F. Kremer

THERMOSETTING POLYMERS

PLASTICS ENGINEERING

Founding Editor

Donald E. Hudgin

Professor
Clemson University
Clemson, South Carolina

1. Plastics Waste: Recovery of Economic Value, *Jacob Leidner*
2. Polyester Molding Compounds, *Robert Burns*
3. Carbon Black-Polymer Composites: The Physics of Electrically Conducting Composites, *edited by Enid Keil Sichel*
4. The Strength and Stiffness of Polymers, *edited by Anagnostis E. Zachariades and Roger S. Porter*
5. Selecting Thermoplastics for Engineering Applications, *Charles P. MacDermott*
6. Engineering with Rigid PVC: Processability and Applications, *edited by I. Luis Gomez*
7. Computer-Aided Design of Polymers and Composites, *D. H. Kaelble*
8. Engineering Thermoplastics: Properties and Applications, *edited by James M. Margolis*
9. Structural Foam: A Purchasing and Design Guide, *Bruce C. Wendle*
10. Plastics in Architecture: A Guide to Acrylic and Polycarbonate, *Ralph Montella*
11. Metal-Filled Polymers: Properties and Applications, *edited by Swapan K. Bhattacharya*
12. Plastics Technology Handbook, *Manas Chanda and Salil K. Roy*
13. Reaction Injection Molding Machinery and Processes, *F. Melvin Sweeney*
14. Practical Thermoforming: Principles and Applications, *John Florian*
15. Injection and Compression Molding Fundamentals, *edited by Avraam I. Isayev*
16. Polymer Mixing and Extrusion Technology, *Nicholas P. Cheremisinoff*
17. High Modulus Polymers: Approaches to Design and Development, *edited by Anagnostis E. Zachariades and Roger S. Porter*
18. Corrosion-Resistant Plastic Composites in Chemical Plant Design, *John H. Mallinson*
19. Handbook of Elastomers: New Developments and Technology, *edited by Anil K. Bhowmick and Howard L. Stephens*
20. Rubber Compounding: Principles, Materials, and Techniques, *Fred W. Barlow*
21. Thermoplastic Polymer Additives: Theory and Practice, *edited by John T. Lutz, Jr.*
22. Emulsion Polymer Technology, *Robert D. Athey, Jr.*
23. Mixing in Polymer Processing, *edited by Chris Rauwendaal*
24. Handbook of Polymer Synthesis, Parts A and B, *edited by Hans R. Kricheldorf*

25. Computational Modeling of Polymers, *edited by Jozef Bicerano*
26. Plastics Technology Handbook: Second Edition, Revised and Expanded, *Manas Chanda and Salil K. Roy*
27. Prediction of Polymer Properties, *Jozef Bicerano*
28. Ferroelectric Polymers: Chemistry, Physics, and Applications, *edited by Hari Singh Nalwa*
29. Degradable Polymers, Recycling, and Plastics Waste Management, *edited by Ann-Christine Albertsson and Samuel J. Huang*
30. Polymer Toughening, *edited by Charles B. Arends*
31. Handbook of Applied Polymer Processing Technology, *edited by Nicholas P. Cheremisinoff and Paul N. Cheremisinoff*
32. Diffusion in Polymers, *edited by P. Neogi*
33. Polymer Devolatilization, *edited by Ramon J. Albalak*
34. Anionic Polymerization: Principles and Practical Applications, *Henry L. Hsieh and Roderic P. Quirk*
35. Cationic Polymerizations: Mechanisms, Synthesis, and Applications, *edited by Krzysztof Matyjaszewski*
36. Polyimides: Fundamentals and Applications, *edited by Malay K. Ghosh and K. L. Mittal*
37. Thermoplastic Melt Rheology and Processing, *A. V. Shenoy and D. R. Saini*
38. Prediction of Polymer Properties: Second Edition, Revised and Expanded, *Jozef Bicerano*
39. Practical Thermoforming: Principles and Applications, Second Edition, Revised and Expanded, *John Florian*
40. Macromolecular Design of Polymeric Materials, *edited by Koichi Hatada, Tatsuki Kitayama, and Otto Vogl*
41. Handbook of Thermoplastics, *edited by Olagoke Olabisi*
42. Selecting Thermoplastics for Engineering Applications: Second Edition, Revised and Expanded, *Charles P. MacDermott and Aroon V. Shenoy*
43. Metallized Plastics: Fundamentals and Applications, *edited by K. L. Mittal*
44. Oligomer Technology and Applications, *Constantin V. Uglea*
45. Electrical and Optical Polymer Systems: Fundamentals, Methods, and Applications, *edited by Donald L. Wise, Gary E. Wnek, Debra J. Trantolo, Thomas M. Cooper, and Joseph D. Gresser*
46. Structure and Properties of Multiphase Polymeric Materials, *edited by Takeo Araki, Qui Tran-Cong, and Mitsuhiro Shibayama*
47. Plastics Technology Handbook: Third Edition, Revised and Expanded, *Manas Chanda and Salil K. Roy*
48. Handbook of Radical Vinyl Polymerization, *Munmaya K. Mishra and Yusuf Yagci*
49. Photonic Polymer Systems: Fundamentals, Methods, and Applications, *edited by Donald L. Wise, Gary E. Wnek, Debra J. Trantolo, Thomas M. Cooper, and Joseph D. Gresser*
50. Handbook of Polymer Testing: Physical Methods, *edited by Roger Brown*
51. Handbook of Polypropylene and Polypropylene Composites, *edited by Harutun G. Karian*
52. Polymer Blends and Alloys, *edited by Gabriel O. Shonaike and George P. Simon*
53. Star and Hyperbranched Polymers, *edited by Munmaya K. Mishra and Shiro Kobayashi*
54. Practical Extrusion Blow Molding, *edited by Samuel L. Belcher*

55. Polymer Viscoelasticity: Stress and Strain in Practice, *Evaristo Riande, Ricardo Díaz-Calleja, Margarita G. Prolongo, Rosa M. Masegosa, and Catalina Salom*
56. Handbook of Polycarbonate Science and Technology, *edited by Donald G. LeGrand and John T. Bendler*
57. Handbook of Polyethylene: Structures, Properties, and Applications, *Andrew J. Peacock*
58. Polymer and Composite Rheology: Second Edition, Revised and Expanded, *Rakesh K. Gupta*
59. Handbook of Polyolefins: Second Edition, Revised and Expanded, *edited by Cornelia Vasile*
60. Polymer Modification: Principles, Techniques, and Applications, *edited by John J. Meister*
61. Handbook of Elastomers: Second Edition, Revised and Expanded, *edited by Anil K. Bhowmick and Howard L. Stephens*
62. Polymer Modifiers and Additives, *edited by John T. Lutz, Jr., and Richard F. Grossman*
63. Practical Injection Molding, *Bernie A. Olmsted and Martin E. Davis*
64. Thermosetting Polymers, *Jean-Pierre Pascault, Henry Sautereau, Jacques Verdu, and Roberto J. J. Williams*

Additional Volumes in Preparation

THERMOSETTING POLYMERS

JEAN-PIERRE PASCAULT
HENRY SAUTEREAU
Institut National des Sciences Appliquées
Villeurbanne, France

JACQUES VERDU
Ecole Nationale Supérieure d'Arts et Métiers
Paris, France

ROBERTO J. J. WILLIAMS
University of Mar del Plata and
National Research Council (CONICET)
Mar del Plata, Argentina

MARCEL DEKKER, INC. NEW YORK • BASEL

ISBN: 0-8247-0670-6

This book is printed on acid-free paper.

Headquarters
Marcel Dekker, Inc.
270 Madison Avenue, New York, NY 10016
tel: 212-696-9000; fax: 212-685-4540

Eastern Hemisphere Distribution
Marcel Dekker AG
Hutgasse 4, Postfach 812, CH-4001 Basel, Switzerland
tel: 41-61-261-8482; fax: 41-61-261-8896

World Wide Web
http://www.dekker.com

The publisher offers discounts on this book when ordered in bulk quantities. For more information, write to Special Sales/Professional Marketing at the headquarters address above.

Current printing (last digit):
10 9 8 7 6 5 4 3 2 1

PRINTED IN THE UNITED STATES OF AMERICA

Preface

The subject of thermosetting polymers receives very brief consideration in most books covering the fundamentals of polymer science. Usually the chemistry is represented by the structure of a phenolic network of the resol type, and some statistical calculations based on Flory's derivations are presented. Therefore, anyone trying to get a first approach to the subject finds only books with chapters written by different authors and aimed at specialists in the field.

The aim of this book is to present a unified coverage of the field of thermosetting polymers, written for readers who are making their first contact with this area. The analysis, however, is carried out to a derivation of concepts and equations useful for practical purposes. The book will be of value for undergraduate and graduate students as well as for people involved in R&D activities in the industrial sector. Some of the material has been used in undergraduate and graduate courses given at ENSAM (Paris), INSA (Lyon), and INTEMA (Mar del Plata, Argentina).

The scientific basis of different aspects of thermosetting polymers was well established in the 1980s and '90s. It was a period of great excitement for people working in the field (the Gordon Conferences were devoted to this subject; a Polymer Network Group was created; several symposia were organized by different research groups; and new, specific journals were launched). Most of the developments during this period are now well settled and constitute the heart of the book. Among the subjects that may be considered firmly established are mean-field theories of network formation; transformation diagrams used to follow the evolution of the material along polymerization; the interpretation of experimental data obtained by differential scanning calorimetry; rheology and dielectrical spectroscopy; ways to synthesize rubber-modified thermosets and thermoplastic-thermoset blends; the influence of crosslinks on physical, elastic, and viscoelastic properties of polymer networks, and factors affecting the yielding and fracture of neat and modified thermosetting polymers. Subjects that are still controversial or less established, such as the inhomogeneity of some thermosetting polymers and aspects related to their durability, are also discussed in the book. Topics related to processing and to composite materials are analyzed only in the contexts of temperature and conversion profiles developed during the cure.

The core of research activities in the field has evolved to a more sophisticated level. The development of advanced materials based on thermosetting polymers is a very active field. It includes nano-structured materials, organic–inorganic hybrid materials, multicomponent blends, polymer-dispersed liquid crystals and networks with liquid crystalline behavior, and materials with a very low dielectric constant, etc. New analytical techniques are available such as microcalorimetry and atomic-force microscopy. Sophisticated computer simulations give a more realistic approach to modeling the build-up of the polymer network or following the evolution of temperature and conversion during the cure in a complex mold. Most of these subjects are outside the scope of this book. A list of selected references, suggested to complement and expand on the material presented in the book, is available at the end of many chapters.

This book is the result of almost two decades of scientific cooperation among the authors that resulted in a significant number of joint publications. In spite of the official grant codes, we always designated our joint research activities as the Asado (typical Argentine meat) – Beaujolais (typical French wine) Program. And some of the more fruitful discussions took place in the course of a Beaujolais tour or during the ritual of preparing the asado.

Over all these years we have learned much from discussions with many colleagues working in this field. The list would be so large—and the possibility of forgetting someone so great—that we prefer to present a general acknowledgment to many friends who contributed significantly to the ideas and concepts developed in this book.

Jean-Pierre Pascault
Henry Sautereau
Jacques Verdu
Roberto J. J. Williams

Contents

THERMOSETTING POLYMERS

1

Introduction

1.1 THERMOPLASTIC AND THERMOSETTING POLYMERS

Based on their response to temperature, plastic materials may be classified into two main categories: thermoplastics and thermosets. A thermoplastic behaves like a fluid above a certain temperature level, but the heating of a thermoset leads to its degradation without its going through a fluid state. In fact, this reflects the behavior of plastic materials under the action of fire, providing the simplest identification test. This classification is not restricted to plastic materials but may also be extended to the behavior of coatings, adhesives, and several other categories. This is why we find it better to use the term *thermosetting polymers*, which implies the different ways in which these materials are used and adds the fact that a constitutional repeating unit (CRU) is present in their chemical structures. These materials are also referred to as *thermosetting resins*, which is a vaguer definition that may be applied to the starting monomers or oligomeric precursors, as well as to the final materials.

Typical examples of thermosetting polymers are phenolic and urea–formaldehyde resins, unsaturated polyesters, and epoxy resins. Typical thermoplastics are polyethylene, polypropylene, polystyrene, and poly(vinyl chloride). The thermoplastic or thermosetting character of a polymer depends on whether it can reach a fluid state by heating. This state can be attained only if individual macromolecules separate to produce flow. This is in principle possible for materials composed of linear or branched macromolecules held together by secondary forces (van der Waals, dipole–dipole, hydrogen bonding). These materials may be either amorphous (e.g., atactic polystyrene) or semicrystalline (e.g., polyethylene, polypropylene). For amorphous thermoplastics, flow will take place by heating beyond the glass transition temperature (after devitrification). For semicrystalline thermoplastics, it is necessary to heat beyond the melting temperature to reach a fluid state.

For high-molar-mass thermoplastics (with molar masses much higher than the entanglement limit), the fluid state is characterized by very large relaxation times so that processing by conventional methods is not possible. In some cases the synthesis and shaping of a thermoplastic polymer is produced in the same process as for thermosetting polymers (reactive processing). For example, some poly(amide-imides) are processed by combining the injection molding of prepolymers with reactive chain ends, with the subsequent cure in a heated mold, to obtain a part made of a high-molar-mass linear polymer that cannot be processed by conventional methods. Cast poly(methyl methacrylate) sheets are produced from a solution of the final polymer in methyl methacrylate. The dilution of the initial monomer with the polymerization product is required to reduce both the temperature increase due to the heat of reaction and the final shrinkage of the part. In poly(tetrafluoroethylene) or poly(acrylonitrile), the density of secondary forces is so high that degradation of the chemical structure takes place before a fluid state is reached. All these materials are classified as thermoplastics because individual molecules can in principle be separated by flow (even though unrealistic times are required) or through the action of an adequate solvent.

A distinctive characteristic of a thermosetting polymer is that one giant macromolecule consisting of covalently bonded repeating units is formed during the polymerization process. This giant macromolecule that percolates throughout the sample is called a gel. Unless the covalent bonds present in the gel are destroyed by the action of temperature or a reactive solvent, the thermosetting polymer will not flow. Therefore, the thermosetting nature of a polymer is due to the presence of a network formed by covalent bonds. A good solvent will swell the polymer but will not dissolve it. However, not every polymer network is a thermosetting polymer, because networks may be also be produced by physical crosslinks among individual chains, as in gelatin. In this case, heating dissociates the individual polymer molecules and converts the material from a gel to a fluid state (a sol). And the gel may be reversibly regenerated during a subsequent cooling step.

Therefore, from a more fundamental point of view, thermosetting polymers may be defined as polymer networks formed by the chemical reaction of monomers, at least one of which has 3 or more reactive groups per molecule (a functionality equal to or higher than 3), and that are present in relative amounts such that a gel is formed at a particular conversion during the synthesis. In a symbolic form, it may be stated that a thermosetting polymer is obtained by the homopolymerization of an A_f molecule ($f \geq 3$ is the number of functional groups per molecule), or the polymerization of A_f by reaction with B_g (where f and/or g are ≥ 3) and they are present in a particular ratio such that a gel will be formed. In commercial formulations, mixtures of several monomers differing in chemical structure and functionality may be used and the gel may be formed through competitive reactions among the different monomers.

Because the polymer network is produced in an irreversible way, the synthesis of a thermosetting polymer is carried out to produce the final material with the desired shape. Therefore, polymerization and final shaping are performed in the same process. The situation is completely different for most of the thermoplastic polymers. The processor gets the final part by heating (to provide flow) and cooling (to set the final shape) in an injection machine, an extruder, etc. The mold used to obtain a part made of a thermosetting polymer is a true chemical reactor, while the machine used to obtain a similar part made of a thermoplastic polymer acts as a heat exchanger (part of the necessary heat may be generated by viscous dissipation). In thermosetting polymers incorrect processing leads to the irreversible loss of the material (it may be reused for other purposes, e.g., as a filler); in thermoplastics the material may be reused many times (although a continuous decrease in the average molar mass is produced by the relatively high temperatures used in the processing machines).

The synthesis of a thermosetting polymer starts from monomers (a sol phase). At a particular conversion of functional groups, gelation takes place, meaning that a giant macromolecule that percolates through the sample appears in the system. This sol–gel critical transition is a distinctive feature of thermosetting polymers. Before gelation the material is a sol with a finite value of viscosity. At gelation, viscosity increases to infinite. After gelation an insoluble fraction (the gel fraction) is present in the system. Eventually, at full conversion of functional groups, in stoichiometric formulations, the sol fraction disappears and the final thermosetting polymer is composed of one giant molecule of a gel.

Although a sol–gel transformation is present in the synthesis of any thermosetting polymer, in some fields it is used in a rather restrictive way. For example, ceramists associate sol–gel processes with the hydrolytic condensation of precursors to obtain inorganic polymer networks (SiO_2, TiO_2, etc.) or hybrid inorganic–organic polymer networks.

Thermosetting polymers are usually amorphous because there is no possibility of ordering portions of the network structure due to the restrictions imposed by the presence of crosslinks. Exceptions are networks obtained from rigid monomers exhibiting a nematic–isotropic transition. In these cases, a polymer network presenting a nematic–isotropic transition may be obtained, provided that the concentration of crosslinks is kept at a low value.

One of the most important parameters characterizing a thermosetting polymer is the location of its glass transition temperature (T_g) with respect to the temperature at which it is used (T_{use}). Most thermosetting polymers are formulated and selected so that their T_g is higher than T_{use}; therefore, they behave as glasses during their use. Materials exhibiting a T_g lower than T_{use} are classified as rubbers, but they can also be regarded as thermosetting polymers operating in the rubbery state and are thus included within the scope of the book.

Thermosetting polymers are well established in areas where thermoplastics cannot compete because of either properties or costs. For example, phenolics constitute a first option when fire resistance is required because they are self-extinguishing and exhibit low smoke emission. Urea–formaldehyde polymers for wood agglomerates and melamine–formaldehyde for furniture coatings give products of excellent quality at low costs. Unsaturated polyesters are extensively used to produce structural parts with a glass-fiber reinforcement. In addition, epoxies, cyanate esters, and polyimides are employed for aeronautical and electronic applications where their excellent properties cannot be matched by thermoplastics. The average percent annual growth in the United States during the decade from 1989 to 1999 was 2% for synthetic rubbers, 4% for thermosetting polymers, and 5% for thermoplastics.

1.2 ORGANIZATION OF THE BOOK

Every chapter has been prepared in a self-consistent form and includes a particular notation that is given at its end (although the most frequently occurring variables and parameters have the same notation throughout the book). Thus, a reader with a preliminary knowledge of the subject can go directly to any of the chapters. However, for a reader approaching the subject for the first time, it will be much more convenient to follow the order in which the book is organized.

Different chemistries for the synthesis of thermosetting polymers are described in Chapter 2. The distinction between step-growth and chain polymerizations is clearly established, and several examples of both types of reaction are presented. The buildup of the network is discussed in Chapter 3 for both types of polymerization reactions. Equations derived from mean-field models or from combinations of kinetic and recursive procedures (fragment approach) enable us to calculate statistical parameters of the network as a function of the conversion of functional groups. In particular, the gel conversion may be predicted and stoichiometric ratios necessary to avoid gelation may be obtained. The evolution of the glass transition temperature with conversion is discussed in Chapter 4. This information, together with the gel conversion, is used to develop transformation diagrams that may be employed to predict the different transitions that will take place during a cure cycle. Chapter 5 presents different equations, either phenomenological or based on reaction mechanisms, used to describe the kinetics of network formation. Some incorrect ways of presenting kinetic results are discussed and examples are provided; the influence of diffusion limitations is analyzed.

Both rheologic and dielectric monitoring of network formation are discussed in Chapter 6, while Chapter 7 tries to answer the question concerning the inhomogeneity of some polymer networks (a controversial subject). Chapter 8 deals with the preparation of rubber-modified thermosets and thermoset–thermoplastic blends, which are frequently used to improve some properties of the neat thermosetting polymer. An introduction to the processing of thermosetting polymers is developed in Chapter 9, with the main emphasis placed on the estimation of temperature and conversion profiles generated during the cure process.

The effect (or lack of effect) of crosslinks on basic physical properties of thermosetting polymers is discussed in Chapter 10, while the effect on elastic and viscoelastic properties is analyzed in Chapter 11. Yielding and fracture of neat and modified thermosetting polymers are discussed in Chapters 12 and 13. Finally, the very important problem of the durability of polymer networks is presented in Chapter 14.

2

Chemistry of Crosslinked Polymer Synthesis

2.1 GENERAL CONSIDERATIONS

A comprehensive classification of both linear and crosslinked polymers may be based on the mechanism of the polymerization process. From the point of view of the polymer growth mechanism, two entirely different processes, step and chain polymerization, are distinguishable.

Step-growth polymerization proceeds via a step-by-step succession of elementary reactions between reactive sites, which are usually functional groups such as alcohol, acid, isocyanate, etc. Each independent step causes the disappearance of two coreacting sites and creates a new linking unit between a pair of molecules. To obtain polymers, the reactants must be at least difunctional; monofunctional reactants interrupt the polymer growth.

In chain-growth polymerization, propagation is caused by the direct reaction of a species bearing a suitably generated active center with a monomer molecule. The active center (a free radical, an anion, a cation, etc.) is generated chainwise by each act of growth; the monomer itself constitutes the feed (reactive solvent) and is progressively converted into the polymer.

For both mechanisms of polymer growth, if one of the reactants has a functionality higher than 2, branched molecules and an infinite structure can be formed. To summarize both mechanisms it may be stated that:

1. A step-growth polymerization (with or without elimination of low-molar-mass products) involves a series of monomer + monomer, monomer + oligomer, monomer or oligomer + macromolecule, and macromolecule + macromolecule reactions. The molar mass of the product grows gradually and the molar mass distribution becomes continuously wider. Functionalities of monomers and the molar ratio between coreactive sites are the main parameters for controlling the polymer structure.
2. A chainwise polymerization proceeds exclusively by monomer + macromolecule reactions. When the propagation step is fast compared to the initiation step, long chains are already formed at the beginning of the reaction. The main parameters controlling the polymer structure are the functionalities of the monomers and the ratios between the initiation and propagation rates and between initiator and monomer concentrations.

Thermosetting polymers may be formed in two ways:

1. By polymerizing (step or chain mechanisms) monomers where at least one of them has a functionality higher than 2.
2. By chemically creating crosslinks between previously formed linear or branched macromolecules (crosslinking of primary chains, as vulcanization does for natural rubber).

In fully reacted polymer networks, practically all constituent units are covalently bonded into an infinite three-dimensional structure. It means that during polymerization or crosslinking the system evolves from a collection of molecules of finite size to an infinite network, proceeding through the gel point at which the infinite network structure appears for the first time. This transformation is called gelation.

As polymer networks are very often prepared in bulk, vitrification, which is the transformation from a liquid or rubbery state to a vitreous state, can also take place. These transformations are discussed later (Chapters 3, 4, and 6), but one question that concerns chemistry is the possible effect of these transformations on the mechanisms and kinetics of the reactions.

2.2 STEP-GROWTH POLYMERIZATIONS: POLYCONDENSATIONS AND POLYADDITIONS

2.2.1 General Aspects

Based on the classical definitions given in organic chemistry, the step-growth polymerization process can involve either condensation steps or addition steps. The former proceeds with elimination of by-products while the latter takes place without elimination of by-products. This is illustrated by Eqs (2.1) and (2.2), for the particular case of difunctional molecules:

$$n\ A{-}R{-}B \rightleftharpoons A{-}\left(RC\right)_{n-1}{-}R{-}B \quad \{+\text{by-products}\} \tag{2.1}$$

$$n\ AR^1A + n\ BR^2B \rightleftharpoons A\left(R^1CR^2C\right)_{2n-2}R^1CR^2B \quad \{+\text{by-products}\} \tag{2.2}$$

Each reaction step causes the disappearance of two reactive sites, A and B, converted into a linking unit C, and leaves two reactive sites at the ends of any growing molecule irrespective of its size. The difference between polycondensation and polyaddition is only the formation of by-products during each reaction step.

The extent of reaction (or conversion) at any stage can be expressed by the fraction of total reactive sites that have been consumed. Reactive sites usually display the same reactivity regardless of the size of the molecule to which they are linked. The polymerization process has the characteristics of a statistical combination of fragments. In this way, a distribution of products from the monomer to a generic n-mer is obtained (Table 2.1), with average molar masses increasing continuously with conversion.

If the initial polymerization system contains a single monomer as in Eq. (2.1), the constitutional repeating unit (CRU) of the polymer will contain only one monomer-based unit and the structure of the CRU will be derived from the monomer (polyaddition case), possibly through the elimination of a small molecule (polycondensation case).

If the initial polymerization system contains two different monomers (Eq. 2.2), the CRU will contain two monomer-based units.

When at least one of the monomers bears more than two reactive functional groups, the formation of a polymer network is possible.

When the concentrations of A and B may be varied independently (Eq. 2.2), the stoichiometric ratio of functionalities is defined by $r = A_0/B_0$, where A_0 and B_0 are the initial concentrations of functional groups A and B. As will be shown in Chapter 3, this ratio is very important in designing and controlling a step-growth polymerization. Statistical parameters at any

TABLE 2.1 The step-growth polymerization process

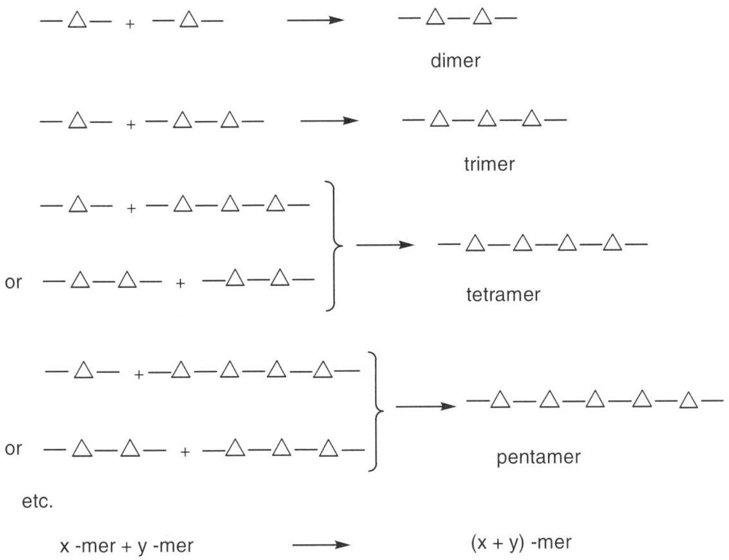

Reactants		Product
—△— + —△—	⟶	—△—△— dimer
—△— + —△—△—	⟶	—△—△—△— trimer
—△— + —△—△—△— or —△—△— + —△—△—	⟶	—△—△—△—△— tetramer
—△— + —△—△—△—△— or —△—△— + —△—△—△—	⟶	—△—△—△—△—△— pentamer
etc.		
x -mer + y -mer	⟶	(x + y) -mer

conversion may be correlated with the initial composition and the number of functional groups per molecule.

Many reactions familiar to organic chemists may be utilized to carry out step polymerizations. Some examples are given in Table 2.2 for polycondensation and in Table 2.3 for polyaddition reactions. These reactions can proceed reversibly or irreversibly. Those involving carbonyls are the most commonly employed for the synthesis of a large number of commercial linear polymers. Chemistries used for polymer network synthesis will be presented in a different way, based on the type of polymer formed (Tables 2.2 and 2.3). Several different conditions may be chosen for the polymerization: in solution, in a dispersed phase, or in bulk. For thermosetting polymers the last is generally preferred.

TABLE 2.2 Typical polycondensations

A and **B** reactive sites	By-product	Linkage **C** formed	Type of polymer
Carboxylic acid + alcohol	H_2O		
Ester of carboxylic acid + alcohol	ROH	$-CO_2-$	Polyester
Anhydride of dicarboxylic acid + alcohol	H_2O		
Carboxylic acid + amine	H_2O	–CONH–	Polyamide
Anhydride of dicarboxylic acid + amine	H_2O	–CO–N–CO– |	Polyimide
Ketone + amine	H_2O	$\rangle$C=N–	Polyazomethine
Phenol + phosgene ($COCl_2$)	HCl	$-OCO_2-$	Polycarbonate
Phenol (+ NaOH) + aryl halide	NaCl	Ar–O–Ar	Polyether
Phenol + formaldehyde	H_2O	$Ar-CH_2-$	"Phenolic resin"
Urea (or melamine) + formaldehyde	H_2O	$-NH\ CH_2-$	"Amino resin"
Isocyanate (2**A**, no **B**)	CO_2	–N=C=N–	Polycarbodiimide
Chlorosilane + H_2O	HCl	$\rangle$SiO–	Polysiloxane

Table 2.3 Typical polyadditions

A and **B** reactive sites	Linkage **C** formed	Type of polymer
Isocyanate + alcohol	$-O-CO-NH$	Polyurethane
Isocyanate + amine	$-NH-CO-NH-$	Polyurea
Epoxy (or oxirane) + amine	$-CH(OH)-CH_2-N<$	Polyepoxy
Epoxy (or oxirane) + isocyanate		Poly(2-oxazolidone)
Cyanate–ester (3**A**, no **B**)		Polycyanurate or triazine
Thiol + double bond	$-CH_2-CH_2S-$	Polysulfide for example, amine + fumarate double bond
Michael-type additions		
Diels–Alder reaction		

2.2.2 Organic Acid Reactions

a. Mechanism: An Example of Polycondensation Involving Carbonyl Groups

Reactions of this type are employed for the synthesis of a large number of commercial (linear) polymers such as polyesters and polyamides. A small molecule, water, is split out during these condensation reactions:

$$R^1-\underset{\underset{O}{\|}}{C}-OH + R^2-OH \rightleftharpoons R^1-\underset{\underset{O}{\|}}{C}-OR^2 + H_2O \tag{2.3}$$

$$R^1-\underset{\underset{O}{\|}}{C}-OH + R^2-NH_2 \rightleftharpoons R^1-\underset{\underset{O}{\|}}{C}-NHR^2 + H_2O \tag{2.4}$$

As a result of reactions (2.3) and (2.4), an equilibrium between reactants and products would be reached in a closed system. The removal of water is necessary in order to allow the reaction to proceed to high conversions.

Instead of carboxylic acids, other carbonyl compounds can be used: acid halides, esters, amides, etc. The commonly accepted general mechanism for these reactions consists of the initial nucleophilic addition of an active hydrogen compound to the electron-poor carbonyl carbon atom of the R^1COOH molecule, with the formation of a metastable intermediate that can undergo a subsequent elimination reaction:

$$R^1-\underset{\underset{O}{\|}}{C}-X + AH \rightleftharpoons R^1-\overset{\overset{O^-}{|}}{\underset{\underset{\overset{+}{A}-H}{|}}{C}}-X \rightleftharpoons R^1-\underset{\underset{O}{\|}}{C}-A + HX \tag{2.5}$$

Such a mechanism can explain why aromatic R^1 acids have a higher reactivity than aliphatic ones. AH is a Lewis base, which carries an unshared pair of electrons on the A atom. Hence an increase in its nucleophilic character should facilitate the addition to the electron-poor carbonyl carbon atom and should make easier the elimination of X as a negative ion.

The use of a catalyst is frequently necessary in order to obtain high-molar-mass products. Strong protonic acids or carboxylic organic acids are frequently employed as catalysts:

$$R^1\text{-}\underset{\underset{O}{\|}}{C}\text{-}X \overset{+H^+}{\rightleftharpoons} R^1\text{-}\overset{+}{\underset{\underset{OH}{|}}{C}}\text{-}X \overset{+AH}{\rightleftharpoons} R^1\text{-}\overset{AH^+}{\overset{|}{\underset{\underset{OH}{|}}{C}}}\text{-}X \overset{-HX}{\rightleftharpoons} R^1\text{-}\overset{A}{\overset{|}{\underset{\underset{OH}{|}}{C}}}+ \overset{-H^+}{\rightleftharpoons} R^1\text{-}\underset{\underset{O}{\|}}{C}\text{-}A \tag{2.6}$$

Different types of metal compounds can also be used as catalysts; for example, zinc acetate, titanium alkoxide, phosphorus derivatives, etc.

Cyclic anhydrides are diacids with one molecule of water eliminated from the condensation of the two acid groups. They can be useful for the synthesis of polyesters. The reaction proceeds in two steps because the free acid formed in the first step (Eq. 2.7) is much less reactive than the original anhydride:

$$R^1\langle(C{=}O)_2\rangle O + R^2OH \longrightarrow HO\text{-}\underset{\underset{O}{\|}}{C}\text{-}R^1\text{-}\underset{\underset{O}{\|}}{C}\text{-}OR^2 \tag{2.7}$$

$$HO\underset{\underset{O}{\|}}{C}\text{-}R^1\text{-}\underset{\underset{O}{\|}}{C}\text{-}OR^2 + R^2OH \rightleftharpoons R^2O\text{-}\underset{\underset{O}{\|}}{C}\text{-}R^1\text{-}\underset{\underset{O}{\|}}{C}\text{-}OR^2 + H_2O \tag{2.8}$$

The reaction of cyclic anhydrides with amines can be different from that with alcohols, because in the case of amines, the amido acid formed during the first step (Eq. 2.9) can close a cycle to give an imide group (Eq. 2.10):

$$R^1\langle(C{=}O)_2\rangle O + R^2NH_2 \longrightarrow R^1\begin{cases}\overset{\overset{O}{\|}}{C}\text{-}NHR^2\\ \underset{\underset{O}{\|}}{C}\text{-}OH\end{cases} \tag{2.9}$$

$$R^1\begin{cases}\overset{\overset{O}{\|}}{C}\text{-}NHR^2\\ \underset{\underset{O}{\|}}{C}\text{-}OH\end{cases} \rightleftharpoons R^1\langle(C{=}O)_2\rangle N\text{-}R^2 + H_2O \tag{2.10}$$

The reaction can be very fast, even at low temperature without a catalyst.

b. Some Examples of Polymer Networks Based on Esterification

Direct polyesterification can be used to prepare polymer networks for coating applications. In this case it is necessary to increase the reactivity of the system by using anhydrides instead of a diacid (glycero-phthalic or glyptal "resins," Eq. 2.11) or activated alcohols (powder coatings, Eq. 2.12).

Phthalic anhydride + glycerol

$$C_6H_4\begin{matrix} CO \\ CO \end{matrix}\!\!>O \; + \; \begin{matrix} CH_2 & - & CH & - & CH_2 \\ | & & | & & | \\ OH & & OH & & OH \end{matrix} \tag{2.11}$$

Diacid + activated alcohol

$$\begin{matrix} HO - & C & - R - & C & - OH \\ & \| & & \| & \\ & O & & O & \end{matrix} \; + \; (CH_2)_4\begin{matrix} \diagup CO\,N\,(CH_2\,CH_2\,OH)_2 \\ \diagdown CO\,N\,(CH_2\,CH_2\,OH)_2 \end{matrix} \tag{2.12}$$

c. Synthesis of Unsaturated Polyesters, UP Oligomers

Maleic anhydride [R^1 equals –CH=CH– in Eq. (2.7); cis isomer] is reacted with aliphatic diols to form low molar mass unsaturated polyesters, UP. For molar masses higher than 1000 g/mol, products are diluted with a liquid vinyl monomer, most often styrene. This reactive mixture, generally called "unsaturated polyester, UP resin," can be transformed into crosslinked polymers through a free-radical chain polymerization (see Sec. 2.3).

Equations (2.13) and (2.14) (Table 2.4) describe the synthesis of UP oligomers. This is usually carried out in bulk at elevated temperatures. During a first step, the temperature is kept in the range of 60–130°C and is increased up to 160–220°C in a second step. During this second step most of the maleate groups (cis isomer) are isomerized into fumarate groups (trans isomer), Eq. (2.15) (Table 2.4). The degree of isomerization is determined by the esterification conditions (temperature, acid content, catalyst, nature of the diol). It must be carefully controlled because the content of fumarate units determines many properties of UP networks.

Since esterification is a reversible process, water must be efficiently removed, especially in the last stages of the reaction. These stages are usually carried out under a vacuum with the difficulty to avoid losses of other volatile reactants such as diols.

Table 2.4 Main reactions occurring during UP synthesis

Monoester formation:

$$\text{(maleic anhydride)} + HO-R^2-OH \longrightarrow HO-\underset{\|}{\overset{}{C}}(=O)-CH=CH-C(=O)-OR^2OH \qquad (2.13)$$

Polycondensation (polyesterification):

$$n\ HO-C(=O)-CH=CH-C(=O)-OR^2OH \rightleftharpoons HO\left[-C(=O)-CH=CH-C(=O)-OR^2O\right]_n-H + (n-1)\,H_2O \qquad (2.14)$$

Maleate–fumarate isomerization:

$$HO-C(=O)-CH=CH-C(=O)-OR^2OH\ (\textit{cis}) \rightleftharpoons HO-C(=O)-CH=CH-C(=O)-OR^2OH\ (\textit{trans}) \qquad (2.15)$$

Ordelt saturation of a monoester by a diol:

$$HO-C(=O)-CH=CH-C(=O)-OR^2OH + HOR^2OH \longrightarrow HO-C(=O)-CH_2-CH(O-R^2-OH)-C(=O)-OR^2OH \qquad (2.16)$$

Ordelt saturation of a monoester by a monoester:

$$HO-C(=O)-CH=CH-C(=O)-OR^2OH + HO-C(=O)-CH=CH-C(=O)-OR^2OH$$

$$\longrightarrow HO-C(=O)-CH_2-CH\left(O-R^2-O-C(=O)-CH=CH-C(=O)OH\right)-C(=O)-OR^2OH \qquad (2.17)$$

Table 2.4 Continued

General UP prepolymer structure (see text):

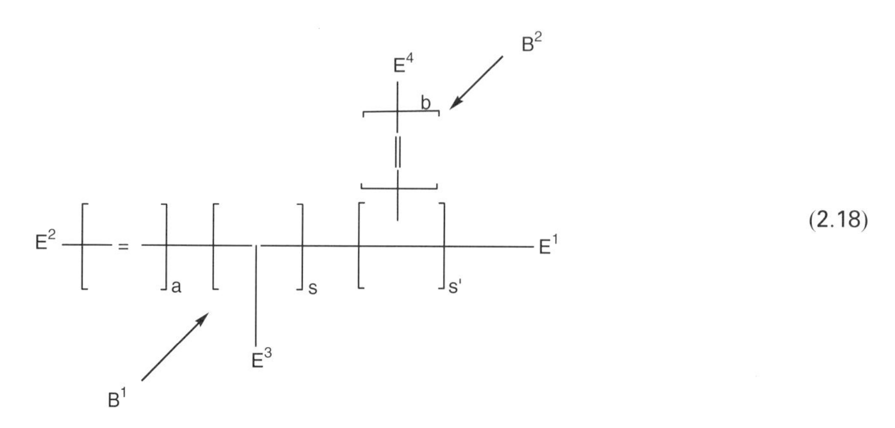

(2.18)

Dehydration of α-diol:

$$2\ HO-R^2-OH \xrightarrow{H^+} HO-R^2-O-R^2-OH + H_2O \qquad (2.19)$$

The chemical structure of UP oligomers is more complex than might be expected from the chemistry of the reactions, Eqs. (2.13) and (2.14). The addition of hydroxyl groups to the activated double bonds is one of the most important side reactions – Eqs. (2.16) and (2.17) (Table 2.4) – called Ordelt reactions. It leads to the formation of side chains and to a modification of the COOH/OH stoichiometry due to diol consumption.

Equation (2.18) gives a general molecular structure for the UP prepolymers. E^1 and E^2 are the main chain end groups, hydroxyl or carboxyl groups; E^3 and E^4 the branch chain end groups, mainly hydroxyl groups; B^1 the short-chain branch; B^2 the long-chain branch; and a, s, s′, b the number of constitutional units, $a > 0$, $s \geq 0$, $s' \geq 0$, $b \geq 0$. The number of chain ends per molecule, including hydroxyl and carboxyl groups, is, in fact, larger than 2 (the theoretical value for linear chains).

Another possible side reaction under experimental polyesterification conditions is the dehydration of α-diols (Eq. 2.19, Table 2.4). It changes the structure of some constitutional units. Depending on the diol used, cyclic ethers and aldehydes can also be formed.

Because of these side reactions, the molar mass distribution of UP prepolymers is larger than expected, with a polydispersity index that can be 10 or more (Fig. 2.1).

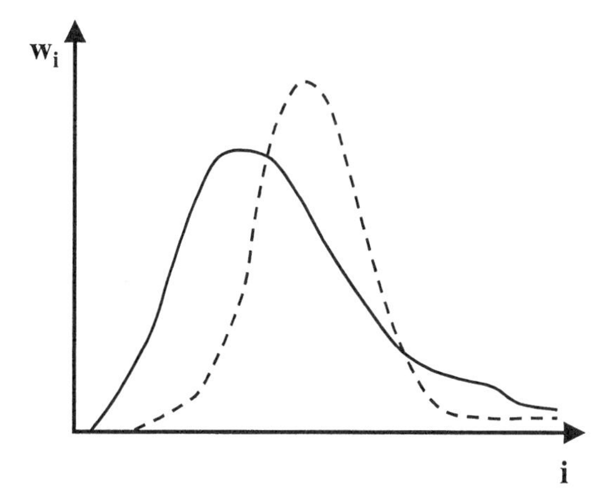

FIGURE 2.1 Molar mass distribution of an UP prepolymer, - - - theoretical curve assuming no side reactions and equal reactivity for –COOH and –OH groups; —— experimental curve.

d. Synthesis of Bismaleimide (BMI) Monomers

When maleic anhydride is reacted with diamines instead of aliphatic diols, intramolecular cyclization occurs and bismaleimides are obtained (Eq. 2.20) (with chloroform, acetone, or toluene as solvent and acetic anhydride for cyclization).

$$2\ (\text{maleic anhydride}) + H_2N-R^2-NH_2 \longrightarrow (\text{maleimide})N-R^2-N(\text{maleimide}) + 2H_2O \qquad (2.20)$$

The intramolecular reaction (Eq. 2.20) is easier with aromatic diamines than with aliphatic ones. Reactions of BMI monomers will be discussed later (Sec. 2.2.7).

2.2.3 Isocyanate Reactions

a. Mechanism: An Example of a Polyaddition Reaction Involving Carbonyl Groups

The mechanism for isocyanate reactions also consists of the nucleophilic addition of an active hydrogen compound, AH, to the electron-poor carbonyl atom:

$$\mathrm{R-N=C=O + AH \rightleftharpoons R-N=\underset{\underset{^{+}A-H}{|}}{C}-O^{-} \dashleftarrow\!\!\!\longrightarrow R-\underset{\underset{H}{|}}{N}-\underset{\underset{O}{\|}}{C}-A} \quad (2.21)$$

A broad spectrum of hydrogen-containing nucleophiles react with both aromatic and aliphatic isocyanates: compounds containing OH groups (H_2O, alcohols, phenols, oximes, acids), SH groups (H_2S, mercaptans), NH groups (NH_3, amines, hydrazines, amides, ureas, urethanes), enolizable compounds such as malonic and aceto acetic esters, etc. Some reactions are given in Table 2.5.

Reactivities of isocyanates depend on their structure. Table 2.6 gives the main isocyanates used for polymer network synthesis. Conjugation with aromatic nuclei makes ArNCO particularly reactive. The reactivity of diisocyanates is well documented in the literature. For symmetric diisocyanates such as diphenylmethane 4,4′-diisocyanate (MDI) or para-phenylene 4,4′-diisocyanate (PPDI), both NCO groups have initially the same reactivity. But as the NCO group itself exhibits an activating effect on isocyanate reactivity, the fact that one NCO group has reacted introduces a substitution effect that usually decreases the reactivity of the second NCO group.

This effect is more pronounced in PPDI than in MDI; the ratio of the rate constants for the reaction with an aliphatic alcohol is $k_1/k_2 = 9$ and $k_1/k_2 = 2$, respectively (at room temperature).

Asymmetric diisocyanates such as 2,4-TDI are more complex because the initial reactivity of the two isocyanate groups is not equivalent and the substitution effect amplifies the difference. The 4-NCO is about 10–20 times more reactive than the 2-NCO, but the reactivity ratio also depends on temperature (see Chapter 5). This difference also explains why the TDI dimer can be prepared quantitatively (Eq. 2.28).

Reaction kinetics with the various reagents becomes faster as their nucleophilicity is increased. The following order of reactivity can be given: primary aliphatic amine > primary aromatic amine > secondary aliphatic amine ~ primary alcohol > secondary alcohol > water > tertiary alcohol ~ phenol > mercaptan.

Catalysts may be Lewis bases like tertiary amines. The catalyst forms an initial coordination complex with the carbonyl carbon atom, with subsequent displacement by the active hydrogen compound:

$$\mathrm{R-N=C=O \overset{B}{\rightleftharpoons} R-N^{-}-\underset{\underset{B^{+}}{|}}{C}=O \overset{AH}{\rightleftharpoons} RNH-\underset{\underset{B^{+}}{|}}{\overset{\overset{A}{|}}{C}}-O^{-} \xrightarrow{-B} RNH-\underset{\underset{O}{\|}}{C}-A} \quad (2.31)$$

Table 2.5 Main reactions of isocyanate groups

Reaction	Product	Eq.
Reaction with water:		
$R^1NCO + H_2O \longrightarrow R^1NHCOOH$ (carbamic acid) $\longrightarrow R^1NH_2 + CO_2$		(22.2)
Amine addition:		
$R^1NCO + R^2NH_2 \rightleftharpoons R^1NHCONHR^2$	urea	(2.23)
Alcohol addition:		
$R^1NCO + R^2OH \rightleftharpoons R^1NHCOOR^2$	urethane	(2.24)
Urethane addition:		
$R^1NCO + R^1NHCOOR^2 \rightleftharpoons R^1NHCON(R^1)COOR^2$	allophanate	(2.25)
Urea addition:		
$R^1NCO + R^1NHCONHR^2 \rightleftharpoons R^1NHCON(R^1)CONHR^2$	biuret	(2.26)
Epoxy addition:		
$R^1NCO + R^2\text{-}CH\text{-}CH_2$ (O) $\rightleftharpoons$ $R^1\text{-}N$, $CH_2\text{-}CH\text{-}R^2$, O, $C{=}O$	oxazolidone	(2.27)
Homopolyaddition (with catalyst like triphenylphosphine):		
$2\,R^1NCO \rightleftharpoons$ $R^1\text{-}N$, $N\text{-}R^1$, $C{=}O$, $C{=}O$	uretdione	(2.28)
Homopolyaddition (base as catalyst):		
$3\,R^1NCO \rightleftharpoons$ $R^1\text{-}N$, $N\text{-}R^1$, $N\text{-}R^1$, $C{=}O$, $C{=}O$, $C{=}O$	isocyanurate	(2.29)
Homopolycondensation:		
$2\,R^1NCO \longrightarrow R^1\text{-}N{=}C{=}N\text{-}R^1 + CO_2$	carbodiimide	(2.30)

Table 2.6 Main commercial isocyanates for polymer synthesis

(a) *Toluene 2,4 and 2,6-diisocyanate, TDI*

CH_3 NCO NCO (2,4-TDI) — OCN CH_3 NCO (2,6-TDI)

- Commercial mixtures of isomers 80/20 or 65/35
- Dimer of 2,4-TDI: cf. Eq. (2.28) ($T_m \sim 180°C$)

(b) *para-Phenylene 4,4′-diisocyanate*, PPDI

Solid ($T_m \sim 90°C$) OCN—C₆H₄—NCO

(c) *Diphenylmethane 4,4′-diisocyanate*, MDI

Solid ($T_m \sim 42°C$) OCN—C₆H₄—CH_2—C₆H₄—NCO

Liquid: 85% MDI + 15% "oligomers"

OCN - R - N / C(=N - R - NCO) / N - R - NCO / C=O — uretonimine

R = —C₆H₄—CH_2—C₆H₄—

- "Polymeric" OCN—C₆H₄—CH_2—C₆H₃(NCO)—CH_2—C₆H₄—NCO

- Isomer mixtures of 4,4′ and 2,4′-MDI

(d) *Dicyclohexylmethane diisocyanate, or hydrogenated MDI*

Liquid OCN—C₆H₁₀—CH_2—C₆H₁₀—NCO cis, trans

Table 2.6 Continued

(e) *Isophorone diisocyanate*, IPDI

Liquid

$$\text{1-(isocyanatomethyl)-1,3,3-trimethyl-5-isocyanatocyclohexane: ring bearing } (CH_3)_2,\ CH_3,\ CH_2NCO,\ NCO$$

(f) *Hexamethylene-1,6-diisocyanate, HDI*

- Solid $OCN{-}\!\left(CH_2\right)_6\!{-}NCO$
- Trimer of HDI: see Eq. (2.29)
- Commercial adducts by reacting an excess HDI with trimethylol propane (TMP) – see Eq. (2.24)

Catalysts based on organometallic compounds of tin (dibutyltin dilaurate), lead, bismuth, mercury, and cobalt are also frequently employed. As a general rule they are not selective; they catalyze the reaction of isocyanates with both hydroxyl groups and water.

The reactions between isocyanates and protic nucleophilic reagents are characterized by an intrinsic reversibility under heating. Typical urethanes tend to decompose to isocyanate and alcohol at temperatures T~ 180–250°C, depending on the substituents. In the same range of temperature or just below, exchange reactions are also possible between two urethanes:

$$R^1NHCOOR^2 + R^3NHCOOR^4 \rightleftharpoons R^1NHCOOR^4 + R^3NHCOOR^2 \tag{2.32}$$

Isocyanates which are reacted with phenols, oximes, dialkylmalonates, lactams, triazole, imidazoline, oxamate, etc., are usually termed blocked isocyanates. They decompose at temperatures which decrease roughly as the isocyanate reactivity and the nucleophilic reagent acidity increase: 50–150°C for aromatic isocyanates, 120–180°C for aliphatic ones. They give back, often in almost quantitative yields, the starting isocyanate and the blocking agent. Blocked isocyanates in blends with polyols or polyamines are employed as crosslinking agents for many applications (e.g., single-package blocked-adduct urethane coatings). It is not clear from the literature if the mechanism can be described by two reaction paths (Eqs (2.33) and (2.34)) or just by one (Eq. (2.35)):

$$R^1\text{-}NH\text{-}\underset{\underset{O}{\|}}{C}\text{-}B \rightleftharpoons R^1\text{-}N=C=O + BH \tag{2.33}$$

$$R^1N=C=O + R^2NH_2 \longrightarrow R^1NH-\underset{\underset{O}{\|}}{C}-NHR^2 \quad (2.34)$$

$$R^1-NH\underset{\underset{O}{\|}}{C}-B + R^2NH_2 \longrightarrow R^1NH-\underset{\underset{O}{\|}}{C}-NHR^2 + BH \quad (2.35)$$

b. Polyurethanes (PU) and Polyureas

The synthesis of PU can be carried out by the reaction described in Eq. (2.24). If the functionality of the hydroxy-containing compounds or the isocyanate is increased beyond 2, branched and possibly crosslinked polymers are produced. Because the nature of the polyol (polyether, polyester, polybutadiene, etc.) and isocyanate components can vary widely, PU are among the most versatile polymers, producing a wide variety of materials such as elastomers, foams, coatings, adhesives, or fibers.

To obtain elastomers, one or two diols can be reacted with the isocyanate. When two diols are used, the first one is a macrodiol with a molar mass in the range 500–10000 g mol^{-1}, and the second one is a short diol, typically 1,4-butanediol. The PU may be prepared by either the one-shot process (three components reacting together), or the prepolymer approach : a prepolymer is prepared first (Eq. 2.36) and then reacted with the short diol (chain extender):

$$\underset{\text{macrodiol, 1 mol}}{HO\sim\sim X\sim\sim OH} + \underset{\text{in excess}}{OCNR^1NCO} \longrightarrow OCN-R^1\left[NHCO-O\sim\sim X\sim\sim OCONHR^1\right]_i NCO + \underset{\text{less excess}}{OCN-R^1NCO} \quad (2.36)$$

At the end of the prepolymer synthesis, the diisocyanate excess depends on the initial excess and also on the reactivity ratio between the two isocyanate groups.

A typical formulation for the reaction–injection molding (RIM) process is shown below (one-shot process):

$$\underset{\text{trifunctional polyol}}{2\ HO\sim\sim \underset{\underset{OH}{\wr}}{X}\sim\sim OH} + \underset{\text{short diol}}{HO-R^2-OH} + \underset{\text{diisocyanate}}{4\ OCN-R^1NCO}$$

At the end of the reaction the crosslinked PU contains hard and soft blocks:

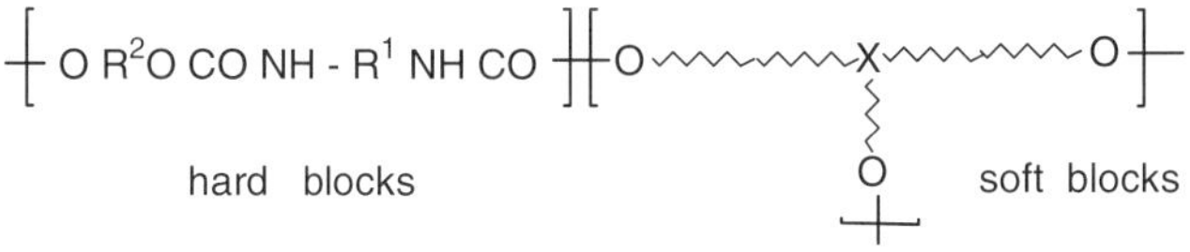

$$\left[O\,R^2O\,CO\,NH - R^1\,NH\,CO \right]\left[O \sim\sim\sim X(\sim\sim O-) \sim\sim\sim O \right]$$

hard blocks soft blocks

Diamines may be introduced as short chain extenders. As the amine reacts rapidly (Eq. 2.23), hindered aromatic amines must often be used to allow a controllable reaction. Polyurethane ureas are formed.

If no polyol is used and only amine-terminated polyethers with a diamine chain extender are employed, a polyurea RIM system is obtained.

With an excess isocyanate in the above systems, allophanate and biuret reactions take place (Eqs (2.25) and (2.26)), resulting in further crosslinking. When increased rigidity and high-temperature performance are desired, further crosslinking may be accomplished via isocyanurate formation (Eq. (2.29)). Base catalysts such as alkoxides, quaternary ammonium or phosphonium, etc., promote this reaction. Aromatic isocyanates give isocyanurates far more easily than aliphatic ones.

Flexible PU foams are prepared from basically the same raw materials as PU elastomers. Instead of a short-chain extender, water is used. Water reacts with isocyanates to form an amine and CO_2 (Eq. (2.22)), which results in foaming. The diamine thus created can then react with isocyanate groups giving polyurea short segments. A surfactant is also introduced as a cell control agent.

c. Use of Isocyanate Reactions to Prepare Reactive Oligomers

When a macrodiol is reacted with a diisocyanate and a monoalcohol is employed instead of a chain extender, oligomers are formed:

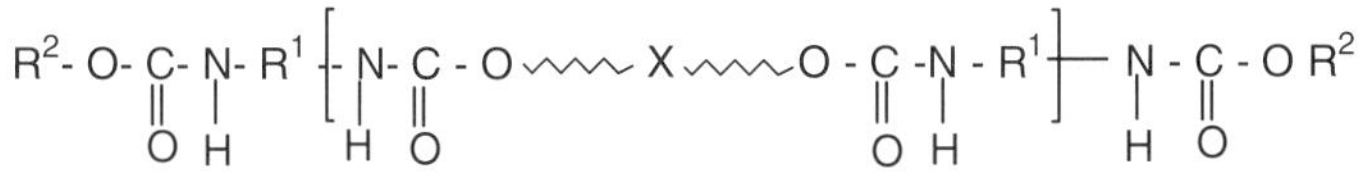

$$R^2-O-\underset{\underset{O}{\|}}{C}-\underset{\underset{H}{|}}{N}-R^1\left[\underset{\underset{H}{|}}{N}-\underset{\underset{O}{\|}}{C}-O\sim\sim X\sim\sim O-\underset{\underset{O}{\|}}{C}-\underset{\underset{H}{|}}{N}-R^1\right]-\underset{\underset{H}{|}}{N}-\underset{\underset{O}{\|}}{C}-O\,R^2$$

If the monoalcohol R^2OH bears some other functional groups, reactive oligomers are obtained. Typically, the monoalcohol is an acrylate or a methacrylate such as hydroxyethylacrylate or methacrylate:

$$CH_2=\underset{\underset{H\,(CH_3)}{|}}{C}-COO\,CH_2\,CH_2\,OH$$

Such oligomers are often used in ultraviolet (UV) cure coatings (free-radical chain polymerization, Sec. 2.3).

2.2.4 Epoxy Reactions

a. A Versatile Monomer

The epoxy or oxirane group is characterized by its reactivity toward both nucleophilic and electrophilic species and it is thus receptive to a wide range of reagents. Epoxy monomers polymerize through step-growth and chain-growth processes. The ionic polymerization initiated by both Lewis bases or acids will be discussed later (Sec. 2.3.4). The case of polyaddition polymerizations is mainly represented by epoxy–amine reactions.

b. Epoxy Monomers ("Resins") Preparation

A major type of epoxy monomers is that derived from the reaction of bis(4-hydroxy phenylene)-2,2 propane (called bisphenol A) and 1-chloropropene 2-oxide (called epichlorohydrin) in the presence of sodium hydroxide (condensation reaction). The structure of the major product, bisphenol A diglycidyl ether (DGEBA) and its condensed forms (Table 2.7a), is dependent upon the stoichiometry of the reactants.

Typically "resins" are marketed with $\bar{n}$ in the range 0.15–10. They are liquid up to $\bar{n} = 0.5$ and vitreous ($T_g \sim 40$–$90°C$) for higher $\bar{n}$.

Another major group of epoxy monomers derived from epichlorohydrin is that comprising monomers synthesized with an aromatic amine, such as aniline (DGA), para-aminophenol (TGpAP) and methylene dianiline (TGMDA) – (Table 2.7e, f, and g). The reaction of epichlorohydrin with an alcohol is more difficult. Liquid monomers based on butanediol, neopentylglycol, and polypropylene oxide (molar mass in the range of 500 g mol^{-1}), are the most common.

Another approach to the formation of the oxirane groups is the peroxidation of a carbon–carbon double bond. Various types of oligomers fall into this category: epoxidized oils, epoxidized rubbers, and cycloaliphatic oxides (Table 2.7h).

c. Epoxy–Amine Reaction

The polyaddition reaction involves epoxy groups reacting with primary and secondary amines. Amines are the most commonly used curing agents/hardeners for epoxides. One epoxy ring reacts with each aminoproton:

$$\mathrm{E\text{-}\underset{\diagdown O \diagup}{CH\text{-}CH_2}} + \mathrm{ANH_2} \xrightarrow{k_1} \mathrm{E\text{-}\underset{OH}{\underset{|}{CH}}\text{-}CH_2\text{-}\underset{H}{\underset{|}{N}}\text{-}A} \tag{2.37}$$

Table 2.7 Main commercial epoxy monomers (resins)

(a) *Diglycidyl ether of bisphenol A, DGEBA*

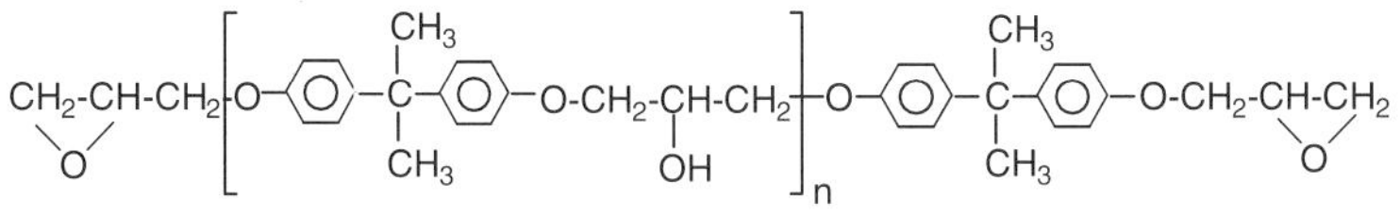

(b) *Diglycidyl ether of bisphenol F, DGEBF*

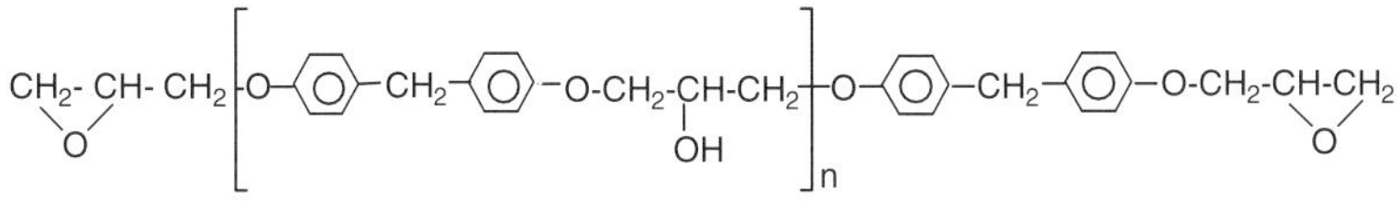

(c) *Epoxy novolac, EN*

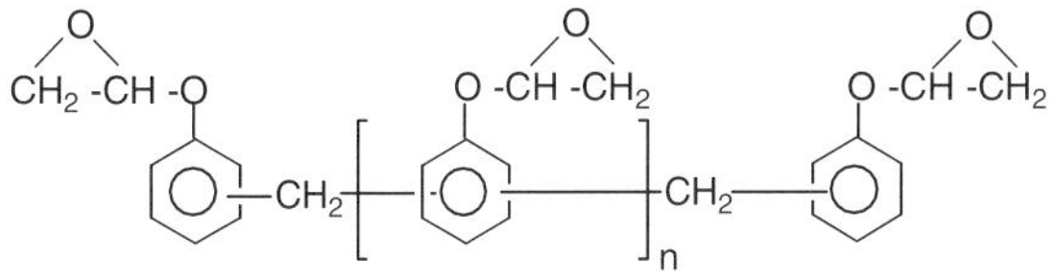

(d) *Triglycidyl ether of tris(hydroxyphenyl)methane*

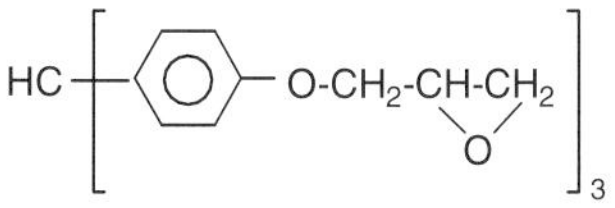

(e) *N,N-diglycidylaniline, DGA*

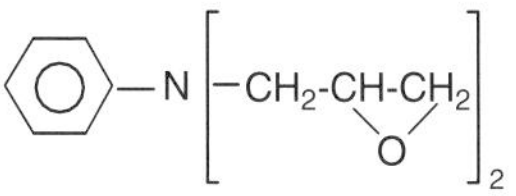

(f) *Triglycidylparaaminophenol, TGpAP*

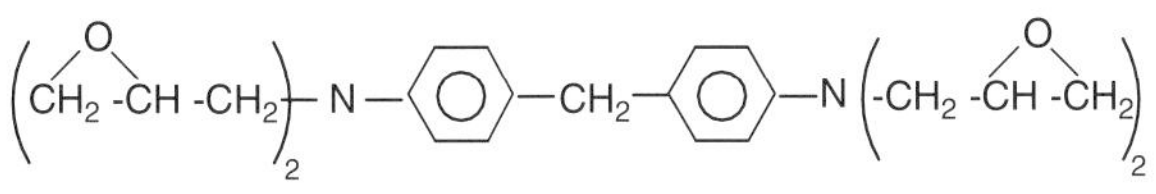

TABLE 2.7 Continued

(g) *N,N,N′,N′-tetraglycidyl-4,4′-methylene dianiline (TGMDA)*

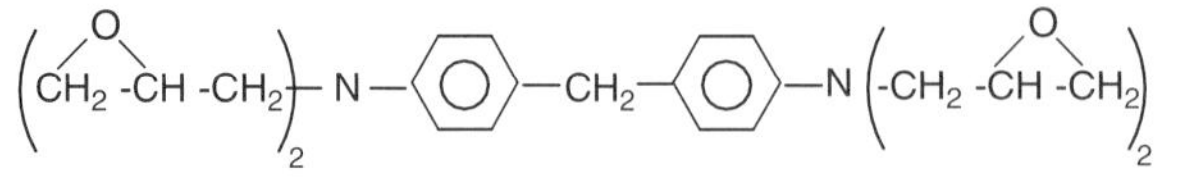

$$\left(\overset{\overbrace{\quad O \quad}}{CH_2\text{-}CH}\text{-}CH_2\right)_2 N-C_6H_4-CH_2-C_6H_4-N\left(\text{-}CH_2\text{-}\overset{\overbrace{\quad O \quad}}{CH\text{-}CH_2}\right)_2$$

(h) *3,4-epoxycyclohexyl methyl 3′,4′-epoxy cyclohexane carboxylate*

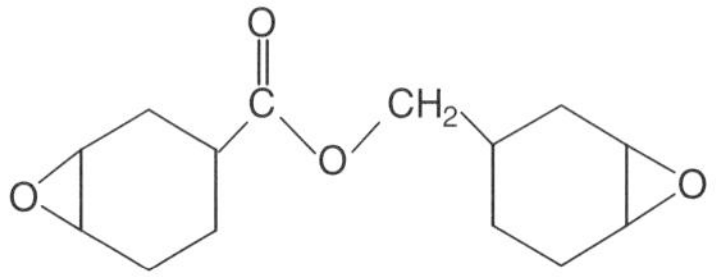

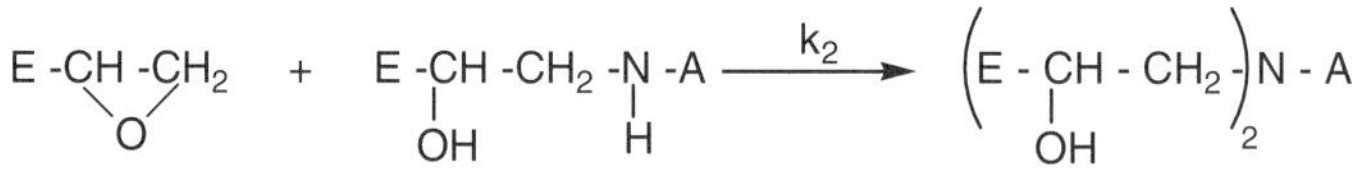

$$\text{(3,4-epoxycyclohexyl)}-\overset{O}{\overset{\|}{C}}-O-CH_2-\text{(3,4-epoxycyclohexyl)}$$

$$E\text{-}\underset{\diagdown O \diagup}{CH\text{-}CH_2} + E\text{-}\underset{OH}{\underset{|}{CH}}\text{-}CH_2\text{-}\underset{H}{\underset{|}{N}}\text{-}A \xrightarrow{k_2} \left(E\text{-}\underset{OH}{\underset{|}{CH}}\text{-}CH_2\text{-}\right)_2 N\text{-}A \tag{2.38}$$

Usually when the concentration of epoxy groups is equal to or lower than the concentration of NH groups, side reactions do not take place. The reactivity of the amine increases with its nucleophilic character: aliphatic > cycloaliphatic > aromatic.

The two amino hydrogens have initially the same reactivity but once the first one has reacted, the secondary amine formed may be less reactive. For aliphatic amines both primary and secondary amino hydrogens have approximately the same reactivity. But for aromatic amines, the reactivity of the secondary amine is typically 2 to 5 times less than the reactivity of a primary amino hydrogen. This change in reactivity is called the "substitution effect".

Hydroxyl groups (water, alcohol, phenol, acid, etc.) catalyze the reaction through the formation of a trimolecular complex, which facilitates the nucleophilic attack of the amino group:

$$
\begin{array}{c}
A \\
| \\
H - \ddot{N} - H \\
\downarrow \\
E - CH - CH_2 \\
\diagdown O \diagup \\
\ddot{} \\
\downarrow \\
HOB
\end{array}
$$

As secondary alcohols are continuously generated (Eqs (2.37) and (2.38)), epoxy-amine reactions are autocatalyzed.

When there is an excess epoxy groups or when the secondary amino groups have a low reactivity, a third reaction can compete with the two previous ones:

$$
E - \underset{\diagdown O \diagup}{CH - CH_2} + \underset{OH}{\overset{\diagdown\ \diagup}{CH}} \xrightarrow{k_3} E - \underset{OH}{CH} - CH_2 - O - \overset{\diagup}{\underset{\diagdown}{CH}} \qquad (2.39)
$$

The epoxy–hydroxyl reaction (etherification) modifies the initial stoichiometric ratio based on epoxy to amino hydrogen groups.

Other factors may also influence the path of the curing reaction, such as the presence of a catalyst, or of an initiator (see Sec. 2.3.4).

Table 2.8 gives the main diamine hardeners used for the crosslinking of epoxides. Aliphatic amines are used for low-temperature curing systems (adhesives, coatings, etc.), and aromatic diamines for composite materials. Cyanoguanidine, or dicyanodiamide (Dicy), is a very versatile hardener widely used in one-pack epoxy formulations for prepregs, laminates, powder coatings, etc. Its latency as a curing agent lies in its high melting point ($T_m = 207°C$) and in its low solubility in epoxy monomers. The reactions of Dicy with epoxy groups are very complex, complicated by the fact that "an accelerator" such as a tertiary amine is often used (this accelerator works like an initiator for anionic chain polymerization – see Sec. 2.3.4 – and the reaction is a mixture of polyaddition and chain polymerization).

Most of the matrices for aerospace composites are based on a combination of TGMDA and DDS and/or Dicy. Again reactions are very complex and different reaction paths could be considered: additions, etherifications, but also cyclizations introduced by the neighboring N,N-diglycidyl groups:

Table 2.8 Main diamine hardeners used in epoxy systems

(a) *Aliphatic*

- $H_2N-(CH_2)_2-[NH-(CH_2)_2]_i-NH_2$ — liquid

Diethylene triamine, DETA, i = 1; triethylene tetramine, TETA, i = 2; tetraethylene pentamine, TEPA, i = 3

- N-aminoethylpiperazine 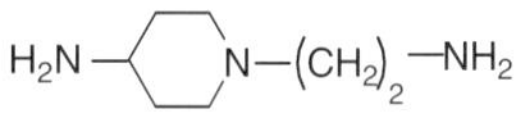 — liquid

(b) *Cycloaliphatic*

- Isophorone diamine, IPD 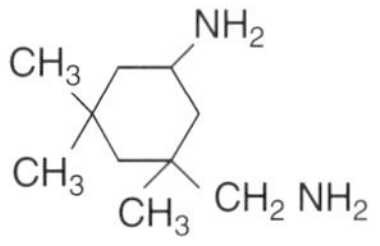 — liquid

- 4,4′-Diamino dicyclohexyl methane, PACM — $T_m \sim 40°C$

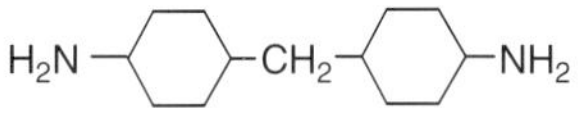

- 4,4′-Diamino 3,3′-dimethyl dicyclohexyl methane, 3 DCM — liquid

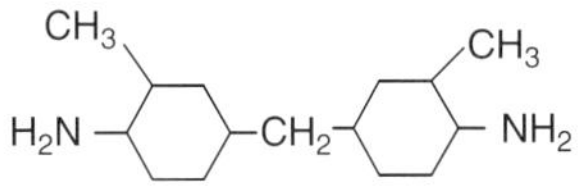

(c) *Aromatic*

- m-Phenylene diamine, mPDA 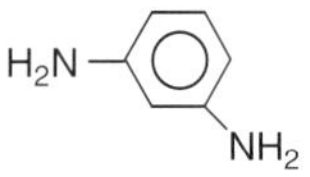 — $T_m \sim 80°C$

- 4,4′-Diamino diphenylmethane, DDM or MDA — $T_m \sim 80°C$

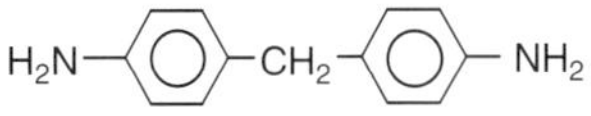

and also aromatic di- or trisubstituted Me, Et, Cl ...

TABLE 2.8 Continued

• 4,4′-Diamino diphenyl sulfone, DDS	$T_m \sim 175°C$
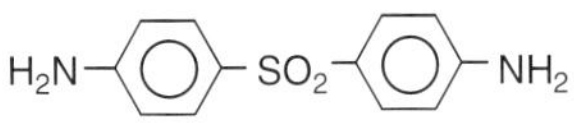 $H_2N-C_6H_4-SO_2-C_6H_4-NH_2$	
(d) *Latent hardener*	
• Cyanoguanidine or dicyanodiamide, Dicy	$T_m \sim 207°C$
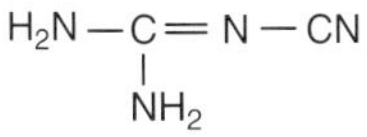 $H_2N-C(NH_2)=N-CN$	

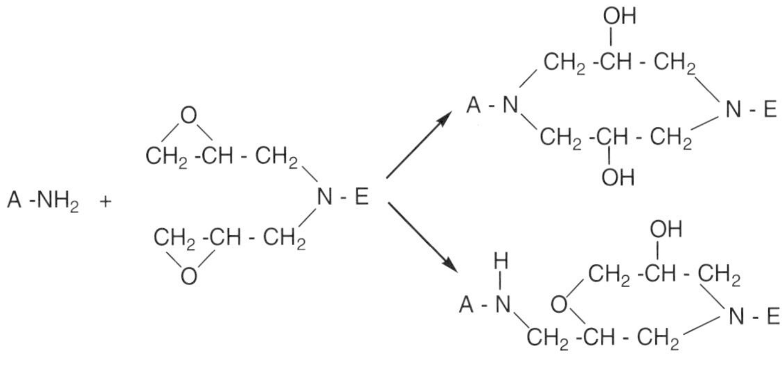

$$A\text{-}NH_2 + (\text{CH}_2\text{-}\overset{O}{\text{CH}}\text{-CH}_2)_2N\text{-E} \rightarrow A\text{-}N(\text{CH}_2\text{-CH(OH)-CH}_2)_2N\text{-E} \quad \text{or} \quad A\text{-}NH\text{-}\text{CH}_2\text{-CH-CH}_2\text{-}N(\text{E})\text{-CH}_2\text{-CH(OH)-CH}_2\text{-O} \tag{2.40}$$

d. Other Polyaddition Reactions of Epoxy Groups

Some of these reactions are listed in Table 2.9, but they are not as clear as they are described in the table because catalysts that can also initiate a chain polymerization (tertiary amines, triphenylphosphine, imidazoles, chromates, etc; see Sec 2.3.4) are practically always used.

2.2.5 Formaldehyde Reactions

a. The Major Use of Formaldehyde

Thermosetting polymers based on the reaction of formaldehyde ($H_2C = O$) with phenol, urea, or melamine are used for many applications, such as

TABLE 2.9 Different addition reactions of the epoxy groups

Reaction			Product
E - CH - CH_2 (epoxide ring, -O-) reacted with			
• phenol	C_6H_5—OH	⟶	E - CH(OH) - CH_2 - O - C_6H_5
• mercaptan	R-SH	⟶	E - CH(OH) - CH_2 - S - R
• isocyanate	R-NCO	⟶	E - CH - CH_2 - N(R) - C(=O) - O - (five-membered ring, O bonded back to CH)
but RNCO also reacts with hydroxyl groups			
• acid	RCOOH	⟶	E - CH(OH) - CH_2 - O - C(=O) - R
but also:			
transesterification:			E - CH(CH_2OH) - O - C(=O) - R
esterification:			>CH - O - C(=O) - R
etherification:			>CH - O - CH_2 - CH(OH) - E

coatings, molding compounds, and matrices for glass fibers or wood composites.

b. Reaction of Formaldehyde with Phenols: "Phenolic Resins"

Phenolic prepolymers are obtained by step-growth polymerization of a difunctional monomer, CH_2O, with phenol or substituted phenols. Phenol has three reactive positions – ortho, ortho′, and para – and behaves as a trifunctional monomer. CH_2O is present in aqueous solution as methyleneglycol, which is the key monomer species in the synthesis:

$$CH_2=O + H_2O \rightleftharpoons HO-CH_2-OH \tag{2.41}$$

Novolacs are obtained by the reaction of phenol and formaldehyde in acidic conditions. Novolac oligomers are linear or slightly branched addition products linked by methylene bridges (molar masses in the range 500–5000 g mol^{-1}). The reaction is usually carried out using a molar ratio CH_2O/PhOH close to 0.8, to avoid gelation in the reactor (see Chapter 3):

$$HOCH_2OH + H^+ \rightleftharpoons {}^+CH_2OH + H_2O \tag{2.42}$$

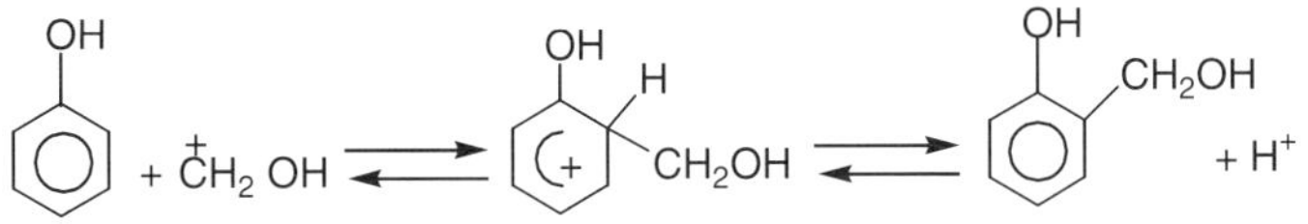

$$\text{PhOH} + {}^+CH_2OH \rightleftharpoons [\text{ring}^+(H)(CH_2OH)\text{-OH}] \rightleftharpoons o\text{-}HOCH_2C_6H_4OH + H^+ \tag{2.43}$$

Hydroxymethylated phenols cannot be isolated, and the reaction continues:

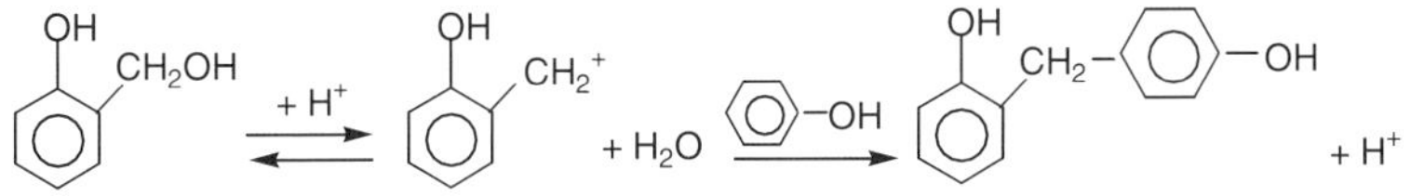

$$o\text{-}HOCH_2C_6H_4OH \xrightarrow{+H^+} o\text{-}{}^+CH_2C_6H_4OH + H_2O \xrightarrow{C_6H_5OH} o\text{-}HOC_6H_4CH_2\text{-}C_6H_4\text{-}OH + H^+ \tag{2.44}$$

The last reaction step, Eq. (2.44), is 10 times faster than the initial one, Eq. (2.43). Strongly acid-catalyzed prepolymers contain 50–75% 2,4′-linkages (Eq. 2.44). At typical molar masses of 500–1000 g mol^{-1}, novolac molecules are essentially linear because of the much lower reactivity of double-reacted phenolic units. Branching is observed for prepolymers of higher molar masses.

In the range of pH 4–7, formaldehyde substitution of the phenolic ring is also possible using particular catalysts such as zinc or calcium acetate. Novolacs produced in these conditions exhibit a high content of 2,2′-methylene units:

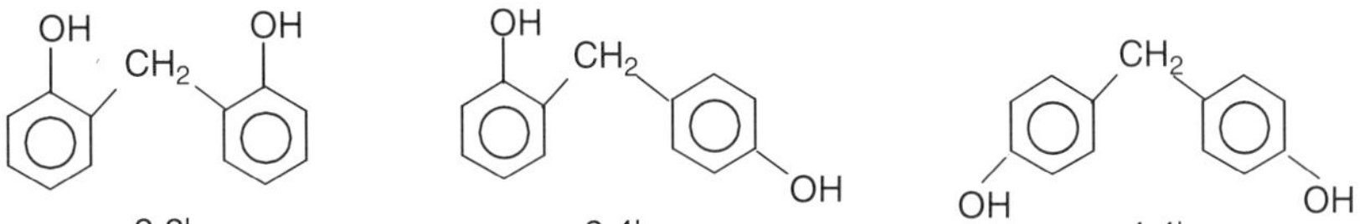

OH CH2 OH — 2,2'

OH CH2 … OH — 2,4'

CH2, OH, OH — 4,4'

These prepolymers are cured by the addition of a hardener, the most common one being hexamethylenetetramine, HMTA (typically 5–15 wt %):

$$6CH_2O + 4NH_3 \rightleftharpoons (CH_2)_6N_4 + 6H_2O \tag{2.45}$$

The reaction mechanism is complex and leads to a polymer network containing as much as 75% of the initial nitrogen chemically bound.

Resols are obtained by phenol–formaldehyde reaction under alkaline conditions. The phenoxide anion is the reactive species:

$$C_6H_5O^- + \overset{\delta_+}{CH_2} = \overset{\delta_-}{O} \longrightarrow C_6H_5(=O)(H)(CH_2O^-) \longrightarrow {}^-O\text{–}C_6H_4\text{–}CH_2OH \tag{2.46}$$

The phenoxide anion is substituted via the electrophilic formaldehyde species in ortho and para positions. Phenol is a weak acid ; alkyl phenols are slightly less acidic and hydroxymethylphenols are more acidic than phenols. In contrast to the reaction under acidic conditions (novolac synthesis), the substituted phenols can be isolated in this alkali-catalyzed reaction.

An excess of formaldehyde is used, generally from 1.5–3. Oligomers consist of mono or polynuclear hydroxymethylphenols, which are not stable at room temperature.

Due to the presence of reactive CH_2OH groups, resol oligomers may be converted into highly crosslinked products without the addition of hardeners. Heat curing is conducted at $T \sim 130$–$200°C$. The polycondensation mechanisms are complex and different bridges are possible: $–CH_2–O–CH_2–$ and $–CH_2–$. The latter is thermodynamically the most stable. Therefore the methylene bridges are the prevalent crosslinks in cured resols.

$$R^1\text{–}C_6H_3(OH)\text{–}CH_2OH + R^2\text{–}C_6H_4OH \longrightarrow R^1\text{–}C_6H_3(OH)\text{–}CH_2\text{–}C_6H_3(OH)\text{–}R^2 + H_2O \tag{2.47}$$

Equation (2.47) is quite similar to Eq. (2.44), which means that acidic curing of resols is possible and may be achieved at room temperature by a variety of strong organic and inorganic acids.

c. Reaction of Formaldehyde with Urea and Melamine: "Amino Resins"

Compounds with > NH groups can react with aldehydes and ketones to form addition and condensation products. It is possible to divide the addi-

TABLE 2.10 Main reactions during the synthesis of UF or MF oligomers and networks

- Urea = $H_2N{-}C({=}O){-}NH_2$ melamine = 2,4,6-triamino-1,3,5-triazine (ring of alternating C and N, each C bearing NH_2)

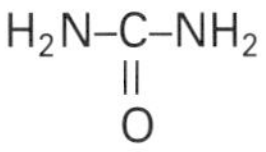

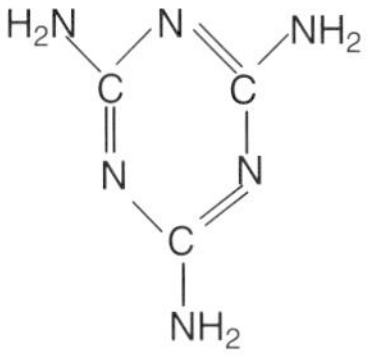

- Addition (T < 100°C)

$$-U-NH_2 + CH_2O \xrightarrow{H^+ \text{ or } B^-} -U-HN-CH_2OH \longrightarrow \cdots \qquad (2.48)$$

with $-U- = \;>C{=}O$ or $>C{=}N-$

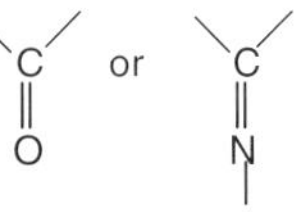

- Condensation

$$-U-NH-CH_2OH + -U-NH_2 \longrightarrow -U-NH-CH_2-NH-U- + H_2O \qquad (2.49)$$

$$2\,-U-NH-CH_2OH \longrightarrow -U-NH-CH_2O-CH_2-NH-U- + H_2O \qquad (2.50)$$

$$-U-NH-CH_2O-CH_2-NH-U- \xrightarrow{T>120^\circ C} -U-NH-CH_2-NH-U- + CH_2O \qquad (2.51)$$

---

tion/condensation of urea and melamine with formaldehyde into three steps (Table 2.10):

1. Formation of hydroxymethyl compounds under acidic or basic conditions at low temperature, 50–100°C (Eq. 2.48). A complicated reaction mixture can result.
2. Condensation of these hydroxymethyl compounds to form methylene or dimethylene ether bridges, leading to oligomers (Eqs (2.49) and (2.50)). At T < 100°C, the reactions are relatively slow.
3. As urea and melamine have a functionality higher than 2, a network is rapidly formed at T > 100°C. As methylene groups are thermodynamically more stable, some rearrangements are also possible at these temperatures (Eq. (2.51)).

Similar structures result when CH_2O is co-condensed with both phenol and melamine or urea.

Molding compounds were among the earliest applications for solid phenolic resins. The molded articles exhibit high-temperature, flame, and chemical resistance; retention of modulus at elevated temperatures; and hardness. Systems with good electrical properties can be formulated at low costs. Adhesive materials and friction materials (brakes) are made from molding compounds.

Laminate manufacture involves impregnation of glass cloths with a liquid "resin" in a dip coating operation. The treated glass cloths are dried in an oven and the "resin" pre-cured. Even if most of the products are applied as aqueous solutions of the prepolymers, powdered materials are also available. The advantage of the powdered products is their better chemical stability compared to the aqueous solutions. The "resin" content of the final products lies between 30 and 70%.

Another application for these thermosetting polymers is the production of particle boards where the polymers are used as adhesives and binders for wood particles.

2.2.6 Silanol Condensation

a. Synthesis of Linear Polysiloxanes

Polysiloxanes are another class of linear or cyclic oligomers that are prepared by a nucleophilic substitution. The first reaction takes place between an organohalosilane (typically dichlorosilane) and water (Eq. (2.52)). To increase the molar mass it is necessary to introduce a catalyst (Eq. (2.53)):

$$y\ X-\underset{R^2}{\overset{R^1}{Si}}-X + (y+1)\,H_2O \longrightarrow HO\left(\underset{R^2}{\overset{R^1}{Si}}-O\right)_y H + 2y\,HX \tag{2.52}$$

$$n\ HO\left(\underset{R^2}{\overset{R^1}{Si}}-O\right)_y H \xrightarrow[\text{or } B^-]{H^+} HO\left(\underset{R^2}{\overset{R^1}{Si}}-O\right)_{ny} H + (n-1)\,H_2O \tag{2.53}$$

Silanols are stronger acids than their hydrocarbon homologues. Condensation reactions are governed by the polarity of the Si –O– and –OH bonds. The most widely used product is polydimethylsiloxane, PDMS, $R^1 = R^2 = -CH_3$. The formation of cyclic oligomers always competes with that of linear chains.

Another method of preparing PDMS is the ring-opening polymerization of cyclic oligomers (Sec. 2.3).

b. Polysiloxane Networks

Labile silane groups, e.g., alkoxysilane groups, react through a two-stage mechanism. The first stage is hydrolysis (Eq. (2.54)). The second stage is condensation, either with elimination of water, as in Eq. (2.53), or of an alcohol, as in Eq. (2.55):

$$\equiv \mathrm{Si-OR} + \mathrm{H_2O} \xrightarrow[\text{or } B^-]{H^+} \equiv \mathrm{Si-OH} + \mathrm{ROH} \tag{2.54}$$

$$\equiv \mathrm{Si-OR} + \equiv \mathrm{Si-OH} \xrightarrow[\text{or } B^-]{H^+} \equiv \mathrm{Si-O-Si} \equiv + \mathrm{R\,OH} \tag{2.55}$$

The polymerization of silicon alkoxides, $Si(OR)_4$, to produce inorganic glasses proceeds through these three reactions and is commonly referred to as "the sol–gel" process. Reactions of trialkoxysilane-terminated molecules $R^1Si(OR^2)_3$ are similar and give hybrid organic–inorganic networks, $R^1\text{-}\overset{|}{\underset{|}{Si}}\text{-O-}\overset{|}{\underset{|}{Si}}\text{-}R^1$. These reactions can be generalized to any labile ≡Si–X group (Table 2.11).

TABLE 2.11 Different labile ≡Si–X groups used in silicone formulations

–X groups	Formula
Alkoxy	≡Si–OR
Acyloxy	≡Si–O–COR
Amino	$\equiv Si\text{–}NR^1R^2$
Amido	$\equiv Si\text{–}NR^1\text{–}COR^2$
Enoxy	$\equiv Si\text{–}OCR^1{=}CHR^2$
Aminoxy	$\equiv Si\text{–}O\text{–}NR^1R^2$
Ketiminoxy	$\equiv Si\text{–}O\text{–}N{=}CR^1R^2$

Most of the formulations of silicone elastomers that can be crosslinked at room temperature contain an α, ω-dihydroxylated polysiloxane and a crosslinking agent with more than two labile Si–X groups:

$$\mathrm{HO}\left(\underset{R^2}{\overset{R^1}{\mathrm{Si}}}\mathrm{-O}\right)_n\mathrm{H} + \mathrm{Si\,X_4}\ (\text{or } R^3\text{-}\mathrm{Si\,X_3}) \xrightarrow[\text{and condensation } (H^+ \text{ or } B^-)]{\text{hydrolysis}} \leftarrow \mathrm{Si}\text{-}\mathrm{O}\left(\underset{R^2}{\overset{R^1}{\mathrm{Si}}}\mathrm{-O}\right)_n\mathrm{Si} \rightarrow$$

(Arrows indicate a structure going to infinity) (2.56)

Additives like silica and other inorganic fillers are also introduced. One-pack or two-pack silicone-elastomer formulations find their uses as flowing materials for applications in coatings, adhesive bonding, molding, and electrical insulation or as non-flowing thixotropic materials in the fields of adhesive bonding and sealing.

Instead of α,ω-dihydroxylated polysiloxanes, other α,ω-dihydroxylated organic polymers such as polyether, polyester, or polybutadiene can be used to prepare hybrid organic–inorganic networks.

Another method for preparing polysiloxane networks is to use the hydrosilylation reaction: an SiH group is added to a double bond; e.g., an allyl group with the help of a catalyst (platinum catalyst) :

$$H\left(-\overset{Me}{\underset{Me}{Si}}-O\right)-\overset{Me}{\underset{Me}{Si}}-H + 2\,R\left(CH_2-CH=CH_2\right)_3 \xrightarrow{\text{catalyst}} \;\; >R-CH_2-CH_2CH_2\left(-\overset{Me}{\underset{Me}{Si}}-O\right)-\overset{Me}{\underset{Me}{Si}}-CH_2-CH_2-CH_2-R< \tag{2.57}$$

2.2.7 Miscellaneous Systems

a. Aceto–Acetic Esters

Some possible reactions of the aceto–acetoxy (or acetylacetoxy) group are described in Table 2.12. Ways to synthesize aceto–acetoxy functional polymers or oligomers are shown in Table 2.13.

b. Azlactone Chemistry for Coatings

Vinyl azlactone can be also copolymerized with acrylic monomers to give functionalized polymers. The azlactone-functional copolymers, P-, can be reacted with amine-, thiol-, and hydroxyl-functional molecules (Heilmann *et al.*, 1984):

$$P\text{-azlactone} + RH \xrightarrow[RT]{H^+} P-C(=O)-NH-C(CH_3)_2-C(=O)-R \tag{2.66}$$

TABLE 2.12 Some reactions of the aceto–acetoxy group

(1) *Reactions of the keto form* $Ⓟ\text{-O-C(=O)-CH}_2\text{-C(=O)-CH}_3$ with

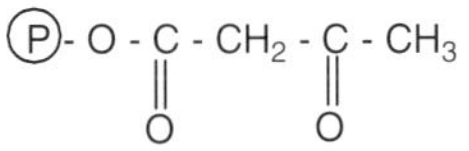

• melamine–formaldehyde oligomers end-capped with methanol

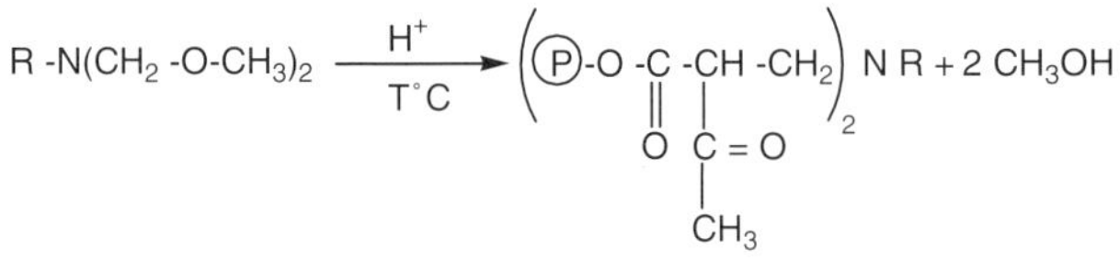

$$\text{R-N(CH}_2\text{-O-CH}_3)_2 \xrightarrow[T^\circ C]{H^+} \left(Ⓟ\text{-O-C(=O)-CH(C(=O)CH}_3\text{)-CH}_2\right)_2 \text{NR} + 2\,\text{CH}_3\text{OH} \tag{2.58}$$

• isocyanate

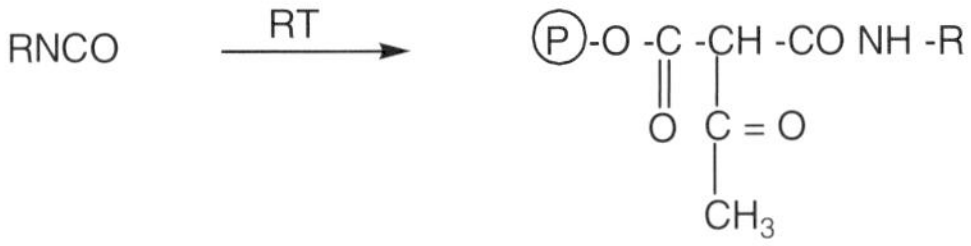

$$\text{RNCO} \xrightarrow{RT} Ⓟ\text{-O-C(=O)-CH(C(=O)CH}_3\text{)-CONH-R} \tag{2.59}$$

• acrylate (Michael reaction)

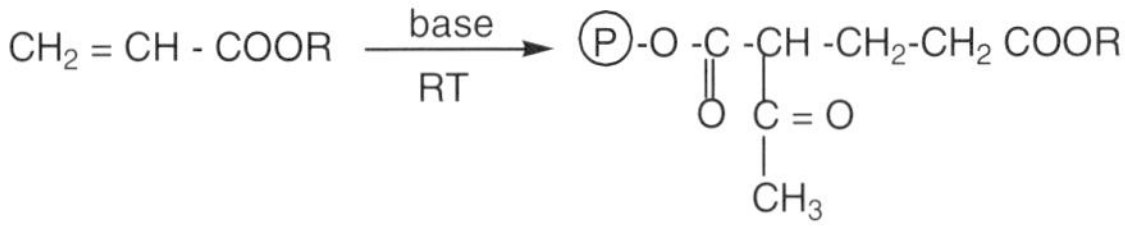

$$\text{CH}_2=\text{CH-COOR} \xrightarrow[RT]{base} Ⓟ\text{-O-C(=O)-CH(C(=O)CH}_3\text{)-CH}_2\text{-CH}_2\text{COOR} \tag{2.60}$$

• aldehydes

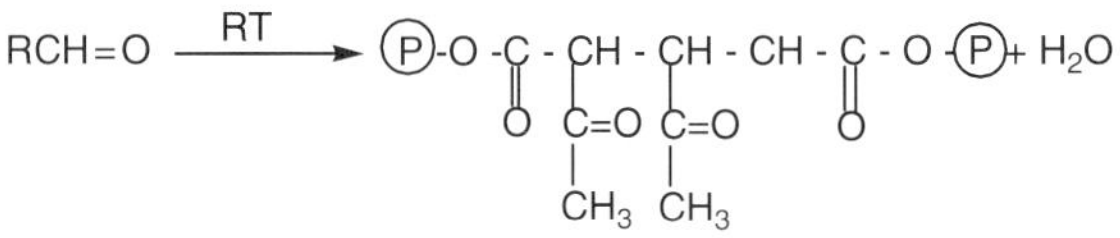

$$\text{RCH=O} \xrightarrow{RT} Ⓟ\text{-O-C(=O)-CH(C(=O)CH}_3\text{)-CH(C(=O)CH}_3\text{)-CH-C(=O)-O-}Ⓟ + \text{H}_2\text{O} \tag{2.61}$$

(2) *Reactions of the enol form* $Ⓟ\text{-O-C(=O)-CH=C(OH)-CH}_3$ with

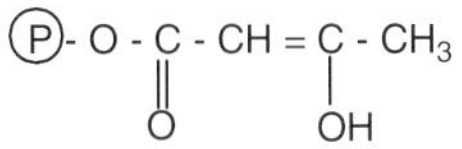

• amine

$$\text{RNH}_2 \xrightarrow{RT} Ⓟ\text{-O-C(=O)-CH=C(CH}_3\text{)-NHR} + \text{H}_2\text{O} \tag{2.62}$$

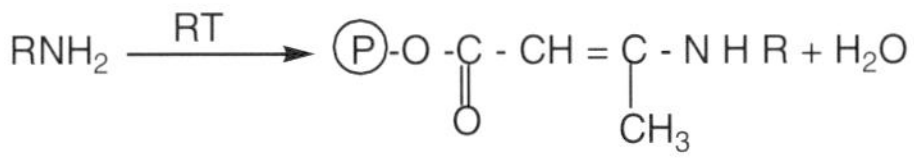

• metal ions chelation

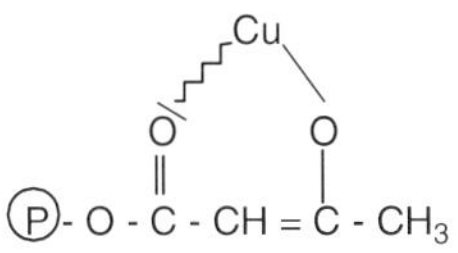

$$Ⓟ\text{-O-C(=O}\cdots\text{Cu)-CH=C(O-Cu)-CH}_3$$

Table 2.13 Some synthesis of aceto–acetoxy functional polymers or oligomers, Ⓟ

(1) *Radical chain-growth copolymerization with aceto-acetoxy ethyl methacrylate AAEM*

$$CH_2 = C(CH_3)\text{-}COOR + CH_2 = C(CH_3)\text{-}C(=O)\text{-}O(CH_2)_2\text{-}O\text{-}C(=O)\text{-}CH_2\text{-}C(=O)\text{-}CH_3 \tag{2.63}$$

(2) *Use of diketene*

$$Ⓟ\text{-}OH + \underset{\text{diketene}}{\overset{O=}{C}\text{—}O\text{—}C(=CH_2)\text{—}CH_2} \longrightarrow Ⓟ\text{-}O\text{-}C(=O)\text{-}CH_2\text{-}C(=O)\text{-}CH_3 \tag{2.64}$$

(3) *Use of methyl- or ethyl-acetoacetate*

$$Ⓟ\text{-}OH + R\text{-}O\text{-}C(=O)\text{-}CH_2\text{-}C(=O)\text{-}CH_3 \longrightarrow Ⓟ\text{-}O\text{-}C(=O)\text{-}CH_2\text{-}C(=O)\text{-}CH_3 + ROH \tag{2.65}$$

c. Cyanate–Ester Reactions

All the cyanated phenols polymerize by a trimerization mechanism to give cyanurate ring systems:

$$3\ Ar\text{-}OC\equiv N \xrightarrow[\text{catalyst ArOH}]{\Delta} (ArO\text{-}C{=}N)_3 \text{ (cyanurate ring, 2,4,6-tris(aryloxy)-1,3,5-triazine)} \tag{2.67}$$

The extreme sensitivity of cyanate chemistry to experimental conditions and minor traces of impurities makes the reaction of cyanate not as simple as described by Eq. (2.67).

Polycyanurate networks are used for electronics or structural materials applications.

d. Bismaleimide, BMI Reactions

Aromatic maleimides have rather high melting points. The maleimide double bond is strongly electrophilic and can react with nucleophilic reagents such as amines and thiols:

$$\text{maleimide(N-Ar)} + RXH \longrightarrow R\text{-}X\text{-succinimide(N-Ar)} \qquad X = NH \text{ or } S \tag{2.68}$$

Many side reactions can occur because when BMIs are heated above 200°C, double bonds can also polymerize by a chain polymerization mechanism with the formation of a crosslinked polymer. A typical formulation contains a molar excess of BMI with respect to a diamine. Linear maleimide end-capped aspartimides are formed, which can then be crosslinked by further heating with or without an initiator ("a catalyst").

The maleimide C=C double bond can undergo a Diels–Alder reaction with dienes: for example, reaction with o,o′-diallylphenols is thought to proceed initially by ene-synthesis, followed by Diels–Alder reaction:

$$\text{R-C}_6\text{H}_3(\text{OH})\text{-CH}_2\text{-CH=CH}_2 + \text{maleimide(N-Ar)} \longrightarrow \text{R-C}_6\text{H}_3(\text{OH})\text{-CH=CH-CH}_2\text{-succinimide(N-Ar)} \xrightarrow{+\text{maleimide}} \text{Diels–Alder adduct (OH; two N-Ar imide rings)} \tag{2.69}$$

e. Other Thermally Stable Polymers

Polycyanurates and polyimides can be classified as thermally stable polymers. Other groups that react at high temperatures (higher than the melting temperature for crystalline monomers) can be used (Table 2.14). The reaction may be classified as an addition polymerization, but the mechanisms are very complex and not always well known.

Monomers bearing these functional groups must also be thermally stable. The main types of chemical compounds used are imides, aromatic ether sulfones, aromatic sulfides, aromatic esters, aromatic ketones or

TABLE 2.14 Some reactive groups for preparing thermostable networks

Cyanamide	$-NH-C\equiv N$
Nitrile	$-C\equiv N$
Nadimide	$-N(CO)_2$ (norbornene dicarboximide)
Acetylene	$-C\equiv CH$
Propargyl	$-CH_2-C\equiv CH$
Phthalonitrile	$-C_6H_3(CN)_2$

heterocyclic compounds. Nucleophilic (Eqs (2.70), (2.71), (2.72)) or electrophilic (Eq. (2.73)) substitution reactions in polar aprotic solvents can be used to prepare these monomers/oligomers:

$$Ar^1-ONa + Cl-CH_2Ar^2 \longrightarrow Ar^1-O-CH_2-Ar^2 + NaCl \qquad (2.70)$$

$$Ar^1-ONa + ClAr^2 \longrightarrow Ar^1-O-Ar^2 + NaCl \qquad (2.71)$$

$$n\,X-Ar-SH \longrightarrow X{\left[Ar-S\right]}_n-H + (n-1)\,HX \qquad (2.72)$$

$$Ar^1 + 2\,X-\underset{\underset{O}{\|}}{C}-Ar^2 \xrightarrow[AlCl_3]{} Ar^2-\underset{\underset{O}{\|}}{C}-Ar^1-\underset{\underset{O}{\|}}{C}-Ar^2 + 2HX \qquad (2.73)$$

2.3 CHAIN POLYMERIZATIONS

2.3.1 Description of the Different Stages

a. Three Stages

An addition chain polymerization has three clearly defined reaction stages. The chain reaction must be started by first providing a suitable active center, I*, capable of reacting with the monomer, M (initiation stage). The addition reaction of the active center with the monomer is the second stage of the reaction. The active center is regenerated at every step, allowing the reaction to continue (propagation stage). The third and last stage is the termination of the growing chain through bimolecular termination or transfer reaction (termination stage).

When one monomer has two or more reactive sites (double bonds, cycles, etc.), thermosetting polymers may be obtained.

b. Initiation and Types of Active Centers

The active center I*, which first initiates and then propagates the chain, can be a free radical, an anion, a cation or a transition-metal based initiator (Ziegler–Natta systems or metathesis reactions).

Initiation can be described by

$$I^* + M \longrightarrow IM^* \qquad (2.74)$$

The rate of this reaction is generally very fast compared to the rate of the reaction giving I*.

Free radicals are certainly the most commonly used type of active centers. They are produced by decomposition of a suitable molecule, an initiator, either thermally or by irradiation. Typical initiators are given in Table 2.15. Homolytic scission of a σ bond leads to two free radicals. Depending on the selected initiator, two identical or two different radicals can be formed.

For each initiator there is a useful temperature range for which the initiator decomposition rate constant, k_d, will produce radicals at suitable rates for polymerization. The initiation rate is usually controlled by the decomposition rate of the initiator, which depends directly on its concentration (first-order reaction). The temperature window can be enlarged by the use of catalysts such as a tertiary amine (Eq. (2.80)), or an organometallic compound in a redox reaction (Eqs (2.81) and (2.82)).

Another effective method to initiate a radical chain polymerization is to use a photosensitive initiator which absorbs UV radiation and decomposes to produce radicals. This method has the advantage that a specific temperature is not needed. Radical production starts immediately when the system is exposed to UV radiation and stops when it is switched off.

Another possibility is to irradiate the monomer with high-energy ionizing radiation such as X-, or β-, or γ-rays, a procedure used in some industrial processes even if the mechanisms by which active centers are generated are not well known.

Ionic initiators are much more specific toward the type of monomer and can initiate polymerization in some cases where radical initiation is ineffective (cyclic, vinyl ether, or allylic monomers). Monomers suitable for anionic initiation are those with an electron-withdrawing substituent attached to the double bond (phenyl, carbonyl, etc.). The reaction consists

Table 2.15 Some radical initiators for chain polymerization

- Peroxides: benzoyl peroxide (BPO)

$$Ph-\underset{\underset{O}{\|}}{C}-O-O-\underset{\underset{O}{\|}}{C}-Ph \longrightarrow 2\,Ph-\underset{\underset{O}{\|}}{C}-O^{\bullet} \longrightarrow 2Ph^{\bullet}+2CO_2 \quad (2.75)$$

- Hydroperoxides: cumyl peroxide

$$Ph-\underset{Me}{\overset{Me}{C}}-O-OH \longrightarrow Ph-\underset{Me}{\overset{Me}{C}}-O^{\bullet}+OH^{\bullet} \quad (2.76)$$

- Peresters: tertbutyl perbenzoate

$$Ph-\underset{\underset{O}{\|}}{C}-O-O-tBu \longrightarrow Ph-\underset{\underset{O}{\|}}{C}-O^{\bullet}+tBuO^{\bullet} \quad (2.77)$$

- Azocompounds: 2,2′-azobisisobutyronitrile (AIBN)

$$(Me)_2\underset{CN}{C}-N=N-\underset{CN}{C}(Me)_2 \longrightarrow Me_2-\underset{CN}{C^{\bullet}}+N_2 \quad (2.78)$$

- Organometallics

$$Mt+RCl \longrightarrow Mt^{+}Cl^{-}+R^{\bullet} \quad (2.79)$$

- Catalyzed peroxide decomposition

$$Ph-\underset{\underset{O}{\|}}{C}-O-O-\underset{\underset{O}{\|}}{C}-Ph+R_3N \longrightarrow R_3\overset{+}{N}^{\bullet}\ {}^{-}O-\underset{\underset{O}{\|}}{C}-Ph+Ph-\underset{\underset{O}{\|}}{C}-O^{\bullet} \quad (2.80)$$

$$Ph-\underset{\underset{O}{\|}}{C}-O-O-\underset{\underset{O}{\|}}{C}-Ph+Co^{2+} \xrightarrow[\text{medium}]{\text{organic}} Co^{3+}+Ph-\underset{\underset{O}{\|}}{C}-O^{-}+Ph-\underset{\underset{O}{\|}}{C}-O^{\bullet} \quad (2.81)$$

and

$$Ph-\underset{\underset{O}{\|}}{C}-O^{-}+Co^{3+} \longrightarrow Co^{2+}+Ph-\underset{\underset{O}{\|}}{C}-O^{\bullet} \quad (2.81\text{ bis})$$

$$H_2O_2+Fe^{2+} \xrightarrow[\text{medium}]{\text{water}} Fe^{3+}+HO^{-}+HO^{\bullet} \quad (2.82)$$

- Photosensitive initiator: benzoin

$$Ph-\underset{\underset{O}{\|}}{C}-\underset{OH}{CH}-Ph \xrightarrow{h\nu} Ph-\underset{\underset{O}{\|}}{C^{\bullet}}+Ph-\underset{OH}{CH^{\bullet}} \quad (2.83)$$

TABLE 2.16 Some examples of ionic initiation reactions

Anionic initiation:

$$Bu^- Li^+ + CH_2{=}CH(C_6H_5) \longrightarrow Bu\text{-}CH_2\text{-}CH(C_6H_5)^- Li^+ \quad (2.84)$$

$$RO^- K^+ + CH_2\text{-}CH_2\ (\text{epoxide, -O-}) \longrightarrow RO\text{-}CH_2CH_2O^- K^+ \quad (2.85)$$

Cationic initiation:

$$BF_3, H_2O + CH_2 = C(CH_3)_2 \longrightarrow CH_3\text{-}C^+(CH_3)_2, BF_3OH^- \quad (2.86)$$

of a nucleophilic attack of the monomer by the carbanion or the oxanion (Eqs (2.84) and (2.85), Table 2.16).

Monomers with electron-donating substituents (methyl, ether, etc.) are the most susceptible to initiation by carbonium or oxonium ions as they are prone to electrophilic attack on the double bond (Eq. (2.86), Table 2.16). In both anion- and cation-initiated polymerizations a counterion is present.

c. Chain Growth

Once initiation has taken place, growth of the polymer chain is effected by the repetitive addition of a monomer to the active center:

$$IM^* + nM \xrightarrow{k_p} IM\left[M\right]_{n-1}M^* \quad (2.87)$$

The propagation step occurs many times during the (very short) lifetime of a particular radical center. This leads to a polymer chain comprising 10^2–10^6 monomer units before termination takes place.

For an ionic center, the propagation rate is generally not so fast as for a radical center. In this case the propagation mechanism involves the insertion of a monomer between the counterion and the carbanion/carbocation on the terminal unit of the growing chain, followed by nucleophilic/electrophilic attack of the anion/cation on the monomer double bond. Thus, the position of the counterion can assist the control of the orientation of the incoming monomer and stereoregular polymer chains can be formed.

d. Termination Reactions

Two quite different mechanisms can occur for termination of the growing chain:

1. Bimolecular termination: two active centers react together and disappear.
2. Transfer reactions.

Chain transfer can occur with molecular species that are not active molecules: solvent molecules, monomers, another polymer chain backbone, or any molecule, generically called TA, susceptible to an abstraction reaction:

$$\sim\!\sim CH_2-\underset{\displaystyle R}{\underset{|}{C}}H^* + TA \longrightarrow \sim\!\sim CH_2-\underset{\displaystyle R}{\underset{|}{C}}H\,A + T^* \tag{2.88}$$

T* can be sufficiently reactive to reinitiate chain growth.

Anionic reactions have no bimolecular termination mechanism and in the absence of impurities (water, alcohols, etc.) or transfer agents, the end remains active indefinitely (living center). Termination reactions are significant for both cationic and free-radical polymerizations.

The bimolecular termination of radical active centers involves combination (Eq. (2.89 a)) or disproportionation reactions (Eq. (2.89 b)):

$$2\ I\sim\!\sim CH_2-\underset{\displaystyle R}{\underset{|}{C}}H^{\bullet} \begin{cases} \xrightarrow{(a)} I\sim\!\sim CH_2-\underset{\displaystyle R}{\underset{|}{C}}H-\underset{\displaystyle R}{\underset{|}{C}}H-CH_2\sim\!\sim I \\ \xrightarrow{(b)} I\sim\!\sim CH_2-\underset{\displaystyle R}{\underset{|}{C}}H_2 + \underset{\displaystyle R}{\underset{|}{C}}H=CH\sim\!\sim I \end{cases} \tag{2.89}$$

As radicals are very active, many transfer reactions can also occur. For linear polymers, such side reactions can lead to branching or even network formation. If during the propagation there is a chain transfer from a growing chain to an inactive chain (hydrogen abstraction, for example) a new radical site is created that is capable of initiating further chain growth (branching) or coupling with another radical site (branching or crosslinking).

A typical effect observed in the synthesis of linear polymers by a free-radical mechanism is the auto-acceleration process. At a particular conversion, when sufficient polymer has accumulated in the system for the viscosity to reach a certain level, the rate of the bimolecular termination reaction begins to fall because of diffusional restrictions to the encounter of two chain ends. However, the initiation and growth rates are hardly affected.

This produces a rise in the concentration of active centers and a corresponding increase in the propagation rate. Chains produced at this stage are longer, and this leads to a broadening of the molar mass distribution. The term "gel effect" is widely used to describe this effect, although no gel is actually formed in the system. The effect is also called the "Trommsdorff effect" (see Chapter 5).

e. Competition between the Different Stages. Control of the Polymerization Process

The competition between the different stages of the chain polymerization is mainly a kinetic problem (see Chapter 5). There are two extreme cases:

1. A case classically associated with radical chain polymerization for which a (pseudo)steady state is assumed for the concentration of active centers; this condition is attained when the termination rate equals the initiation rate (the free-radical concentration is kept at a very low value due to the high value of the specific rate constant of the termination step). The propagation rate, is very much faster than the termination rate, so that long chains are produced from the beginning of the polymerization. For linear chains, the polydispersity of the polymer fraction varies between 1.5 and 2.
2. A case associated with anionic polymerization with no bimolecular termination reaction, for which the initiation rate is generally much faster than the propagation rate. This means that the growth of every chain begins at the same time and that there are

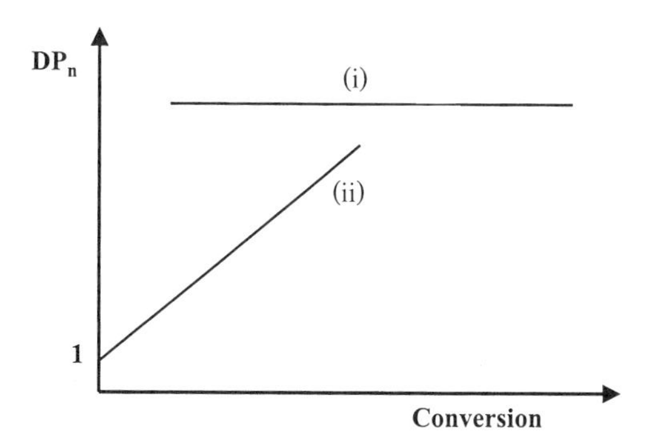

FIGURE 2.2 Evolution of the degree of polymerization, DP_n, of primary chains for (i) free-radical chainwise polymerization and (ii) anionic (living) polymerization.

no long chains formed at the beginning of the polymerization. For linear polymers the polydispersity remains close to 1 (monodisperse polymer).

Figure 2.2 shows the evolution of the degree of polymerization of primary chains for (i) free-radical and (ii) living (anionic) polymerizations.

To obtain a behavior similar to case 2 for free-radical polymerizations, it is necessary to decrease the influence of the bimolecular termination reaction by decreasing the concentration of active centers. This is possible by producing a reversible combination between a growing chain (a polymer radical P*) with a stable radical N* to form an adduct P–N, which behaves as a "dormant" species:

$$P^* + N^* \rightleftharpoons P-N \tag{2.90}$$

N* is active only toward P*, while P* is active not only toward N* and the monomer (propagation, Eq. (2.87)) but also toward P′* (irreversible bimolecular termination, Eq. (2.89)) and toward neutral molecules (chain transfer, Eq. (2.88)). When the last two reactions are unimportant compared with the first two, the system may be viewed as a living (controlled) polymerization.

Several systems have been proposed to fulfill these conditions. Among them is the one mediated by stable nitroxyl radicals such as 2, 2, 6, 6-tetramethylpiperidinyl-1-oxy (TEMPO):

I∼∼CH$_2$-CH$^\cdot$ + $^\cdot$O-N ⇌ ∼∼ CH$_2$-CH-O-N

active chain end dormant chain end

(2.91)

TEMPO can also be introduced by the use of alkoxyamines. These compounds are readily prepared by decomposing an alkyl radical source such as AIBN in the presence of a nitroxide:

N-O-C(Me)(Me)-CN

The alkoxyamines work as iniferters (*ini*tiator, trans*fer* agent, chain *ter*minator). Their C–O bond undergoes reversible homolysis on heating to afford an alkyl radical and a stable nitroxide. The alkyl radical initiates the polymerization and the nitroxide combines with the propagating radical to form the dormant species (Eq. (2.91)).

f. Copolymerization and Networks

The problem of predicting copolymer composition and sequence in the case of chain copolymerizations is determined by a set of differential equations that describe the rates at which both monomers, M_A and M_B, enter the copolymer chain by attack of the growing active center. This requires a kinetic model of the copolymerization process. The simplest one is based on the assumption that the reactivity of a growing chain depends only on its active terminal unit. Therefore when the two monomers M_A and M_B are copolymerized, there are four possible propagation reactions (Table 2.17).

The reactivity ratios, r_A and r_B, express the relative preference of a growing chain for one of the two monomers. In free-radical copolymerizations they are roughly independent of solvent, initiator, and temperature. In ionic copolymerization, however, they depend strongly on the counterion and the solvent.

Special cases can be discussed:

(i) $r_A \sim r_B \sim 1 (k_{AA} \sim k_{AB}$ and $k_{BB} \sim k_{BA})$

Under these conditions, the growing chains find the monomers equally attractive, so that the relative rates of monomer consumption are determined only by the relative monomer concentration in the feed mixture.

TABLE 2.17 Propagation reactions and reactivity ratios for the copolymerization of two monomers

$$\sim\!\sim M_A^* + M_A \xrightarrow{k_{AA}} \sim\!\sim M_A M_A^*$$

$$\sim\!\sim M_A^* + M_B \xrightarrow{k_{AB}} \sim\!\sim M_A M_B^*$$

$$\sim\!\sim M_B^* + M_A \xrightarrow{k_{BA}} \sim\!\sim M_B M_A^*$$

$$\sim\!\sim M_B^* + M_B \xrightarrow{k_{BB}} \sim\!\sim M_B M_B^*$$

and $r_A = \frac{k_{AA}}{k_{AB}}$ $\qquad r_B = \frac{k_{BB}}{k_{BA}}$

(ii) $r_A \sim r_B \sim 0(k_{AA} \sim k_{BB} \sim 0)$

When both reactivity ratios are zero, a chain end cannot add its own monomer. This leads to a perfectly alternating copolymer.

(iii) $r_A > 1; r_B > 1(k_{AA} > k_{AB}$ and $k_{BB} > k_{BA})$

Both active centers prefer to add M_A to the chain, so that the copolymer is always enriched in M_A with respect to the feed.

(iv) $r_A r_B = 1$

Both active centers show the same preference for the addition of one of the monomers.

The free-radical copolymerization of a vinyl monomer with a small amount of a divinyl monomer offers one of the simplest routes to the preparation of polymer networks and gels. Examples are

styrene $CH_2 = CH$–C_6H_5 and divinyl benzene $CH_2 = CH$–C_6H_4–$CH = CH_2$

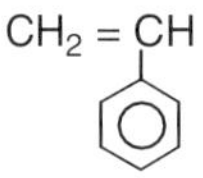

or methyl methacrylate $CH_2 = C(CH_3)COOCH_3$

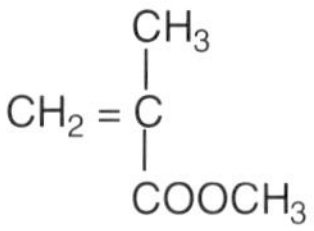

and bis(methacrylate)s $CH_2 = C(CH_3)$–COO - R - OCO–$C(CH_3) = CH_2$

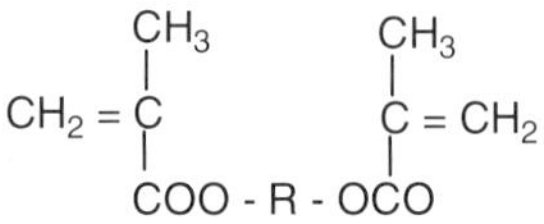

These polymer networks have commercial applications. A number of experimental studies suggest that the network formation by this method proceeds in a highly non-ideal fashion. During copolymerization, a high fraction of pendant vinyls of a primary chain are consumed by intramolecular reactions, causing practically no increase in the molar mass of the system (see Chapter 7):

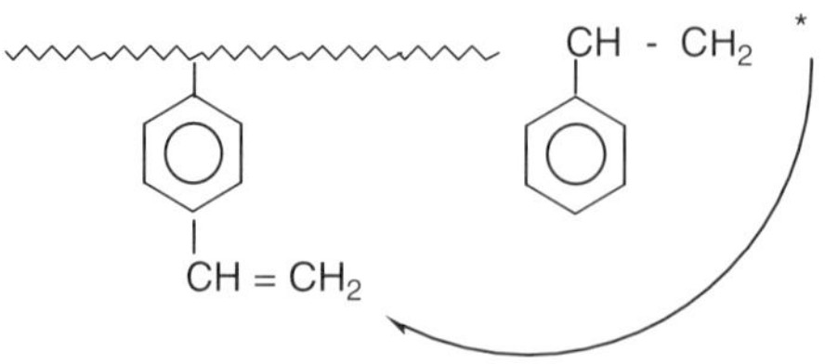

[Scheme: polymer chain bearing a pendant C_6H_4–$CH = CH_2$ group and a radical chain end CH - CH_2* (C_6H_5 substituent); an arrow from the radical end points back to the pendant $CH = CH_2$]

This means that the reactivity of pendant double bonds and the reactivity ratios are modified along the reaction.

2.3.2 Styrene Copolymers and Polyester "Resins"

a. "UP Resins"

In Sec. 2.2.2c the synthesis and the structure of unsaturated polyester oligomers were described. The general UP prepolymer structure is given in Table 2.4, Eq. (2.18). They are solid at room temperature and are usually diluted with styrene (~40 wt%), which behaves as a reactive solvent. The mixture is transformed into crosslinked polymers through a free-radical reaction. Copolymerization occurs between the styrene and the fumarate double bonds. Experimental reactivity ratios are r_F ~0 and r_S ~0.3, which means that there is a strong tendency for alternating copolymerization. UP oligomers are multifunctional monomers, they contain approximately ~ 10 double bonds per molecule depending on the molar mass.

In fact, styrene is generally not a very good solvent for UP oligomers. Figure 2.3 gives a scheme of cloud point curves for UP oligomers dissolved in styrene. The system is initially miscible but during reaction phase separation can occur, leading to UP-rich and styrene-rich phases. This phenomenon amplifies the possibility of intramolecular reactions and the formation of microgels (see Chapter 7).

Depending on the application, the peroxide used can initiate the reaction at room temperature or at a higher temperature, typically 140°C. One problem to solve is the high shrinkage, ~ 10%, occurring during the curing process. The shrinkage is mainly due to styrene polymerization (it amounts

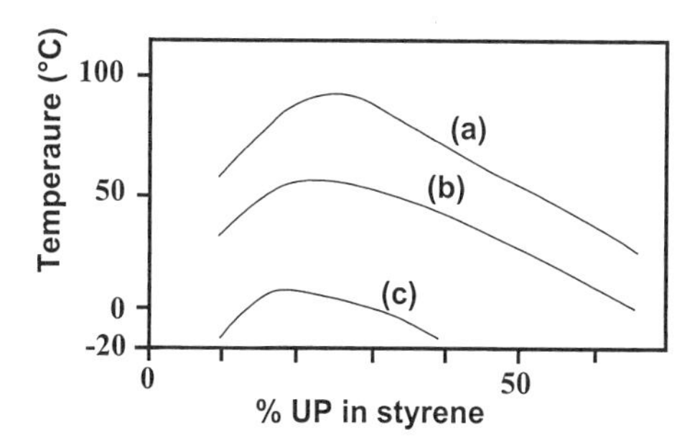

FIGURE 2.3 Cloud point curves of diethanediol-based polyester oligomers (M_n ~ 1 kg mol^{-1}), dissolved in styrene; the diacids used for polyester synthesis were (a) fumaric acid, (b) blend of isophthalic and fumaric acid, and (c) isophthalic acid.

to 17% for pure styrene and 21% for pure methyl methacrylate). To compensate the shrinkage, additives such as poly(vinyl acetate), PVAc, may be introduced in the system. Additives are initially miscible but they phase separate rapidly at the beginning of the reaction (see Chapter 8). This is another reason why the mechanism of this copolymerization is very complex.

Another pratical problem concerns the viscosity of the reactive mixture at low temperatures. For processing purposes it is useful to have a high viscosity at low temperatures but a fast decrease of viscosity when temperature increases. In Sec. 2.2.2c we have indicated that carboxyl and hydroxyl end groups were present in the structure of UP oligomers. One method of increasing the viscosity of the system is to react these carboxyl or hydroxyl groups to increase the average molar mass of the UP oligomer. A usual procedure is to add an oxide such as MgO, which forms a salt by reacting with the terminal COOH groups (Eq. (2.92)). This leads to the formation of ionic clusters that act as physical crosslinks. Another method is to produce a chain extension by reaction of the terminal OH groups with a diisocyanate (Eq. (2.93)).

$$\text{UP-COOH} + \text{MgO} \longrightarrow (\text{UP-COO}^-)(\text{Mg}^{2+})(^-\text{OCO-UP}) + \text{H}_2\text{O} \qquad (2.92)$$

$$\text{UP-OH} + \text{OCN-R-NCO} \longrightarrow \text{UP-OCONH-R-NHCOO-UP} \qquad (2.93)$$

Another method of increasing the viscosity is to use an UP prepolymer that is able to crystallize at room temperature, with a $T_m \sim 50°C$.

A typical UP formulation for sheet molding compounds (SMC) is shown in Table 2.18.

Table 2.18 Typical formulation for SMC

UP prepolymer	100 parts
Styrene	40 parts
Initiator	1–2 parts
TP polymer: typically low-shrinkage additive (PE, PS, etc.) or low-profile additive (PVAc, etc.)	20 parts
Thickening agent: typically MgO	1–3 parts
Fillers (calcium carbonate, alumina, etc.)	150 parts
Short glass fibers	~ 20–40%

b. "Vinyl Ester, VE Resins"

Similarly to UP resins, VE resins consist of an unsaturated oligomer dissolved in styrene. The unsaturated oligomer is based on the epoxy chemistry (Sec. 2.2.4). When one mole of a DGEBA monomer is reacted with 2 moles of (meth)acrylic acid, an α, ω-di(meth)acrylate oligomer is obtained:

$$\underset{\diagdown O \diagup}{CH_2 - CH} - E - \underset{\diagdown O \diagup}{CH - CH_2} + 2\, CH_2 = \overset{R}{\overset{|}{C}} - COOH \longrightarrow$$

$$CH_2 = \overset{R}{\overset{|}{C}} - \underset{O}{\underset{\|}{C}} - O - CH_2 - \underset{OH}{\underset{|}{CH}} - E - \underset{OH}{\underset{|}{CH}} - CH_2 - O\underset{O}{\underset{\|}{C}} - \overset{R}{\overset{|}{C}} = CH_2$$

with

$$E = CH_2 \left[O - \langle\bigcirc\rangle - \underset{CH_3}{\underset{|}{\overset{CH_3}{\overset{|}{C}}}} - \langle\bigcirc\rangle - O - CH_2 - \underset{OH}{\underset{|}{CH}} - CH_2 \right]_n - O - \langle\bigcirc\rangle - \underset{CH_3}{\underset{|}{\overset{CH_3}{\overset{|}{C}}}} - \langle\bigcirc\rangle - OCH_2$$

and

$$R = H \text{ or } -CH_3 \tag{2.94}$$

In this case, the free-radical reaction is a copolymerization between styrene and (meth)acrylate double bonds. The reactivity ratios are $r_S \sim 0.5$ and $r_M \sim 0.5$.

"VE resins" are more expensive than "UP resins." They are used as matrices for glass-fiber composites when a better corrosion resistance is needed.

2.3.3 Photopolymerization of (Meth)acrylates

Photoinitiated polymerization and crosslinking of monomers, oligomers, and polymers constitute the basis of important commercial processes with broad applicability, including photoimaging and UV curing of coatings and inks.

Photoinitiators absorb light in the UV-visible spectral range, 250–450 nm. The photoinitiator, PI, is raised to an electronically excited state, PI* by promotion of an electron to a higher-energy orbital, and then it converts this light energy into chemical energy in the form of reactive intermediates, such as free radicals or cations, which subsequently initiate polymerization of monomers.

$$\mathrm{PI} \underset{\longrightarrow}{\overset{h\nu}{\longleftarrow}} \mathrm{PI}^* \longrightarrow \mathrm{I}^\bullet \text{ or } \mathrm{I}^+ \tag{2.95}$$

(Meth)acrylates are polymerized by free-radical polymerization. Because of their high reactivity, acrylate double bonds are preferred: their polymerization rate is 10 times faster than that of methacrylate monomers.

Photoinitiators are generally aryl alkyl ketones or diaryl ketones (Table 2.19). For aryl alkyl ketones two free radicals are produced by homolytic scission of a C–C bond (Eq. (2.96)). Diaryl ketones are usually mixed with a tertiary amine; the mechanism of production of free radicals involves H abstraction from the tertiary amine by the excited state PI*, via a charge-transfer stabilized exciplex (Eq. (2.97)). The α-amino alkyl radical formed is very reactive and is in fact the true initiator because the cetyl radical disappears rapidly through a coupling reaction (formation of pinacol).

A typical UV-cured coating has three or sometimes four main ingredients: (1) a multifunctional acrylate monomer, (2) a reactive oligomer, (3) possibly a monoacrylate as reactive diluent, and (4) a photoinitiator (Table 2.20).

The reactive oligomer can be any low-molar-mass polymer containing at least a couple of double bonds. It can be based on a polyester, polyether, or polyurethane backbone. One mole of α, ω-OH-terminated polyester or polyether is prereacted with two moles of acrylic acid to obtain an α, ω-diacrylate oligomer. For polyurethanes, 1 mole of α, ω-diisocyanate oligomer is prereacted with 2 moles of hydroxyethylacrylate (Sec. 2.2.3c).

As these reactive oligomers have a high viscosity, they are diluted with a multifunctional and if necessary a monofunctional acrylate monomer. The acrylate monomers reduce the viscosity of the mixture for easier application on substrates. The multifunctional acrylate monomers increase the crosslink density; the monofunctional monomers decrease it.

The thickness of UV-cured samples is generally less than 100 μm. To cure sample thicknesses up to 1 cm or more, it is possible to use a dual initiation process. The heat of reaction developed during the UV cure of the material located close to the surface is used to decompose a peroxide that initiates the polymerization in the bulk.

For large pieces, electron-beam (EB) curing can also be used. Electron-beam curing systems are similar in composition to UV systems.

2.3.4 Epoxy Reactions

a. Homopolymerization

Epoxy groups can react with both nucleophilic and electrophilic species. Thus both Lewis acids and bases are able to initiate the chain polymeriza-

TABLE 2.19 Decomposition reactions of some photoinitiators

- Homolytic scission of C–C bond
  benzoin ether

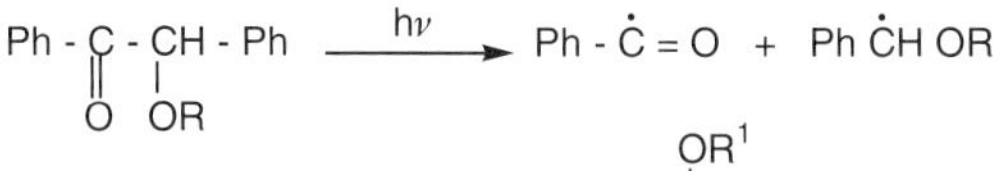

$$\text{Ph-}\underset{\text{O}}{\underset{\|}{\text{C}}}\text{-}\underset{\text{OR}}{\underset{|}{\text{CH}}}\text{-Ph} \xrightarrow{h\nu} \text{Ph-}\dot{\text{C}}\text{=O} + \text{Ph}\dot{\text{C}}\text{H OR} \quad (2.96)$$

or benzylic diacetal $\text{Ph-}\underset{\text{O}}{\underset{\|}{\text{C}}}\text{-}\overset{\text{OR}^1}{\overset{|}{\underset{\text{OR}^2}{\underset{|}{\text{C}}}}}\text{-Ph}$

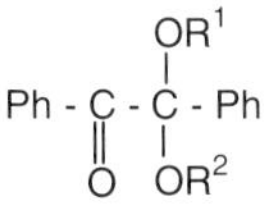

or hydroxy-alkylphenone $\text{Ph-}\underset{\text{O}}{\underset{\|}{\text{C}}}\text{-}\overset{\text{OR}^1}{\overset{|}{\underset{\text{OH}}{\underset{|}{\text{C}}}}}\text{-R}^2$

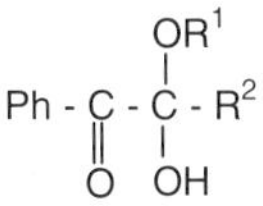

α-amino cetone $\text{O(CH}_2\text{CH}_2)_2\text{N-C}_6\text{H}_4\text{-}\underset{\text{O}}{\underset{\|}{\text{C}}}\text{-}\overset{\text{R}^1}{\overset{|}{\underset{\text{R}^2}{\underset{|}{\text{C}}}}}\text{-N(CH}_3)_2$

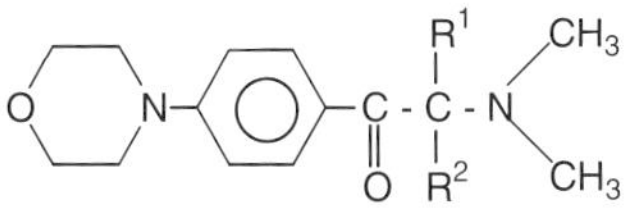

- H abstraction on a co-initiator molecule

  benzophenone, BP

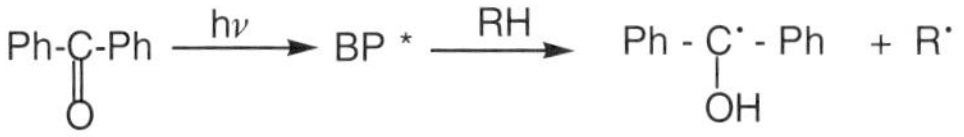

$$\text{Ph-}\underset{\text{O}}{\underset{\|}{\text{C}}}\text{-Ph} \xrightarrow{h\nu} \text{BP}^* \xrightarrow{\text{RH}} \text{Ph-}\underset{\text{OH}}{\underset{|}{\text{C}^\cdot}}\text{-Ph} + \text{R}^\cdot \quad (2.97)$$

or thioxanthone

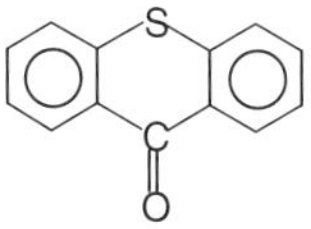

or acylphosphine $\text{C}_6\text{H}_5\text{-}\underset{\text{O}}{\underset{\|}{\text{C}}}\text{-}\underset{\text{O}}{\underset{\|}{\text{P}}}(\text{C}_6\text{H}_5)_2$

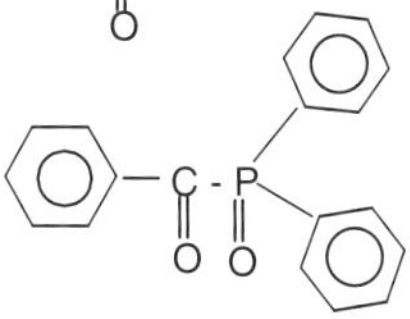

RH is a tertiary amine $\text{R}^1\text{R}^2\text{N-CH}_2\text{R}^3$

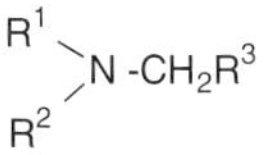

$\text{R}^\cdot$ is very reactive and the effective initiator

TABLE 2.20 Typical UV-cured coating formulations

• Multifunctional acrylates:	
hexane diol diacrylate or	$CH_2{=}CH{-}COO{-}(CH_2)_6{-}OCO{-}CH{=}CH_2$
trimethylol propane triacrylate	
etc.	$CH_3{-}CH_2{-}C(CH_2OCO\ CH{=}CH_2)_3$
• Monoacrylates:	
n-butyl acrylate	$CH_2{=}CH{-}COO{-}C_4H_9$
or 2 ethyl-n-hexyl acrylate	$CH_2{=}CH{-}COO{-}CH_2{-}CH(C_2H_5)\ (CH_2)_3{-}CH_3$
etc.	
• Reactive oligomers:	

Polyester $CH_2{=}CH{-}COO\ R^1{-}\left[O{-}CO\ R^2\ COO\ R^1\right]_n{-}O\ CO\ CH = CH_2$

or polyether $CH_2 = CH{-}COO{-}\left(\underset{CH_3}{CH}{-}CH_2{-}O\right)_x R\left({-}O{-}CH_2{-}\underset{CH_3}{CH}\right)_y{-}OCOCH{=}CH_2$

or polyurethane $CH_2{=}CH{-}COO(CH_2)_2{-}OCO\ NH\ R^1\ NHCO\left[OR^2OCO\ NH\ R^1NH\ CO\right]_n O\ (CH_2)_2\ O\ CO\ CH = CH_2$

tion of epoxy monomers. Commonly used initiators (often named "catalytic" curing agents in the literature) include tertiary amines, imidazoles, ammonium salts for anionic chain polymerization, and boron trifluoride complexes for cationic chain polymerization.

The mechanisms of these reactions have been the subject of considerable controversy for many years: those presented in Tables 2.21 and 2.22 are still questionable. As in the case of classical chain polymerization, many (unknown) transfer reactions with impurities, or with the polymer formed, stop the chain propagation, which explains why degrees of polymerization less than 10 are obtained when a monofunctional epoxy monomer is poly-

TABLE 2.21 Different steps of the anionic polymerization of an epoxy monomer

- Initiation with a tertiary amine:

$$R^1_3N: \longrightarrow \underset{\text{O}\cdots\text{H-OR}^2}{CH_2\text{-}CH\text{-}E} \longrightarrow R^1_3N^+\text{-}CH_2\text{-}\underset{OH}{CH}\text{-}E + R^2O^-$$

or $R^1_3N + R^2OH \longrightarrow R^1_3N^+H + R^2O^-$

- Initiation with an imidazole instead of a tertiary amine:

$$\text{(imidazole, } R^1 \text{ on N, } R^2 \text{ on C2)} + \underset{O}{CH_2\text{-}CH\text{-}E} \longrightarrow \text{(imidazolium } R^1, R^2)N^+\text{-}CH_2\text{-}\underset{E}{CH}\text{-}O^- \ (= I^-)$$

- Propagation:

$$R^2O^- + n\,\underset{O}{CH_2\text{-}CH\text{-}E} \longrightarrow R^2O\left(CH_2\text{-}\underset{E}{CH}\text{-}O\right)_{n-1}CH_2\text{-}\underset{E}{CH}\text{-}O^-$$

(or I^-)

- Transfer reactions?

with $C_6H_5\text{—}O\text{-}CH_2\text{-}CH\text{-}CH_2$ (epoxide ring, O) as monomer and

$C_6H_5\text{—}CH_2\text{-}N(CH_3)_2$ as initiator, $DP_n \sim 6$

merized. The chain polymerization of difunctional epoxy monomers leads to networks (Matejka *et al.*, 1994).

The reactivity of cycloaliphatic epoxy monomers with cations is higher than that of glycidyl ether monomers. For this reason, cycloaliphatic epoxies are used for UV-cure coatings. The most efficient initiators are complex aromatic salts of Lewis acids such as diaryl iodonium, triarylsulfonium, or arene diazonium (Table 2.23) (Crivello, 1999).

When exposed to UV radiation, these salts dissociate and react with a proton (impurities, ROH), to liberate a protonic acid that is able to initiate a cationic chain polymerization of epoxy groups.

Table 2.22 Different steps of the cationic polymerization of an epoxy monomer

- Initiation with BF_3 complexes:

$$F_3B-NH_2-R + CH_2\text{-}CH\text{-}E \text{ (epoxide)} \longrightarrow F_3B^{-}N(R)(E)\text{---}H\text{---}O^{+}(CH_2\text{-}CH\text{-}E)$$

oxonium ion (structure subject to controversy) written, $H-O^{+}(CH_2\text{-}CH\text{-}E)$

- Propagation:

$$H\text{-}O^{+}(CH_2\text{-}CH\text{-}E) + n\, O(CH_2\text{-}CH\text{-}E) \longrightarrow H\left(O\text{-}\underset{E}{CH}\text{-}CH_2\right)_n\text{-}O^{+}(CH_2\text{-}CH\text{-}E)$$

- Transfer reactions ?

with $C_6H_5-O-CH_2-CH-CH_2$ (epoxide, O bridging CH and CH_2) as monomer and

$BF_3, H_2N-C_6H_4-Cl$ as initiator, $DP_n \sim 8\text{–}10$

An advantage of this type of photopolymerizations is that as they are non-radical chain polymerizations, they are insensitive to oxygen. In addition, as the cation is relatively stable, the reaction is able to continue in the dark. Applications of this chemistry may be found in the fields of coatings, adhesives, printing inks, and also for photocurable composites and microelectronic photoresists.

Table 2.23 Decomposition reactions of some cationic photoinitiators

$$Ar_2\, I^+\, BF_4^- + RH \xrightarrow{h\nu} Ar\,I + Ar^{\cdot} + R^{\cdot} + HBF_4$$

$$Ar_3\, S^+\, PF_6^- + RH \xrightarrow{h\nu} Ar_2\,S + Ar^{\cdot} + R^{\cdot} + HPF_6$$

b. Discussion about Dual Polymerization Systems

In practice, epoxy–amine reactions in carbon fiber prepregs, and epoxy–phenol reactions in molding compounds, are often accelerated by the addition of a Lewis acid (typically a BF_3 – amine complex) or a Lewis base (often a tertiary amine), as "catalysts."

Even though these systems exhibit a very good performance, the chemistry of the cure process is rather complex. Both step-growth and chain-growth mechanisms are operative in network formation. But the competition between both pathways depends on the cure temperature. Moreover, they can interfere with one another. For example, the primary amine or the phenol can act as strong transfer agents. Multiple-peak exotherms are usually observed during a temperature scan in differential scanning calorimetry (DSC). The stoichiometry corresponding to the step-growth reaction is usually not respected. Generally, a large excess of epoxy groups is initially introduced in the formulation, but the optimum for processing and properties is only obtained experimentally and cannot be predicted. Some of the initiators ("catalysts") frequently used are given in Table 2.24.

c. Epoxy–Cyclic Anhydride Systems

After the amines, acid anhydrides constitute the next most commonly used reagents for curing epoxy monomers. The epoxy–acid reaction proceeds through a stepwise mechanism (Sec. 2.2.4) while the reaction of epoxides with cyclic anhydrides, initiated by Lewis bases, proceeds through a chainwise polymerization, comprising initiation, propagation, and termination or chain transfer steps. Some of the postulated reactions are shown in Table 2.25 (Matejka *et al.*, 1985b ; Mauri *et al.*, 1997).

Initiation involves the reaction of the tertiary amine (the most widely used Lewis base) with an epoxy group, giving rise to a zwitterion that contains a quaternary nitrogen atom and an alkoxide anion. The alkoxide reacts at a very fast rate with an anhydride group, leading to a species containing a carboxylate anion as the active center. This ammonium salt can be considered as the initiator of the chainwise copolymerization.

Propagation occurs through the reaction of the carboxylate anion with an anhydride group, regenerating the carboxylate anion. The presence of alkoxide anions is not easily detected, a result that is explained by the fact that $k_{p2} \gg k_{p1}$ (Table 2.25). Propagation results then in an alternating chainwise copolymerization of epoxide and anhydride groups. The alternating copolymerization is certainly the reason why in the past this reaction has been erroneously regarded as a stepwise process. The presence of a chainwise mechanism was confirmed by characterizing the

TABLE 2.24 Some "catalysts"/initiators used in epoxy–diamine and epoxy–diphenol formulations

- Tertiary amine: benzyl dimethylamine $C_6H_5-CH_2-N(CH_3)_2$

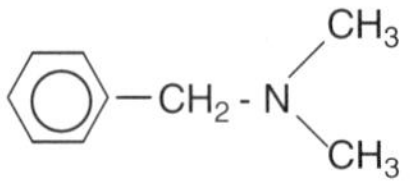

or $HO-C_6H_2(CH_2-N(Me)_2)_3$ called DMP 30

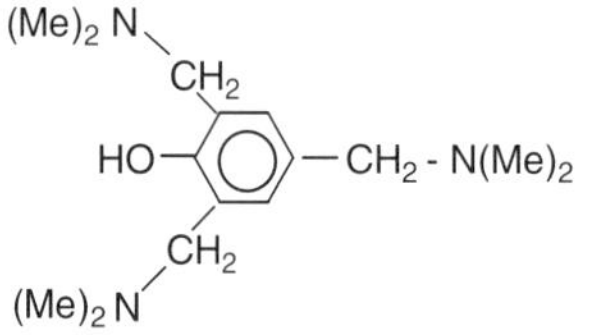

- Blocked isocyanate: 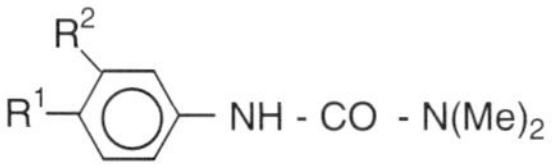 $R^1(R^2)C_6H_3-NH-CO-N(Me)_2$ $R^1 = Cl, R^2 = H, Cl$

- Imidazole:

ethyl-2-methyl-4-imidazole (ring N, NH; substituents CH_3, C_2H_5)

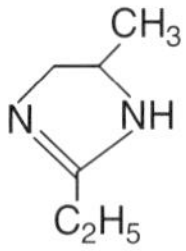

- Ammonium salts:

tetrabutyl ammonium bromide $(C_4H_9)_4N^+ Br^-$

- Boron trifluoride complexes:

$BF_3, H_2NC_2H_5$ or BF_3, MEA

linear copolymer formed in the reaction of phenyl glycidyl ether (PGE) with phthalic anhydride, using matrix-assisted laser desorption/ionization time of flight (MALDI–TOF) mass spectrometry (Leukel and Burchard, 1996). The strictly alternating copolymerization was confirmed, together with the fact that the initiator remained chemically bound during the whole reaction. Molar masses were in the range of 4–80 kg mol^{-1}, depending on the initiator used and on the purity of the starting materials – impurities can be water, acid from anhydride partial hydrolysis, etc. (Steinmann, 1990).

TABLE 2.25 Anhydride–epoxide copolymerization initiated by tertiary amines

- Initiation

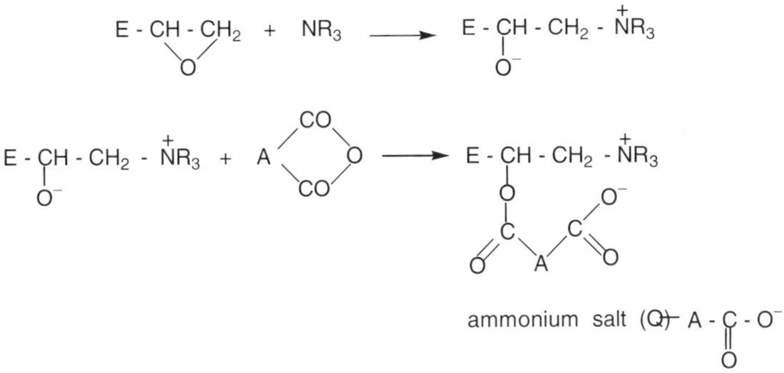

$$\text{E-CH-CH}_2\text{(-O-)} + NR_3 \longrightarrow \text{E-CH(O}^-\text{)-CH}_2\text{-}\overset{+}{N}R_3$$

$$\text{E-CH(O}^-\text{)-CH}_2\text{-}\overset{+}{N}R_3 + \text{A(CO)}_2\text{O} \longrightarrow \text{E-CH(O-C(=O)-A-C(=O)O}^-\text{)-CH}_2\text{-}\overset{+}{N}R_3$$

ammonium salt (Q)-A-C(=O)-O^-

- Propagation

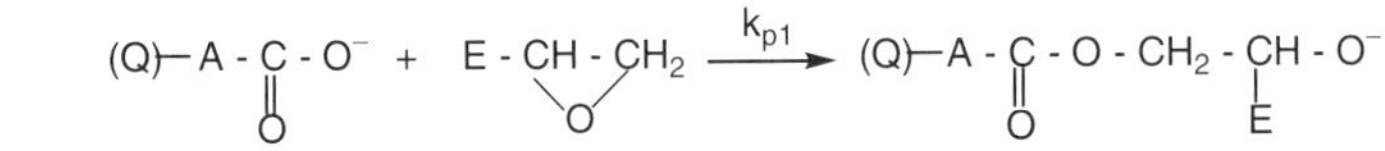

$$\text{(Q)-A-C(=O)-O}^- + \text{E-CH-CH}_2\text{(-O-)} \xrightarrow{k_{p1}} \text{(Q)-A-C(=O)-O-CH}_2\text{-CH(E)-O}^-$$

monoester

$$\text{(Q)-A-C(=O)-O-CH}_2\text{-CH(E)-O}^- + \text{A(CO)}_2\text{O} \xrightarrow{k_{p2}} \text{(Q)-A-COO CH}_2\text{-CH(E)-O-C(=O)-A-C(=O)-O}^-$$

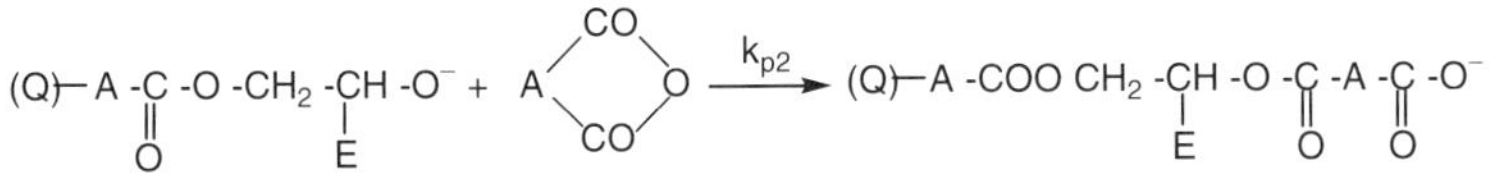

diester

$$k_{p2} >>> k_{p1}$$

- Termination and chain transfer reactions

2.3.5 Other Ring-Opening Polymerizations (ROPs)

a. Synthesis of Reactive Oligomers

The ring-opening polymerization of cyclic monomers can be performed by ionic chain polymerization, as is the case of epoxy monomers. Anionic polymerization of ethylene oxide propylene oxide, and caprolactone can be initiated by alkoxides:

$$RO^-K^+ + CH_2\text{-}CH(CH_3)\text{(epoxide, O)} \longrightarrow RO\text{-}CH_2\text{-}CH(CH_3)\,O^-K^+ \quad (2.98)$$

When a diol is used, an α, ω-dihydroxy-terminated polymer is obtained. With a triol, a branched triol polymer results. Oligomers with molar masses in the range of 250–8000 $g\,mol^{-1}$ are prepared with this reaction for uses in isocyanate or acrylate chemistries.

Cationic initiation is used in the polymerization of tetrahydrofurane and cyclic siloxanes.

b. Synthesis of Linear and Crosslinked Polymers

Anionic polymerization of ε-caprolactam is used to make cast or RIM polyamide-6. Using a premade lactam chain end and a metal catalyst, it proceeds rapidly at 100–160°C, well below the melting temperature of the polymer, $T_m \sim 220°C$. The propagation differs from anionic propagation of most unsaturated monomers because the growth center at the chain end is not represented by an anionically activated group but by a neutral N-acylated lactam, and the anionically activated species is the incoming monomer (Table 2.26).

Dicyclopentadiene (DCP) is another cyclic monomer that can be polymerized very rapidly at room temperature to give a crosslinked polymer. It is also well suited for the RIM process. Metathesis initiators are used, typically WCl_6 + Et_2AlCl. The crosslinked polymer formed has a very complex structure: ~20% of cyclopentene rings are opened, which is enough to have a thermosetting polymer (Matejka *et al.*, 1985a, Table 2.27). The reaction is highly exothermic and the reactivity can be decreased by the use of a chelating agent of the catalyst such as acylacetone or benzonitrile. Because of its olefinic nature, polyDCP has excellent solvent resistance but due to the numerous double bonds, a surface oxidation process takes place after several hours' air exposure.

c. New Cyclic Monomers for ROP

Other cyclic monomers have been prepared and polymerized through fast ROP. The main focus has been first on bisphenol A carbonate oligocyclic monomers (Brunelle *et al.*, 1994). The oligocyclic monomers were prepared using an amine-catalyzed reaction of bisphenol A–bischloroformate, via an interfacial hydrolysis/condensation reaction that also produces linear oligomers and polymers, depending on the structure and concentration of the tertiary amine (Aquino *et al.*, 1994, Table 2.28).

Oligocyclic monomers are useful precursors to high-molar-mass polycarbonates (Brunelle *et al.*, 1994; Otaige, 1997), because they have a low

TABLE 2.26 Initiation and propagation mechanisms for the anionic polymerization of ε-caprolactam

(i) Initiator systems:

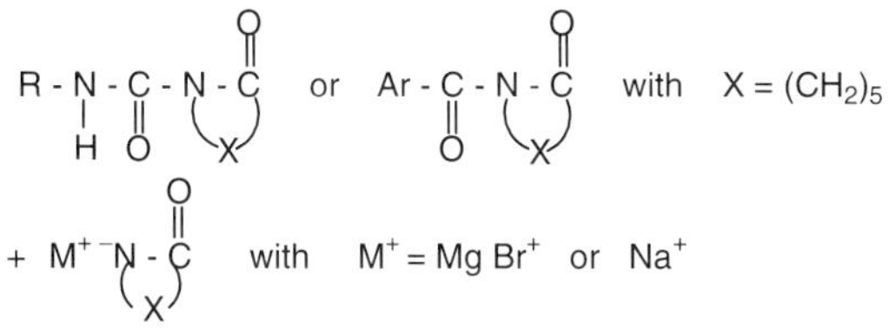

R - N(H) - C(=O) - N - C(=O) (ring X) or Ar - C(=O) - N - C(=O) (ring X) with $X = (CH_2)_5$

$+ M^+ \, ^-N - C(=O)$ (ring X) with $M^+ = Mg\,Br^+$ or Na^+

(ii) Ion attack to add a monomer:

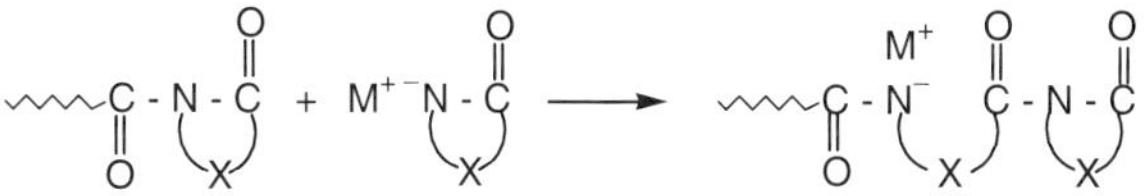

~~~C(=O) - N - C(=O) (ring X) + $M^{+}\,^{-}N$ - C(=O) (ring X) ⟶ ~~~C(=O) - N^- (M^+) (X) C(=O) - N - C(=O) (ring X)

(iii) Hydrogen abstraction:

~~~C(=O) - N^- (M^+) (X) C(=O) - N - C(=O) (ring X) + H - N - C(=O) (ring X) ⟶ ~~~C(=O) - NH (X)CO - N - C(=O) (ring X) + $M^{+}\,^{-}N$ - C(=O) (ring X)

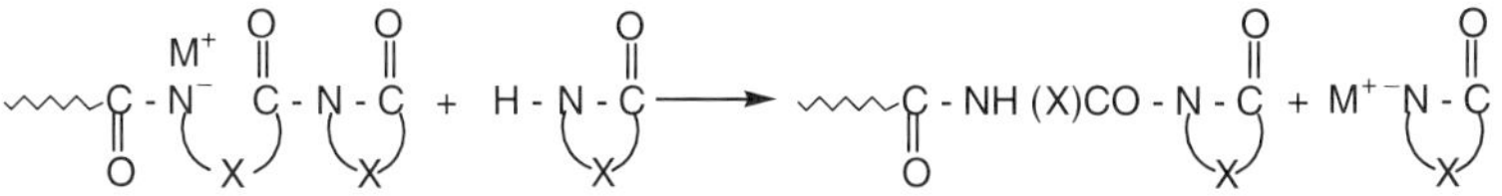

etc.

TABLE 2.27 Scheme of dicyclopentadiene (DCP) ring-opening polymerization

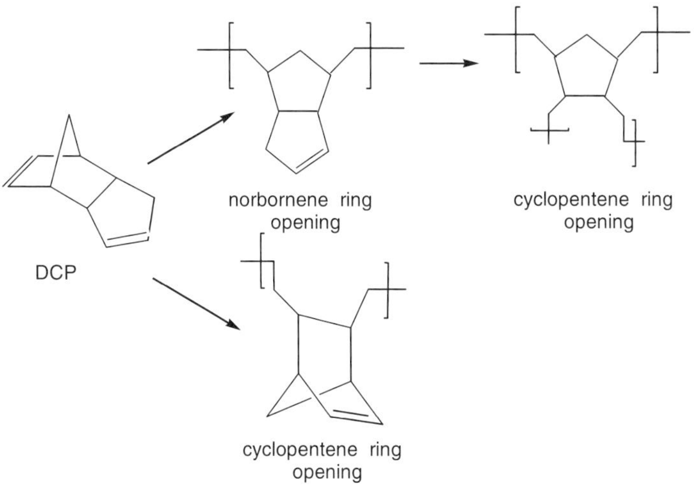

DCP

norbornene ring opening

cyclopentene ring opening

cyclopentene ring opening

TABLE 2.28 Synthesis of cyclic oligomers for ROP

n Cl-C(=O)-O-[bisphenol A]-O-C(=O)-Cl (chloroformate) + n HO-[bisphenol A]-OH (bisphenol A) $\xrightarrow[CH_2Cl_2 / H_2O,\ NaOH]{Et_3N}$ cyclic oligomers + 2n NaCl

viscosity that facilitates reactive processing ("cyclomer technology"). They can be polymerized at high temperatures, 250–300°C, by using an anionic initiator. The reaction is very fast, in the order of 30 s to 5 min. It can be used in low-pressure molding technologies such as resin transfer molding (RTM) for preparing structural composites.

Such a route has been applied to other chemistries: aromatic polyesters (PET or PEN), polyamides, polysulfones, polyphenylene sulfide, etc.

2.3.6 Miscellaneous

a. Diallyl Monomers

Monoallylic compounds do not form homopolymers of high molar mass under free-radical initiation because of the low activation of the double bond and the so-called degradative chain transfer, resulting from the high reactivity of hydrogen atoms, which tends to terminate chain growth.

$$\mathrm{I-CH_2-\dot{C}H(CH_2R) + CH_2{=}CH(CH_2R) \longrightarrow I-CH_2CH_2(CH_2R) + CH_2{=}CH(\dot{C}HR)} \tag{2.99}$$

For this reason, the polymerization of diallyl monomers does not lead to a gel, up to a conversion of double bonds close to 25%.

The most commonly used diallyl monomer is diallyl diethylene glycol carbonate, DADC or CR 39, developed for preparing clear, colorless, abrasion- and heat-resistant polymers (Table 2.29). DADC is obtained by reaction of diethylene glycol bis(chloroformate) with allyl alcohol, or allyl-chloroformate and diethylene glycol.

Sheets, rods, and lens preforms are cast from CR39 or CR39 prepolymer at 60–125°C. A relative high concentration (4–8 wt%) of peroxides or AIBN is needed to obtain complete polymerization. The shrinkage during polymerization is about 14%.

Another important diallyl monomer is diallyl phthalate (DAP), prepared from phthalic anhydride and allyl alcohol.

Triallyl cyanurate (TAC) and triallyl isocyanurate (TAIC) are used as crosslinking agents for methacrylates and UP resins to improve heat and solvent resistance as well as thermooxidative stability.

Table 2.29 Some allyl monomers

Diallyl diethylene glycol carbonate (DADC)

$$CH_2=CH-CH_2-O-\underset{\underset{O}{\|}}{C}-O(CH_2)_2-O-(CH_2)_2-O-\underset{\underset{O}{\|}}{C}-O-CH_2-CH=CH_2$$

Diallyl o-phthalate (DAP)

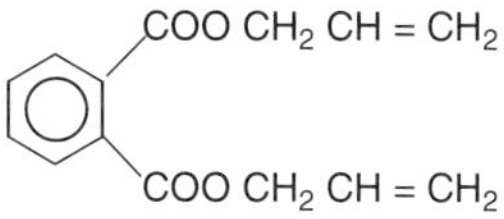

$$C_6H_4(COO\,CH_2\,CH=CH_2)_2$$

Triallyl cyanurate (TAC)

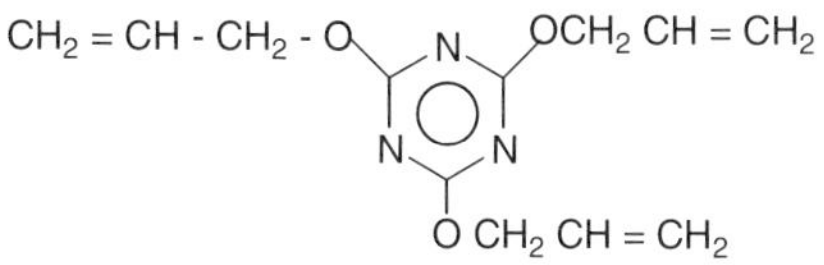

$$CH_2=CH-CH_2-O-C_3N_3(OCH_2\,CH=CH_2)(O\,CH_2\,CH=CH_2)$$

Triallyl isocyanurate (TAIC)

$$(CH_2=CH-CH_2)_3N_3C_3O_3$$

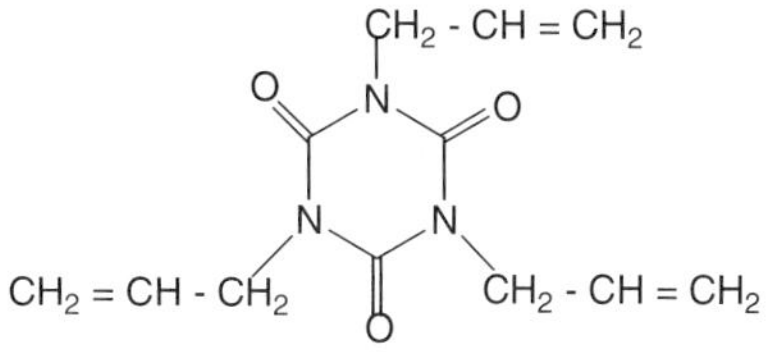

b. Thiol–Olefin Polymerization

The addition of thiols to olefins (thiolene reaction), to form thioethers, is a well-known reaction. The process can occur by either free-radical or ionic mechanisms. The free-radical reaction can be initiated thermally via a peroxide or by UV irradiation with benzophenone. The initiation step involves the formation of a thiyl radical by hydrogen atom abstraction. Both of these species are capable of starting polymer chains (Table 2.30).

The photoinitiated addition of a multifunctional thiol to a difunctional monomer or oligomer gives a crosslinked polymer. The most widely used multifunctional thiol is pentaerythritol tetramercaptopropionate. Ene-monomers can be styrene, acrylates, vinyl ethers, allylic oligomers, etc. For example:

$$(CH_2=CH-CH_2-O-CONH)_2-R \text{ can react with } (HS\,CH_2\,CH_2\,COO\,CH_2)_4\text{-}C \tag{2.100}$$

The ene reactivity depends on the structure of the monomer. In the case of acrylate monomers, a fraction of homopolyacrylate is formed. Norbornyl double bonds exhibit high reactivity.

Although the growth of polymer chains results from a free-radical reaction, the reaction rate is relatively slow, resembling that of typical step-growth polymerizations. When the number of double bonds equals the number of SH groups, the system described by Eq. (2.100) can be classified as an A_2 (ene) + B_4 (thiol) reaction.

TABLE 2.30 Mechanism of the thiolene reaction

- Initiation by a radical, $H^{\cdot}$:

$$H^{\cdot} + CH_2=CH\text{-}R^1 \longrightarrow CH_3\text{-}\dot{C}H\text{-}R^1$$

$$CH_3\,\dot{C}H\,R^1 + R^2SH \longrightarrow CH_3\text{-}CH_2\,R^1 + R^2S^{\cdot}$$

$$\text{or } H^{\cdot} + R^2SH \longrightarrow H_2 + R^2S^{\cdot}$$

- Propagation:

$$R^2S^{\cdot} + CH_2=CH\text{-}R^1 \longrightarrow R^2S\,CH_2\,\dot{C}H\,R^1$$

- Transfer:

$$R^2S\,CH_2\,\dot{C}H\,R^1 + R^2SH \longrightarrow R^2S^{\cdot} + R^2S\,CH_2\,CH_2\,R^1$$

The shrinkage of these polymers during reaction is generally 2–6 %, depending on the ene component and its functionality. This chemistry is used for many applications, such as adhesives, coatings, photoresists, ... etc. One advantage of these formulations is the relatively high amount of S atoms introduced into the polymer; this increases its refractive index, an important factor for optical glasses.

2.4 CONCLUSIONS

The chemistry described in this chapter is the same for the synthesis of both thermoplastic and thermosetting polymers. The transformations occurring during network formation may have a bearing either on the mechanisms (e.g., variation of the reactivity ratios along polymerization) or on the kinetics of network formation (e.g., decrease of reaction rate at the time of vitrification). These transformations and the effects they produce on the buildup of the polymer network will be discussed in the following chapters.

NOTATION

DP_n = number-average degree of polymerization
k_{AA}, k_{AB}, k_{BA}, k_{BB} = specific rate constants for different propagation steps in a chain copolymerization, $m^3\ kmol^{-1}\ s^{-1}$
k_d = specific rate constant for the decomposition of an initiator, s^{-1}
k_p = specific rate constant for the propagation step, $m^3\ kmol^{-1}\ s^{-1}$
M_A = monomer A
M_B = monomer B
M_n = number-average molar mass
r = stoichiometric ratio of functionalities supplied by two monomers
r_A, r_B = reactivity ratios in a chain copolymerization
T_m = melting temperature, K

REFERENCES

Aquino EC, Brittain WJ, Brunelle DJ, *J. Polym. Sci., Polym. Chem.*, 32, 741–746 (1994).
Brunelle DJ, Krabbenhoft HO, Bonauto DK, *Macromol. Symp.* 77, 117–124 (1994).
Crivello JV, *J. Polym. Sci., Polym. Chem.*, 37, 4241–4254 (1999).
Heilmann SM, Rasmussen JK, Palensky FJ, Smith HK, *J. Polym. Sci., Polym.Chem.*, 22, 1179-1186 (1984).
Leukel J, Burchard W, *Macromol. Rapid Commun.*, 17, 359–366 (1996).

Matejka L, Chabanne P, Tighzert L, Pascault JP, *J. Polym. Sci., Polym. Chem.*, 32, 1447–1458 (1994).
Matejka L, Houtman C, Macosko C, *J. Appl. Polym. Sci.*, 30, 2787–2803 (1985a).
Matejka L, Pokorny S, Dusek K, *Makromol. Chem.*, 186, 2025–2036 (1985b).
Mauri AN, Galego N, Riccardi CC, Williams RJJ, *Macromolecules*, 30, 1616–1620 (1997).
Otaige JU, *TRIP*, 5, 17–23 (1997).
Steinmann B, *J. Appl. Polym. Sci.*, 39, 2005–2026 (1990).

3

Gelation and Network Formation

3.1 INTRODUCTION

The aim of this chapter is to describe the process of network formation using qualitative arguments together with simple mathematical tools. The intention is to provide a first approach to the subject that should enable the reader to get acquainted with the basic concepts and definitions of the network structure.

It is evident that the way in which the network structure is developed will depend primarily on the type of polymerization reaction that is involved: stepwise or chainwise. In the former case the network growth occurs smoothly, as schematically represented in Fig. 3.1, for the paradigmatic case of an A_3 homopolymerization – e.g., a molecule with three OH groups that undergoes a polyetherification reaction.

In several types of chainwise polymerizations, species with high molar mass are generated from the beginning of the reaction. This is depicted in Fig. 3.2 for an A_2 (one double bond per molecule) $+A_4$ (two double bonds per molecule) free-radical polymerization – e.g., a vinyl–divinyl system.

Living polymerizations (2.3.1, Fig. 2.2) exhibit a different type of network growth. They are classified among the chainwise polymerizations because it is always the monomer that reacts, adding to growing chains. But the growth of primary chains occurs smoothly, as in the case of stepwise polymerizations.

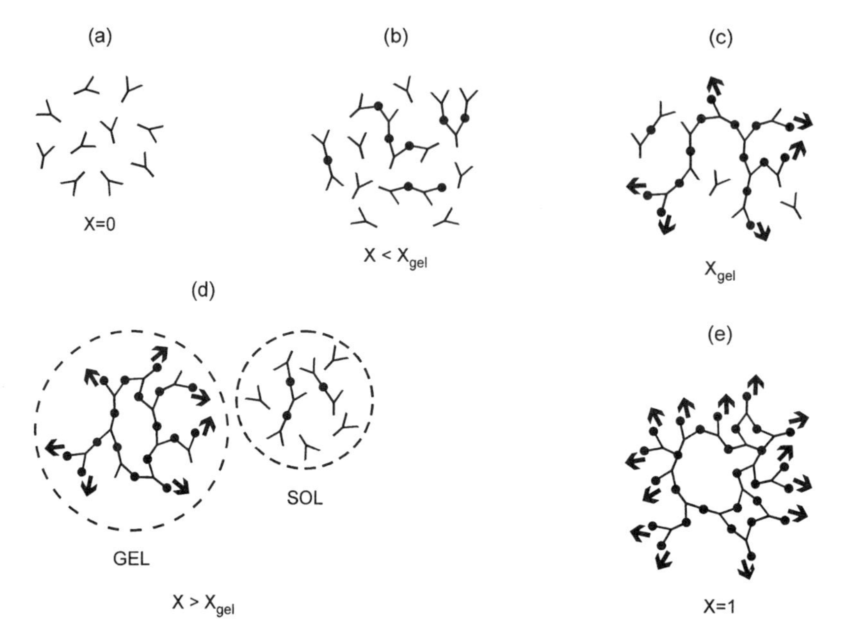

Figure 3.1 Network growth during the stepwise homopolymerization of an A3 monomer.

3.1.1 Gelation

A characteristic feature during network formation is the presence of a critical transition called gelation, which involves an abrupt change from a liquidlike to a solidlike behavior. Figure 3.3 illustrates the evolution of (zero-shear) viscosity, elastic modulus and fraction of soluble material (sol fraction), as a function of the conversion of reactive groups (x). At $x = x_{gel}$, the (zero-shear) viscosity becomes infinite, there is a buildup of the elastic modulus, and an insoluble fraction (gel fraction) suddenly appears.

For conversions lower than x_{gel} the average molar mass of the polymer exhibits a continuous increase. The first two moments of the molar mass distribution are the number-average molar mass, M_n, and the mass-average molar mass, M_w, respectively. M_n is defined in terms of the number contribution of every species to the whole population. The weight factor used to define this average is the molar fraction. M_w is defined in terms of the mass contribution of every species to the whole mass, so that the mass fraction is the weight factor used in its definition.

Gelation occurs when one of the growing molecules reaches a mass so large that it interconnects every boundary of the system. A similar way to

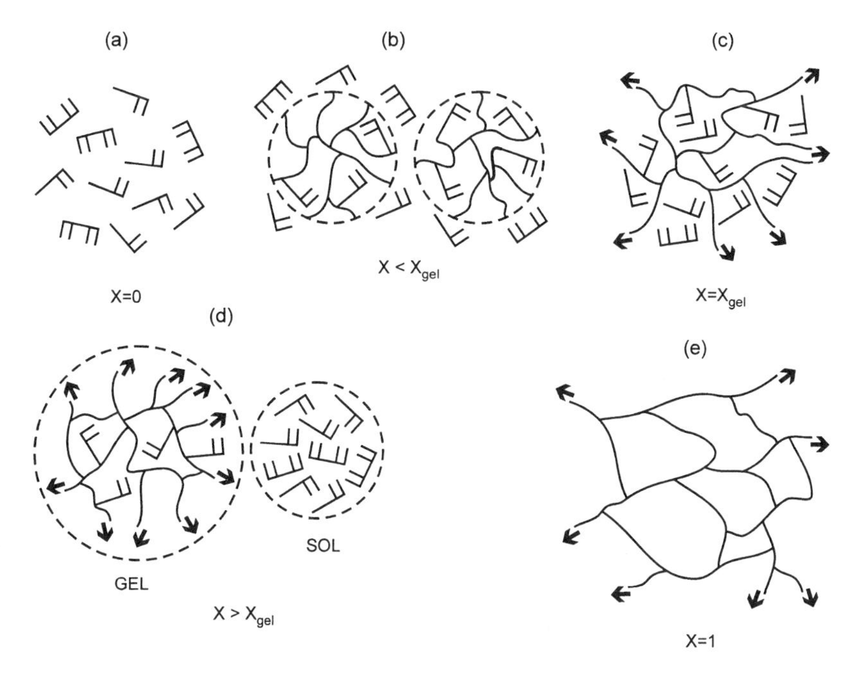

Figure 3.2 Network growth during the chainwise copolymerization of a vinyl (A_2) – divinyl (A_4) system.

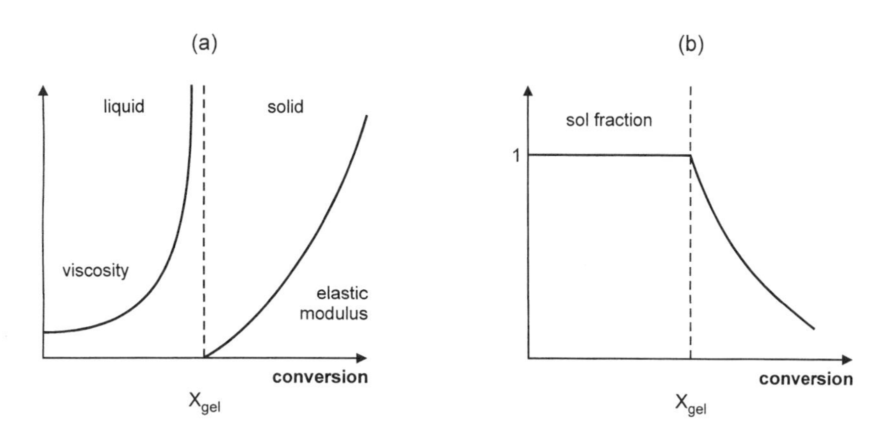

Figure 3.3 Evolution of physical properties of the thermosetting polymer as a function of conversion of reactive groups: (a) zero-shear viscosity and elastic modulus, (b) sol fraction.

describe this critical transition is by stating that gelation is defined by the percolation of the giant macromolecule throughout the system (the arrows in Figs 3.1 and 3.2 indicate the continuation of the structure up to the boundaries of the reaction vessel).

The phenomenon of gelation is not restricted to thermosetting polymers. It may also take place in linear polymers dissolved in specific solvents for a particular range of concentrations and temperature. Physical association of portions of linear chains promoted by crystallization, ionic interactions, triple-helix formation, etc., may lead to the formation of a gel, characterized by the percolation of a giant structure throughout the system. This is a reversible process called physical gelation. A reversion from a gel to a liquid state may be simply produced by increasing temperature (gelatin desserts are a typical example of physical gels). In thermosetting polymers, gelation is an irreversible process. The only way to destroy the gel is by breaking the covalent bonds generated by the chemical reaction.

When gelation takes place during reaction of a thermosetting polymer, there is only one giant macromolecule present in the system. This means that its contribution to the total number of molecules is absolutely negligible. So, the giant macromolecule does not contribute to the value of the number-average molar mass of the polymer. This is reflected by the fact that M_n is unaffected by gelation. The situation is completely different for M_w. As the mass fraction of the giant species is significant, its contribution to the M_w value largely prevails over contributions of the (much) smaller species. This means that when $x = x_{gel}$, M_w diverges. Thus, a mathematical definition of gelation states that $M_w \rightarrow \infty$ at $x = x_{gel}$.

We feel it necessary to warn readers that on occasion one may find in the literature the statement (ascribed to Carothers) that gelation is produced when $M_n \rightarrow \infty$. This statement is incorrect and devoid of physical meaning.

Practical ways to determine gelation are frequently related to the fact that the viscosity of the reaction mass becomes infinite at the gel conversion. For example, a laboratory test of gelation consists of periodically tilting the tube where reaction is taking place to visualize the manifestation of a liquid behavior. When gelation takes place, the tube may be turned over and the material will not drop out. Alternatively, a glass rod may be cyclically introduced and removed from the reaction mass. When gelation takes place, an attempt to remove the glass rod from the tube will lift the tube together with the rod.

Gelation has important practical consequences for the production of polyfunctional oligomers that are precursors of thermosetting polymers: e.g., for one-stage phenolic resins (resols). The polymerization is advanced to a particular conversion $x < x_{gel}$, and the resulting product is discharged from the reactor. Depending on the particular application (e.g., embedding fibers, filling a mold, coating a surface, etc.), a particular viscosity is required. Therefore, the polymerization must be arrested at a particular conversion before the gel point. The time to gel (t_{gel}) may be predicted from the knowledge of x_{gel}, the polymerization kinetics, and the reaction temperature. Trying to carry out the polymerization in an industrial reactor without a previous knowledge of t_{gel} may be dangerous. The problem is that the increase in viscosity is relatively sharp at conversions close to gelation. When a high viscosity level is detected there may not be enough time to discharge the reactor. Should this happen, the polymerization must be continued (without agitation) to obtain a glassy material after cooling to room temperature. Then, a pneumatic hammer must be used to remove the product and recuperate the reaction vessel.

With the definition of the critical conversion, x_{gel} , the network formation is clearly divided into two parts: the pregel and the postgel stages. Stepwise and chainwise polymerizations, schematically depicted in Figs 3.1 and 3.2, exhibit a different behavior in the pregel stage. While the former involves a homogeneous system, the latter may develop inhomogeneities on a nanoscale, called microgels (dashed-line circles in Fig. 3.2b). These microgels are the result of the polymerization mechanism. A growing primary chain incorporates monomers of both kinds (A_2 and A_4) at a fast rate. Incorporation of an A_4 monomer leads to the presence of pendant double bonds in the primary chain. Occasionally, the active center of the growing chain may attack a pendant double bond belonging to the same molecule (intramolecular reaction) or to another polymer chain (intermolecular reaction). Because of the dilution of the growing chain in the solution of monomers, intramolecular reactions prevail at low conversions. Once the active center becomes trapped inside the polymer coil, the probability of reacting with neighboring pendant double bonds is very high. This leads to the formation of highly crosslinked polymer coils swollen by unreacted monomers (microgels). As polymerization continues, both the dimensions and the concentration of microgels increase. Eventually, an interconnected structure that percolates the system results, leading to (macro)gelation.

For some particular formulations (e.g., unsaturated polyesters formulated with a high styrene concentration), the primary chains that are first generated are not miscible with the unreacted monomers. In this case, there is a phase separation phenomenon characterized by the appearance of relatively large polymer-rich particles. These microgels are formed by a thermodynamic driving force and their sizes are large enough to be detected in both the course of polymerization and the final materials.

The presence of inhomogeneities in polymer networks will be analyzed in more detail in Chapter 7.

When the fraction of the A_4 monomer in the formulation is very small, intramolecular reaction with pendant double bonds may be neglected, and the picture again becomes that of a homogeneous system.

The mathematical model of network formation in the pregel stage will focus on the prediction of the gel conversion and the evolution of number- and mass-average molar masses, M_n and M_w, respectively. For chainwise polymerizations, calculations will be restricted to the limit of a very low concentration of the polyfunctional monomer (A_4 in the previous example). Thus, homogeneous systems will always be considered.

3.1.2 Postgel Stage

In the postgel stage the mass of the system is divided between a gel fraction and a sol fraction. As polymerization continues, the gel fraction increases at the expense of the sol, and at full conversion there will be practically no sol fraction remaining in the system. In chainwise polymerizations the sol is mostly composed of free monomers. In stepwise polymerizations the sol consists of a mixture of oligomers of different sizes. But since larger oligomers have more free functionalities, they react with a high probability with available functionalities in the gel. Then, the sol is continuously enriched in the low-molar-mass fraction, meaning that the average molar mass of the sol will decrease with conversion.

It is important to realize that a thermosetting polymer reacted to high (but not full) conversion contains a small fraction of free monomers: if the monomers are volatile, their emissions may produce forbidden contamination levels, particularly for indoor applications. The decline in the use of urea–formaldehyde resins in agglomerated wood panels resulted from contamination problems associated with formaldehyde emission.

Apart from the evolution of sol and gel fractions, there are important structural parameters that may be characterized during the postgel stage. Figure 3.4 shows part of the gel structure formed in the reaction of a trifunctional monomer (A_3) with a bifunctional one (B_2). A_3 molecules with three reacted functionalities are called branching units. Crosslinks (or crosslinking units) are branching units that exhibit continuity to the boundaries of the system when moving away in the three possible directions (they go to infinity in three directions). For a generic A_f molecule, branching units and crosslinks of degree m ($3 \leq m \leq f$) may be found during the polymerization in the postgel stage.

In Fig. 3.4, **b**, **c**, **d**, and **e** are branching units but only **b**, **c**, and **d** are crosslinks. The finite chain attached to **e** is called a pendant chain.

The crosslink concentration is a very important structural parameter because it is directly related to the elastic modulus of the network in the

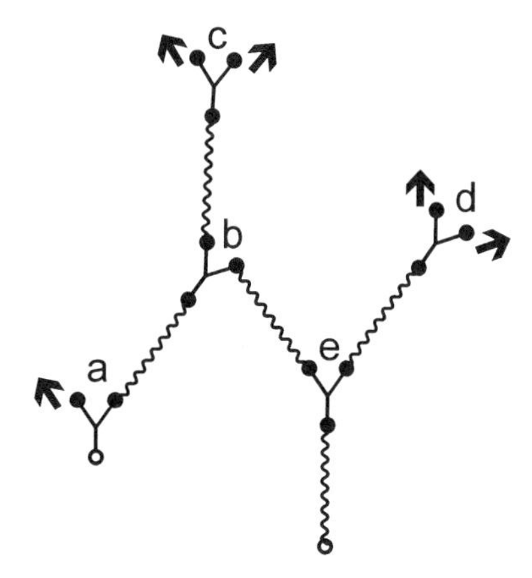

Figure 3.4 Part of the gel structure formed by reacting a trifunctional monomer (A_3) with a bifunctional monomer (B_2).

rubbery state. Chains located between two crosslinks are called elastically active network chains (EANC). In Fig. 3.4, the chain located between **b** and **c** is an EANC, while another EANC bearing an attached pendant chain is present between crosslinks **b** and **d**. For a trifunctional monomer there are three ends of EANC associated to one crosslink: then, the EANC concentration is 3/2 of the crosslink concentration (each EANC has two ends).

The situation may be different if it is possible for the A_3 monomer to generate additional elastic chains through its own structure, as shown in Fig. 3.5. When this monomer becomes a crosslink, apart from the 3/2 elastic chains that have to be counted, there are three more elastic chains contributed by the monomer itself. Therefore, one must be very careful when

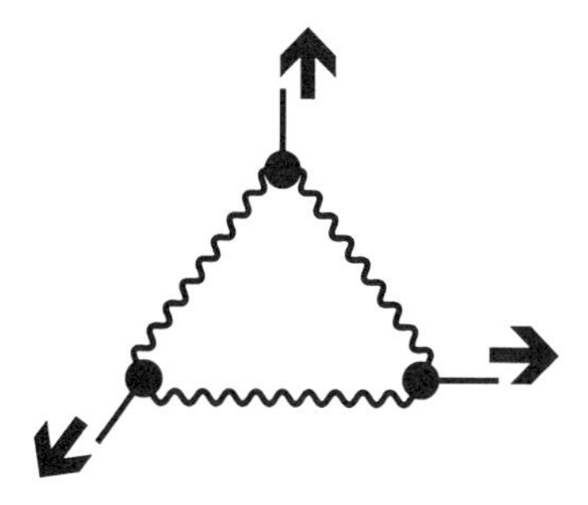

Figure 3.5 Trifunctional monomer with a structure capable of activating internal elastic chains.

defining the elastic chains to be considered for a given chemical system. This is particularly important for thermosetting polymers that originate short elastic chains.

For an A_f ($f \geq 3$) that is polymerized with a B_2, crosslinks of functionality 3,4,. . ., f may be present. The higher the functionality of a crosslink, the larger the contribution to the elastic modulus in the rubbery state.

A simple mathematical description of the postgel stage will be presented for stepwise and free-radical chainwise polymerizations (in this case, the description will be limited to the range of low concentrations of the polyfunctional monomer leading to a homogeneous system). Calculations will be restricted to the evolution of sol and gel fractions, the mass fractions of pendant and elastic chains, and the concentration of crosslinks and EANC as a function of conversion.

3.1.3 Models of Network Formation

Flory (1941a,b, 1953) and Stockmayer (1943, 1944) laid out the basic relations for establishing the evolution of structure with conversion in nonlinear polymerizations. Their analysis is based on the following assumptions defining an ideal network:

1. Functional groups are equally reactive.
2. Reactivities are not affected by the state of reaction of neighboring groups (substitution effects) or by the size of the species to which they belong.
3. Intramolecular cycles are absent in finite species.

Using combinatorial arguments, Flory and Stockmayer derived expressions for the size distribution of the finite molecules as a function of conversion. A compilation of these expressions is given by Peebles (1971).

Instead of the entire size distribution of finite molecules, what is usually necessary is to know the evolution of particular average values of the population as a function of conversion. Although these averages may be obtained from the size distribution by somewhat tedious calculations, a direct procedure to generate the averages using statistical arguments is highly desirable. Moreover, this enables the extension of the analysis to nonideal polymerizations, where the use of distribution functions appears to be prohibitive.

Gordon (1962) showed that different averages of the population could be calculated directly using the theory of stochastic branching processes (cascade substitution). The original method or related recursive procedures was used by a large number of authors to generate most of our present knowledge of the evolution of structure with conversion in nonlinear polymerizations. One popular method was proposed by Macosko and Miller (Macosko and Miller, 1976; Miller and Macosko, 1976; Miller *et al.*, 1979).

In this chapter, one of the recursive methods developed to calculate statistical averages of the population will be used. The first step is the definition of the structures (monomers, fragments of monomers, and reaction products in different states of reaction) to be considered in the statistical analysis. The second step is the calculation of the concentration of selected fragments as a function of conversion. This step requires a simple statistical analysis for ideal polymerizations (e.g., those following the Flory–Stockmayer assumptions), or the solution of a set of kinetic equations for the general case of nonideal polymerizations. The third and final step is the calculation of statistical averages using the known population of fragments as a function of conversion. The fragment approach is particularly useful for analyzing unequal reactivities or substitution effects (Aranguren *et al.*, 1984; Vázquez *et al.*, 1984; Riccardi and Williams, 1986).

The statistical generation of branched and crosslinked structures from the selected fragments leads to a pseudo–most probable distribution. This consists of a population in which the distribution of fragments is deterministic (determined by the set of kinetic equations) and the rest of the distribution is of the most-probable type. But nonidealities may easily lead to distributions that are completely different from the most-probable one. Therefore, recursive procedures must be regarded as approximate solutions for nonideal polymerizations. The rigorous solution can be obtained by stating a set of kinetic differential equations that describe the evolution of every species along the reaction and solving for the averages by applying the method of moments or related procedures (Dušek, 1985). However, by including larger fragments in the analysis, the fragment approach may still be used for highly nonideal polymerizations (Williams *et al.*, 1987,1991; Riccardi and Williams, 1993).

On- and off-lattice Monte Carlo simulations are among the best available methods to analyze complex nonlinear polymerizations, particularly those presenting a high extent of intramolecular cyclization (Šomvársky and Dušek, 1994; Anseth and Bowman, 1994).

In the rest of the chapter we will consider separately stepwise and chainwise polymerizations. A small separate section will be devoted to the hydrolytic condensation of alkoxysilanes. This system exhibits such a large departure from the usual assumptions involved in the description of network formation that it merits particular consideration.

3.2 STEPWISE POLYMERIZATIONS

3.2.1 A_3 Homopolymerization

The ideal homopolymerization of a monomer with three functional groups (functionalities) that may react among themselves will be considered. The ideal case means that the three functionalities are equally reactive, there are no substitution effects, and there are no intramolecular cycles in finite species.

a. Definition of Fragments

Selected fragments are shown in Fig. 3.6. They represent the different possible reaction states for an A_3 molecule (α, unreacted monomer; β, monoreacted fragment; γ, bireacted fragment; δ, trireacted fragment).

b. Concentration of Fragments as a Function of Conversion

For an ideal polymerization the concentration of different fragments may be obtained by a simple statistical analysis. For example, the concentration of unreacted monomer at any conversion x of functional groups is equal to the simultaneous probability that the three functionalities remain unreacted. If x is the probability that a functionality selected at random has reacted, $(1 - x)$ is the probability that it remains unreacted at the particular conversion level. Then, the simultaneous probability that three functionalities have not reacted is $(1 - x)^3$ (the product of the three individual probabilities). So, the fraction of the initial monomer A_{30} that remains unreacted is given by

$$\alpha = A_{30}(1 - x)^3 \tag{3.1}$$

Similarly,

$$\beta = A_{30} 3x(1 - x)^2 \tag{3.2}$$

$$\gamma = A_{30} 3x^2(1 - x) \tag{3.3}$$

$$\delta = A_{30} x^3 \tag{3.4}$$

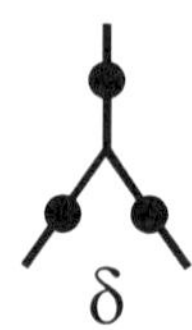

FIGURE 3.6 Different reaction states of an A_3 monomer undergoing a stepwise homopolymerization.

The fraction β arises from the simultaneous probability that one functionality has reacted and the other two remain unreacted. The factor 3 takes into account the three undistinguishable possible elections of the reacted functionality. Fractions γ and δ were obtained using similar arguments. It may be verified that

$$\alpha + \beta + \gamma + \delta = A_{30} \tag{3.5}$$

c. Statistical Averages in the Pregel Stage

In the pregel stage, if one selects a particular fragment β, γ, or δ, the mass attached to it through the reacted functionalities will be finite. We may imagine the experiment depicted in Fig. 3.7. A reacted functionality is introduced into the polymerization mass at a particular conversion x. This reacted functionality will necessarily join to another reacted functionality belonging to species β, γ, or δ. If it becomes joined to β, only a small "fish" will be captured with a mass equal to the mass of fragment β. But if we capture γ by one of its reacted functionalities we have to add the mass that is attached to the other reacted functionality. And if we pick up δ, we have to add the masses attached to the two remaining reacted functionalities. In different castings we will get different masses. But we may ask the question in terms of statistical arguments: Which is the average mass that we expect to get at a particular conversion x? Let us call W this average mass. It may be defined as

$$W = \Sigma_i \text{ (probability of capturing fragment i)} \\ \text{(mass attached to fragment i)} \tag{3.6}$$

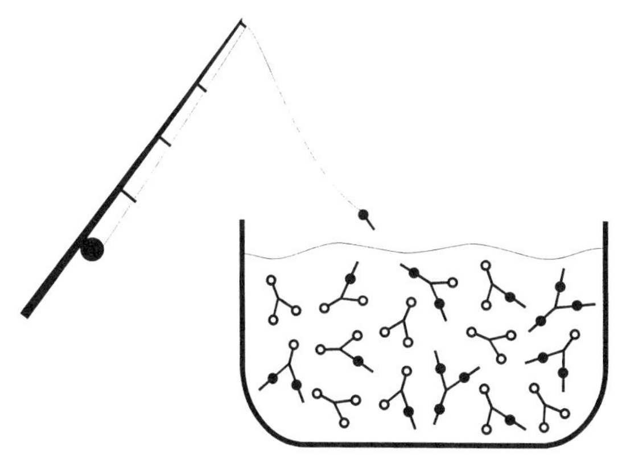

FIGURE 3.7 "Fishing" with a reacted functionality in the reaction medium formed by the stepwise homopolymerization of an A_3 monomer.

The probability of getting β is equal to the fraction of reacted functionalities that belong to this particular fragment:

$$(\text{Probability of capturing } \beta) = \beta/(\beta + 2\gamma + 3\delta) \tag{3.7}$$

Similarly,

$$(\text{Probability of capturing } \gamma) = 2\gamma/(\beta + 2\gamma + 3\delta) \tag{3.8}$$

$$(\text{Probability of capturing } \delta) = 3\delta/(\beta + 2\gamma + 3\delta) \tag{3.9}$$

Let us call M the molar mass of the A_3 monomer and assume, for reasons of clarity, that no small condensation products (such as water) originate in the polymerization. Then, the masses attached to different fragments when they are captured by one of their reacted functionalities are

$$(\text{Mass attached to fragment } \beta) = M \tag{3.10}$$

$$(\text{Mass attached to fragment } \gamma) = M + W \tag{3.11}$$

$$(\text{Mass attached to fragment } \delta) = M + 2W \tag{3.12}$$

where we have again used the definition of W as the average mass joined to a reacted functionality.

Inserting Eqs (3.7)–(3.12) into Eq. (3.6), using Eqs (3.1)–(3.4), and solving, we obtain

$$W = M/(1 - 2x) \tag{3.13}$$

The mass-average molar mass is defined as

$$M_w = \Sigma_i \text{ (mass fraction of fragment i)(mass of fragment i)} \tag{3.14}$$

As the total mass of the system is

$$M_{tot} = A_{30}M = M(\alpha + \beta + \gamma + \delta), \tag{3.15}$$

the respective mass fractions are

$$(\text{Mass fraction of } \alpha) = (1 - x)^3 \tag{3.16}$$

$$(\text{Mass fraction of } \beta) = 3x(1 - x)^2 \tag{3.17}$$

$$(\text{Mass fraction of } \gamma) = 3x^2(1 - x) \tag{3.18}$$

$$(\text{Mass fraction of } \delta) = x^3 \tag{3.19}$$

Then,

$$M_w = (1-x)^3 M + 3x(1-x)^2(M+W) + 3x^2(1-x)(M+2W) + x^3(M+3W) \tag{3.20}$$

Inserting Eq. (3.13) into (3.20) and solving leads to

$$M_w = M(1+x)/(1-2x) \tag{3.21}$$

The gel condition may be obtained by making $M_w \to \infty$. This gives $x_{gel} = 0.5$. The gel conversion could have also been obtained by making $W \to \infty$ in Eq. (3.13).

The number-average molar mass may simply be calculated by

$$M_n = (\text{total mass})/(\text{total number of moles}) \tag{3.22}$$

The total number of moles is the initial number (A_{30}) less the number of moles lost by reaction. When two functionalities belonging to different molecules react, the number of molecules in the system is reduced in one unit. At any conversion $x < x_{gel}$, the number of reacted functionalities is $3A_{30}x$ and the number of moles that have disappeared is $(3/2)A_{30}x$. Then,

$$M_n = A_{30}M/A_{30}(1-3x/2) = M/(1-3x/2) \tag{3.23}$$

Figure 3.8 shows the variation of the number-average ($DP_n = M_n/M$) and mass-average ($DP_w = M_w/M$) degrees of polymerization with conversion, in the pregel stage. While M_w diverges at x_{gel}, M_n remains finite; this is because the total number of moles is still high at the gel conversion.

Equation (3.23) has no physical sense for $x > x_{gel}$, because intramolecular reactions in the gel cannot be neglected; so, the relationship between the number of moles in the system and the conversion is no longer valid. For an actual system, the departure of the theoretical prediction of M_n and the experimental value in the pregel stage may be ascribed to the formation of intramolecular cycles. This constitutes the usual way of determining the significance of cyclization for a particular system.

d. Statistical Averages in the Postgel Stage

Properties of the system in the postgel stage may be obtained in a similar way, using a simple statistical analysis. Now we change the question and ask for the probability of obtaining a finite species in the hypothetical experiment shown in Fig. 3.7.

Let us call Z the probability of having a finite continuation from a reacted functionality. It is defined by

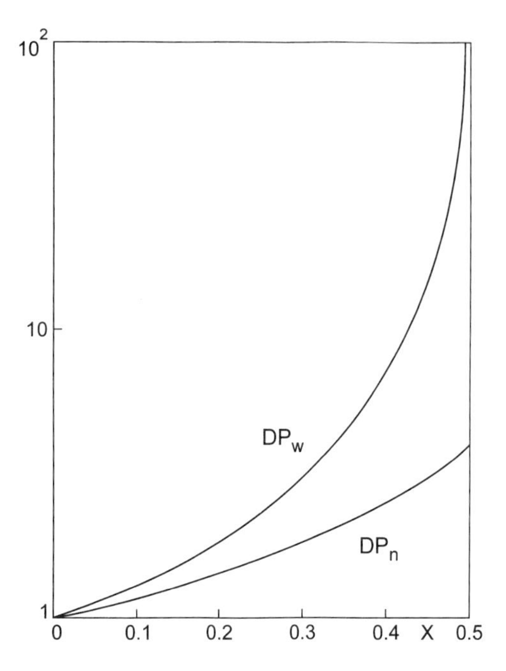

FIGURE 3.8 Evolution of the number-average and mass-average degrees of polymerization, DP_n and DP_w, as a function of conversion in the pregel stage, during the stepwise homopolymerization of an A_3 monomer.

$$Z = \Sigma_i \text{ (probability of capturing fragment i)} \text{ (probability to have a finite continuation from fragment i)} \quad (3.24)$$

The last factor of Eq. (3.24) is equal to 1 if we capture β (as there is no continuation from this species), it equals Z if the captured fragment is γ; and it equals Z^2 if the attached fragment is δ (as there are two possible continuations from this species and the probability of having a finite continuation is equal to the simultaneous probability that both directions lead to a finite species). Then,

$$Z = [\beta/(\beta + 2\gamma + 3\delta)] + [2\gamma/(\beta + 2\gamma + 3\delta)]Z + [3\delta/(\beta + 2\gamma + 3\delta)]Z^2 \quad (3.25)$$

Using Eqs (3.1)–(3.4) and solving for Z leads to the following two roots:

$$Z = 1 \quad (3.26)$$

$$Z = (1 - x)^2/x^2 \tag{3.27}$$

Equation (3.26) is the root with physical meaning in the pregel stage, as all continuations are finite. Equation (3.27) is the root valid in the postgel stage (for $x = x_{gel} = 0.5$, both roots are coincident). It is observed that for $x = 1$, $Z = 0$, which means that all fragments are interconnected; i.e., at full conversion all the initial mass is contained in only one giant molecule.

The mass fraction of finite species, w_s (sol fraction), is obtained from

$$w_s = (\alpha/A_{30}) + (\beta/A_{30})Z + (\gamma/A_{30})Z^2 + (\delta/A_{30})Z^3 \tag{3.28}$$

Equation (3.28) is based on the fact that only those fragments that have finite continuation in every direction contribute to the sol. Inserting Eqs (3.1)–(3.4) and (3.27), and using some algebra, we get

$$w_s = (1 - x)^3/x^3 \tag{3.29}$$

Figure 3.9 shows the rapid decrease of the sol fraction with conversion in the postgel stage. The contribution of every fragment to the sol is also depicted. At $x = x_{gel}$, β and γ fragments are the predominant species. The contribution of γ to the sol fraction decreases at a faster rate than that of β,

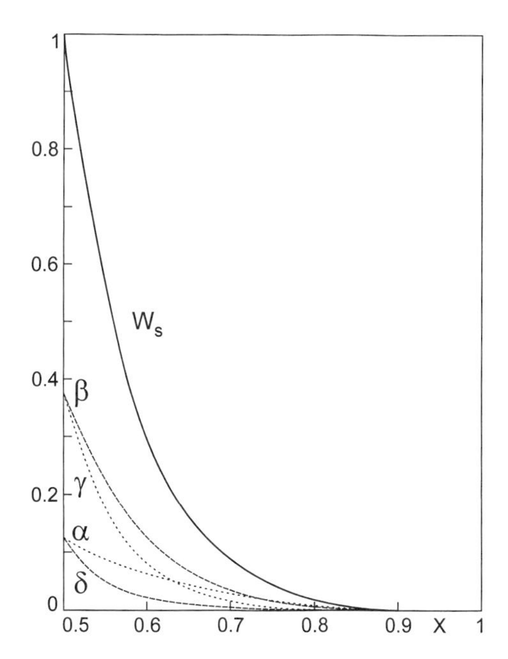

FIGURE 3.9 Contribution of different fragments to the sol fraction for an A_3 stepwise homopolymerization.

due to the presence of two reacted functionalities and the consequent probability that it becomes part of the gel. Both the free monomer (α) and the branching units (δ) have the same concentration at x_{gel}, but while the contribution of δ to the sol fraction decreases rapidly with conversion, the free monomer constitutes the major contribution to the sol fraction at high conversions in the postgel stage.

We will now analyze the evolution of branching units (δ) and crosslinks, along the polymerization. Crosslinks are δ fragments that have an infinite continuation (to the system's boundaries) in the three possible directions. The concentration of crosslinks per unit mass is given by

$$n_c = (\delta/A_{30})[A_{30}](1 - Z)^3 = (2 - 1/x)^3[A_{30}] \tag{3.30}$$

where $[A_{30}]$ is the initial concentration of monomer per unit mass.

Figure 3.10 shows the evolution of the fraction of branching units, and crosslinks with conversion. Before gelation there are no crosslinks in the system. At full conversion their concentration is equal to that of branching units, because every initial A_3 molecule has become a crosslink of the polymer network.

Assuming that no internal elastic chains are activated, the concentration of elastically active network chains per unit mass, ν_e, may be calculated as

$$\nu_e = (3/2)n_c = (3/2)(2 - 1/x)^3[A_{30}] \tag{3.31}$$

The concentration of elastic chains increases from zero at x_{gel} to $(3/2)[A_{30}]$ at full conversion.

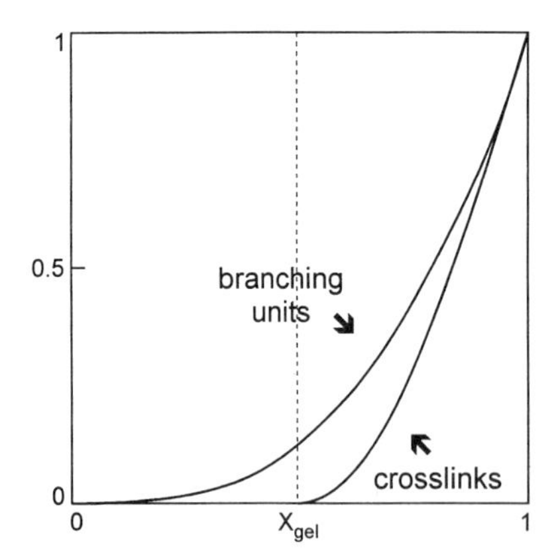

FIGURE 3.10 Evolution of the fraction of branching units (δ/A_{30}) and crosslinks, $n_{c,3}/[A_{30}]$, for the A_3 stepwise homopolymerization.

Pendant chains in the gel may be visualized as network elements that do not follow the overall deformation of the whole material during a mechanical test, thus contributing to viscous dissipation. Pendant chains are formed by fragments β, γ and δ that are linked to the gel by only one arm; e.g., there is an infinite continuation in only one direction.

The mass fraction of pendant chains is given by

$$w_p = (\beta/A_{30})(1 - Z) + (\gamma/A_{30})2Z(1 - Z) + (\delta/A_{30})3Z^2(1 - Z) \quad (3.32)$$

Using Eqs (3.2)–(3.4) and (3.27), we get

$$w_p = 3(1 - x)^2(2x - 1)/x^3 \quad (3.33)$$

The mass fraction of pendant chains is zero at $x_{gel} = 0.5$ and at $x = 1$, being a maximum at an intermediate conversion.

The mass fraction of elastic chains may be obtained as

$$w_e = 1 - w_s - w_p \quad (3.34)$$

Figure 3.11 shows the evolution of w_s, w_p, and w_e as a function of conversion in the postgel stage.

3.2.2 $A_4 + B_2$ Stepwise Polymerization

a. Epoxy – Amine Networks

A good example of an $A_4 + B_2$ stepwise polymerization is the formation of an epoxy – amine network, starting from a difunctional epoxy molecule (B_2)

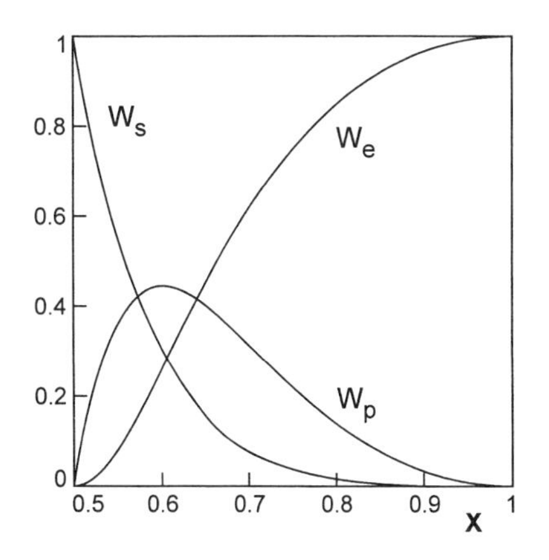

Figure 3.11 Evolution of the mass fraction of sol (w_s), pendant chains (w_p) and elastic chains (w_e), for the A_3 stepwise homopolymerization.

as the diglycidyl ether of bisphenol A and a tetrafunctional diamine (A_4). The following main reactions take place:

$$RNH_2 + \underset{\diagdown O \diagup}{CH_2 - CH}- \rightarrow RNH - CH_2 - CH(OH)- \tag{3.35}$$

$$RNH - CH_2 - CH(OH) - + \underset{\diagdown O \diagup}{CH_2 - CH}- \rightarrow RN(CH_2 - CH(OH)-)_2 \tag{3.36}$$

Under particular conditions, e.g., an excess of epoxy groups or slow reactivity of the secondary amine, the homopolymerization of epoxy groups may take place. This is a chainwise homopolymerization initiated by either the secondary hydroxyls or the tertiary amines (Chapter 2). It may be described by the following equation:

$$-CH(OH) - + \underset{\diagdown O \diagup}{CH_2 - CH}- \rightarrow -CH(O - CH_2 - CH(OH)-)- \tag{3.37}$$

In this section we will assume that this third reaction may be neglected so that a neat stepwise polymerization takes place.

The stoichiometric ratio of amine/epoxy equivalents is given by

$$r = A_0/B_0 \tag{3.38}$$

where A_0 and B_0 are, respectively, the initial concentration of amine and epoxy equivalents. The conversion of A and B functionalities is then related through

$$x_B = r x_A \tag{3.39}$$

b. Definition of Fragments

Figure 3.12 shows the structural fragments that may be used to analyze the network buildup. Fragments F1, F3, and F4 represent half of the diamine molecule with different states of reaction. Both amine hydrogens of F1 (represented by unfilled squares) remain unreacted; F3 and F4 have, respectively, one and two reacted amine hydrogens. F2 represents half of the diepoxide molecule bearing an unreacted epoxy group (represented by an unfilled circle). Reacted epoxies are joined to reacted amine hydrogens in fragments F3 and F4. The network structure is generated by randomly joining half-diamine segments among themselves as well as half-epoxy arrows among themselves. The stoichiometric ratio may be also expressed as

$$r = 2\ F10/F20 \tag{3.40}$$

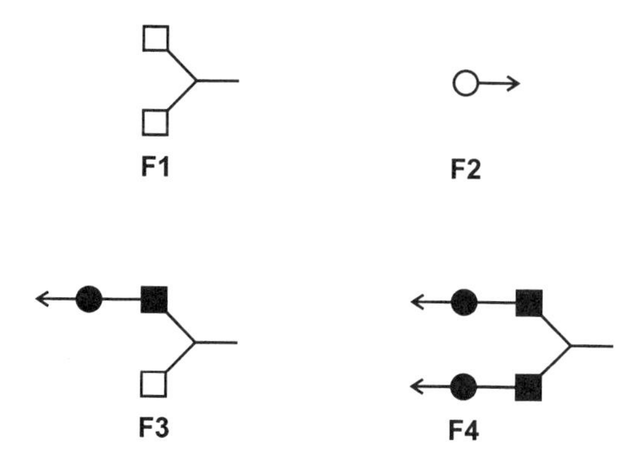

FIGURE 3.12 Different reaction states of half of an A_4 monomer and half of a B_2 monomer in the course of an $A_4 + B_2$ stepwise polymerization.

where F10 and F20 are, respectively, the initial concentrations of fragments F1 and F2.

c. Concentration of Fragments as a Function of Conversion

The evolution of the concentration of different fragments along polymerization depends on the reaction probability of primary and secondary amine hydrogens. If both react at the same rate, the concentration of fragments may be obtained using only probabilistic arguments (as those used to obtain the concentration of fragments α–δ in the previous section). In this case,

$$F1 = (1 - x_A)^2 F10 \tag{3.41}$$

$$F2 = (1 - x_B) F20 \tag{3.42}$$

$$F3 = 2x_A(1 - x_A) F10 \tag{3.43}$$

$$F4 = x_A^2 F10 \tag{3.44}$$

It may be corroborated that F1 +F3 +F4 = F10 and F2 +F3 +2 F4 = F20.

If, on the other hand, primary and secondary amine hydrogens react at different rates, it is necessary to use a kinetic scheme to obtain the evolution of the concentration of different fragments along the reaction. The epoxy – amine reaction may take place both by a noncatalytic path (specific rate constant k′) and by a reaction catalyzed by OH groups (specific rate constant k). The secondary amine hydrogen is usually less reactive than the primary amine hydrogen. Specific rate constants for the secondary amine hydrogen, k_2 and k'_2, are lower than rate constants for the primary amine hydrogen, k_1 and k'_1. A reactivity ratio is usually defined as

$$R_{21} = k_2/k_1 = k'_2/k'_1 \tag{3.45}$$

As a rough approximation, R_{21} may be assumed constant and independent of temperature.

The rate of disappearance and formation of the fragments depicted in Fig. 3.12 may be written as

$$-dF1/dt = 2[k'_1 + k_1(OH)]F1\ F2 \tag{3.46}$$

$$-dF2/dt = \{2[k'_1 + k_1(OH)]F1 + [k'_2 + k_2(OH)]F3\}F2 \tag{3.47}$$

As rate constants are defined on the basis of one hydrogen bonded to an N atom, $R_{21} = 1$ for a system without substitution effects (when rate constants are defined on the basis of the amine group, $R_{21} = 0.5$ in the absence of substitution effects).

The following balances and definitions may be written:

$$F3 = 2(F10 - F1) - (F20 - F2) \tag{3.48}$$

$$(OH) = (OH)_0 + x_B F20 \tag{3.49}$$

$$F4 = F10 - F1 - F3 \tag{3.50}$$

$$x_B = (F20 - F2)/F20 = (F3 + 2F4)/F20 \tag{3.51}$$

Equation (3.48) arises from Eqs (3.50) and (3.51).

Solving Eqs (3.46)–(3.51) gives the concentration of the different fragments as a function of the conversion of epoxy groups x_B related to the conversion of amine hydrogens x_A, through Eq. (3.40).

d. Statistical Averages in the Pregel Stage

With the knowledge of the concentration of different fragments along the reaction, either for the ideal system through Eqs (3.41)–(3.44) or for the system with substitution effects, Eqs (3.46)–(3.51), average statistical parameters of the polymer network may be calculated.

Two average masses may be defined in the pregel stage: W = average mass joined to a segment (in F1, F3, and F4) and Y = average mass joined to an arrow (in F2, F3, and F4). Using arguments similar to those in the previous section, these two averages can be calculated as:

$$\begin{aligned} W = {} & (F1/F10)(M_A/2) + (F3/F10)[(M_B + M_A)/2 + Y] \\ & + (F4/F10)[M_B + M_A/2 + 2Y] \end{aligned} \tag{3.52}$$

$$Y = (F2/F20)(M_B/2) + (F3/F20)[(M_B + M_A)/2 + W] + 2(F4/F20)[M_B + M_A/2 + W + Y] \quad (3.53)$$

where (F1/F10), (F3/F10), and (F4/F10) are the probabilities of capturing fragments F1, F3, and F4, respectively, using a segment; and (F2/F20), (F3/F20), and (2F4/F20) are the probabilities of capturing fragments F2, F3, and F4, respectively, using an arrow. M_A and M_B are the molar masses of both monomers.

For the ideal case, an analytical solution may be obtained. Substituting Eq. (3.53) into (3.52) and using Eqs (3.41)–(3.44), we get

$$W = M_A/2 + x_A M_B + 2x_A Y \quad (3.54)$$

$$Y = [x_B M_A + M_B(0.5 + x_B/2 + x_A x_B)]/(1 - 3x_A x_B) \quad (3.55)$$

The mass-average molar mass may be calculated with the aid of Eq. (3.14):

$$M_W = w_1(M_A/2 + W) + w_2(M_B/2 + Y) + w_3[(M_A + M_B)/2 + Y + W] + w_4(M_B + M_A/2 + W + 2Y) \quad (3.56)$$

where w_1–w_4 are the mass fractions of fragments F1–F4, respectively, which may be readily calculated from the concentrations and particular masses of each fragment.

The system gels when $M_W \to \infty$, a condition that is satisfied when $Y, W \to \infty$ (both diverge simultaneously). From Eqs (3.54) and (3.55), this condition is fulfilled when

$$(x_A x_B)_{gel} = 1/3 \quad (3.57)$$

For a stoichiometric system, $x_A = x_B = x$, leading to

$$x_{gel} = (1/3)^{1/2} = 0.577 \quad (3.58)$$

This value is close to experimental values of gel conversion reported for several diepoxide – diamine systems.

If a diamine excess is used in the initial formulation, meaning that $r > 1$, the limiting reactant is the epoxy. Replacing x_A by Eq. (3.39) in Eq. (3.57), results in

$$x_{B,gel} = (r/3)^{1/2} \quad (3.59)$$

If an epoxy excess is used ($r < 1$), the limiting reactant is the amine. Replacing x_B by Eq. (3.39) in Eq. (3.57), we get

$$x_{A,gel} = (1/3r)^{1/2} \quad (3.60)$$

Figure 3.13 shows the variation of the gel conversion of the limiting reactant as a function of the stoichiometric ratio. For $r > 3$, no gel is formed and the polymer remains in the liquid state after complete reaction of epoxy groups. If the amount of epoxy monomer necessary to obtain a stoichiometric system is added in a second step, polymerization restarts, leading to gelation and the formation of a network. The two-step polymerization is the basis of several commercial thermosetting polymers.

For example, novolacs are phenolic resins obtained by the condensation of phenol (trifunctional monomer) and formaldehyde (bifunctional monomer), using a stoichiometric excess of phenol so that formaldehyde is completely consumed without leading to gelation. In a second step, instead of directly adding the necessary formaldehyde, the polymer network is formed by reaction with hexamethylenetetramine (a condensation product of formaldehyde and ammonia usually called hexa), through a complex set of chemical reactions (Chapter 2).

For $r < 1/3$, no gel can be produced by epoxy–amine reactions. However, the homopolymerization of the epoxy excess (Eq. 3.37) may take place, finally leading to gelation. So, it is not convenient to use an epoxy excess to synthesize stable epoxy–amine prepolymers. Commercial epoxy–amine adducts are based on an amine excess.

For the general case when substitution effects are present, it is necessary to obtain Y and W from Eqs (3.52) and (3.53). Inserting Y from Eq.

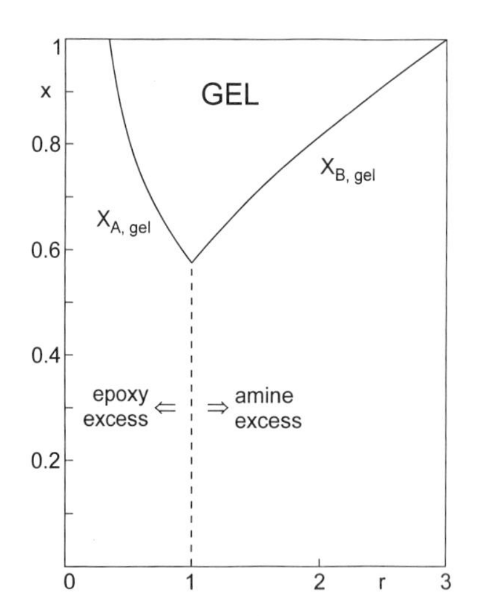

FIGURE 3.13 Gel conversion of the limiting reactant as a function of the stoichiometric ratio.

(3.53) into Eq. (3.52) and solving for W, we can obtain the condition for which both Y and W become infinite (gel conversion). This is given by

$$F10(F20 - 2\ F4) = (F3 + 2\ F4)^2 \qquad (3.61)$$

Substituting F10 from Eq. (3.40) and using Eq. (3.51) results in

$$(x_{B,gel})^2 = (r/2)(1 - 2F4/F20) \qquad (3.62)$$

For a given set of specific rate constants, F4 and, consequently $x_{B,gel}$ may be obtained by solving Eqs (3.46)–(3.51). For the ideal case, F4 may be replaced by Eq. (3.44) and, using Eqs (3.39) and (3.40), the gel condition given by Eq. (3.57) is again obtained. Two limiting cases of gelation in systems with substitution effects may be considered. For the case where the secondary amine hydrogen reacts at a much slower rate than the primary amine hydrogen, $R_{21} \rightarrow 0$ in Eq. (3.45), meaning that all F1 is first converted to F3 and only then does F4 begin to be formed. From this time on, it is verified that

$$F4 = F10 - F3 \qquad (3.63)$$

Inserting Eqs (3.51) and (3.40) into Eq. (3.63), using Eq. (13.62), Eq. (3.62), and solving for $x_{B,gel}$ gives

$$x_{B,gel} = \{(r/2)[1 + 2r - (3r^2 + 2r)^{1/2}]\}^{1/2} \qquad (3.64)$$

For a stoichiometric system (r = 1), Eq. (3.64) leads to

$$x_{B,gel} = 0.618 \qquad (3.65)$$

Substitution effects produce only a slight shift in the gel conversion from 0.577 to 0.618 in the limit of $R_{21} \rightarrow 0$. Most of the reported experimental values of gel conversion in stoichiometric epoxy–amine systems are close to 0.60. Within the experimental error of the measurement of gel conversion, it is difficult to quantify the value of the reactivity ratio from this single determination.

The critical gelation ratio, r_c, in formulations containing an excess of amine groups, constitutes a better way to quantify substitution effects. The system does not gel when $r > r_c$. For the ideal case ($R_{21} = 1$), $r_c = 3$. For the limiting case of $R_{21} = 0$, Eq. (3.64) leads to $x_{B,gel} = 1$ for $r_c = 2$. This means that a much lower amine excess than in the ideal case prevents gelation. The experimental determination of r_c constitutes a good approach to obtaining the value of R_{21}. Solving Eqs (3.46)–(3.51) leads to r_c as a function of R_{21}, as plotted in Fig. 3.14.

The limiting case of $R_{21} \rightarrow \infty$ may also be considered. Although this situation does not occur for an epoxy–amine system, it is of interest to determine the effect of an instantaneous activation of the second available

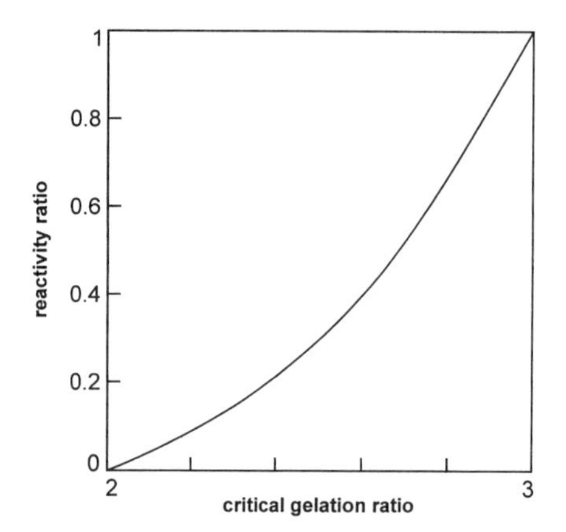

FIGURE 3.14 Critical gelation ratio (r_c) in mixtures with an amine excess for different values of the reactivity ratio between secondary and primary amine hydrogens (R_{21}).

functionality in the F1 fragment, after reaction of the first one. In this case, the concentration of fragment F3 is negligible (F3 → 0), because once the first functionality of F1 reacts, the second functionality reacts immediately. Then, from Eq. (3.51),

$$x_B = 2F4/F20 \tag{3.66}$$

Substituting Eq. (3.66) into Eq. (3.62) and solving, we obtain

$$x_{B,gel} = 0.5(0.25r^2 + 2r)^{1/2} - 0.25r \tag{3.67}$$

For a stoichiometric system ($r = 1$), Eq. (3.67) leads to

$$x_{B,gel} = 0.5 \tag{3.68}$$

Because of the effective generation of branched fragments (F4) from the very beginning of polymerization, the gel conversion is advanced with respect to the value for the ideal case.

e. Statistical Averages in the Postgel Stage

Properties of the system in the postgel stage may be obtained in a similar way to that for the A_3 homopolymerization. Let us denote Z_A and Z_S as the probabilities of finding a finite chain when looking out from an arrow or a segment, respectively.

$Z_A = \Sigma_{fragments}$ (fraction of total arrows belonging to a particular fragment) × (probability that all exits of the fragment captured by an arrow, are finite)

(3.69)

Then,

$$Z_A = (F2/F20) + (F3/F20)Z_S + (2F4/F20)Z_AZ_S \quad (3.70)$$

Similarly,

$$Z_S = (F1/F10) + (F3/F10)Z_A + (F4/F10)Z_A^2 \quad (3.71)$$

The concentration of fragments as a function of conversion is obtained from Eqs (3.41)–(3.44) for an ideal system, or by numerical solution of the set of Eqs. (3.46)–(3.51) for the general case. Eqs. (3.70) and (3.71) are then solved, looking for the roots comprised between zero and 1. This gives Z_A and Z_S as a function of conversion.

The sol fraction may be calculated as

$$w_S = w_1Z_S + w_2Z_A + w_3Z_AZ_S + w_4Z_A^2Z_S \quad (3.72)$$

The mass fraction of pendant chains is given by

$$w_p = w_1(1 - Z_S) + w_2(1 - Z_A) + w_3[Z_A(1 - Z_S) + Z_S(1 - Z_A)] + w_4[2Z_AZ_S(1 - Z_A) + Z_A^2(1 - Z_S)] \quad (3.73)$$

The mass fraction of material pertaining to elastically active network chains (EANC) is obtained from Eq. (3.34).

The concentration of crosslinks per unit mass is given by the concentration of F4 fragments issuing to infinity in the three possible directions:

$$n_C = (F4/F10)(1 - Z_A)^2(1 - Z_S)[F10] \quad (3.74)$$

where [F10] is the initial concentration of amine groups per unit mass.

The concentration of EANC per unit mass is given by

$$\nu_e = (3/2)n_C \quad (3.75)$$

Equation (3.75) assumes that the chain located between both amine groups may become an elastic chain of the polymer network.

Figure 3.15 shows the fraction of elastic chains, ν_e/[F10], calculated numerically as a function of conversion, for a stoichiometric system ($r = 1$) with the following kinetic parameters: $R_{21} = 0.4$, $(OH)_0 / F20 = 0.061$, and $k'_1 / (k_1F20) = 0.0576$. At full conversion, n_C = [F10] (all the amine groups act as crosslinks) and ν_e / [F10] = 1.5.

The three elastic chains issuing from the crosslink are not equivalent. One of them is the chain joining both amine groups in the diamine mono-

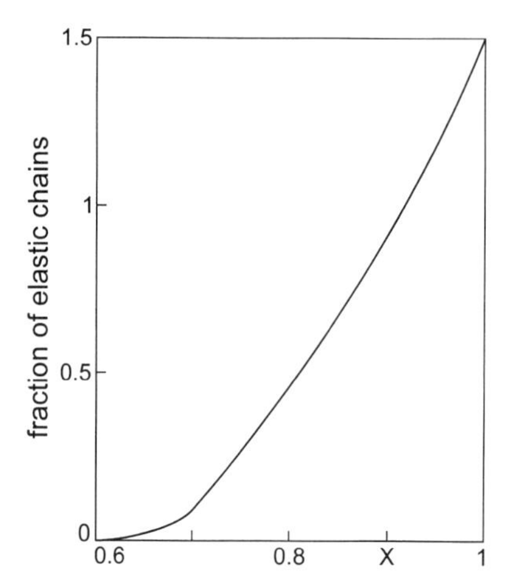

Figure 3.15 Fraction of elastic chains, $\nu_e/[F10]$, as a function of conversion, for a diepoxide (B_2) – diamine (A_4) stoichiometric system, with the following kinetic parameters: $R_{21} = 0.4$, $(OH)_0/F20 = 0.061$, and $k'_1/(k_1F20) = 0.0576$.

mer. If this chain is very much shorter than the one of the diepoxide, the calculation of EANC is better carried out taking the whole diamine unit as a single fragment. This gives crosslinks with three and four branches going to infinity.

f. Limitations of the Analysis

The example developed in this section should enable the reader to analyze any particular system undergoing stepwise polymerization. The first step is to make a suitable definition of fragments; then a kinetic scheme taking unequal reactivities and substitution effects (if any) into account must be stated and solved to obtain the concentration of fragments as a function of conversion. Statistical averages may be calculated in both pregel and postgel stages, as in the example of epoxy–amine networks. There are still two approximations in these calculations:

1. Intramolecular cycles are absent in the pregel stage (although short cycles may be introduced in the definition of fragments).
2. The network structure at any conversion may be obtained by joining the different fragments at random, with a probability given by the concentration of every fragment in the mixture. This is a mean-field approach and is not valid when nonidealities are present. Unequal reactivities, substitution effects, and intramolecular cycles give place to preferred nonrandom combinations.

The error introduced by the second assumption may be decreased by adding larger fragments to describe the system. Usually, it is verified that the mean-field approach gives a very good approximation, and it is not necessary to increase the number of fragments. In most cases the presence of intramolecular cycles is responsible for departures between experimental results and theoretical predictions.

3.2.3 Extension to A_f and (A_f + B_g) Stepwise Polymerizations

Several approaches can be used to obtain statistical parameters of these general systems under ideal conditions (equal reactivities, and absence of substitution effects and intramolecular cycles). In this section we will discuss some of these results. The reader is referred to the papers of Macosko and Miller (1976), Miller and Macosko (1976), and Miller *et al.* (1979) for the deductions of the equations used in this section.

a. Pregel Stage (Monomers with a Single Value of Functionality)

For an A_f homopolymerization, number- and mass-average molar masses are given by

$$M_n = M_{Af}/(1 - x\, f/2) \tag{3.76}$$

$$M_w = M_{Af}(1 + x)/[1 - x(f - 1)] \tag{3.77}$$

Equations (3.76) and (3.77) reduce to Eqs (3.21) and (3.23) for $f = 3$. They also reduce to the known equations for the stepwise polymerization of bifunctional monomers, by making $f = 2$.

Gelation occurs when $M_w \rightarrow \infty$, a condition that is fulfilled for

$$x_{gel} = 1/(f - 1) \tag{3.78}$$

The higher the functionality of a monomer, the lower the gel conversion.

For an $A_f + B_g$ stepwise polymerization without evolution of condensation products, M_n and M_w are given by

$$M_n = (M_{Af}A_f + M_{Bg}B_g)/(A_f + B_g - x_A f A_f) \tag{3.79}$$

$$\begin{aligned} M_w = {} & [x_B(M_{Af}^2/f) + x_A(M_{Bg}^2/g)]/[x_B(M_{Af}/f) + x_A(M_{Bg}/g)] \\ & + \{x_A x_B[x_A(f - 1)M_{Bg}^2 + x_B(g - 1)M_{Af}^2 + 2M_{Af}M_{Bg}]\}/ \\ & \{[x_B(M_{Af}/f) + x_A(M_{Bg}/g)][1 - x_A x_B(f - 1)(g - 1)]\} \end{aligned} \tag{3.80}$$

Equation (3.79) was obtained by dividing the total mass of the system by the total number of moles (A_f and B_g represent the initial number of moles of each one of the monomers). The factor $x_A f A_f$ in the denominator represents the moles of type A functionalities that have reacted, which is equal to the number of moles that are lost by reaction provided that no intramolecular cycles are formed. As 1 mole of A reacts with 1 mole of B, it must be verified that

$$x_A f A_f = x_B g B_g \tag{3.81}$$

or

$$x_B = r x_A \tag{3.82}$$

where r is the stoichiometric ratio,

$$r = f A_f / g B_g \tag{3.83}$$

The factor $x_A f A_f$ may be then replaced by $x_B g B_g$ in Eq. (3.79). Gelation takes place when $M_w \rightarrow \infty$. This leads to

$$(x_A x_B)_{gel} = 1/[(f-1)(g-1)] \tag{3.84}$$

Equation (3.84) reduces to Eq. (3.57) for $f = 4$ and $g = 2$.

To calculate the critical gelation ratio, r_c, let us call A_f the monomer which is present in excess over the stoichiometric value. Substituting Eq. (3.39) in Eq. (3.84), results in

$$(x_{B,gel})^2 = r/[(f-1)(g-1)] \tag{3.85}$$

The critical gelation ratio is obtained for $x_{B,gel} = 1$, leading to

$$r_c = (f-1)(g-1) \tag{3.86}$$

For $r > r_c$, the system will not gel (provided it behaves ideally).

Let us apply Eq. (3.86) to phenol – formaldehyde polymers synthesized in an acid medium with a phenol excess (novolacs). Phenol is a trifunctional reactant (A_3), the functional groups being the aromatic hydrogens located in positions 2, 4, and 6 of the phenolic ring. Formaldehyde acts as a bifunctional monomer (B_2), forming methylene bridges between the reactive positions of phenol. Novolacs are synthesized with a phenol excess, such that gelation does not occur at full formaldehyde conversion. From Eqs (3.83) and (3.86), we obtain

$$r_c = 2 = (3\ A_3)/(2B_2) \tag{3.87}$$

For $r > 2$, gelation will not occur. This leads to $B_2/A_3 < 0.75$, to avoid gelation. However, many commercial novolacs are synthesized using a formaldehyde/phenol molar ratio in the range 0.80–0.85, and gelation does not occur in the reactor as industrial practice has convincingly proved for about a century. The reason is that in this system there is a significant substitution effect for the reaction of the third functionality of a phenolic ring when the two other functionalities have reacted. Similarly than for $A_4 + B_2$ polymerization, this reduces r_c and increases the formaldehyde/phenol molar ratio that may be safely introduced into the reactor to avoid gelation (Aranguren *et al.*, 1984). In fact, under the normal conditions of the synthesis, novolacs are better described as linear oligomers with a very small fraction of branching (trireacted phenolic rings).

b. Pregel Stage (Monomers with a Distribution of Functionalities)

A further generalization arises if one considers the system $A_{fi} + B_{gj}$,

$$A_{f1} + A_{f2} + \cdots + A_{fi} + \cdots + B_{g1} + B_{g2} + \cdots + B_{gj} + \cdots,$$

where reactant A is a mixture of molecules of different functionality (f_1, f_2, $\cdots$, f_i, $\cdots$), bearing the same type of functional groups, that reacts with B, which is also a mixture of molecules of different functionality (g_1, g_2, $\cdots$, g_j, $\cdots$), with the same functional groups. This system has practical applications in the formulations of coatings. Mixtures containing bifunctional and polyfunctional monomers may be used to regulate the crosslink density and the corresponding final properties of the polymer network.

The number-average molar mass is calculated by the following generalization of Eq. (3.79):

$$M_n = \left(\sum A_{fi}M_{Afi} + \sum B_{gj}M_{Bgj}\right)/\left(\sum A_{fi} + \sum B_{gj} - x_A \sum A_{fi}M_{Afi}\right) \tag{3.88}$$

Both conversions are related through

$$x_A \sum f_i A_{fi} = x_B \sum g_j B_{gj} \tag{3.89}$$

where

$$r = (\sum f_i A_{fi})/(\sum g_j B_{gj}) \tag{3.90}$$

The mass-average molar mass is given by

$$\begin{aligned} M_w = {} & \{(x_B \sum M_{Afi}^2 a_{fi}/f_i + x_A \sum M_{Bgj}^2 b_{gj}/g_j)/(x_B \sum M_{Afi} a_{fi}/ \\ & f_i + x_A \sum M_{Bgj} b_{gj}/g_j) + \{x_A x_B [x_A (f_e - 1)(\sum M_{Bgj} b_{gj})^2 \\ & + x_B (g_e - 1)(\sum M_{Afi} a_{fi})^2 + 2(\sum M_{Afi} a_{fi})(\sum M_{Bgj} b_{gj})]\}/ \\ & \{(x_B \sum M_{Afi} a_{fi}/f_i + x_A \sum M_{Bgj} b_{gj}/g_j)[1 - x_A x_B (f_e - 1)(g_e - 1)]\} \end{aligned} \tag{3.91}$$

where

$$a_{fi} = f_i A_{fi} / \sum f_i A_{fi} \tag{3.92}$$

$$b_{gj} = g_j\, B_{gj} / \sum g_j B_{gj} \tag{3.93}$$

are, respectively, the mole fraction of all A's on A_{fi} and of all B's on B_{gj} molecules.

M_w depends on functionality-average functionalities defined by

$$f_e = (\sum f_i^2\, A_{fi}) / (\sum f_i A_{fi}) \tag{3.94}$$

$$g_e = (\sum g_j^2 B_{gj}) / (\sum g_j\, B_{gj}) \tag{3.95}$$

The mass-average molar mass goes to infinity when

$$(x_A\, x_B)_{gel} = 1/[(f_e - 1)(g_e - 1)] \tag{3.96}$$

An example of the application of Eq. (3.96) to a practical case will be analyzed. Let us consider a system based on glycerin (A_3), phthalic anhydride (B_2), and an unsaturated fatty acid, such as linoleic acid (B_1). The reaction of the triol with the anhydride and the monoacid gives rise to polyesters that form the basis of some coatings. The polyester is usually an oligomer that is soluble in a suitable solvent. After application, a cross-linking reaction takes place through the unsaturations of the fatty acid (e.g., through oxidation catalyzed by organometallic salts), leading to a polymer network. We want to know what is the minimum B_1/B_2 ratio in a stoichiometric system for which the polyesterification reaction will not lead to gelation at full conversion. In this case,

$$f_e = 3 \tag{3.97}$$

$$g_e = (4B_2 + B_1)/(2B_2 + B_1) = [(B_1/B_2) + 4]/[(B_1/B_2) + 2] \tag{3.98}$$

From Eq. (3.96), as $x_A = x_B = x$, the necessary condition to avoid gelation at full conversion is that

$$(f_e - 1)(g_e - 1) < 1 \tag{3.99}$$

Substituting Eqs (3.97) and (3.98) in Eq. (3.99) leads to

$$B_1/B_2 > 2 \tag{3.100}$$

Similar calculations may be carried out for nonstoichiometric mixtures. For example, the minimum amount of B_1 to avoid gelation of a stoichiometric $A_3 + B_2$ mixture may be determined. In this case there will be an excess of acid groups, so that when $x_A = 1$, $x_B = r = 3A_3/(B_1 + 2B_2)$. From Eq. (3.96), gelation will be avoided if

$$(f_e - 1)(g_e - 1) < 1/r \tag{3.101}$$

Substituting Eqs (3.97), (3.98), and the definition of r in Eq. (3.101), and taking into account that $3A_3 = 2B_2$ (stoichiometric $A_3 + B_2$ mixture), we obtain

$$B_1/B_2 > 0.83 \tag{3.102}$$

Results expressed by Eqs (3.100) and (3.102) are strictly valid for ideal systems. Unequal reactivities, substitution effects and the formation of intramolecular cycles will affect them. The first two nonidealities may be conveniently taken into account using the fragment approach described in the previous sections.

c. Crosslinking of Polymer Chains (Vulcanization)

Consider the case of polymer chains with an arbitrary distribution of chain lengths and with a number of potentially reactive sites equal to the degree of polymerization (every monomeric unit of the polymer chain constitutes one potentially reactive site). The random crosslinking (vulcanization) of these linear primary chains may be considered a stepwise homopolymerization of A_{fi} species, where the functionality of every species is directly proportional to its molar mass,

$$M_{Afi} = f_i\ M_c \tag{3.103}$$

M_c is the mass between crosslinkable sites.

By assuming that the crosslinking reaction takes place by coupling of chains without any mass change, Eq. (3.91) may be used and adapted to the case of a single type of monomer:

$$M_w = [(\sum M_{Afi}^2 a_{fi}/f_i)/(\sum M_{Afi} a_{fi}/f_i)] + x(\sum M_{Afi} a_{fi})^2/\{(\sum M_{Afi} a_{fi}/f_i)[1 - x(f_e - 1)]\} \tag{3.104}$$

Substituting Eq. (3.103) into Eq. (3.104), and taking into account the definitions of f_e and a_{fi}, gives

$$DP_w = M_w/M_c = f_e(1 + x)/[1 - x(f_e - 1)] \tag{3.105}$$

Putting Eq. (3.103) into Eq. (3.94), gives

$$f_e = \sum M_{Afi}^2 A_{fi}/[M_c(\sum M_{Afi} A_{fi})] = DP_w^0 \tag{3.106}$$

where DP_w^0 is the mass-average degree of polymerization of the initial distribution of primary chains.

Substituting Eq. (3.106) into Eq. (3.105) leads to

$$DP_w = DP_w^0(1 + x)/[1 - x(DP_w^0 - 1)] \tag{3.107}$$

Gelation takes place when $DP_w \to \infty$, a condition that is fulfilled when

$$x_{gel} = 1/(DP_w^0 - 1) \tag{3.108}$$

Low values of the gel conversion are expected for the usual case of long primary chains.

d. Postgel Stage (Monomers with a Single Value of Functionality)

We will now analyze some of the statistical predictions in the postgel stage. They are usually expressed in terms of the following parameters:

Z_A^{out} = probability of having a finite continuation from an A functionality (either reacted or unreacted),

Z_B^{out} = probability of having a finite continuation from a B functionality (either reacted or unreacted).

For an A_f homopolymerization, Z_A^{out} is the root between 0 and 1 of

$$x(Z_A^{out})^{f-1} - Z_A^{out} + 1 - x = 0 \tag{3.109}$$

Solutions for f = 3 and f = 4 are, respectively,

$$Z_A^{out} = (1 - x)/x, \qquad x \geq 1/2 \text{(for f = 3)} \tag{3.110}$$

$$Z_A^{out} = (1/x - 0.75)^{1/2} - 0.5, \qquad x \geq 1/3 \text{ (for f = 4)} \tag{3.111}$$

The sol fraction is calculated by the probability that all possible continuations are finite:

$$w_S = (Z_A^{out})^f \tag{3.112}$$

For f = 3, substituting Eq. (3.110) into Eq. (3.112) gives the same expression calculated in Sec. 3.2.1 (Eq. 3.29). Our definition of Z in that section was equivalent to the probability of finding a finite continuation when looking out from a *reacted* A functionality. This gives $Z = (Z_A^{out})^2$, for the A_3 homopolymerization.

As Z_A^{out} (between 0 and 1) decreases when f increases, Eq. (3.112) shows that the sol fraction decreases more rapidly with conversion for molecules of higher functionality; this is because high-functional molecules have a greater probability of being bound to the gel.

For a generic A_f molecule ($f \geq 3$), crosslinks of degree m ($3 \leq m \leq f$) may be formed along the polymerization in the postgel stage. A crosslink of degree m is generated by m functionalities with an infinite continuation, an event with probability $(1 - Z_A^{out})^m$, and (f – m) functionalities with a finite continuation, an event with probability $(Z_A^{out})^{f-m}$. The concentration of crosslinks of degree m is given by

$$n_{c,m} = \{f!/[(f-m)!m!]\}(Z_A^{out})^{f-m}(1 - Z_A^{out})^m[A_{f0}] \qquad (3.113)$$

where the first factor in the right-hand side of the equation accounts for the number of independent possibilities of selecting m elements from a set of f elements, and $[A_{f0}]$ is the initial number of moles of A_f per unit mass.

For f = 3, the only possible value for m is 3. The resulting expression for $n_{c,3}$ reduces to Eq. (3.30), obtained in Sec. 3.2.1.

Figure 3.16 shows the evolution of trifunctional and tetrafunctional crosslinks for an A_4 homopolymerization. While the fraction of trifunctional crosslinks goes through a maximum, the fraction of tetrafunctional crosslinks increases continuously. At full conversion all A_4 molecules are converted into tetrafunctional crosslinks.

As (m/2) elastic chains issue from a crosslink of degree m, the concentration of elastic chains is given by

$$\nu_e = \sum(m/2)n_{c,m} \quad (\text{for } m = 3 \text{ to } f) \qquad (3.114)$$

The elastic modulus in the rubbery state is directly related to ν_e. It has been proposed that the network stiffness under a mechanical solicitation depends on the functionality of the crosslinks (Duiser and

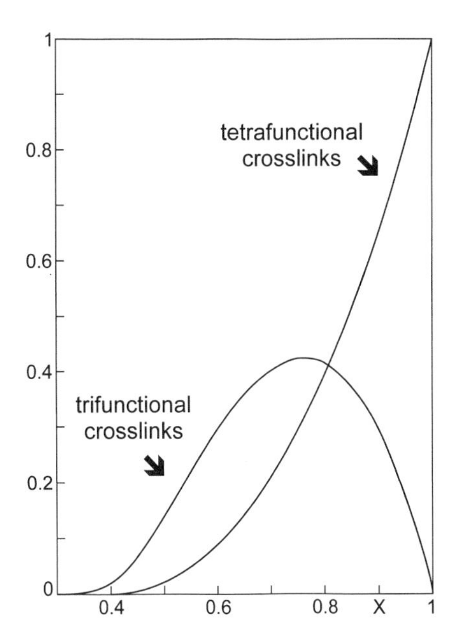

FIGURE 3.16 Evolution of the fraction of trifunctional and tetrafunctional crosslinks as a function of conversion, for an A_4 homopolymerization.

Staverman,1965; Graessley,1975). The higher the value of m, the more constrained the chain movements and the higher the elastic modulus. A weight factor equal to (m - 2)/m has been suggested. It varies from 1/3 for $m = 3$, to 1 for $m \to \infty$.

e. Postgel Stage (Monomers with a Distribution of Functionalities)

For an $A_{fi} + B_{gj}$ step polymerization, the following expressions for Z_A^{out} and Z_B^{out} have been derived:

$$Z_A^{out} = x_A \sum b_{gj}[x_B \sum a_{fi}(Z_A^{out})^{f_i-1} + 1 - x_B]^{g_j-1} + 1 - x_A \quad (3.115)$$

$$Z_B^{out} = x_B \sum a_{fi}[x_A \sum b_{gj}(Z_B^{out})^{g_j-1} + 1 - x_A]^{f_i-1} + 1 - x_B \quad (3.116)$$

The numerical solution of these two equations gives the two roots, between 0 and 1, which are necessary to obtain relevant parameters of the postgel stage. For example, the sol fraction is calculated by

$$w_S = \sum w_{Afi}(Z_A^{out})^{fi} + \sum w_{Bgj}(Z_B^{out})^{gj} \quad (3.117)$$

where the w's are the mass fractions of the different monomers.

3.2.4 Cyclotrimerization Reactions

An example of cyclotrimerization reactions is the homopolymerization of aromatic cyanate esters, forming triazine rings (Fig. 3.17). As these particular stepwise polymerizations cannot be described by the equations developed so far, we will consider them in this section with the usual assumptions for an ideal polymerization.

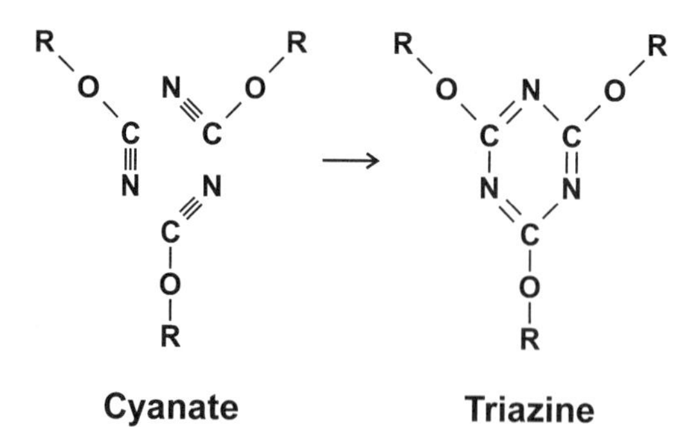

FIGURE 3.17 Cyclotrimerization of cyanates leading to triazine rings.

Consider the fragments shown in Fig. 3.18 for a bifunctional monomer, where α is the unreacted monomer and β and γ are mono- and bireacted fragments. The fraction of initial monomer A_{20} that is present in one of these states at any conversion x is given by

$$\alpha = A_{20}(1 - x)^2 \tag{3.118}$$

$$\beta = A_{20}2x(1 - x) \tag{3.119}$$

$$\gamma = A_{20}x^2 \tag{3.120}$$

Cycles are formed by three reacted functionalities, pertaining to fragments β or γ.

In the pregel stage we want to know the average mass (W) that is bound to a reacted functionality. This reacted functionality will be part of a cycle that necessarily includes two other reacted functionalities (2 β's, one β and one γ or two γ's). Then,

W $=\sum$(probability of completing the cycle with one possible combination) (mass attached to the particular cycle) (3.121)

The probability of completing the cycle with two reacted functionalities of β fragments is given by

$$P(2\beta) = [\beta/(\beta + 2\gamma)]^2 \tag{3.122}$$

The probability of completing the cycle with one β and one γ fragment is obtained from

$$P(\beta, \gamma) = 2\beta(^2\gamma)/(\beta + 2\gamma)^2 \tag{3.123}$$

Similarly,

$$P(2\gamma) = 4\gamma^2/(\beta + 2\gamma)^2 \tag{3.124}$$

Substituting Eqs (3.118)–(3.120) into Eqs (3.122)–(3.124), and solving, results in

$$P(2\beta) = (1 - x)^2 \tag{3.125}$$

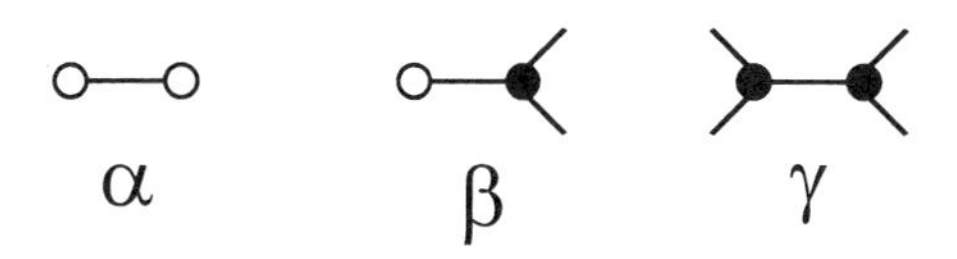

FIGURE 3.18 Different states of reaction of a monomer in a cyclotrimerization reaction.

$$P(\beta, \gamma) = 2x(1 - x) \quad (3.126)$$

$$P(2\gamma) = x^2 \quad (3.127)$$

Equations (3.125)–(3.127) could have also been obtained by realizing that when a cycle is completed with two β fragments, two unreacted functionalities remain attached to the cycle (probability $= (1 - x)^2$); when the cycle is completed with one β and one γ fragment, one reacted and one unreacted functionality are joined to the cycle (probability $= 2x(1 - x)$); and when it is completed with two γ fragments, there are two reacted functionalities joined to the cycle (probability $= x^2$).

Substituting Eqs (3.125)–(3.127) in Eq. (3.121), we obtain

$$W = (1 - x)^2 2M + 2x(1 - x)(2M + W) + x^2(2M + 2W) \quad (3.128)$$

which leads to

$$W = 2M/(1 - 2x) \quad (3.129)$$

where M is the molar mass of the monomer.

The mass-average molar mass, expressed by Eq. (3.14), is given by

$$M_w = (1 - x)^2 M + 2x(1 - x)(M + W) + x^2(M + 2W) \quad (3.130)$$

Substituting Eq. (3.129), and rearranging, gives

$$M_w = M(1 + 2x)/(1 - 2x) \quad (3.131)$$

The system gels when $x = 0.5$, the same conversion as for the A_3 homopolymerization. However, the evolution of the mass-average degree of polymerization, DP_w, is different for both systems, as depicted in Fig. 3.19 (Eqs (3.21) and (3.131)).

For every three reacted functionalities (forming one cycle), 3 moles are converted into 1 mole, implying that 2 moles are lost. Then, for 1 reacted functionality, 2/3 moles are lost. Therefore, M_n is given by

$$M_n = (A_2 M)/[A_2 - (2/3)2A_2 x] \quad (3.132)$$

which reduces to

$$M_n = M/[1 - (4/3)x] \quad (3.133)$$

In the postgel stage we have to calculate the probability of having a finite continuation when completing the cycle starting from a reacted functionality. We define

$Z = \sum$ (probability of completing the cycle with one possible combination) × (probability of having a finite continuation from the particular cycle) (3.134)

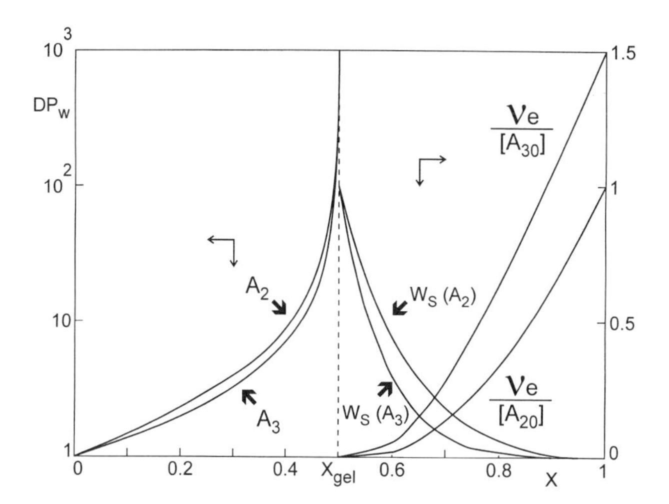

FIGURE 3.19 Comparison of the evolution of different statistical parameters during a homopolymerization of an A_3 and a cyclotrimerization of an A_2.

This leads to

$$Z = P(2\beta) + P(\beta, \gamma)Z + P(2\gamma)Z^2 \tag{3.135}$$

Using Eqs (3.125)–(3.127), and solving for Z, we obtain

$$Z = 1 \tag{3.136}$$

$$Z = (1 - x)^2/x^2 \tag{3.137}$$

Equation (3.136) is the root with physical meaning in the pregel stage, because all continuations are finite. Equation (3.137) is the root valid in the postgel stage.

The sol fraction is given by

$$w_s = (\alpha/A_{20}) + (\beta/A_{20})Z + (\gamma/A_{20})Z^2 \tag{3.138}$$

Substituting Eqs (3.118)–(3.120) and rearranging, results in

$$w_s = (1 - x)^2/x^2 \tag{3.139}$$

In Fig. 3.19, the evolution of the sol fraction for the A_2 cyclotrimerization (Eq. (3.139)) is compared with the case of the A_3 homopolymerization.

Crosslinks can be defined as cycles composed of three γ fragments that exhibit an infinite continuation in the three directions. Their concentration per unit mass is given by

$$n_c = P(3\gamma)(1 - Z)^3[C] \tag{3.140}$$

where [C] represents the concentration of cycles per unit mass at any conversion x.

The probability of simultaneously selecting three reacted functionalities that belong to γ fragments is given by

$$P(3\gamma) = (2\gamma)^3/(\beta + 2\gamma)^3 \tag{3.141}$$

Substituting Eqs (3.119) and (3.120) in Eq. (3.141), and putting this in Eq. (3.140), together with Eq. (3.137) for Z, gives

$$n_c = [(2x - 1)/x]^3[C] \tag{3.142}$$

As three reacted functionalities form one cycle, [C] is one-third of the moles of reacted functionalities:

$$[C] = (1/3)\ 2[A_{20}]x \tag{3.143}$$

Therefore, the concentration of crosslinks per unit mass is given by

$$n_c = (2/3)[A_{20}](2x - 1)^3/x^2 \tag{3.144}$$

This leads to the following expression for the concentration of EANC:

$$\nu_e = (3/2)\ n_c = [(2x - 1)^3/x^2][A_{20}] \tag{3.145}$$

The evolution of the concentration of EANC per initial mole of monomer is shown in Fig. 3.19 for both the A_3 homopolymerization (Eq. (3.31)) and the A_2 cyclotrimerization (Eq. (3.145)).

3.3 CHAINWISE POLYMERIZATIONS

As stated in Sec. 3.1, only ideal systems will be considered in this section. This definition implies that there is no intramolecular reaction, a condition which is satisfied in practice for very low concentrations of A_f monomers ($f > 2$), in the $A_2 + A_f$ chainwise polymerization. To take into account intramolecular reactions it would be necessary to introduce more advanced methods to describe network formation, such as dynamic Monte Carlo simulations.

We will first analyze the case of the pure A_2 homopolymerization that leads to a linear polymer; e.g., a vinyl polymerization. This will enable the reader to get acquainted with the usual parameters that are necessary to describe a chainwise polymerization. Then, we will consider the general $A_f + A_2$ copolymerization, leading to the formation of a polymer network.

3.3.1 A_2 Homopolymerization

We consider styrene homopolymerization by a free-radical mechanism. Styrene, like any other vinyl monomer, is bifunctional, because the double bond opens two arms in the polymerization processs (one of the carbons is attacked by a free radical, activating the other carbon atom, which may continue to propagate the chain).

Once initiated, a chain propagates at a fast rate (r_P = propagation rate), until termination takes place by one of the following events:

1. Chain transfer with a rate r_T,
2. Chain disproportionation with a rate r_D .
3. Chain combination with a rate r_C .

In a particular system, one, two, or the three termination mechanisms may be simultaneously present. But we assume that a termination step is always present (the treatment of living polymerizations is beyond our scope). The first two termination events have the same consequence: the "dead" chain that is formed keeps the same length that it had at the time the termination event took place. In this sense, these termination mechanisms are statistically equivalent. On the contrary, the combination mechanism leads to a "dead" chain that has a length equal to the sum of the lengths of the two chains that were combined.

The following ratios of kinetic events determine the average size of polymer chains:

$$q = r_P/(r_P + r_T + r_D + r_C) \tag{3.146}$$

$$\xi = r_C/(r_T + r_D + r_C) \tag{3.147}$$

The parameter q gives the probability that the active end of the propagating chain adds another monomer to the growing chain. For free-radical vinyl polymerizations, q attains values close to 1 (usual values are larger than 0.99). However, for some nonliving anionic or cationic polymerizations, such as the cationic polymerization of epoxy groups, values of q may be lower.

The parameter ξ gives the probability that the termination of active species takes place by combination (for $\xi = 0$, termination takes place by chain transfer or disproportionation; for $\xi = 1$, termination takes place by combination exclusively).

At a particular conversion of double bonds (or of any generic bifunctional group), the fragments depicted in Fig. 3.20 may be distinguished; α is the unreacted monomer and β is a monomer that participated in a propagation step (($-$) bonds must be linked to ($+$) bonds). Looking in the direction

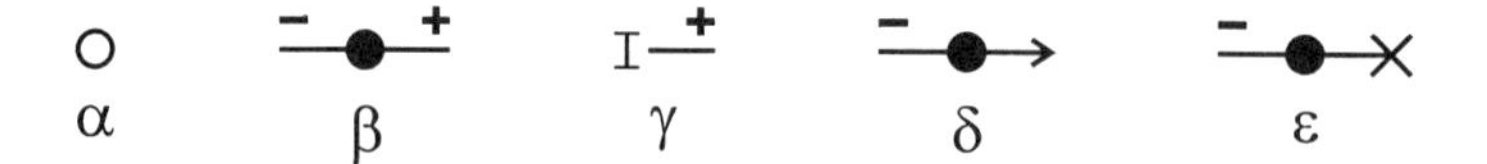

Figure 3.20 Polymer fragments during the chainwise homopolymerization of an A_2 monomer.

of the (−) sign, we go up along the chain through various β units and, at the end, we find the activated initiator denoted by γ. Looking in the direction of the (+) sign, we go down along the chain through various β units and arrive either at a fragment δ or ε. The former represents a reacted monomer that participated in a combination step (an arrow must be linked to another arrow when regenerating the polymer chain); the latter is a reacted monomer present at a chain end that took part in a termination by chain transfer or disproportionation.

Taking into account Eqs (3.146) and (3.147), the concentrations of different fragments are given by

$$\alpha = A_2(1 - x) \tag{3.148}$$

$$\beta = A_2 x q \tag{3.149}$$

$$\delta = A_2 x(1 - q)\xi \tag{3.150}$$

$$\varepsilon = A_2 x(1 - q)(1 - \xi) \tag{3.151}$$

$$\gamma = \delta + \varepsilon = A_2\, x\, (1 - q) \tag{3.152}$$

Equation (3.148) is just the definition of conversion in the system. The concentration of β given by Eq. (3.149), arises from the simultaneous probability that a monomer had reacted (probability x) and propagated (probability q). Similarly, δ and ε give the concentration of monomer that has reacted (probability x) and terminated by combination (probability ξ), or by chain transfer or disproportionation (probability (1 − ξ)), respectively. As the total concentration of (−) bonds must be equal to the concentration of (+) bonds, γ must be equal to δ + ε.

We will call Y(+), Y(−) and W, the average masses attached to a (+) bond (joined to the (−) bonds of different fragments), a (−) bond (joined to the (+) bonds of different fragments), and an arrow (joined to other arrows), respectively. They are given by

$$Y(+) = [\beta/(\beta + \delta + \varepsilon)][M + Y(+)] + [\delta/(\beta + \delta + \varepsilon)][M + W] + [\varepsilon/(\beta + \delta + \varepsilon)]M \tag{3.153}$$

$$Y(-) = [\beta/(\beta + \gamma)][M + Y(-)] + [\gamma/(\beta + \gamma)]M_I \tag{3.154}$$

$$W = M + Y(-) \tag{3.155}$$

where M and M_I are, respectively, the mass of monomer and the mass of initiator. For long chains, the mass of the initiator fragment may be neglected; thus arbitrarily we make $M_I = 0$.

Solving Eqs (3.153)–(3.155) with the aid of Eqs (3.149)–(3.152), we obtain

$$Y(+) = M(1 + \xi)/(1 - q) \tag{3.156}$$

$$Y(-) = Mq/(1 - q) \tag{3.157}$$

$$W = M/(1 - q) \tag{3.158}$$

Average masses obtained when looking up (toward the initiator) or down (toward the monomer that participated in the termination reaction), are different. The misunderstanding of this fact has been the origin of some incorrect statistical calculations reported in the literature.

The mass-average molar mass of polymer chains is calculated by

$$\begin{aligned} M_w = {} & [\beta/(\beta + \delta + \varepsilon)][M + Y(+) + Y(-)] + [\delta/(\beta + \delta + \varepsilon)] \\ & [M + Y(-) + W] + [\varepsilon/(\beta + \delta + \varepsilon)][M + Y(-)] \end{aligned} \tag{3.159}$$

Substituting Eqs (3.149)–(3.152) and Eqs (3.156)–(3.158), results in

$$M_w = M(1 + q + \xi)/(1 - q) \tag{3.160}$$

The number-average molar mass of polymer chains is defined as

$$M_n = (\text{mass of reacted monomers})/[(1/2) \text{ of chain ends}] \tag{3.161}$$

$$M_n = M\ A_2\ x/[(\gamma + \varepsilon)/2] \tag{3.162}$$

Substituting Eqs (3.151) and (3.152) leads to

$$M_n = 2\ M/[(1 - q)(2 - \xi)] \tag{3.163}$$

The polydispersity of polymer chains is given by

$$M_w/M_n = (1 + q + \xi)(2 - \xi)/2 \tag{3.164}$$

For the case of long chains ($q \to 1$), the polydispersity is equal to 2 for termination by chain transfer or disproportionation ($\xi = 0$), and 1.5 for termination by combination ($\xi = 1$).

It is interesting to compare these equations with those corresponding to the stepwise polymerization of a bifunctional monomer. For the particular case of $\xi = 0$, equations giving M_n and M_w for the chainwise polymerization may be generated from those derived by the stepwise polymerization of bifunctional monomers (Eqs (3.76) and (3.77)), by replacing x by q. This is just a mere formalism, because x is an independent variable and q is a parameter that is both a function of conversion and temperature.

The polydispersity for the stepwise polymerization of bifunctional monomers may be obtained by dividing Eq. (3.77) by Eq. (3.76), both written for $f = 2$:

$$M_w/M_n = 1 + x \tag{3.165}$$

For the limit of $x \to 1$, the polydispersity is (almost) the same as the one obtained for chainwise polymerizations, provided that the termination mechanism is chain transfer or disproportionation and primary chains are long enough so that $q \to 1$. However, in the former case polydispersity increases continuously with conversion, while in the latter it gets a value close to 2 from the very beginning of reaction.

Temperature may have a very significant effect in the case of chainwise polymerizations, because it may affect the values of both q and ξ, thus modifying the molar-mass distribution.

3.3.2 A_2 $+A_f$ Chainwise Polymerization

An example of this case is a vinyl (A_2) – divinyl (A_4) polymerization. The assumption of an ideal polymerization means that we consider equal initial reactivities, absence of substitution effects, no intramolecular cycles in finite species, and no phase separation in polymer- and monomer-rich phases. These restrictions are so strong that it is almost impossible to give an actual example of a system exhibiting an ideal behavior. An $A_2 + A_4$ copolymerization with a very low concentration of A_4 may exhibit a behavior that is close to the ideal one. But, in any case, the example developed in this section will show some of the characteristic features of network formation by a chainwise polymerization.

a. Concentration of Fragments

Figure 3.21 shows the new fragments, apart from those depicted in Fig. 3.20, that must be considered in this case. Fragment γ_1 is the skeleton of an A_f molecule, where bonds *(+) must be linked to bonds *(−). The states of the (f/2) functional groups (e.g., double bonds) of the A_f monomer may be the following ones: unreacted (α_1), propagated (β_1), terminated by combination (δ_1), or terminated by chain transfer or disproportionation (ε_1).

Taking into account Eqs (3.146) and (3.147), the concentration of different fragments is given by

$$\gamma_1 = A_f \tag{3.166}$$

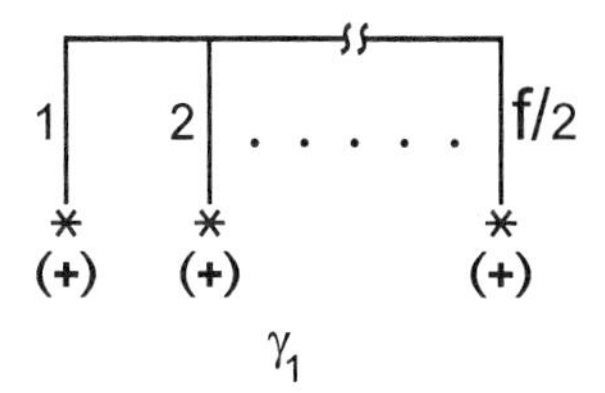

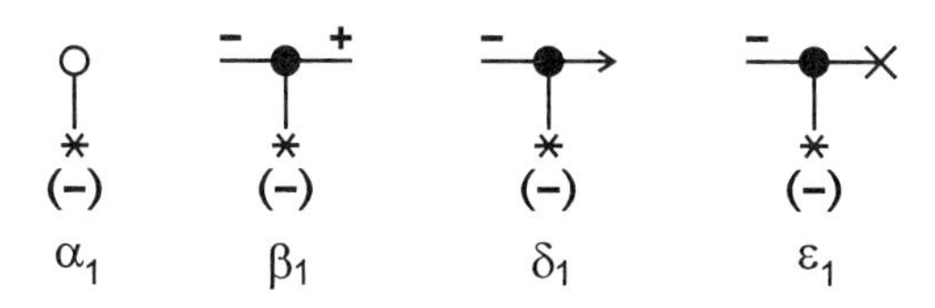

Figure 3.21 Polymer fragments of an A_f monomer participating in a chain-wise copolymerization with an A_2 monomer.

$$\alpha_1 = (f/2)A_f(1 - x) \tag{3.167}$$

$$\beta_1 = (f/2)A_f\ xq \tag{3.168}$$

$$\delta_1 = (f/2)A_f\ x(1 - q)\xi \tag{3.169}$$

$$\varepsilon_1 = (f/2)A_f x(1 - q)(1 - \xi) \tag{3.170}$$

Now, the concentration of activated initiator is obtained from

$$\gamma = \delta + \delta_1 + \varepsilon + \varepsilon_1 = x(1 - q)[A_2 + (f/2)A_f] \tag{3.171}$$

It is verified that the concentration of *(+) bonds, $(f/2)A_f$, is equal to the concentration of *(−) bonds.

b. Pregel Stage

Average masses attached to (+) bonds, Y(+), (−) bonds, Y(−), *(+) bonds, Y*(+), *(−) bonds, Y*(−), and arrows, W, may be obtained in a similar way as for the A_2 homopolymerization. The following equations result:

$$Y(+) = (1 + \xi)M^*/F(x, q, \xi) \tag{3.172}$$

$$Y(-) = qM^*/F(x, q, \xi) \tag{3.173}$$

$$Y^*(+) = (2/f)M_f + x(2q + \xi)M^*/F(x, q, \xi) \tag{3.174}$$

$$Y^*(-) = (1 - 2/f)M_f + x(2q + \xi)(f/2 - 1)M^*/F(x, q, \xi) \tag{3.175}$$

$$W = M^*/F(x, q, \xi) \tag{3.176}$$

where M_f is the mass of the A_f monomer and

$$M^* = (1 - a_f)M_2 + a_f M_f \tag{3.177}$$

$$a_f = (fA_f)/(2A_2 + fA_f) \tag{3.178}$$

$$F(x, q, \xi) = 1 - q - x(2q + \xi)a_f(f/2 - 1) \tag{3.179}$$

Gelation is attained when all the average masses become infinite. This condition is verified when $F(x_{gel}, q, \xi) = 0$, or

$$x_{gel} = (1 - q)/[a_f(f - 2)(q + \xi/2)] \tag{3.180}$$

As in most chainwise polymerizations, $q \rightarrow 1$, the first observation arising from Eq. (3.180) is the very low value of the gel conversion. In actual systems, intramolecular cyclization and microgel formation produce an increase in the gel conversion. But reported values of x_{gel} for the free-radical polymerization of systems containing multifunctional monomers are usually below 0.10. This is the case for the crosslinking of unsaturated polyesters (A_f) with styrene (A_2).

The second observation is that, for $q \rightarrow 1$,

$$x_{gel}(\xi = 1)/x_{gel}(\xi = 0) = 2/3 \tag{3.181}$$

Then, when termination takes place by combination ($\xi = 1$), the gel conversion decreases by a factor 2/3 with respect to the case where termination occurs by chain transfer or disproportionation.

Equation (3.180) constitutes a particular case of a general equation derived by Stockmayer (1944) to predict gelation in chainwise polymerization. This equation may be written as

$$x_{gel} = 2/[a_f(f - 2)(DP_w - 1)] \tag{3.182}$$

where DP_w is the mass-average degree of polymerization of primary chains (produced by single activation and termination steps). From Eq. (3.160), DP_w of primary chains is given by

$$DP_w = (1 + q + \xi)/(1 - q) \tag{3.183}$$

Substituting Eq. (3.183) into Eq. (3-182) leads to Eq. (3.180).

The concentration of unreacted monomer A_f, at any conversion x, is given by $A_f(1 - x)^{f/2}$. This gives the simultaneous probability that none of the f/2 double bonds has reacted. Then, mass fractions of unreacted monomers and polymer are given by

$$w_{Af} = A_f\, M_f(1 - x)^{f/2}/(A_2M_2 + A_fM_f) \tag{3.184}$$

$$w_{A2} = A_2M_2(1 - x)/(A_2M_2 + A_fM_f) \quad (3.185)$$

$$w_{pol} = (A_2M_2x + A_fM_f[1 - (1 - x)^{f/2}])/(A_2M_2 + A_fM_f) \quad (3.186)$$

The mass-average molar mass of the whole system, i.e., including free monomers, is defined as

$$\begin{aligned} M_w = \textstyle\sum&(\text{mass fraction of a particular fragment})\times \\ &(\text{mass attached to the fragment}) \end{aligned} \quad (3.187)$$

Mass fractions are obtained from the concentrations of the different fragments (Eqs (3.148)–(3.151) and Eqs (3.166)–(3.171)). In particular, fragments γ and δ_1 are considered devoid of mass. This means that the contribution of the initiator mass is neglected and the mass of A_f is distributed among the (f/2) functional groups, with individual masses equal to $(2/f)M_f$, attached to a skeleton devoid of mass. For example, using Eq. (3.169), we can calculate the mass fraction of fragment δ_1 as

$$w_{\delta 1} = (f/2)A_f\ x(1 - q)\xi\ (2/f)M_f/(A_2M_2 + A_fM_f) \quad (3.188)$$

The mass attached to this fragment is given by

$$(\text{Mass attached to }\ \delta 1) = (2/f)M_f + Y(-) + Y^*(+) + W \quad (3.189)$$

Substituting in Eq. (3.187) for every fragment, using Eqs (3.172)–(3.176), and rearranging, leads to

$$\begin{aligned} M_w = (A_2M_2^2 + A_fM_f^2 &+ (2q + \xi)x\ W[A_2M_2 \\ &+ (f/2)A_fM_f])/(A_2M_2 + A_fM_f) \end{aligned} \quad (3.190)$$

The mass-average molar mass of the polymer fraction may be obtained from

$$M_w = w_{pol}M_{w,pol} + w_{A2}M_2 + w_{Af}M_f \quad (3.191)$$

Using Eqs (3.184)–(3.186) and (3.190), and rearranging, we obtain

$$\begin{aligned} M_{w,pol} = (A_2M_2^2x + A_fM_f^2[1 - (1 - x)^{f/2}] &+ (2q + \xi)xW[A_2M_2 \\ + (f/2)A_fM_f])/(A_2M_2x &+ A_fM_f[1 - (1 - x)^{f/2}]) \end{aligned} \quad (3.192)$$

The number-average molar mass of the polymer fraction may be obtained by dividing the polymer mass by the number of moles in the polymer fraction. This leads to (Williams and Vallo, 1988):

$$\begin{aligned} M_{n,pol} = (A_2M_2x + A_fM_f&[1 - (1 - x)^{f/2}])/((x/2)(1 - q) \\ (2 - \xi)[A_2 + (f/2)A_f] &- A_f[(f/2)x - (1 - (1 - x)^{f/2})]) \end{aligned} \quad (3.193)$$

Apart from the condition defining an ideal polymerization, the equations derived in this section assume that both q and ξ remain constant in the pregel stage; e.g., no drift occurs. This is usually a good assumption in the pregel period due to the extremely low value of the gel conversion. However, if drift occurs, integral equations must be used (Dotson *et al.*,1988).

c. Postgel Stage

To derive statistical parameters in the postgel stage, we have to determine the probability of having a finite continuation when leaving a fragment from (+) bonds, (−) bonds, *(+)bonds, *(−)bonds and arrows. We will call these probabilities Z(+), Z(−), Z*(+), Z*(−) and Z(A), respectively. For example, Z(−) is defined as

$$Z(-) = \sum(\text{probability of joining a particular fragment having } (+)\text{bonds})\times(\text{probability that all extra branches leaving the fragment are finite}) \tag{3.194}$$

The total concentration of (+) bonds is equal to x $[A_2 + (f/2)A_f]$. Then, Z(−) is given by:

$$Z(-) = (A_2 x q Z(-) + x(1-q)[A_2 + (f/2)A_f] + (f/2)A_f x q Z(-) Z^*(-))/(x[A_2 + (f/2)A_f]) \tag{3.195}$$

Similar expressions may be derived for the other probabilities. Solving the system of five equations in five unknowns leads to

$$Z(-) = 1 - q\text{Root} \tag{3.196}$$

$$Z(+) = (1 - \text{Root})(1 - \xi\text{Root}) \tag{3.197}$$

$$Z^*(-) = 1 - [(1-q)\text{Root}]/[a_f(1 - q\text{Root})] \tag{3.198}$$

$$Z^*(+) = (1 - [(1-q)\text{Root}]/[a_f(1 - q\text{Root})])^{2/(f-2)} \tag{3.199}$$

$$Z(A) = 1 - \text{Root} \tag{3.200}$$

where Root is the root between 0 and 1 of the following equation:

$$(1-[(1-q)\mathrm{Root}]/[a_f(1-q\mathrm{Root})])^{2/(f-2)} = 1 - x + x(1-q\mathrm{Root})^2 (1-\xi\mathrm{Root}) \quad (3.201)$$

Root = 0 is always a root of Eq. (3.201). It is the only root with physical meaning for $x \le x_{gel}$. For $x > x_{gel}$, the only root with physical meaning is the one lying between 0 and 1.

We may now calculate different statistical parameters in the postgel stage. For example, crosslinks of degree f are A_f molecules with infinite continuation from its f branches. This condition is satisfied only if the f/2 branches of γ_1 are joined to β_1 and δ_1 fragments and, additionally, both β_1 and δ_1 exhibit an infinite continuation in both additional directions.

The simultaneous probabiity of selecting (f/2) β_1 or δ_1 fragments with infinite continuation in (+) and (−) directions for β_1, and (−) and the arrow direction for δ_1, is given by

$$[(f/2)A_f x(q[1-Z(+)][1-Z(-)] + (1-q)\xi[1-Z(-)][1-Z(A)])]^{f/2} / [(f/2)A_f]^{f/2} \quad (3.202)$$

where the denominator gives the simultaneous probability of selecting (f/2) possible fragments (α_1, β_1, δ_1 and ε_1), that may be joined to the A_f skeleton. Replacing Z(−), Z(+) and Z(A) by Eqs. (3.196), (3.197) and (3.200), and rearranging, we get

$$n_{c,f} = (xq\mathrm{Root}^2[q(1-\xi\mathrm{Root}) + \xi])^{f/2}[A_{f0}] \quad (3.203)$$

By a similar procedure we can obtain the concentration of crosslinks of degree (f − 1). Now, one of the branches attached to the A_f skeleton must be a pendant chain. The resulting expression is

$$n_{c,f-1} = (f/2)(xq\mathrm{Root}^2[q(1-\xi\mathrm{Root}) + \xi])^{(f/2)-1} x\mathrm{Root}(1-q\mathrm{Root}) [2q(1-\xi\mathrm{Root}) + \xi][A_{f0}] \quad (3.204)$$

d. Crosslinks Generated in an $A_2 + A_4$ Chainwise Polymerization

Let us consider an $A_2 + A_4$ chainwise polymerization; e.g., a vinyl – divinyl system with $a_f = 0.01$ (a very small concentration of the crosslinker in order to keep ideal conditions), and q = 0.999. Figure 3.22 shows the fraction of tetrafunctional crosslinks, $n_{c,4}/[A_{40}]$, as a function of conversion, for two limiting values of ξ. Termination by combination (ξ = 1) increases the crosslink concentration with respect to termination by chain transfer or disproportionation (ξ = 0). At full conversion, termination by combination leads

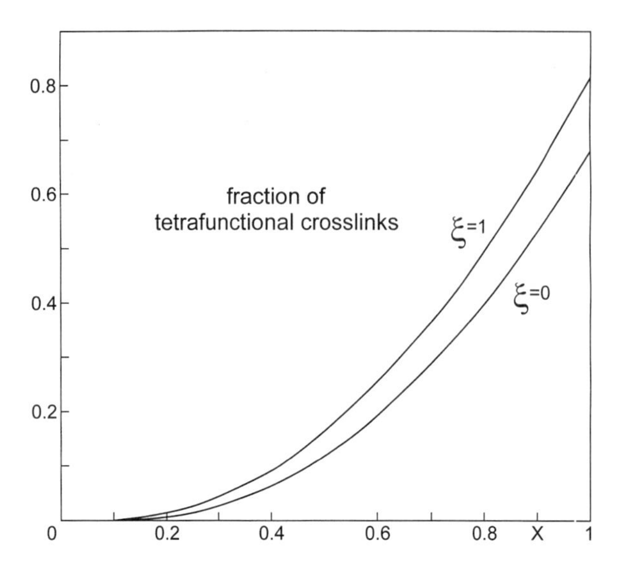

FIGURE 3.22 Fraction of tetrafunctional crosslinks formed in the course of an $A_2 + A_4$ chainwise copolymerization ($a_f = 0.01$, $q = 0.999$), with two different termination mechanisms ($\xi = 0$ represents the case of a termination by chain transfer or disproportionation, while $\xi = 1$ represents the case of a termination by combination).

to a fraction of tetrafunctional crosslinks equal to 0.813 and a fraction of trifunctional crosslinks equal to 0.176; therefore, 98.9% of the initial A_4 monomers end as crosslinks of degree 3 or 4. For the same conditions, termination by chain transfer or disproportionation leads to a fraction of tetrafunctional crosslinks equal to 0.680 and a fraction of trifunctional crosslinks equal to 0.275; therefore, 95.5% of the initial A_4 monomers end as crosslinks of degree 3 or 4. The ratio of the concentration of elastically active network chains produced by both mechanisms is given by

$$\nu_e(\xi = 1)/\nu_e(\xi = 0) = (1.5n_{c,3} + 2n_{c,4})(\xi = 1)/(1.5n_{c,3} + 2n_{c,4})(\xi = 0)$$
$$= 1.066 \tag{3.205}$$

3.4 HYDROLYTIC CONDENSATION OF ALKOXYSILANES

In this section we take a brief look at inorganic polymer networks produced by the acid-catalyzed hydrolytic condensation of tetraalkoxysilanes; e.g., tetraethoxysilane (TEOS). The hydrolytic condensation may be represented by the following reactions:

$$\equiv Si(OC_2H_5) + H_2O \rightarrow \equiv SiOH + C_2H_5OH \quad (3.206)$$

$$\equiv Si(OC_2H_5) + \equiv SiOH \rightarrow \equiv Si{-}O{-}Si\equiv + C_2H_5OH \quad (3.207)$$

$$2 \equiv SiOH \rightarrow \equiv Si{-}O{-}Si\equiv + H_2O \quad (3.208)$$

TEOS is a tetrafunctional monomer that homopolymerizes, forming Si–O–Si bonds by a stepwise mechanism. According to the statistics of stepwise polymerizations discussed in Sec. 3.1, the theoretical gel conversion is

$$x_{gel}\ (\text{theor.}) = 1/(f - 1) = 1/3 \quad (3.209)$$

However, ^{29}Si nuclear magnetic resonance shows experimental gel conversions,

$$x_{gel}\ (\text{exp}) = 0.83 \pm 0.02 \quad (3.210)$$

over a wide range of initial acid ($10^{-5} - 10^{-1}$ mol l^{-1}), water ($4 - 22$ mol l^{-1}), and TEOS ($1.3 - 2.5$ mol l^{-1}) concentrations (Ng *et al.*, 1995).

The key to understanding the extremely large departure of experimental from theoretical values is to take into account the flexibility of Si–O–Si bonds that makes the formation of cyclic and polyhedral species possible. In such cases, the generation of intramolecular cycles is the mechanism by which the inorganic polymer network is built up. Cagelike intermediates and ladder structures are formed in a first stage and interconnected in later stages, leading to gelation.

To provide a rough estimation of the gel conversion arising from such a mechanism, let us assume a very simple model of network formation in which a first stage involves the hydrolytic condensation of the monomer to produce octahedra as shown in Fig. 3.23. In the formation of octahedra,

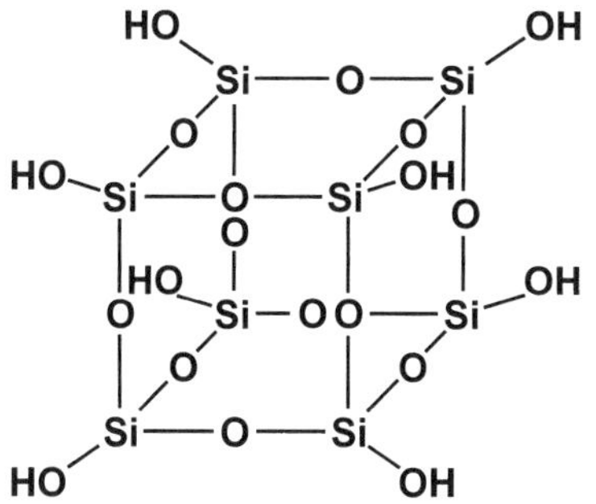

Figure 3.23 Octahedron produced by the hydrolytic condensation of tetraethoxysilane (TEOS).

75% of the total SiOH groups were condensed. In a second stage, it may be assumed that these octahedra are randomly interconnected. For the stepwise homopolymerization of an A_8(8-functional monomer), gelation is attained when 1/7 of the initial free functional groups are reacted. Then, the model predicts an overall gel conversion,

$$x_{gel}\ (\text{theor.}) = 0.75 + 0.25/7 = 0.786 \tag{3.211}$$

which is of the order of experimental values.

In fact, the polymerization generates a distribution of cyclic and cage-like intermediates that continue to react in a nonrandom way, producing larger cycles and finally leading to gelation.

This simple model provides an explanation of the large values of gel conversion found in practice.

NOTATION

A_f = f-functional monomer
$[A_f]$ = concentration of the f-functional monomer per unit mass, kmol kg^{-1}
a_{fi} = mole fraction of all A's on Afi molecules
B_g = g-functional monomer
b_{gj} = mole fraction of of all B's on Bgj molecules
$[C]$ = concentration of cycles per unit mass, kmol kg^{-1}
DP_n = number-average degree of polymerization
DP_w = mass-average degree of polymerization
$DP_w{}^0$ = mass-average degree of polymerization of the initial distribution of primary chains
f = monomer functionality
f_e = functionality-average functionality of A_{fi} molecules
f_i = functionality of monomer A_{fi}
F1, F2, ... = denotes a particular fragment
[F1], [F2], ... = concentration of a particular fragment per unit mass, kmol kg^{-1}
g = monomer functionality
g_e = functionality-average functionality of B_{gj} molecules
g_j = functionality of monomer B_{gj}
i = generic fragment
k, k′, k_i = specific rate constants, s^{-1} or $(kmol\ m^3)^{1-n}\ s^{-1}$
m = number of branches issuing to infinity from an f-functional crosslink ($3 \leq m \leq f$)
M = molar mass of a monomer, kg $kmol^{-1}$
M_c = mass between crosslinkable sites, kg $kmol^{-1}$
M_i = molar mass of a generic fragment, kg $kmol^{-1}$
M_I = molar mass of initiator, kg $kmol^{-1}$
M_n = number-average molar mass, kg $kmol^{-1}$

M_{tot} = total mass of the system, kg
M_w = mass-average molar mass, kg $kmol^{-1}$
n_c = concentration of crosslinks per unit mass, kmol kg^{-1}
$n_{c,m}$ = concentration of crosslinks of degree m per unit mass, kmol kg^{-1}
0 = initial value
P(i,j) = probability of closing a cycle with fragments i and j
q = probability that the active end of the propagating chain adds another monomer to the growing chain
r = stoichiometric ratio of functionalities supplied by two monomers
r_c = critical gelation ratio
r_C = combination rate, kmol m^{-3} s^{-1}
r_D = disproportionation rate, kmol m^{-3} s^{-1}
r_P = propagation rate, kmol m^{-3} s^{-1}
r_T = chain-transfer rate, kmol m^{-3} s^{-1}
R_{21} = reactivity ratio of secondary to primary amine hydrogens
t_{gel} = time to gel, s
x = conversion of functional groups
x_i = conversion of functional groups of class i
x_{gel} = gel conversion
W = average mass attached to a particular bond, kg $kmol^{-1}$
w_e = mass fraction of elastic chains
w_i = mass fraction of fragment i
w_p = mass fraction of pendant chains
w_s = mass fraction of finite species (sol fraction)
Y, Y(+), Y(−), Y*(+), Y*(−) = average mass attached to a particular bond, kg $kmol^{-1}$
Z, Z_A, Z_B, Z(+), Z(−), Z*(+), Z*(−), Z(A) = probability of having a finite continuation from a reacted functionality
Z_i^{out} = probability of having a finite continuation from an i functionality (either reacted or unreacted)

Greek Letters

α = identifies a particular fragment
β = identifies a particular fragment
γ = identifies a particular fragment
δ = identifies a particular fragment
ε = identifies a particular fragment
ν_e = concentration of elastically active network chains per unit mass, kmol kg^{-1}
ξ = probability that the termination of active species takes place by combination

Subscripts

0 = initial value

REFERENCES

Anseth KS, Bowman CN, *Chem. Eng. Sci.*, 49, 2207–2217 (1994).
Aranguren MI, Borrajo J, Williams RJJ, *Ind. Eng. Chem. Prod. Res. Dev.*, 23, 370–374 (1984).
Dotson NA, Galván R, Macosko CW, *Macromolecules*, 21, 2560–2568 (1988).
Duiser JA, Staverman AJ. In *Physics of Non-Crystalline Solids*, Prins JA (ed), North-Holland, Amsterdam, 1965, p. 376.
Dušek K, *Br. Polym. J.*, 17, 185–189 (1985).
Flory PJ, *J. Am. Chem.* Soc., 63, 3083–3090 (1941a).
Flory PJ, *J. Am. Chem. Soc.*, 63, 3097–3099 (1941b).
Flory PJ, *Principles of Polymer Chemistry*, Cornell University Press, Ithaca, N.Y., 1953.
Gordon M, *Proc. R. Soc. London, Ser. A*, 268, 240–259 (1962).
Graessley WW, *Macromolecules*, 8, 186–190 (1975).
Macosko CW, Miller DR, *Macromolecules*, 9, 199–206 (1976).
Miller DR, Macosko CW, *Macromolecules*, 9, 206–211 (1976).
Miller DR, Vallés EM, Macosko CW, *Polym. Eng. Sci.*, 19, 272–283 (1979).
Ng LV, Thompson P, Sanchez J, Macosko CW, *Macromolecules*, 28, 6471–6476 (1995).
Peebles LH, *Molecular Weight Distributions in Polymers*, Wiley, New York, 1971.
Riccardi CC, Williams RJJ, *Polymer*, 27, 913–920 (1986).
Riccardi CC, Williams RJJ, *J. Polym. Sci.*, B: Polym. Phys., 31, 389–393 (1993).
Šomvársky J, Dušek K, *Polym. Bull.*, 33, 369–376 (1994).
Stockmayer WH, *J. Chem. Phys.*, 11, 45–55 (1943).
Stockmayer WH, *J. Chem. Phys.*, 12, 125–131 (1944).
Vazquez A, Adabbo HE, Williams, RJJ, *Ind. Eng. Chem. Prod. Res. Dev.*, 23, 375–379 (1984).
Williams RJJ, Riccardi CC, Dušek K, *Polym. Bull.*, 17, 515–521 (1987).
Williams RJJ, Vallo CI, *Macromolecules*, 21, 2571–2575 (1988).
Williams RJJ, Riccardi CC, Dušek K, *Polym. Bull.*, 25, 231–237 (1991).

FURTHER READING

Dotson NA, Galván R, Laurence RL, Tirrel M, *Polymerization Process Modeling*, VCH, New York, 1996.
Dušek K. In *Developments in Polymerisation – 3*, Haward RN (ed), Applied Science Publishers, Barking, UK, 1982, pp. 143–206.
Stepto RFT (ed), *Polymer Networks: Principles of Their Formation, Structure and Properties*, Blackie Academic & Professional, London, 1998.

4

Glass Transition and Transformation Diagrams

4.1 INTRODUCTION

Two main transitions may take place during the formation of a polymer network: gelation, a critical transition defined by the conversion at which the mass-average molar mass becomes infinite (Chapter 3); glass transition, or vitrification, characterized by the conversion at which the polymer begins to exhibit the typical properties of a glass.

Gelation and vitrification produce significant transformations in the physical properties of the thermosetting polymer. When gelation occurs, the system is transformed from a liquid into a covalently crosslinked gel. When vitrification takes place, the system is transformed from a liquid or gel (if gelation has already taken place) into a glass (ungelled or gelled). The cause of the glass transition is the reduction of the system's mobility through the formation of covalent bonds, up to a point where the cooperative movements of large portions of the polymer, which are characteristic of rubbery and liquid states, are no longer possible.

The polymerization temperature, often called the cure temperature, affects both transitions in different ways. In Chapter 3 it was shown that the gel conversion does not depend on temperature for ideal stepwise polymerizations but may show a small dependence on temperature for the case where unequal reactivity of functional groups or substitution effects vary with temperature. For chainwise polymerizations the effect of temperature on the gel conversion is restricted to its influence on the ratios of reaction rates of different elementary steps. But in most cases it is a good approximation to consider the gel conversion for a particular formulation as a constant value independent of temperature.

The situation is completely different for the glass transition. The possibility of producing cooperative movements of fragments of the thermosetting polymer must increase with temperature. So, the conversion at which vitrification takes place increases with the cure temperature.

In the glassy state relaxation times become very large, making the continuation of the polymerization reaction difficult. Moreover, once in the glassy state, small advances in the conversion of functional groups produce a further increase in the already large relaxation times. Thus, the reaction becomes autoretarded and rapidly ceases for practical purposes. Only an increase in the cure temperature can restart the polymerization reaction.

Gelation and vitrification may be represented together to generate different types of transformation diagrams, the most popular of which is the time–temperature transformation (TTT) diagram proposed by Gillham and coworkers (Enns and Gillham, 1983). Another transformation diagram that is particularly important to follow nonisothermal polymerizations is the conversion–temperature transformation (CTT) diagram (Adabbo and Williams, 1982). Different regions of these diagrams may be ascribed to the different states of the thermosetting polymer; e.g., liquid, rubber, ungelled glass, and gelled glass.

Other transitions such as degradation and phase separation may be also observed during the formation of the polymer network. Degradation is usually present when high temperatures are needed to get the maximum possible conversion. Phase separation may take place when the monomers are blended with a rubber or a thermoplastic, to generate rubber-modified or thermoplastic-modified polymer networks. In these cases, formulations are initially homogeneous but phase-separate during the polymerization reaction. This process is discussed in Chapter 8.

In this chapter the main characteristics of the glass transition are analyzed and equations relating the glass transition temperature with the conversion of the thermosetting polymer are discussed. Then, CTT and TTT transformation diagrams are derived and examples of their practical use provided.

4.2 GLASS TRANSITION

4.2.1 Phenomenological Aspects

From the practical point of view, the glass transition is a key property since it corresponds to the short-term "ceiling" temperature above which there is a catastrophic softening of the material. For amorphous polymers in general, and thus for thermosets, one can consider that the glass transition temperature, T_g, is related to the conventional heat deflection temperature (HDT) (usually, HDT is 10–15°C below T_g, depending on the applied stress and the criterion selected to define T_g).

In the field of processing, the glass transition is rather a "floor" temperature because the polymerization (cure) exhibits a very slow or even negligible rate in the glassy state (see Chapter 5). Many important properties, such as the yield stress or the fracture toughness at a temperature T are sharply linked to $(T_g - T)$. Some qualitative and important quantitative differences between the glassy and rubbery states are listed in Table 4.1.

The glass transition is usually characterized as a second-order thermodynamic transition. It corresponds to a discontinuity on the first derivative of a thermodynamic function such as enthalpy (dH/dT) or volume (dV/dT) (A first-order thermodynamic transition, like melting, involves the discontinuity of a thermodynamic function such as H or V). However, T_g cannot be considered as a true thermodynamic transition, because the glassy state is out of equilibrium. It may be better regarded as a boundary surface in a tridimensional space defined by temperature, time, and stress, separating the glassy and rubbery (or liquid) domains.

TABLE 4.1 Main characteristics of glassy and rubbery states

Property	Glass	Rubber
Thermodynamic equilibrium	No	Yes
Molecular mobility	Local	Cooperative
Influence of crosslink density on properties	Low	High
Temperature effect on properties	Arrhenius law	WLF equation
Apparent activation energy (kJ mol^{-1})	≤ 150	≥ 500
Expansion coefficient (K^{-1})	1.5×10^{-4}–3.5×10^{-4}	4×10^{-4}–8×10^{-4}
Heat capacity (kJ kg^{-1} K^{-1})	1–1.6	1.6–2.2
Young's modulus (GPa)	1–5	10^{-3}–10^{-1}
Poisson's ratio	0.30–0.46	0.49–0.50

Most of the physical properties of the polymer (heat capacity, expansion coefficient, storage modulus, gas permeability, refractive index, etc.) undergo a discontinuous variation at the glass transition. The most frequently used methods to determine T_g are differential scanning calorimetry (DSC), thermomechanical analysis (TMA), and dynamic mechanical thermal analysis (DMTA). But several other techniques may be also employed, such as the measurement of the complex dielectric permittivity as a function of temperature. The shape of variation of corresponding properties is shown in Fig. 4.1.

Several comments arise from the plots of Fig. 4.1. The most important one is that vitrification (when cooling) or devitrification (when heating), takes place over a temperature range so that it is always necessary to indicate the way in which T_g is defined. For example, Fig. 4.1b represents a typical determination of T_g from a DSC thermogram. The shift in the baseline represents the variation of c_p during the glass transition (the direction of the shift may be upward or downward, depending on the arbitrary placement of the capsule containing the sample; but in every case the c_p of the rubber (or liquid) is higher than the c_p of the glass). T_g may be defined as the onset value of the baseline shift when increasing temperature at a constant heating rate (point M), or at the temperature corresponding to half of the baseline shift (point N), or at the end of the shift (point O). The definition is arbitrary but it should always be explicitly mentioned. Onset and medium values are the most often employed.

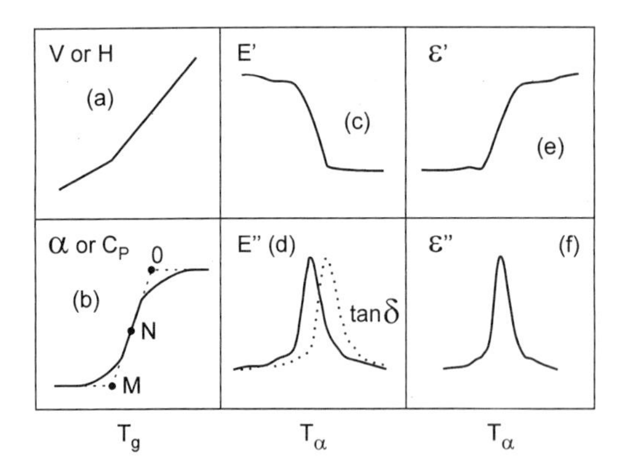

FIGURE 4.1 Variation of physical properties vs temperature, used to determine the glass transition: (a) volume (V) or enthalpy (H); (b) expansion coefficient (α) or heat capacity (c_p); (c) storage modulus (E'); (d) dissipation modulus (E") and dumping factor (tan δ); (e) real part of the complex dielectric permittivity (ε'); (f) imaginary part of the complex dielectric permittivity (ε'').

Another important fact related to the determination of T_g is that the experimental value depends on the rate of temperature variation, as will be discussed in next section. However, from a practical point of view the small kinetic effect on T_g can be neglected, at least in a first approach. But more important than the rate of temperature variation is the way in which the T_g determination is performed: either during a heating test or during a cooling test. As the glass is intrinsically a nonequilibrium state of matter, its physical properties depend on time (see next section). So, in a heating test there may be no unequivocal definition of the starting material and, rigorously speaking, the value of T_g may exhibit a dependence on not only the heating rate but also on the temperature–time history of the material. The starting point of a cooling test is a rubber (or a liquid) that may be considered in thermodynamic equilibrium. When cooling this material, a nonequilibrium glass is obtained. But its state is only determined by the cooling rate, so that if the small kinetic effect is neglected, T_g arising from a cooling test may be regarded as a material property as, for instance, the melting point.

Unfortunately, in the case of thermosetting polymers, physical properties in the rubbery (or liquid) state vary continuously due to the polymerization reaction. Only when full conversion is attained an equilibrium state may be obtained. So, the glass transition of partially reacted thermosetting polymers is usually determined using heating tests. A possible way to make the experimental determination independent of the state of the starting material is to "erase" the time–temperature history by heating to a temperature beyond T_g, rapidly cooling at a constant rate to the starting temperature, and rescanning again. This procedure requires a negligible advance in the conversion of the thermosetting polymer during the thermal cycle.

As indicated in Fig. 4.1, dynamic methods lead to a transition temperature T_α that depends on the frequency employed in the experimental test: it is usually higher than T_g and the shift increases with frequency. For low frequencies, typically ≤ 10 Hz, the difference between both values is less than 20 K. As indicated in Fig. 4.1, there is a systematic shift between the maxima of E'' and tan δ.

Both the changes in the heat capacity, Δc_p, and in the expansion coefficient, $\Delta\alpha$, at the glass transition, are a decreasing function of the crosslink density. So, the experimental determination of T_g by DSC or TMA may be not possible for highly crosslinked thermosets, due to the lack of an adequate sensitivity. In these cases T_g may be conveniently determined using dynamic mechanical methods because it is usually verified that $E'_{glass} / E'_{rubber} \geq 10$. DMTA is particularly useful to determine the T_α of composites containing a low volumetric fraction of thermosetting polymer.

From a more fundamental point of view, the glass transition phenomenon has been explained using different types of physical models, as shown in Table 4.2. The most popular approaches are based on the variation of the free volume fraction (f) and configurational entropy (S) with temperature, as shown in Fig. 4.2. Both parameters are "frozen in" at T_g and become zero at the extrapolated temperature T_∞. Both theories predict that the material must obey a time–temperature equivalence principle expressed by the WLF equation (see Chapter 10), but they differ in some quantitative predictions.

4.2.2 Kinetic Aspects

Let us consider an experiment in which a polymer, initially in the rubbery state, is cooled at a constant rate $q = dT/dt$. In the initial state the material exhibits the equilibrium distribution of possible conformations' states at the starting temperature T_0. Let us call $g_{eq}(T)$ the concentration of the less stable conformation (equivalent to the gauche conformation of some linear polymers), at any temperature T. According to the equilibrium distribution, as temperature decreases $g_{eq}(T)$ must also decrease. But the new equilibrium is not reached instantaneously. The rate at which the system evolves to the new equilibrium situation is determined by a particular kinetic law. Assuming that a first-order kinetic law holds:

$$dg/dt = -k[g - g_{eq}(T)] \tag{4.1}$$

Solving Eq. (4.1) leads to:

$$g - g_{eq}(T) = [g_{eq}(T_0) - g_{eq}(T)]\exp(-kt) \tag{4.2}$$

A characteristic time to reach equilibrium may be defined as

$$\tau_c = 1/k = \tau_{c0}\exp(E_c/RT) \tag{4.3}$$

where an Arrhenius dependence with an activation energy E_c has been assumed for the specific rate constant k.

A characteristic time for the rate of temperature variation may be defined as

$$\tau_q = T_g/q \tag{4.4}$$

Table 4.2 Physical approaches to the glass transition

Physical approach to the glass transition	Scientific domain	Main author
Free volume	Conventional thermodynamics	Fox *et al.* (1955)
Configurational entropy	Statistical thermodynamics	Di Marzio (1964)
Coupling model	Molecular dynamics	Ngai *et al.* (1986)

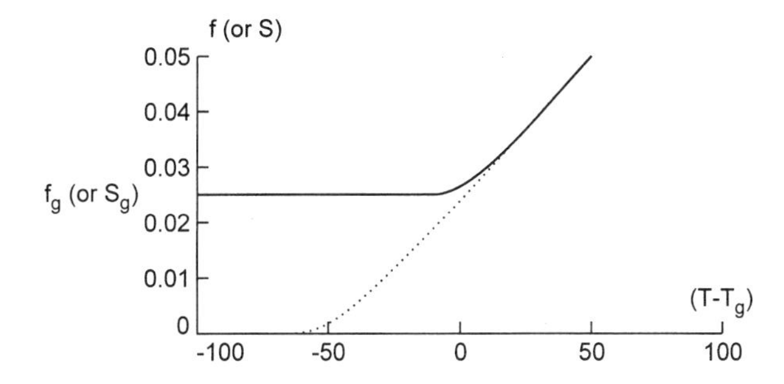

Figure 4.2 Shape of variation of the free volume fraction or the entropy with $(T - T_g)$.

Figure 4.3 shows a plot of both characteristic times as a function of 1/T. When $\tau_c < \tau_q$, the polymer is able to reach, continuously, the equilibrium distribution of conformations. So it remains in the rubbery (or liquid) state. But when $\tau > \tau_q$, the polymer cannot reach equilibrium in the time-scale of the experiment and it behaves as a glass. In the frame of this kinetic model, the glass transition may be defined as the temperature at which $\tau_c = \tau_q$ (Fig. 4.3).

It is also evident from Fig. 4.3 that as the value of τ_q depends on q, the experimental value of T_g will be a function of the cooling rate. The faster the cooling rate the lower the τ_q value and the higher the T_g value. This means that the polymer is not allowed to reach its equilibrium state due to the high rate of temperature variation; so it vitrifies at higher temperatures. But due to the high value of E_c, only very important changes of the cooling rate, typically several decades, can produce a significant shift in the T_g value. This is the reason why T_g may be considered as a material property independent of the cooling (or heating) rate.

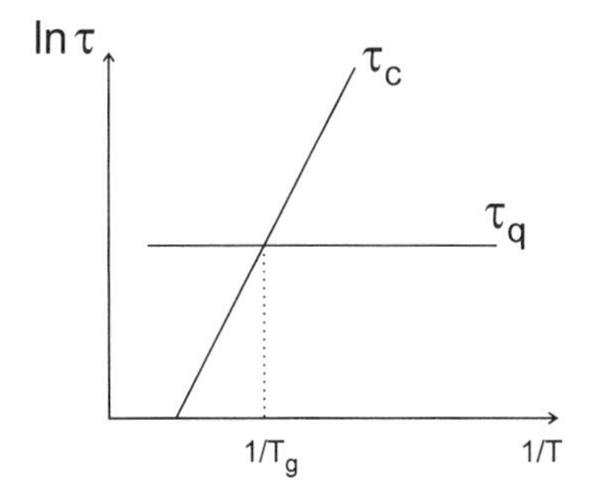

Figure 4.3 Characteristic times for conformational changes (τ_c) and temperature variation (τ_q), as a function of 1 / T.

The kinetic character of the glass transition and the resulting non-equilibrium character of the glassy state are responsible for the phenomena of structural relaxation, glass transition hysteresis, and physical aging (Kovacs, 1963; Struik, 1978).

Figure 4.4 shows a dilatometric or calorimetric experiment to show structural relaxation (physical aging) and glass transition hysteresis. The sample is cooled from T_0 to T_1; it is kept at T_1 for a certain time and heated again to T_0. During the cooling step, the material vitrifies at B, resulting in an abrupt decrease in both the expansion coefficient and the specific heat.

When the material reaches point C at $T = T_1$, there is an excess of both free volume and enthalpy with respect to the equilibrium value (represented with a dashed line in Fig. 4.4). During the storage time at T_1 the polymer evolves toward equilibrium through a densification process associated with the decrease of the volume and enthalpy excesses. This structural relaxation is usually called physical aging. But, as aging occurs, there is a significant decrease in the residual mobility due to the continuous reduction in free volume. This is then an autoretarded phenomenon and the consequence is that equilibrium is never attained except in a temperature region very close to T_g. After a certain storage time at T_1, point D is reached.

When the material is heated starting from point D, devitrification cannot take place at F, placed in the equilibrium line. The previous physical aging decreased the mobility of the sample and it is not able to respond in the timescale of the temperature program. After passing through the equilibrium value the thermal energy continues to increase until it attains a temperature high enough (point G) to promote a catastrophic return to equilibrium (point H).

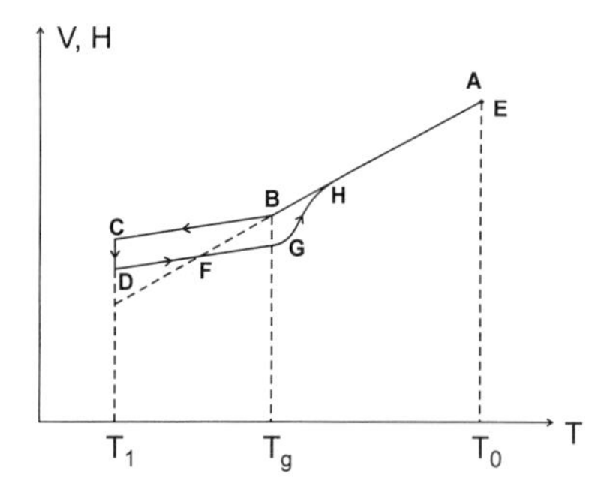

FIGURE 4.4 Dilatometric or calorimetric experiment to show the structural relaxation (physical aging) and the glass transition hysteresis (the sample is cooled from T_0 to T_1; it is kept at T_1 for a certain time and heated again to T_0).

Figure 4.5 shows the variation of the expansion coefficient or the heat capacity in the same experiment described in Fig. 4.4. The area of the peak (which is endothermic in the calorimetric experiment) can be considered as a quantitative measurement of the extent of physical aging. If, having attained point H, the sample is cooled again to C and rapidly reheated, the manifestation of physical aging (the peak shown in Fig. 4.5) is erased and the sample is said to be rejuvenated.

Kinetic models of physical aging are available in the literature (Kovacs *et al.*, 1979). During this process, volume changes are very low, typically less than 1 $m^3\ kg^{-1}$, but they affect a component of free volume that plays a crucial role in creep, relaxation, yielding and fracture (see Chapter 12).

4.2.3 Glass Transition Temperature as a Function of Conversion

a. Experimental Determination of the T_g vs x Relationship

The solution of the precursors of the thermosetting polymer (mixture of monomers or oligomers with or without initiators, catalysts and different additives) is usually a liquid at room temperature; e.g., unsaturated polyester–styrene, some epoxy–anhydride and epoxy–amine formulations, cyanate esters, one-stage phenolics, etc. Cooling any of these solutions below room temperature leads to a glass. The temperature at which the glass–liquid transition of the initial formulation takes place is denoted as T_{g0}. Some other particular formulations, such as two-stage phenolics (novolac–hexa mixtures), some epoxy–amine systems, etc., exhibit a T_{g0} above room temperature.

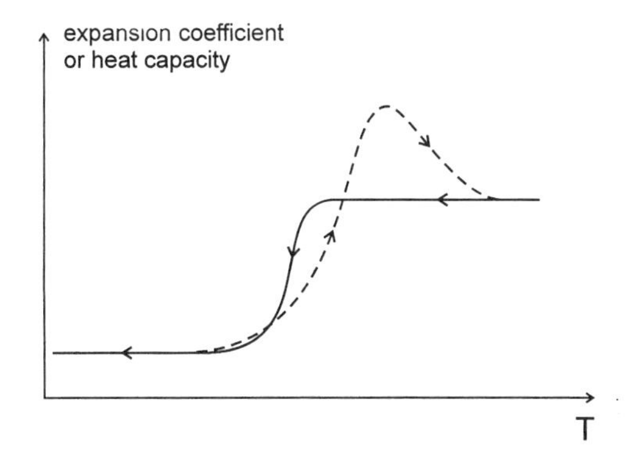

Figure 4.5 Variation of the expansion coefficient or the heat capacity following the thermal cycle of Fig. 4.4.

For certain formulations one of the components may crystallize upon cooling (e.g., certain bismaleimides, cyanates, etc.). In these cases, a thermal quench in liquid nitrogen may avoid crystallization, enabling the determination of T_{g0} in a subsequent scan.

After completing the polymerization, the material attains a glass transition temperature, $T_{g\infty}$, that is significantly higher than T_{g0}. The difference between $T_{g\infty}$ and T_{g0} may be of several hundreds °C, which reflects the dramatic change of the material structure during polymerization.

The increase of the glass transition temperature with the conversion of the thermosetting polymer may be followed by DSC. Figure 4.6 (Montserrat, 1992), shows DSC thermograms in the heating mode, for a system based on diglycidyl ether of bisphenol A (DGEBA)–phthalic anhydride and a commercial tertiary amine as initiator, cured at 80°C (a) and 130°C (b), for different times, t_c (h). The initial formulation exhibits a shift in the baseline at $T_{g0} = -23.2$°C (defined at the midpoint of the shift). T_g increases with cure time at constant temperature. The completely cured

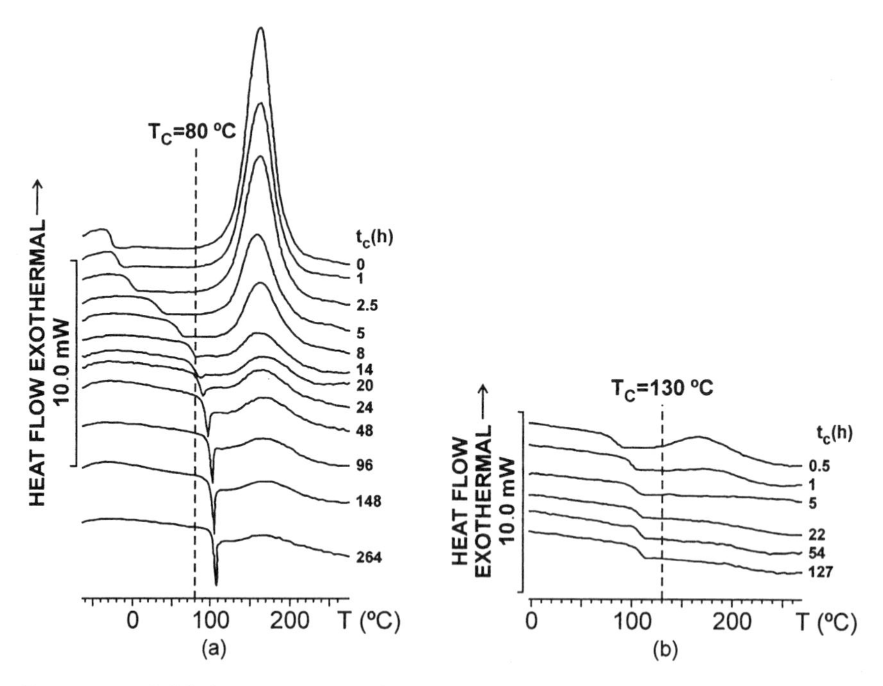

FIGURE 4.6 DSC thermograms of an epoxy–anhydride system cured at 80°C (a) and 130°C (b), for different times, t_c (h). (Montserrat, 1992 – Copyright 2001 – Reprinted by permission of John Wiley & Sons, Inc.)

material, represented by DSC thermograms obtained after cure times equal to or higher than 22 h at 130°C, shows a value of $T_{g\infty} = 109°C$.

The conversion of partially reacted samples is defined by

$$x = 1 - (-\Delta H_R)/(-\Delta H) \quad (4.5)$$

where $(-\Delta H)$ is the reaction heat determined for the initial formulation (exothermic peak for $t_c = 0$ h), and $(-\Delta H_R)$ is the residual reaction heat obtained from the exothermic peak of partially reacted samples.

The cure at $T_c < T_{g\infty}$ produces different DSC thermograms than the cure at $T_c > T_{g\infty}$. For $T_c = 80°C$ ($< T_{g\infty}$), vitrification is attained when T_g equals T_c. This occurs somewhere between 14 and 20 h. From this time on, polymerization takes place at a very slow rate. A residual reaction heat is measured even after 264 h at 80°C.

A characteristic feature of the vitrification process is the endothermic peak superimposed on the baseline shift, observed when the partially cured sample evolves through the glass transition. The origin of this endothermic peak was qualitatively explained in Sec. 4.2.2. When the partially reacted thermoset is kept at constant temperature in the glassy state (meaning that T_g has surpassed the cure temperature), two different processes occur: physical aging and a very slow increase in conversion. The latter becomes practically negligible when T_g exceeds T_c by more than about 20–30°C (although in some particular systems this difference may attain higher values). Cooling such a sample and rescanning, leads to an enthalpy relaxation in the transition from an aged glass to a liquid or rubber. This produces the endothermic peaks observed in Fig. 4.6 for $t_c \geq 20$ h (after vitrification). Physical aging may also shift the observed T_g to higher temperatures, due to the higher thermal energy required to devitrify a densified glass.

In an isothermal cure process, vitrification will be always observed if T_c is selected in the range between T_{g0} and $T_{g\infty}$. So, vitrification was not observed when the epoxy–anhydride formulation was cured at $T_c = 130°C (> T_{g\infty})$.

From Fig. 4.6a and b, couples of T_g and x values may be obtained for both cure temperatures. A similar procedure may be performed at different cure temperatures. In general, a single curve representing T_g as a function of conversion is generated from data obtained at different cure temperatures. Figure 4.7 represents such a plot for the epoxy–anhydride–tertiary amine formulation cured at $T_c = 30, 40, 50,\ldots, 130°C$. Represented values correspond to $T_g < T_c$ to avoid the influence of physical aging. This behavior is observed for most network-forming polymers, such as epoxy–amine and epoxy–anhydride systems, polycyanates, polyimides, polyurethanes, vinyl esters, and unsaturated polyesters, etc. This unique relationship between T_g and conversion is striking because one should expect that different net-

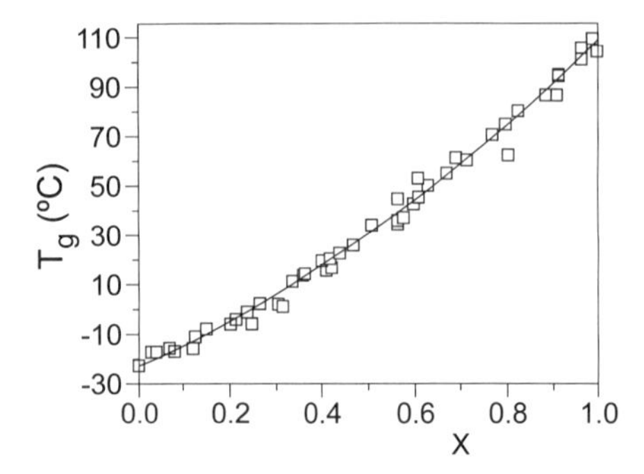

FIGURE 4.7 T_g as a function of conversion for an epoxy–anhydride system cured at $T_c = 30, 40, 50, \ldots, 130°C$. Represented values correspond to the range $T_c > T_g$ to avoid the influence of physical aging. (Montserrat, 1992 – Copyright 2001 – Reprinted by permission of John Wiley & Sons, Inc.)

work structures result from the cure at different temperatures. For example, the rate of termination steps in the epoxy–anhydride chainwise copolymerization should vary with temperature, producing different network structures at the same conversion level. However, this effect is not reflected in large T_g variations, possibly because of the low sensitivity of T_g to the resulting structural modifications.

The existence of a single relationship between T_g and x enables one to use the experimental measurement of T_g to follow the conversion of the thermosetting polymer. This is particularly important at high conversions where T_g is much more sensitive to follow small increases in conversion than the measurement of the residual reaction heat or the direct determination of free functional groups using IR spectroscopy.

b. Prediction of the T_g vs x Relationship

Usually, the T_g vs x relationship shows an upward curvature related to the shape of the curves describing the increase in the concentration of branching points and high-functional crosslinks as a function of conversion (Chapter 3). Theoretical equations relating the T_g increase to structural parameters of the polymer network have been proposed. But, in every case, adjustable parameters are required to fit experimental results.

A different approach to obtaining an expression to describe the T_g vs x relationship has been proposed by Pascault and Williams (1990). Their derivation is based on an expression obtained by Couchman (1987) to express the T_g of a solution with randomly mixed components:

$$\sum \phi_i \int_{T_{gi}}^{T_g} (\Delta c_{pi}/T)\, dT = 0 \tag{4.6}$$

where T_g and T_{gi} are, respectively, the glass transition temperature of the solution and of a generic component i; $\Delta c_{pi} = c_{pi}(l) - c_{pi}(g)$ is the change in the isobaric heat capacity of component i between liquid and glassy states, and ϕ_i is the concentration of component i in the solution, in units matching those of Δc_{pi}.

To integrate Eq. (4.6), it is necessary to express Δc_{pi} as a function of temperature. It is observed that $c_{pi}(l)$ increases less with temperature than $c_{pi}(g)$, meaning that Δc_{pi} decreases with an increase in temperature (Wunderlich, 1960). Assuming that Δc_{pi} varies inversely with temperature, we obtain

$$\Delta c_{pi}(T) = \Delta c_{pi}(T_{gi})T_{gi}/T \tag{4.7}$$

Integrating Eq. (4.6) with $\Delta c_{pi}(T)$, given by Eq. (4.7), gives

$$T_g = \sum \phi_i \Delta c_{pi}(T_{gi})T_{gi} / \sum \phi_i \Delta c_{pi}(T_{gi}) \tag{4.8}$$

A partially reacted polymer network is modeled as a random solution of reacted functionalities with a concentration equal to conversion ($\phi_i = x, T_{gi} = T_{g\infty}, \Delta c_{pi}(T_{g\infty}) = \Delta c_{p\infty}$), and unreacted functionalities with a concentration equal to $(1-x)$ ($\phi_i = 1-x, T_{gi} = T_{g0}$, $\Delta c_{pi}(T_{g0}) = \Delta c_{p0}$). Substituting in Eq. (4.8) with

$$\Delta c_{p\infty}/\Delta c_{p0} = \lambda \tag{4.9}$$

we obtain

$$(T_g - T_{g0})/(T_{g\infty} - T_{g0}) = \lambda x/[1-(1-\lambda)x] \tag{4.10}$$

An equation similar to Eq. (4.10) was used by Adabbo and Williams (1982) to correlate experimental results from the literature. The origin of this equation was incorrectly assigned by these authors to a previous derivation performed by Di Benedetto. So, Eq. (4.10) has been popularized as the Di Benedetto equation.

Equation (4.10), with λ given by Eq. (4.9), was used with success to correlate experimental T_g vs x results for several thermosetting polymers (Pascault and Williams, 1990; Jordan *et al.*, 1992; Georjon *et al.*, 1993). As an example, Fig. 4.8 shows the experimental T_g vs x curve obtained for the cyclotrimerization of 4,4′-dicyanate 2,2′-diphenylpropane. Experimental points were obtained at the following cure temperatures: $T_c = 150$, 165, 200, 225, and 250°C. The formulation showed an initial glass transition temperature, $T_{g0} = -48$°C (determined by thermal quench-

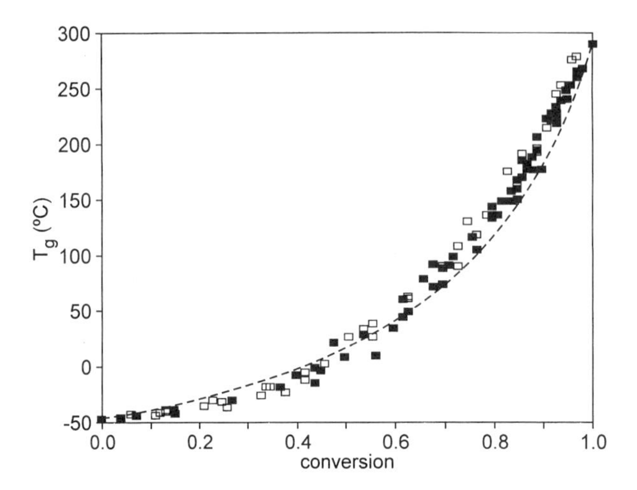

FIGURE 4.8 T_g as a function of conversion for the cyclotrimerization of 4,4′-dicyanate 2,2′-diphenylpropane. The curve is the prediction resulting from Eq. (4.10) with λ given by Eq. (4.9). (Georjon *et al.*, 1993 – Copyright 2001 – Reprinted by permission of John Wiley & Sons, Inc.)

ing to -100°C to avoid crystallization), and $\Delta c_{p0} = 0.56\,J\,g^{-1}K^{-1}$. The final cured material exhibited the following values: $T_{g\infty} = 292$°C and $\Delta c_{p\infty} = 0.13\ J\ g^{-1}\ K^{-1}$. The curve in Fig. 4.8 shows the fitting of experimental results with Eq. (4.10) using $\lambda = \Delta c_{p\infty}/\Delta c_{p0} = 0.23$. Equation (4.10) shows a reasonable predictive capability using experimental calorimetric data of the initial formulation and the final material.

The T_g vs x relationship shown in Fig. 4.7 was fitted using $\lambda = 0.69$ as an adjustable parameter (Montserrat, 1995). The same author reported the following correlation for the Δc_p values of partially reacted epoxy–anhydride samples:

$$\Delta c_p(J\ g^{-1}K^{-1}) = 0.089 + 107.2/T_g(K) \tag{4.11}$$

From the reported values of T_{g0} (250 K) and $T_{g\infty}$ (382 K), a value of $\lambda = \Delta c_{p\infty}/\Delta c_{p0} = 0.71$, results. This is in excellent agreement with the value adjusted to obtain the best fit.

The inverse proportionality of Δc_p with T_g constitutes a general experimental observation for thermosetting polymers, and leads to $\lambda < 1$, which is a necessary condition to obtain an upward curvature in the T_g vs x function defined by Eq. (4.10). Low values of λ are typical of large $(T_{g_\infty} - T_{g0})$ values.

For some highly crosslinked thermosetting polymers with rigid backbones, topological restrictions limit the maximum attainable conversion. For the reaction between epoxidized novolacs and cresol-based novolacs, a maximum conversion, $x_{max} = 0.80$ was reported (Hale *et al.*, 1991). A similar value of x_{max} was reported for the cure of an epoxidized novolac with 4,4′-diaminodiphenylsulfone (DDS) (Oyanguren and Williams, 1993a). In these cases, partially reacted networks may be modeled as a random solution of reacted functionalities with a concentration equal to x/x_{max} ($\phi_i = x/x_{max}$, $T_{gi} = T_{gmax}$, $\Delta c_{pi}(T_{gmax}) = \Delta c_{pmax}$), and unreacted functionalities with a concentration equal to $(1 - x/x_{max})$ ($\phi_i = 1 - x/x_{max}$, $T_{gi} = T_{g0}$, $\Delta c_{pi}(T_{g0}) = \Delta c_{p0}$). Substituting into Eq. (4.10) and rearranging, gives

$$(T_g - T_{g0})/(T_{gmax} - T_{g0}) = \lambda' x/[x_{max} - (1 - \lambda')x] \tag{4.12}$$

where

$$\lambda' = \Delta c_{pmax}/c_{p0} \tag{4.13}$$

Figure 4.9 shows the fit of experimental results obtained for the epoxidized novolac–DDS system, using Eq. (4.12) with $\lambda' = 0.15$ (Oyanguren *et al.*, 1993b). As Δc_{pmax} could not be determined with enough precision, λ' was used as an adjustable coefficient.

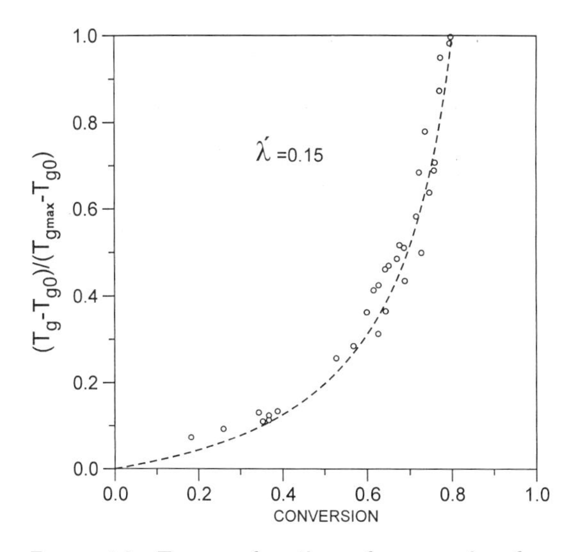

FIGURE 4.9 T_g as a function of conversion for an epoxidized novolac–DDS system ($x_{max} = 0.80$). The dashed line is the curve predicted using Eq. (4.12) with $\lambda' = 0.15$ (Oyanguren and Williams, 1993b – Copyright 2001 – Reprinted by permission of John Wiley & Sons, Inc.)

4.3 DEGRADATION AND PHASE SEPARATION

When the polymerization reaction takes place at high temperatures, side reactions including degradation of the polymer network may take place. The rate of these reactions depends on the composition of the gaseous phase; e.g., air or an inert atmosphere, and on the rate at which volatiles are removed from the sample. If degradation takes place from the glassy state of the polymer network, it will produce a devitrification process followed, at later times, by revitrification due to char formation (Chan et al., 1984). In the rubbery state only char formation will take place. All these transitions are better described using a TTT diagram (Sec. 4.5).

Several commercial formulations based on thermosetting polymers also include an elastomer or a thermoplastic that is initially soluble in the mixture of monomers. As polymerization takes place, the solubility of this component continuously decreases due to the unfavorable variation of the entropy of mixing (Chapter 8). The variation of the enthalpy of mixing may also contribute to the decrease in solubility. At a particular conversion that depends on the cure temperature, demixing of a phase rich in the elastomer or thermoplastic begins to take place. This transition is called the cloud point of the system, because the intensity of transmitted light decreases due to the light scattering produced by the heterogeneous medium. From this time on, the thermosetting polymer is fractioned between both phases, which exhibit different values of conversion and glass transition temperatures. Rubber- and thermoplastic-modified thermosets are used in commercial formulations of adhesives, coatings, and advanced composites (Chapter 8).

4.4 THE CONVERSION–TEMPERATURE TRANSFORMATION (CTT) DIAGRAM

4.4.1 Regions of the CTT Diagram

The CTT diagram showing gelation, vitrification, and degradation curves is represented in Fig. 4.10. The degradation curve is the boundary of a high-temperature region where side reactions, including chemical degradation, may occur. It is assumed that the stability of the thermoset to thermal degradation increases with conversion. Obviously, the diagram may be only used for systems in which the gel conversion is independent of temperature and a unique relationship between T_g and conversion may be stated. In certain thermosetting polymers, the reaction pathway and the resulting network structure are strongly dependent on the thermal cycle selected for the polymerization, such as for formulations based on epoxy–dicyandiamide where no unique T_g vs x relationship is found.

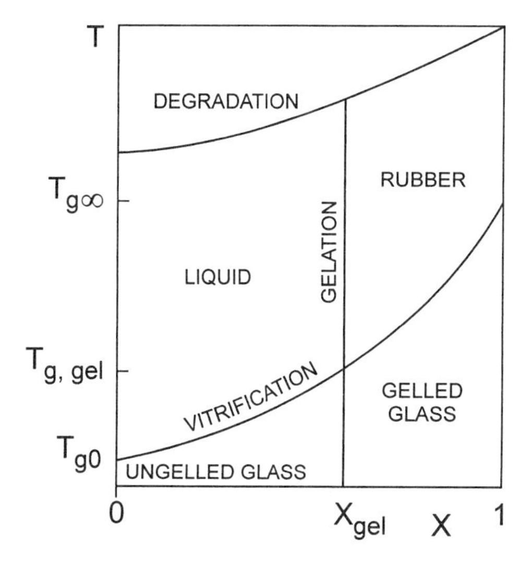

FIGURE 4.10 CTT diagram showing gelation, vitrification, and degradation curves.

Different states in which the thermosetting polymer may be found are indicated in Fig. 4.10. The initial formulation ($x = 0$) may be either present as a liquid or as an ungelled glass, depending on the value of T_{g0} with respect to the storage temperature. It is obviously much safer to store the initial formulation in the glassy state to prevent an undesired advance of the polymerization reaction.

The cure is always started at $T > T_{g0}$. Gelation produces the transition from a liquid to a rubber or, eventually, from an ungelled glass to a gelled glass. At the particular temperature $T_{g,gel}$, the liquid is directly transformed into a gelled glass upon gelation. At the right of the vitrification curve, either an ungelled or a gelled glass is obtained. The state at which a cured thermoset is present in normal use is that of a gelled glass. Physical properties of thermosetting polymers in the glassy state are, then, of practical importance.

4.4.2 Isothermal Trajectories

Figure 4.11 shows three possible isothermal trajectories in the CTT diagram. Curing at $T_g > T_{g_\infty}$ (trajectory **a**) leads to complete conversion. However, due to the high values of both the reaction heat and the polymerization rate, it is usually not possible to keep isothermal conditions when curing at high temperatures. The temperature of the sample continuously increases as it is

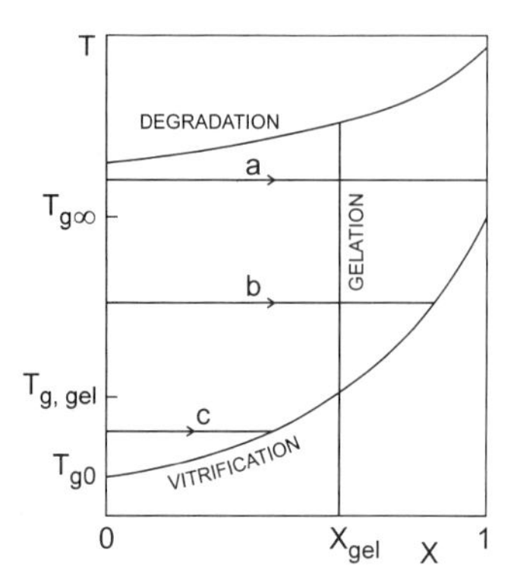

FIGURE 4.11 CTT diagram showing isothermal trajectories.

impossible to dissipate the reaction heat at the rate at which it is generated. This leads rapidly to thermal degradation of the thermosetting polymer.

Trajectory **b**, located at an intermediate temperature between T_{g0} and T_{g_∞}, represents the usual condition of an isothermal cure. The reaction rate at this generic temperature is low enough to enable the dissipation of heat generated and to keep isothermal conditions (this also requires the use of small amounts of samples or thin specimens). The final conversion of these trajectories is limited by vitrification. So, to obtain a complete cure, it is often necessary to use several isothermal trajectories with increasing temperatures, with a final step carried out at $T > T_{g_\infty}$. In particular, the safest way to cure very large structures, avoiding exothermic excursions, is to increase the temperature in small steps, arresting the reaction by vitrification in each one of them. Within this limitation, the cure proceeds very slowly along the vitrification curve.

As trajectory **b** produces a gelled glass, the part must have the desired shape before the cure takes place; e.g., by curing in a heated mold. Trajectories located between T_{g0} and $T_{g,gel}$, as trajectory **c**, may be used to advance the conversion of the initial formulation without reaching gelation. Resulting products are usually called B-stage formulations and may be used as the starting materials for the final cure process.

4.4.3 Polymerization at a Constant Heating Rate

Polymerizations performed at a constant heating rate are typical of characterization techniques such as DSC. In these cases, a small mass of material is employed so that the temperature remains uniform in the whole specimen but varies linearly with time:

$$dT/dt = q \tag{4.14}$$

Usual scanning rates vary from 2 to 20°C min^{-1}.

Polymerization rates are in general described by a set of differential equations giving the individual rates of consumption and generation of species and functional groups participating in the kinetic scheme (Chapter 5). For some particular cases, the general system of differential equations may be simplified to give a single kinetic equation of the type

$$dx/dt = r(x, T) \tag{4.15}$$

where the function r(x,T) is called the polymerization rate.

For systems where Eq. (4.15) is valid, dividing Eq. (4.14) by Eq. (4.15) leads to

$$dT/dx = q/r(x, T) \tag{4.16}$$

Integration of Eq. (4.16) for a particular r(x,T) expression leads to the particular trajectory followed in the CTT diagram. For some thermosetting polymers, the polymerization rate may be expressed as the product of a factor that has an Arrhenius dependence on temperature and a factor that depends exclusively on conversion:

$$r(x, T) = A \exp(-E/RT) f(x) \tag{4.17}$$

Inserting Eq. (4.17) in Eq. (4.16), and integrating, gives

$$\int_{T_0}^{T} \exp(-E/RT)dT = (q/A) \int_{0}^{x} dx/f(x) \tag{4.18}$$

Figure 4.12 shows three possible trajectories followed by the thermosetting polymer cured at a constant heating rate. For the lowest q value the trajectory intercepts the vitrification curve. But now the reaction is not arrested, because heat is continuously supplied to increase T at the desired heating rate. So, once the vitrification curve is intercepted, the system evolves close to it until full conversion is attained. For the intermediate q value, the system evolves from the liquid to the rubbery state without vitrifying. But for the highest q value, degradation occurs before attaining full conversion.

Usually, experimental DSC curves obtained at a constant heating rate are processed to obtain the polymerization kinetics (the inverse problem). In this case it is advisable to plot the experimental trajectories in the CTT diagram and to use only the portions of the trajectories that do not intercept the vitrification or degradation curves.

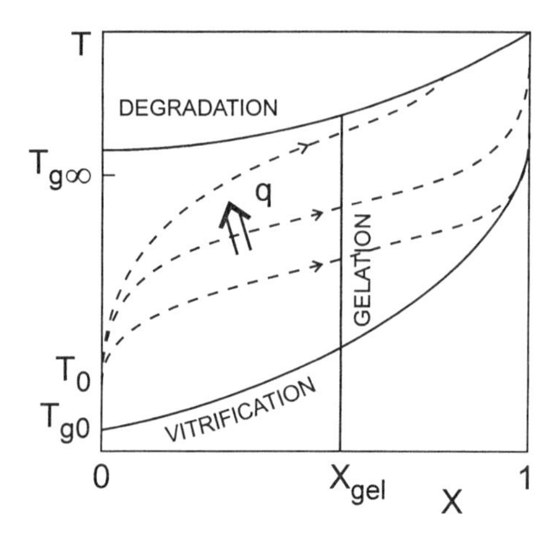

FIGURE 4.12 CTT diagram showing trajectories at a constant heating rate.

4.4.4 Adiabatic Trajectories

Adiabatic trajectories are typical of the cure of large parts, where the rate of heat diffusion is much lower than the rate of heat generation. A thermal energy balance may be stated as follows:

$$c_p dT/dt = (-\Delta H) r(x, T) \tag{4.19}$$

where c_p and $(-\Delta H)$ are, respectively, the isobaric specific heat capacity and the heat of reaction per unit mass.

Dividing Eq. (4.19) by Eq. (4.15) and integrating, assuming that both c_p and $(-\Delta H)$ are constant, leads to

$$(T - T_0)/\Delta T_{ad} = x \tag{4.20}$$

where

$$\Delta T_{ad} = (-\Delta H)/c_p \tag{4.21}$$

Figure 4.13 shows three possible adiabatic trajectories in the CTT diagram. For the trajectory with the lowest $\Delta T_{ad} (< T_{g\infty} - T_0)$, the straight line is intercepted by the vitrification curve. But when the system vitrifies, the reaction practically ceases, and there is no more possibility of getting out from the glassy state because the only source of temperature increase is the chemical reaction; therefore, full conversion cannot be attained, and a post-cure step at $T > T_{g_\infty}$ is necessary to complete the polymerization.

For the intermediate ΔT_{ad}, full conversion is attained without problems of vitrification or degradation. But for the highest value of ΔT_{ad}, the degradation of the thermosetting polymer cannot be avoided. The heat of

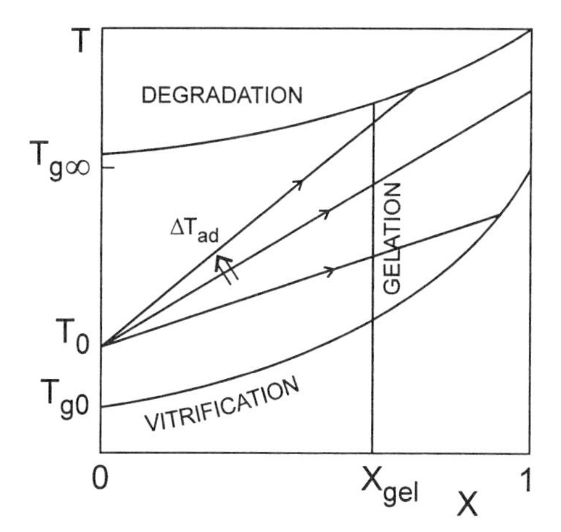

Figure 4.13 CTT diagram showing adiabatic trajectories.

reaction of neat thermosetting polymers is in the range of several hundreds of J g^{-1} while the c_p is in the order of 2 J g^{-1} K^{-1}. This leads to ΔT_{ad} values comprised in the range of 100–300°C. The use of fillers, fibers, and any other inert component in the initial formulation decreases the value of ΔT_{ad} and facilitates processing.

4.5 TIME–TEMPERATURE TRANSFORMATION (TTT) DIAGRAM

The use of a timescale instead of a conversion one requires a previous definition of the cure schedule; e.g., isothermal, constant heating rate, etc. Usually, isothermal conditions are selected to define the timescale; i.e., only trajectories at constant temperature have a physical meaning. This leads to the TTT diagram.

The TTT diagram may be either built from the experimental determination of t_{gel} and t_{vit} under isothermal conditions or generated from the CTT diagram using the polymerization rate r(x,T). Equation (4.15) may be integrated for $T = T_c$, yielding

$$t = \int_0^x dx/r(x, T_c) \tag{4.22}$$

Times to gel (t_{gel}) and times to vitrify (t_{vit}) may be obtained by making $x = x_{gel}$ and $x = x_{vit}$, respectively:

$$t_{gel} = \int_0^{x_{gel}} dx/r(x, T_c) \quad (4.23)$$

$$t_{vit} = \int_0^{x_{vit}(T_c)} dx/r(x, T_c) \quad (4.24)$$

Figure 4.14 shows a typical TTT diagram. The location of the different states of the thermosetting polymer is indicated, including a char region that is characteristic of the thermal degradation occurring at high temperatures. As the reaction rate varies exponentially with temperature, times are better represented in a logarithmic scale.

The shape of the gelation curve prior to interception of the vitrification curve arises from the Arrhenius dependence of the polymerization rate on temperature. For example, assuming the validity of Eq. (4.17), the time to gel may be obtained from

$$t_{gel} = (1/A)\exp(E/RT_c) \int_0^{x_{gel}} dx/f(x) \quad (4.25)$$

If x_{gel} does not depend on temperature, $\ln(t_{gel})$ is a linear function of $1/T_c$. The activation energy of the chemical reaction may be obtained from such a plot.

Close to the vitrification curve and particularly after its interception, the reaction rate becomes diffusionally controlled. It decelerates with an autoretardation effect and eventually becomes negligible for practical pur-

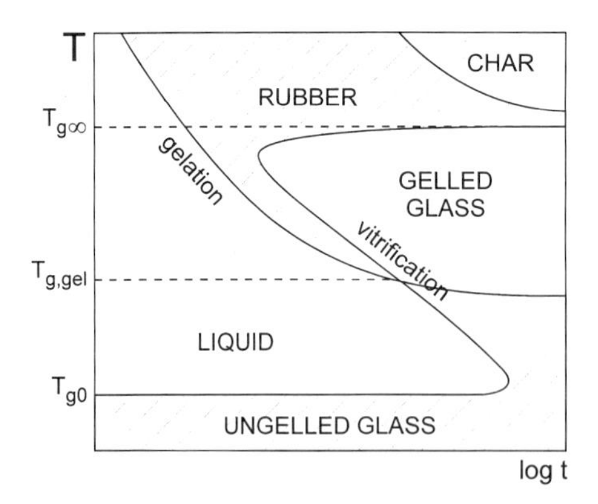

FIGURE 4.14 Time–temperature transformation (TTT) diagram.

poses. This makes the gelation curve turn into a practically horizontal line at long times. It is, however, possible to gel the system after vitrification by operating at $T_c < T_{g,gel}$ but not so distant from it. For example, Verchère *et al.* (1990) reported the TTT diagram for a DGEBA-based epoxy cured with a cycloaliphatic diamine. The experimental value of $T_{g,gel}$ was close to 50°C. The cure at 29°C led to vitrification after 12.5 h, but gelation could be observed after a cure of 50 days in the glassy state.

The S shape of the vitrification curve can also be explained. This arises from the presence of T_c both in the rate equation and in the upper limit of the integral defined in Eq. (4.24). For $T_c = T_{g0}$, x_{vit} $(T_{g0}) = 0$; so, $t_{vit} = 0$. Increasing T_c, starting from T_{g0}, leads to an initial increase in t_{vit} because a finite conversion must be attained to vitrify the system. But, on further increasing T_c, the exponential decrease of vitrification times arising from the Arrhenius dependence of the reaction rate predominates. The trend is reversed again when T_c gets close to $T_{g\infty}$. High conversion values must be attained to vitrify the system. But in the high-conversion range, r(x,T) becomes very low and this produces an increase in vitrification times. An S-shaped vitrification curve results.

Times to reach any constant value of conversion may be calculated in the same way as t_{gel}, which means isoconversional lines can be included in the TTT diagram; the most important one is the full conversion line. The TTT diagram shown in Fig. 4.15 includes such a line (Peng and Gillham, 1985). While gelation and vitrification curves were obtained using experimental points, the full cure curve was estimated from the time necessary to reach $T_{g\infty}$ (about 168°C for the particular system represented in Fig. 4.15), when curing at $T < T_{g\infty}$.

High T_g thermosets require a postcure step at a temperature close to $T_{g\infty}$ to reach full conversion. In these conditions thermal degradation may become competitive with network formation. Figure 4.16 shows a TTT diagram for the cure of a trifunctional epoxy monomer, triglycidyl ether of tris(hydroxyphenyl)methane, with DDS, in a helium atmosphere (Chan *et al.*, 1984). A maximum $T_{g\infty} = 352$°C was estimated as an extrapolated value. However, attempts to get close to this value by curing at high temperatures led to a devitrification branch, followed by revitrification due to char formation. The best T vs time cure schedule that maximizes the final T_g must be determined experimentally.

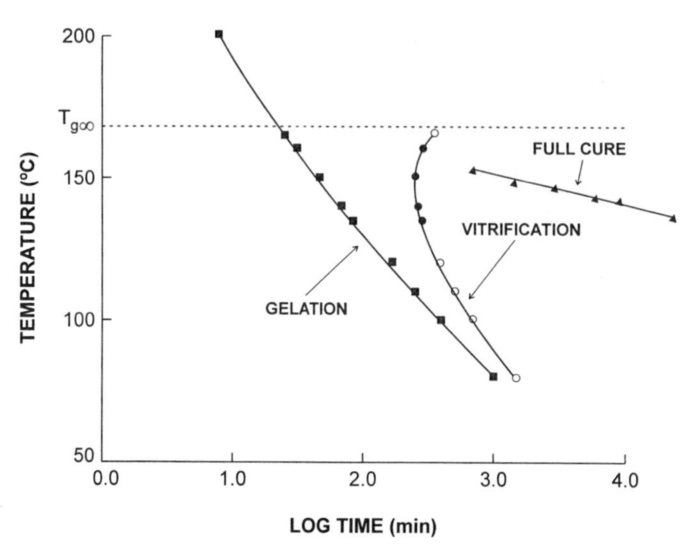

Figure 4.15 TTT diagram for a DGEBA-based diepoxide cured with trimethylene glycol di-p-aminobenzoate (TMAB). (Peng and Gillham, 1985 – Copyright 2001 – Reprinted by permission of John Wiley & Sons, Inc.)

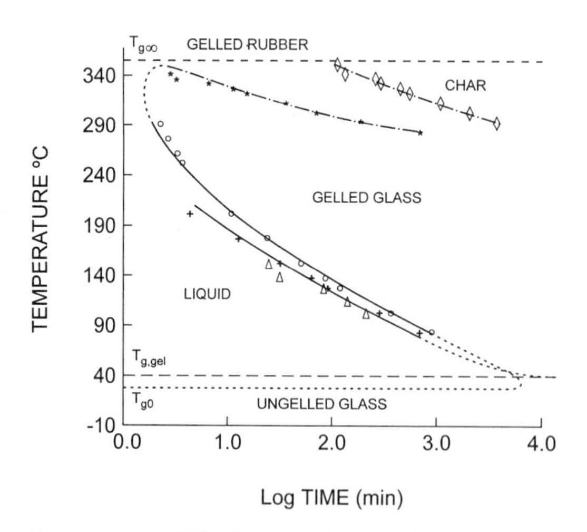

Figure 4.16 TTT diagram for a trifunctional epoxy monomer, triglycidyl ether of tris(hydroxyphenyl)methane, cured with 4,4′-diaminodiphenylsulfone (DDS), under He. (Chan *et al.*, 1984 – Copyright 2001 – Reprinted by permission of John Wiley & Sons, Inc.)

NOTATION

A	= pre-exponential factor of the specific rate constant, s^{-1}
c_p	= isobaric heat capacity, kJ kg^{-1} K^{-1}
DMTA	= dynamic mechanical thermal analysis
DSC	= differential scanning calorimetry
E	= activation energy of the polymerization rate, kJ $kmol^{-1}$
E'	= storage modulus determined in a dynamic mechanical test, GPa
E''	= loss modulus determined in a dynamic mechanical test, GPa
E_c	= activation energy characteristic of the rate to achieve an equilibrium situation, kJ mol^{-1}
f	= free volume fraction
g	= concentration of the less stable conformation states, kmol m^{-3}
g_{eq}	= concentration of the less stable conformation states at equilibrium, kmol m^{-3}
H	= enthalpy, J
HDT	= heat deflection temperature, K
k	= specific rate constant, s^{-1}
q	= rate of temperature variation, dT/dt, K s^{-1}
$r(x,T)$	= polymerization rate, s^{-1}
R	= gas constant = 8.314 kJ $kmol^{-1}$ K^{-1}
S	= entropy, J K^{-1}
t	= time, s
t_c	= cure time, s
t_{gel}	= time to gel, s
t_{vit}	= time to vitrify, s
T	= temperature, K
T_0	= initial temperature, K
T_1	= particular temperature during a thermal cycle, K
T_α	= glass transition temperature determined in a dynamic mechanical test, K
T_c	= cure temperature, K
T_g	= glass transition temperature, K
T_{gi}	= glass transition temperature of a generic component i, K
T_{g0}	= glass transition temperature of the initial formulation, K
$T_{g\infty}$	= glass transition temperature at full conversion, K
$T_{g,gel}$	= glass transition temperature corresponding to the transformation of a liquid into a gelled glass, K
T_{gmax}	= maximum attainable glass transition temperature, K
T_∞	= temperature at which the free volume fraction or configurational entropy become zero, K
$\tan \delta$	= loss tangent = E''/E'
TMA	= thermal mechanical analysis
V	= volume, m^3
x	= conversion
x_{max}	= maximum attainable conversion

Greek Letters

α = expansion coefficient, K^{-1}

Δc_{p0} = change in the isobaric heat capacity of the initial formulation between liquid and glassy states, kJ kg^{-1} K^{-1}

$\Delta c_{p\infty}$ = change in the isobaric heat capacity of the fully reacted thermoset between rubbery and glassy states, kJ kg^{-1} K^{-1}

$\Delta c_{pi} = c_{pi}(l) - c_{pi}(g)$ = change in the isobaric heat capacity of component i between liquid and glassy states, kJ kg^{-1} K^{-1}

Δc_{pmax} = change in the isobaric heat capacity of the thermoset reacted to the maximum attainable conversion, between rubbery and glassy states, kJ kg^{-1} K^{-1}

$(-\Delta H)$ = total reaction heat per unit mass, kJ kg^{-1}

$(-\Delta H_R)$ = residual reaction heat per unit mass, kJ kg^{-1}

ΔT_{ad} = adiabatic temperature rise at complete conversion, K

ε' = real part of the complex dielectric permitivitty, C^2 N^{-1} m^{-2}

ε'' = imaginary part of the complex dielectric permitivitty, C^2 N^{-1} m^{-2}

λ = $\Delta c_{p\infty}/\Delta c_{p0}$

λ' = $\Delta\ c_{pmax}/\Delta c_{p0}$

τ_c = characteristic time to reach equilibrium, s

τ_q = characteristic time for the rate of temperature variation, s

ϕ_I = concentration of component i in units matching those of Δc_{pi}

REFERENCES

Adabbo HE, Williams RJJ, *J. Appl. Polym. Sci*, 27, 1327–1334 (1982).

Chan LC, Naé HN, Gillham JK, *J. Appl. Polym. Sci.*, 29, 3307–3327 (1984).

Couchman PR, *Macromolecules*, 20, 1712–1717 (1987).

Di Marzio EA, *J. Res. NBS*, 68A, 611–617 (1964).

Enns JB, Gillham JK, *J. Appl. Polym. Sci.*, 28, 2567–2591 (1983).

Fox TG, Gratch S, Loshaek S, *J. Polym. Sci.*, 15, 371–390 (1955).

Georjon O, Galy J, Pascault JP, *J. Appl. Polym. Sci.*, 49, 1441–1452 (1993).

Hale A, Macosko CW, Bair HE, *Macromolecules*, 24, 2610–2621 (1991).

Jordan C, Galy J, Pascault JP, *J. Appl. Polym. Sci.*, 46, 859–871 (1992).

Kovacs AJ, *Adv. Polym. Sci.*, 3, 394–507 (1963).

Kovacs AJ, Aklonis JJ, Hutchinson JM, Ramos AR, *J. Polym. Sci., Polym. Phys. Ed.*, 17, 1097–1162 (1979).

Montserrat S, *J. Appl. Polym. Sci.*, 44, 545–554 (1992).

Montserrat S, *Polymer*, 36, 435–436 (1995).

Ngai KL, Rendell RW, Rajagopal AK, Teitler S, *Annals NY Acad. Sci.*, 484, 150–184 (1986).

Oyanguren PA, Williams RJJ, *J. Appl. Polym. Sci.*, 47, 1361–1371 (1993a).

Oyanguren PA, Williams RJJ, *J. Appl. Polym. Sci.*, 47, 1373–1379 (1993b).

Pascault JP, Williams RJJ, *J. Polym. Sci. B: Polym. Phys.*, 28, 85–95 (1990).

Peng X, Gillham JK, *J. Appl. Polym. Sci.*, 30, 4685–4696 (1985).
Struik LCE, *Physical Ageing in Amorphous Polymers and Other Materials*, Elsevier, Amsterdam, 1978.
Verchère D, Sautereau H, Pascault JP, Riccardi CC, Moschiar SM, Williams RJJ, *Macromolecules*, 23, 725–731 (1990).
Wunderlich B, *J. Phys. Chem.*, 64, 1052–1056 (1960)

FURTHER READING

Gillham JK, Enns JB, *Trends Polym. Sci.*, 2, 406–419 (1994).
Williams RJJ, In *Developments in Plastics Technology* – 2, Whelan A, Craft J (eds), Elsevier Applied Science, London, 1985, pp. 339–379.
Williams RJJ, In *Polymer Networks: Principles of Their Formation, Structure and Properties*, Stepto RFT (ed.), Blackie Academic & Professional, London, 1998, pp. 93–124.

5

Kinetics of Network Formation

5.1 INTRODUCTION

An impressive number of papers on the polymerization kinetics of thermosets have been published since the 1970s. This kind of sport of reporting kinetic results is possibly based on the simplicity with which they can usually be obtained. All one needs is a differential scanning calorimeter (DSC) and some centigrams of a commercial formulation. The task is even facilitated if the software for kinetic calculations, provided by most commercial DSC devices, is used to fit a phenomenological rate expression.

After three decades of accumulating experimental results, we should be expected to have an almost complete knowledge of the rate equations that describe the most important thermosetting polymerizations. Unfortunately, the situation seems to be quite different: on the one hand, some authors persist in using intrinsically incorrect methodologies to analyze kinetic data; on the other hand, even for the most studied systems – e.g., the epoxy–amine reaction – no general kinetic schemes are universally accepted.

The first aim of this chapter is to analyze the significant implications associated with the mere statement of a rate equation. Limitations of phenomenological kinetic equations are discussed and more rigorous analysis based on the reaction pathway, for both stepwise and chainwise polymerizations, is presented. The effect of vitrification on polymerization rate is discussed. Some experimental techniques frequently used in kinetic analysis are overviewed briefly and, in a final section, the use of constitutive equations and mass and thermal energy balances is discussed.

5.2 KINETIC EQUATIONS

In this section, constitutive equations describing the polymerization kinetics when the system is in the liquid or rubbery state are analyzed. The influence of vitrification on reaction rate is considered in a subsequent section. First, phenomenological kinetic equations are analyzed; then, the use of a set of kinetic equations based on a reaction model is discussed in separate subsections for stepwise and chainwise polymerizations.

5.2.1 Phenomenological Kinetic Equations

a. Limitations of the Phenomenological Analysis

Many kinetic studies for thermosetting polymers begin by stating a generic rate equation of the form

$$dx/dt = f(x, T) \tag{5.1}$$

Particular expressions derived from Eq. (5.1) are also used as the starting point of kinetic analysis. Examples are

$$dx/dt = A\exp(-E/RT)f(x) \tag{5.2}$$

$$dx/dt = A\exp(-E/RT)(1-x)^n \tag{5.3}$$

$$dx/dt = [A_1\exp(-E_1/RT) + A_2\exp(-E_2/RT)x^m](1-x)^n \tag{5.4}$$

Equation (5.4) is usually referred to as the Kamal equation (Kamal, 1974; Sourour and Kamal, 1976), and is one of the most popular phenomenological kinetic equations used in the literature to fit experimental data for thermosetting polymers.

The specific rate constant A has time^{-1} units in all the previous kinetic equations. Except for first-order rate equations ($n = 1$ in Eq. 5.3), A must be regarded as the product of a specific rate constant times the adequate power of the initial concentration of functional groups, c_0; e.g., for Eq. (5.3), $A = k\ c_0^{n-1}$.

Despite its innocent appearance, Eq. (5.1), as well as every particular equation derived from it, implicitly incorporates a very strong assumption: i.e., for a given set of conversion and temperature values there is unique value of the reaction rate. But this is, in general, not true. In most cases, the reaction rate depends not only on the concentration of remaining functional groups, as measured by the conversion x, but also on the concentration of other species. These other species may be some intermediates produced in the reaction; e.g., the concentration of the secondary amine produced in the epoxy–primary amine reaction, the concentration of pendant double bonds generated in a vinyl–divinyl polymerization, or the concentration of active species (free radicals, anions, etc.) in chainwise polymerizations. And, in general, the concentration of these species evolves following different kinetic laws, which means that for given values of x and T it depends on the particular thermal history followed by the sample.

In addition, Eq. (5.1) should not be applied in systems that exhibit different initial reactivities of functional groups, such as in the copolymerization of double bonds of a particular unsaturated polyester with styrene or in the formation of a polyurethane starting from 2,4-toluene diisocyanate.

Equation (5.1) becomes a rigorous equation only in the case of stepwise polymerizations with equal initial reactivities, absence of substitution effects, and following a single reaction path. It may be also used when the reactivity ratio does not vary with temperature. This is, fortunately, the case of epoxy–amine reactions where the reactivity ratio of the secondary to the primary amine is approximately constant in a broad temperature range. But, even in this case, the parallel polyetherification of epoxy groups must be negligible to keep a single reaction path.

If Eq. (5.1) is used under conditions where it cannot provide a rigorous representation of the polymerization kinetics, the result is that different fittings are obtained when using isothermal and nonisothermal conditions. On occasions, a reasonable fitting is obtained with a single kinetic equation, such as Eq. (5.4), for a restricted range of temperatures and heating rates. But the equation should not be extrapolated beyond this range, because large departures of model predictions from experimental results are highly probable.

b. Is There an Influence of the Heating Rate on the Kinetic Equation?

As stated in the previous section, the use of a phenomenological kinetic equation derived from Eq. (5.1), for a system that does not verify the required restrictions for its use, may lead to different kinetic expressions when trying to fit experimental results obtained under isothermal and nonisothermal conditions. In particular, it may be observed that different kinetic parameters result by varying the heating rates in nonisothermal experiments.

But this observation is just the result of using an incorrect equation to describe the polymerization kinetics. When a correct analysis is performed (see Secs 5.2.2 and 5.2.3), it is found that reaction rates are constitutive equations that depend exclusively on concentrations and temperature but do not depend on the cure history or on the rate at which the temperature may be varying. For a given set of concentrations and temperature, a unique value of the reaction rate may be defined. This rather simple concept is violated by some authors, possibly starting from a publication appearing in *Nature* (McCallum and Tanner, 1970), where the authors speculated incorrectly about the possibility that under nonisothermal conditions the kinetics should include a term accounting for the rate of temperature variation. Although this incorrect statement was rapidly refuted (for example, see the paper by Sesták, 1979), it is amazing to see how often it reappears in the literature.

An important corollary of this analysis is related to the inverse problem. We may state that a constitutive rate equation for a particular system will be adequate if and only if it may be used to fit experimental results obtained under isothermal and nonisothermal conditions.

c. Isoconversional Methods of Kinetic Analysis

The basic idea of this type of analysis is that the reaction rate at a constant conversion depends only on temperature, as stated by Eq. (5.1). So, values of reaction rates (dx/dt) obtained at different temperatures or heating rates, may be compared for the same conversion (x), using the following equation derived from Eq. (5.2):

$$d\ln(dx/dt)/d(1/T)]x = -E/R \tag{5.5}$$

For the very restricted conditions where Eq. (5.2) provides a rigorous description of the reaction kinetics, the activation energy, E, is a constant independent of conversion. But in most cases it is found that E is indeed a function of conversion, E (x). This is usually attributed to the presence of two or more mechanisms to obtain the reaction products: e.g., a catalytic and a noncatalytic mechanism. However, the problem is in general associated to the fact that the statement in which the isoconversional method is based, the validity of Eq. (5.1), is not true. Therefore, isoconversional methods must be only used to infer the validity of Eq. (5.2) to provide a rigorous description of the polymerization kinetics. If a unique value of the activation energy is found for all the conversion range, Eq. (5.2) may be considered valid. If this is not true, a different set of rate equations must be selected.

5.2.2 Kinetic Equations Based on a Reaction Model: Stepwise Polymerizations

General concepts will be analyzed by stating kinetic equations for a diepoxide–diamine system in stoichiometric proportions, and under conditions that the chainwise homopolymerization of epoxy groups may be considered negligible. The main reactions taking place are

$$RNH_2 + \underset{\diagdown O \diagup}{CH_2 - CH}- \rightarrow RNH - CH_2 - CH(OH)- \tag{5.6}$$

$$RNH - CH_2 - CH(OH) - + \underset{\diagdown O \diagup}{CH_2 - CH}- \rightarrow RN[CH_2 - CH(OH)-]_2 \tag{5.7}$$

An adequate set of kinetic equations must describe the rate at which functional groups (epoxides, primary and secondary amines, hydroxy groups), rather than individual species (monomers, dimers, i-mers), evolve during reaction. This assumes that the rate at which a functional group reacts does not depend on the size (finite or infinite) of the molecule to which it is attached. The implication of this hypothesis may be understood if we write Eqs (5.6) and (5.7) in terms of the formation of an activated complex.

Let us call e = epoxy group and a_1, a_2, and a_3 = primary, secondary, and tertiary amine groups. We may assume that the following reactions take place:

$$a_1 + e <=> a_1 \cdots e \rightarrow a_2 \tag{5.8}$$

$$a_2 + e <=> a_2 \cdots e \rightarrow a_3 \tag{5.9}$$

where $a_1 \cdots e$ and $a_2 \cdots e$ are activated complexes that may either be decomposed into the products (a_2 and a_3) or the reactants. An activated complex is visualized as a transient association of two reactants and not as a defined chemical species. If the steps of formation of the activated complex and its dissociation to regenerate the reacting groups are both very fast, its concentration rapidly attains an equilibrium value with those of reactants. In this case, the rate-determining step is the decomposition of the activated complex into reaction products. The reaction proceeds in the chemical-controlled regime and the assumption that the reaction rate does not depend on the sizes of reactants is correct.

On the other hand, if the kinetics of any of the reactions given by Eqs (5.8) and (5.9) are determined by the rate at which the activated complex is formed, the rate is said to proceed in the diffusional-controlled regime. In this case, the reaction rate may eventually depend on the sizes of the reacting species: e.g., an epoxy group belonging to the monomer will diffuse at a faster rate than a similar epoxy group attached to the gel. However, the required diffusion must occur over a very short path (the necessary distance to approach both reactants to form the activated complex). Thus, diffusional effects are in general not observed in stepwise polymerizations until the system begins to vitrify.

A very important concept arising from the previous discussion is that gelation does not alter the polymerization rate in stepwise polymerizations; i.e., a single set of kinetic equations may be used from the beginning of reaction up to vitrification.

The rates at which epoxy and primary amine groups are consumed, in units of moles (equivalents) per unit volume and time, are described by the following constitutive equations:

$$r_e = f_1([e], [a_1], [a_2], [OH], T) \tag{5.10}$$

$$r_{a1} = f_2([e], [a_1], [OH], T) \tag{5.11}$$

where the fact that OH groups, initially present and generated during polymerization, may catalyze the epoxy–amine reaction, is taken into account. For a stoichiometric system, $[a_1]_0 = [e]_0/2$.

The concentration of a_2 and OH groups may be obtained from the following balances (Eqs (3.48) and (3.49) in Chapter 3):

$$[a_2] = 2([a_1]_0 - [a_1]) - ([e]_0 - [e]) \tag{5.12}$$

$$[OH] = [OH]_0 + [e]_0 - [e] \tag{5.13}$$

The statement of kinetics in terms of Eqs (5.10)–(5.13) assumes that all initial, intermediate, and final species participating in the reaction scheme have been identified. This is not necessarily true if there are stable complexes formed by association of OH groups with e, a_1, or a_2 groups. In this case, constitutive equations giving the rate of generation of these new intermediates have to be added to the kinetic description.

A popular reaction model assumes that the reaction may take place by two different mechanisms: (a) noncatalytic; (b) catalyzed by OH groups (Smith, 1961; Horie *et al.*, 1970). For a batch reactor; e.g., a DSC pan, the following mass balances may be stated:

$$-d[e]/dt = [e](2(k_1' + k_1[OH])[a_1] + (k_2' + k_2[OH])[a_2]) \tag{5.14}$$

$$-d[a_1]/dt = [e]2(k_1' + k_1[OH])[a_1] \tag{5.15}$$

with $[a_2]$ and [OH] given, respectively, by Eqs (5.12) and (5.13).

Other possibilities of writing the rate equations take into account the possible formation of complexes between hydroxy groups and any of the functional groups or between epoxy and amine groups themselves (Flammersheim, 1997, 1998).

The best way to validate the proposed kinetic model (or to state an alternative one) is to measure the evolution of the concentration of both epoxy and primary amine groups during polymerization. These concentrations may be conveniently measured by Fourier transformed infrared spectroscopy (FTIR) in the near-IR region (NIR) (e.g., see Min *et al.*, 1993; Paz-Abuín *et al.*, 1997, 1998; López-Quintela *et al.*, 1998). However, most often, only the conversion of epoxy groups is inferred from DSC results, and the experimental data are fitted to a kinetic model. In this case, the most important variable is the conversion of epoxy groups, defined as

$$x = (e_0 - e)/e_0 \tag{5.16}$$

where e_0 and e are the initial and actual number of moles of epoxy groups.

To write conversion in terms of molar concentrations, the volume contraction produced by the polymerization reaction must be taken into account. Usually this effect is ignored (often without even mentioning it). Assuming that volume variations may be neglected, conversion may be also written as

$$x = ([e]_0 - [e])/[e]_0 \tag{5.17}$$

Using Eq. (5.17) and defining $[a_1]/[e]_0 = \alpha_1$ (where $\alpha_{10} = 0.5$), $[a_2]/[e]_0 = \alpha_2$, $K_1 = k_1[e]_0^2$, $K'_1 = k'_1[e]_0$, $K_2 = k_2[e]_0^2$, $K'_2 = k'_2[e]_0$, the set of kinetic equations (5.14) and (5.15), together with the balances (Eqs (5.12) and (5.13)), may be written as

$$dx/dt = (1 - x)[2(K'_1 + K_1[OH]/[e]_0)\alpha_1 + (K'_2 + K_2[OH]/[e]_0)\alpha_2] \tag{5.18}$$

$$-d\alpha_1/dt = (1 - x)2(K'_1 + K_1[OH]/[e]_0)\alpha_1 \tag{5.19}$$

$$\alpha_2 = 1 - 2\alpha_1 - x \tag{5.20}$$

$$[OH]/[e]_0 = [OH]_0/[e]_0 + x \tag{5.21}$$

From Eqs (5.18)–(5.21), it is obvious that for any couple of values of conversion and temperature, the rate of consumption of epoxy groups, (dx/dt), will depend on the particular value of α_1. And this, in turn, depends on the particular cure schedule; e.g., for a particular couple (x,T), a different value of α_1 will result from isothermal runs or from runs at constant heating rate that intercept the particular point (x,T). Therefore, Eq. (5.1) has no general validity for this case.

Kinetic equations may be further simplified by postulating that the reactivity ratio of secondary to primary amine hydrogens is constant and the same for both catalytic and noncatalytic mechanisms,

$$K_2/K_1 = K'_2/K'_1 = R_{21} \tag{5.22}$$

Dividing Eq. (5.18) by Eq. (5.19) and integrating with the aid of Eq. (5.20), leads to

$$x = 1 - [2\alpha_1(1 - R_{21}) + (2\alpha_1)^{0.5R_{21}}]/(2 - R_{21}) \tag{5.23}$$

For this particular case, both α_1 and α_2 are unique functions of conversion, meaning that dx/dt depends only on conversion and temperature; i.e., the polymerization kinetics may be described by the phenomenological Eq. (5.1). Moreover, if one of the mechanisms (e.g., the catalytic) predominates over the other one (e.g., the noncatalytic), Eq (5.2) may be used to correlate experimental results and the activation energy may be obtained using isoconversional methods.

If $R_{21} = 1$, Eq. (5.18) may be written as

$$dx/dt = [K_1' + K_1([OH]_0/[e]_0 + x)](1 - x)^2 \tag{5.24}$$

In this case, a correct fitting is also provided with Eq (5.4), using $m = 1$ and $n = 2$.

However, for several epoxy–amine systems, the simple kinetic model expressed by the set of Eqs (5.18)–(5.21) does not provide a good fitting with experimental results. Reaction mechanisms, including the formation of different kinds of complexes, have been postulated to improve the kinetic description (Flammersheim, 1998). Also, a more general treatment of the kinetics of epoxy–amine reactions would have to include the possibility of the homopolymerization of epoxy groups in the reaction path. Sets of kinetic equations including this reaction have been reported (Riccardi and Williams, 1986; Chiao, 1990; Cole, 1991).

5.2.3 Kinetic Equations Based on a Reaction Model: Chainwise Polymerizations

The kinetic description of chainwise polymerizations requires a set of rate equations accounting for inhibition, initiation, propagation, termination, and transfer steps. The polymerization rate is usually given by the propagation steps (one or more), because they consume functional groups much more frequently than any other step involving them; e.g. initiation or transfer steps. However, it is always necessary to consider the overall set of kinetic equations to determine the evolution of the concentration of active species (free radicals, ions, etc.) that participate in the propagation step.

Therefore, for particular values of the conversion of functional groups and temperature, the rate of chainwise polymerizations depends on the concentration of active species which, in turn, depends on the particular thermal history. Thus, phenomenological equations derived from Eq. (5.1), or isoconversional methods of kinetic analysis, should not be applied for this case.

Polymerizations taking place by a free-radical mechanism usually require a much more complex kinetic description than those proceeding through anionic or cationic mechanisms. Therefore, they are analyzed separately in the following subsections.

a. Free-Radical Polymerizations

The free-radical mechanism frequently leads to long primary chains. Extensive cyclization reactions in these primary chains lead to the appearance of a distribution of microgels, each one composed of one or more primary chains (Sun *et al.*, 1997; see Chapter 8). This may lead to different concentrations of monomers and initiator molecules inside and outside the microgels. This partition effect implies that elementary steps may occur with a different rate in both regions.

Another important consideration for a kinetic analysis is that several of the elementary steps may be affected and even controlled by diffusion in the rubbery or liquid states (previous to vitrification). The most affected step is the termination of growing primary chains by combination or disproportionation. This requires the encounter of two large species, an event with very low probability for highly crosslinked systems. As translational diffusion depends on chain length, the specific rate constant for a termination event must be chain-length dependent. And as gelation occurs at very low conversions (Chapter 3), it may be expected that termination steps are (almost) negligible soon after gelation. It has been proposed that the only effective method for termination is for the radicals on the chain ends to move by the propagation process itself until the propagating chain is within proximity of another propagating chain, a process called reaction diffusion (Russell *et al.*, 1988a,b; Anseth *et al.*, 1995).

To complicate matters still more, propagation reactions are also affected by diffusion which means that the specific rate constant for propagation must be considered as a decreasing function of conversion. The reduction in initiator efficiency as polymerization proceeds is another factor to take into account for an accurate kinetic description (Russell *et al.*, 1988b).

The previous concepts may be illustrated with the experimental determination of the evolution of reaction rate, measured by DSC at T = 60°C, for the copolymerization of methyl methacrylate (MMA) with variable amounts of ethylene glycol dimethacrylate (EGDMA), a vinyl–divinyl system (Sun *et al.*, 1997). The reaction was initiated with 2,5-dimethyl-2,5-bis(2-ethylhexanoyl)peroxy hexane.

Figure 5.1 shows isothermal scans for formulations containing different amounts of EGDMA, and Fig. 5.2 shows the polymerization rate as a function of conversion for the same compositions.

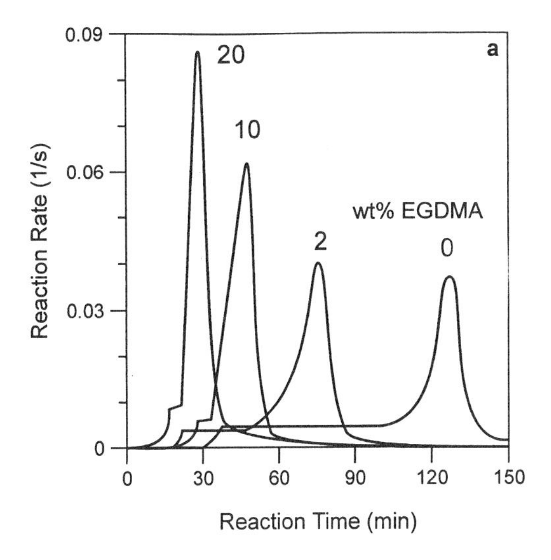

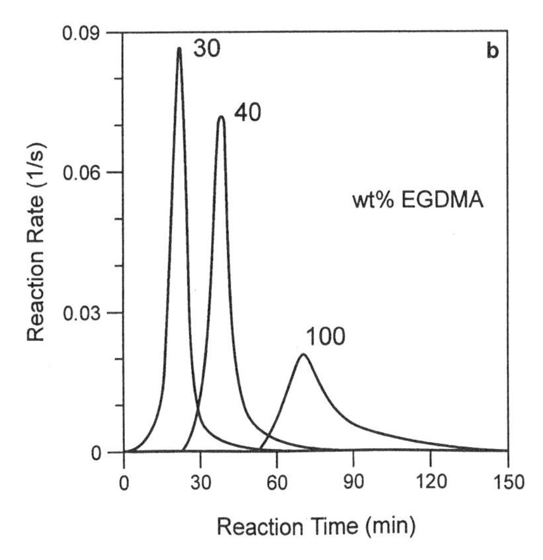

FIGURE 5.1 Polymerization rate as a function of time, at T = 60°C, for MMA–EGDMA formulations containing different wt% EGDMA and initiated with 2,5-dimethyl-2,5-bis(2-ethylhexanoyl)peroxy hexane. (Reprinted with permission from Sun *et al.*, 1997. Copyright 2001 American Chemical Society)

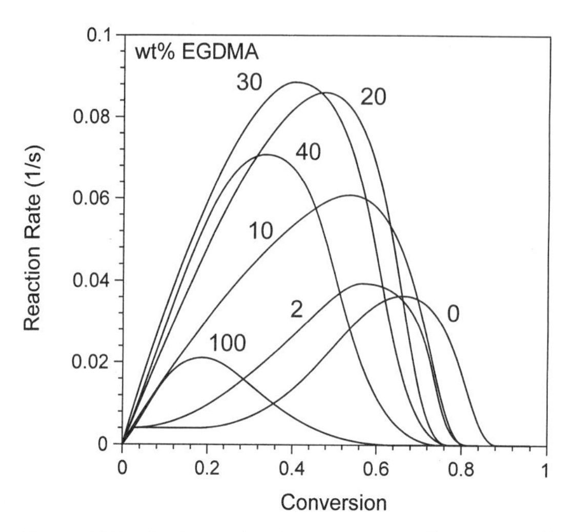

Figure 5.2 Polymerization rate as a function of conversion for the same formulations shown in Fig. 5.1. (Reprinted with permission from Sun *et al.*, 1997. Copyright 2001 American Chemical Society)

Curves for the MMA homopolymerization show the characteristic regions of a linear free-radical polymerization. For about 30 min no reaction is observed. During this induction period, free radicals generated by the decomposition of the initiator are consumed by the inhibitor present in the formulation until it is depleted. At this time there is a steep increase in the reaction rate but very rapidly a plateau is reached. During this period of almost constant reaction rate, the concentration of free radicals remains approximately constant due to fact that the initiation rate becomes equal to the termination rate (pseudo steady-state period). At conversions of about 30%, the reaction rate begins to increase at a fast rate due to the decrease in the specific rate constant of the termination step, which becomes controlled by diffusion, and the corresponding increase in the free-radicals concentration. This is usually called the gel effect or the Trommsdorff effect (a detailed analysis of this effect may be found in the review by O'Neil and Torkelson, 1997). At conversions of about 65%, the reaction rate goes through a maximum: this results from the depletion of monomer and a possible effect of diffusional restrictions on the specific rate constant for propagation. Finally, at a conversion of 95%, the system vitrifies and the reaction rate is no longer measurable.

The EGDMA addition produces a significant decrease in the conversion at which the Trommsdorff effect makes its appearance. For 30 wt% EGDMA, the pseudo steady-state period has completely disappeared, which means that the concentration of free radicals increases continuously after the inhibition period. There is a corresponding increase in the maximum reaction rate up to a 30 wt% EGDMA; but, with further additions, a significant decrease in the maximum reaction rate is observed. This may result from the lower reactivity of pendant double bonds of EGDMA (substitution effect), increase of diffusional restrictions on the propagation rate, decrease of the initiator efficiency, etc.

The above example gives us an idea of the difficulties in stating a rigorous kinetic model for the free-radical polymerization of formulations containing polyfunctional monomers. An example of efforts to introduce a mechanistic analysis for this kind of reaction, is the case of (meth)acrylate polymerizations, where Bowman and Peppas (1991) coupled free-volume derived expressions for diffusion-controlled k_p and k_t values to expressions describing the time-dependent evolution of the free volume. Further work expanded this initial analysis to take into account different possible elemental steps of the kinetic scheme (Anseth and Bowman, 1992/93; Kurdikar and Peppas, 1994; Scott and Peppas, 1999). The analysis of these mechanistic models is beyond our scope. Instead, one example of models that capture the main concepts of a rigorous description, but include phenomenological equations to account for the variation of specific rate constants with conversion, will be discussed.

The kinetic model proposed by Ng and Manas-Zloczower (1989) for the copolymerization of unsaturated polyester (UP) with styrene (S) is one example of this type of hybrid model. Relevant species considered in the analysis are C=C double bonds of the UP (E); styrene (S); initiators (I_n: 1,2, . . ., N); and an inhibitor (Z). The model is based on the following hypotheses:

1. The polymerization reaction does not start until the inhibition reaction goes to completion. The amount of residual initiators after this step may be calculated from the rate equations for the decomposition reactions and the experimental determination of the induction time (period during which no reaction is observed).
2. Termination steps are negligible from the beginning of reaction. This is based on the idea that radicals become trapped in highly crosslinked polymer coils and on the observed continuous increase of free radicals' concentration by electron spin resonance (ESR).
3. The propagation step can be characterized by a single specific rate constant that decreases continuously with conversion due to diffusional restrictions.

By letting $[M] = [E] + [S]$ and $[R.] = [E.] + [S.] + \Sigma[I_n.]$, the following rate equations can be written:

$$d[I_n]/dt = -k_{dn}[I_n] \quad (5.25)$$

$$d[R.]/dt = 2\,\Sigma f_n k_{dn}[I_n] \quad (5.26)$$

$$d[M]/dt = -k_p[M][R.] \quad (5.27)$$

where f_n is the efficiency of the n-th initiator.

The specific constant k_p is affected by diffusion and this is taken into account by the following phenomenological equation:

$$k_p = k_{p0}(1 - x/x_f)^n \quad (5.28)$$

where x_f is the maximum attainable conversion under the particular cure conditions.

This extremely simple model provided a reasonable fit of experimental DSC scans performed in the 3–7°C/min range. Better mechanistic models should consider separately the consumption of E and S double bonds in the frame of typical copolymerization models.

For formulations rich in styrene, phase separation takes place in the course of polymerization (Hsu and Lee, 1991). In this case it would be necessary to account for the different compositions of both phases and the partition of the initiators and radicals between them.

b. Anionic Polymerizations

An example of chainwise polymerizations proceeding by an anionic mechanism is the epoxy–anhydride reaction initiated by tertiary amines. Figure 5.3 shows the reaction mechanism proposed for the alternating copolymerization of phenyl glycidyl ether (PGE) and phthalic anhydride (PA), initiated by imidazoles (Leukel *et al.*, 1996). This is an anionic chainwise copolymerization with a strictly alternating sequence of co-monomers – as proved by matrix-assisted laser desorption/ionization time-of-flight mass spectrometry (MALDI-TOF MS; Leukel *et al.*,1996) – which results from activated epoxides (alkoxides) at chain ends reacting at a much faster rate with anhydrides than epoxides.

A simplified model of this system is based on the assumption that an almost simultaneous addition of an epoxy/anhydride couple to the growing chain takes place. Therefore, the propagation step may be described by a single rate constant.

The reaction is often described as a living polymerization (Matejka *et al*, 1983), but neither the distribution of molar masses nor the observed gel conversion corresponds to a pure living mechanism (Mauri *et al.*, 1997). Experimental observations may be explained by assuming the presence of a chain transfer step that regenerates the active initiator. This step deter-

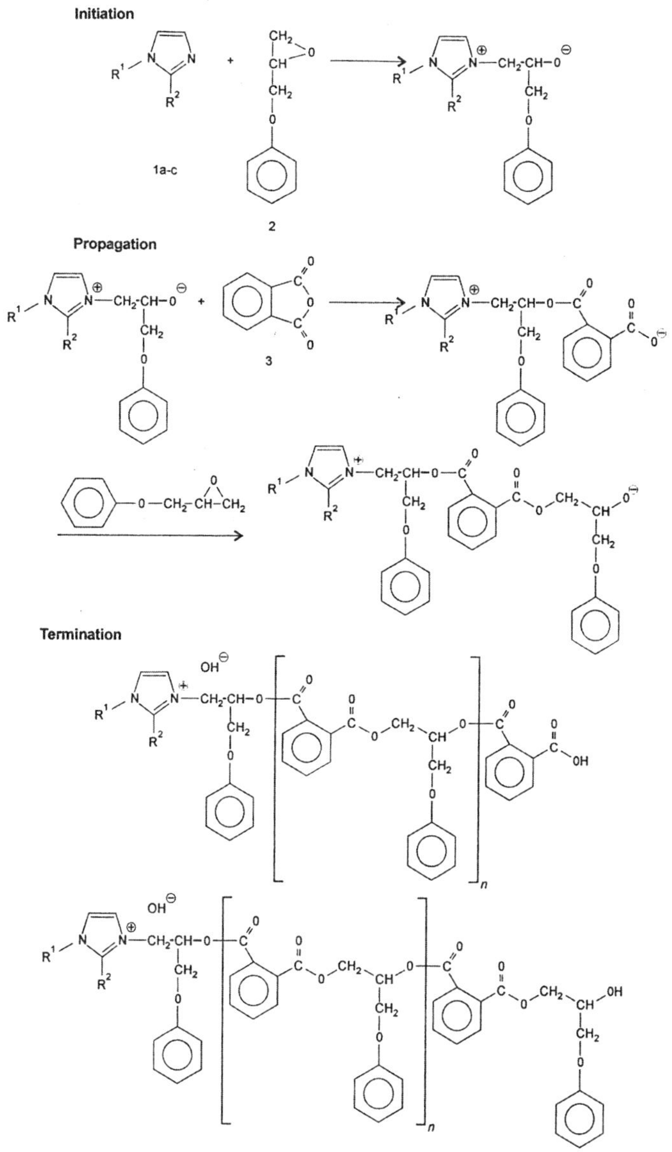

Initiation

1a-c

2

Propagation

3

Termination

R^1	R^2
$CH_2=CH$	H
CH_3	H
$CH_2=CH$	CH_3

FIGURE 5.3 Reaction mechanism of the strictly alternating copolymerization of phenyl glycidyl ether (PGE, **2**) and phthalic anhydride (PA, **3**) initiated by imidazoles (**1a–c**). (Leukel *et al.*, 1996. Copyright 2001. Reprinted by permission of Wiley-VCH)

mines the length of primary chains (and consequently the gel conversion) but does not influence the reaction kinetics.

The initiation step is a rather slow process that generates an induction period when polymerizing at low temperatures.

A mechanistic model of the polymerization kinetics must include an initiation step and a single propagation step.

A survey of the extensive literature of kinetic results reported for epoxy–anhydride–tertiary amine systems is surprising. Both nth order and autocatalytic expressions have been reported for the same system. As an example, we analyze the results reported for the copolymerization of a diepoxide based on diglycidyl ether of bisphenol A (DGEBA) with methyltetrahydrophthalic anhydride (MTHPA), initiated by benzyldimethylamine (BDMA).

Isothermal and nonisothermal DSC runs for this particular system could be fitted by the following autocatalytic equation (Montserrat and Málek, 1993):

$$dx/dt = A\exp(-E/RT)x^{0.49}(1-x)^{1.42} \tag{5.29}$$

with $E = 73$ kJ mol^{-1}. A reasonable fitting was observed for isothermal runs in the 30–110°C range, up to the vitrification time, and for dynamic DSC runs carried out in the 2.5–20 K min^{-1} range. However, the pre-exponential factor showed a decreasing trend with increasing heating rates. At 2.5 K min^{-1}, the fitting was performed with $A = 9.82\times10^6$ s^{-1}, whereas at 20 K min^{-1} the best value was $A = 7.27\times10^6$ s^{-1}. Although in this limited range of heating rates Eq. (5.29) may be reasonably used, it would be risky to employ it for predicting temperature profiles in the cure of large parts, where the rate of temperature increase in the core may be significantly higher (Chapter 9).

Isothermal kinetic runs for the same system were also analyzed by Mauri *et al.* (1997) and Galante *et al.* (1999). An excellent fit of a first-order kinetics after an induction period was reported, with E lying in the range of 70–75 kJ mol^{-1}. Obviously, the overall behavior, including the induction period, is autocatalytic. However, for conversions higher than about 0.15, a first-order kinetics provided an excellent agreement, particularly at low cure temperatures and up to vitrification. This behavior is partially explained by Eq. (5.29): the function f(x) exhibits a behavior close to $(1 - x)$ in the 0.4–0.8 conversion range.

But a striking behavior was observed when analyzing dynamic DSC runs at a constant heating rate of 10 K min^{-1}. In these runs, the first-order behavior was found over the whole conversion range, but the activation energy was $E = 110$ kJ mol^{-1} (35–40 kJ mol^{-1} higher than the isothermal value). Therefore, the simple first-order kinetics seems to be an apparent behavior of a more complex kinetic scheme. This is why it is important to check the validity of rate equations under both isothermal and dynamic conditions, in a broad range of temperatures and heating rates.

Other authors observed the same inconsistent results for other epoxy–anhydride–tertiary amine systems. For example, Peyser and Bascom (1977) observed first-order kinetics under isothermal and dynamic conditions; however, the activation energy for dynamic runs was $E = 104.2 \text{ kJ mol}^{-1}$, much larger than the value for isothermal runs, $E = 58.6 \text{ kJ mol}^{-1}$.

To provide an explanation for the observed experimental behavior under both isothermal and dynamic conditions, the following kinetic model was proposed (Riccardi *et al.*, 1999):

$$I \underset{k_{-1}}{\overset{k_1}{\rightleftharpoons}} I^* \tag{5.30}$$

$$I^* + M \overset{k_p}{\rightarrow} I^* \tag{5.31}$$

where I and I* are, respectively, inactive and active forms of the initiator that are linked by a reversible chemical reaction; M is the monomer that represents a couple of epoxy and anhydride groups (the consumption of monomers in the activation step is neglected).

The following kinetic equations may be written:

$$dI^*/dt = k_1 I - k_{-1} I^* = k_1 I_0 - (k_1 + k_{-1}) I^* \tag{5.32}$$

$$dx/dt = k_p I^* (1 - x) \tag{5.33}$$

where I_0 is the initial concentration of tertiary amine.

By defining $\beta = I^*/I_0$, Eqs (5.32) and (5.33) may be written as

$$d\beta/dt = k_1 - (k_1 + k_{-1})\beta \tag{5.34}$$

$$dx/dt = k_p I_0 \beta (1 - x) \tag{5.35}$$

Under isothermal conditions, a $\beta(x)$ relationship may be analytically obtained by dividing Eq. (5.34) by Eq. (5.35) and integrating. This leads to:

$$\ln[1/(1 - x)] = (\beta_{eq}^2 k_p I_0 / k_1)[\ln[1/(1 - \beta/\beta_{eq})] - (\beta/\beta_{eq})] \tag{5.36}$$

where

$$\beta_{eq} = K/(1 + K) \tag{5.37}$$

$$K = k_1/k_{-1} = (I^*/I)_{eq} \tag{5.38}$$

This simple kinetic model predicts the presence of an induction period in isothermal runs, where β gets close to β_{eq}, followed by a first-order kinetics, as experimentally observed.

Model parameters were obtained by fitting dynamic DSC scans at 10 K min^{-1} for a formulation with $I_0 / M_0 = 0.043$. The following set of values was obtained:

$$k_1(s^{-1}) = 5.882 \times 10^3 \exp(-E_1/RT) \quad (5.39)$$

$$k_{-1}(s^{-1}) = 2.347 \times 10^{-2} \exp(-E_{-1}/RT) \quad (5.40)$$

$$k_p I_o(s^{-1}) = 3.0 \times 10^6 \exp(-E_p/RT) \quad (5.41)$$

where $E_1 = 55$ kJ mol^{-1}, $E_{-1} = 5$ kJ mol^{-1}, and $E_p = 55$ kJ mol^{-1}.

Model predictions may be now analyzed. Figure 5.4 shows the relative approach of β to β_{eq} as conversion increases during isothermal runs. At low temperatures, β becomes equal to β_{eq} after the induction period, and a true first-order kinetics will be observed. At high temperatures, β increases with conversion during the whole run and remains distant from the equilibrium value. However, as shown in Fig. 5.5, an apparent first-order kinetics may be derived from isothermal kinetic runs. The activation energy for the apparent first-order kinetics lies in the 70–80 kJ mol^{-1} range, depending on the particular temperature range covered by experimental runs.

Figure 5.6 shows the evolution of the concentration of active species predicted by the numerical solution of Eqs. (5.34) and (5.35), for runs carried out at constant heating rate. The lower the heating rate, the closer the

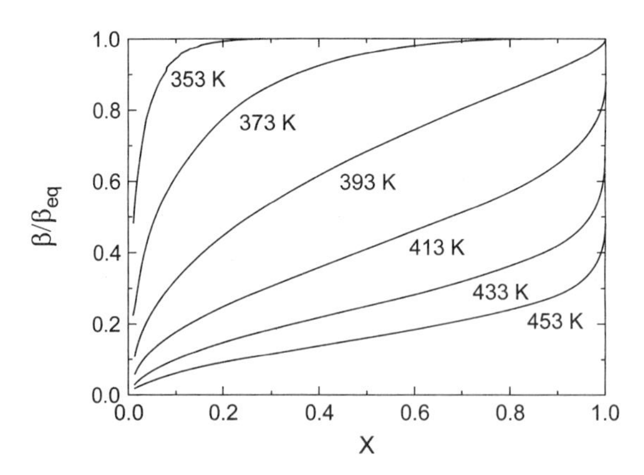

Figure 5.4 Relative approach of the concentration of active species to the equilibrium value for isothermal runs, plotted as a function of conversion. (Riccardi *et al.*, 1999 – Copyright 2001 – Reprinted by permission of John Wiley & Sons, Inc.)

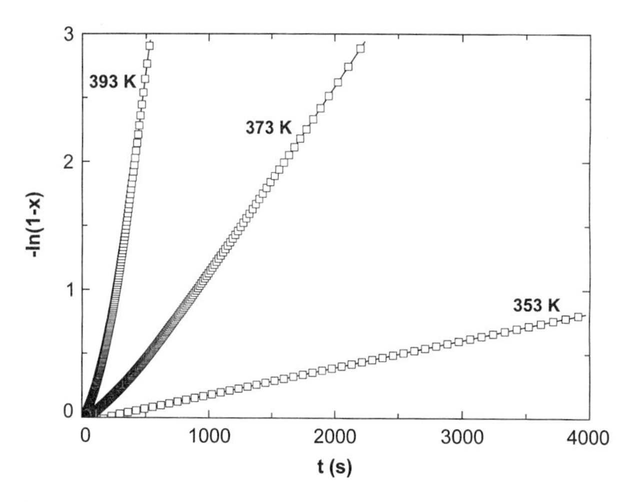

FIGURE 5.5 First-order kinetic fitting of isothermal runs. (Riccardi *et al.*, 1999 – Copyright 2001 – Reprinted by permission of John Wiley & Sons, Inc.)

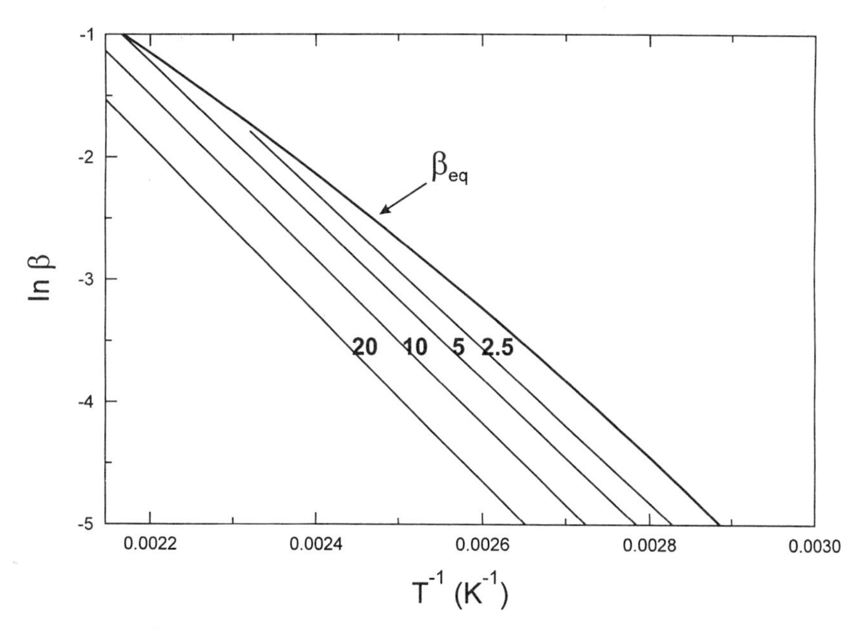

FIGURE 5.6 Concentration of active species as a function of temperature during runs carried out at constant heating rate, in the range 2.5–20 K min^{-1}. (Riccardi *et al.*, 1999 – Copyright 2001 – Reprinted by permission of John Wiley & Sons, Inc.)

approach to β_{eq}. For a first-order kinetics, the apparent rate constant is calculated as

$$\ln k = \ln[(dx/dt)/(1 - x)] = \ln(A_p I_0) - E_p/RT + \ln \beta \quad (5.42)$$

The apparent activation energy is obtained from

$$d\ln k/d(1/T) = -E_p/R + d\ln \beta/d(1/T) \quad (5.43)$$

The model predicts a linear variation of ln β as a function of (1 / T), as shown in Fig. 5.6. Then, an apparent first-order kinetics is verified over the whole conversion range, with an activation energy close to 110 kJ mol^{-1} for every heating rate, which corresponds to the sum of the activation energies of the generation of the active initiator and the propagation step. However, the specific rate constant is higher for the lower heating rate, meaning that the first-order kinetics is just an apparent behavior.

When the isoconversional method is applied to the set of dynamic runs, an activation energy lying in the 69–73 kJ mol^{-1} range is obtained, without any definite trend with conversion. The value is very close to the one reported by Montserrat and Málek (1993) using this method: again, this is an apparent value without any physical meaning.

Therefore, the mechanistic kinetic model gives a consistent explanation of the different rate equations reported in the literature for epoxy–anhydride–tertiary amine systems.

5.3 EFFECT OF VITRIFICATION ON THE POLYMERIZATION RATE

When vitrification sets in, there is an overall diffusion control that affects the rate of both stepwise and chainwise polymerizations, because segmental motions are considerably slowed down.

5.3.1 Experimental Manifestation of the Effect of Vitrification

Let us analyze a particular example where the influence of vitrification on polymerization rate has been clearly established. Wisanrakkit and Gillham (1990) reported the polymerization kinetics for a diepoxide (DGEBA) reacted with a stoichiometric amount of an aromatic diamine (trimethylene glycol di-p-aminobenzoate, TMAB). The reaction rate, measured outside the vitrification region, could be expressed by the following equations:

$$dx/dt = k_c([OH]_0/[e]_0 + x)(1 - x)^2 \quad (5.44)$$

$$k_c(min^{-1}) = 7.60 \times 10^6 \exp(-E/RT) \quad (5.45)$$

with E = 63.7 kJ mol^{-1}.

Equation (5.44) constitutes a particular case where Eq. (5.2) holds. Integrating the generic Eq. (5.2) at constant temperature and taking natural logarithms, leads to

$$\ln\left[\int_0^x dx/f(x)\right] = \ln A - E/RT + \ln t \tag{5.46}$$

As discussed in Chapter 4, for many thermosetting polymers a unique relationship may be established between conversion and glass transition temperature, as was also verified for this particular diepoxide-diamine system (Wisanrakkit and Gillham, 1990). So, the left-hand side of Eq. (5.46) may be written as a unique function of T_g, $F(T_g)$

$$F(T_g) = \ln A - E/RT + \ln t \tag{5.47}$$

With the aid of Eq. (5.47), experimental T_g vs t curves obtained at different cure temperatures may be horizontally shifted to obtain a master curve. Equation (5.47) may be written for a reference temperature, T_r:

$$F(T_g) = \ln A - E/RT_r + \ln t(T_r) \tag{5.48}$$

To obtain the same value of $F(T_g)$, it is necessary to shift the experimental curves plotted as T_g vs ln t, using the following shift factor:

$$a_T = \ln t(T_r) - \ln t(T) = E/RT_r - E/RT \tag{5.49}$$

Figure 5.7 shows the superposition of T_g vs ln t data for the diepoxide (DGEBA)–aromatic diamine (TMAB) system, to form a master curve at 140°C (Wisanrakkit and Gillham, 1990). Vitrification times, defined as the time at which T_g equals the cure temperature, are marked by arrows (T_g was defined as the midpoint of the baseline change during a DSC scan).

The polymerization is kinetically controlled up to the vitrification time, for every cure temperature. Moreover, as in this range of temperatures, gelation arrives before vitrification, the passage through the gel point does not have any influence on the reaction rate: this is a general experimental observation for stepwise polymerizations. After vitrification, a significant decrease in the reaction rate occurs, leading to the observed departure of experimental curves from the master curve.

5.3.2 Kinetic Models Including Diffusion Rate Constants

As the diffusion of reactants to form the activated complex occurs in series with chemical reaction, the total resistance for the generation of products may be written as the sum of diffusional and chemical resistances. Thus,

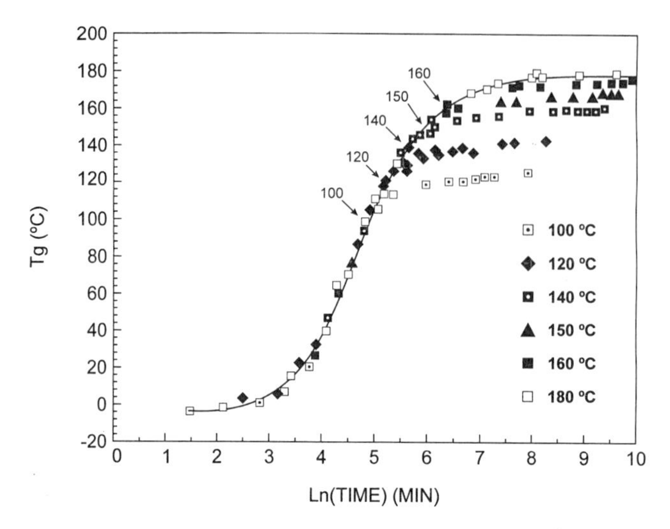

Figure 5.7 Superposition of T_g vs ln (time) data for a diepoxide (DGEBA)–aromatic diamine (TMAB) system, to form a master curve at 140°C. (Wisanrakkit and Gillham, 1990 – Copyright 2001 – Reprinted by permission of John Wiley & Sons, Inc.)

$$1/k_o = 1/k_c + 1/k_d \tag{5.50}$$

where k_o is the overall rate constant, k_c is the chemical rate constant, and k_d is the diffusion rate constant, all in time^{-1} units.

Equation (5.50) is usually called the Rabinowitch model (Rabinowitch, 1937). Although in the particular selected example there is only one chemical rate constant, in the general case Eq. (5.50) must be written for every reaction step. This means that the faster chemical steps will be more affected by diffusional resistances than the slower chemical reactions.

Although different models of diffusion rate constants have been proposed in the literature, one must bear in mind that it is very difficult to capture the evolution of the system in the glassy state with a simple mathematical model. The main problem is the physical aging that occurs in parallel with the advance in conversion. Physical aging produces a densification of the glass, which brings two consequences:

1. A significant decrease of segmental motions at the same values of x and T:
2. An increase in the molar concentration of functional groups, due to volume contraction, that increases the reaction rate.

The influence of the first factor is presumably much more important than the second factor; nevertheless, semiempirical models using two adjustable parameters have been successful in fitting the experimental kinetics in a broad temperature range.

One of these models was proposed by Wisanrakkit and Gillham (1990). They modified the Williams–Landel–Ferry (WLF) equation (Williams *et al.*, 1955), to permit its application both above and below T_g:

$$\ln[\tau(T)/\tau(T_g)] = \ln[k_d(T_g)/k_d(T)] = -c_1(T - T_g)/[c_2 + |T - T_g|] \tag{5.51}$$

where τ represents a characteristic relaxation time of the polymer structure. The use of the absolute value $|T - T_g|$ in the denominator of Eq. (5.51) allows the application of the WLF equation below T_g.

The parameter c_2 was taken equal to 51.6 K, as in the original WLF equation, while $k_d(T_g)$ and c_1 were left as adjustable parameters.

For the particular epoxy–amine system of Fig. 5.7, experimental curves could be fitted with the following equation:

$$k_d(\text{min}^{-1}) = 30.64 \exp[42.61(T - T_g)/(51.6 + |T - T_g|)] \tag{5.52}$$

where $T_g = T_g(x)$.

Figure 5.8 shows the comparison between experimental and predicted values using k_o, calculated using Eqs (5.45), (5.50), and (5.52), rather than k_c in Eq. (5.44).

For $T = T_g$, the value of k_c increases with the selected cure temperature while the value of k_d remains constant. This means that diffusional restrictions are observed earlier at high temperatures. But in this particular system, they become significant only when $T_g > T$.

A simpler semiempirical relationship was proposed by Chern and Poehlein (1987), based on free volume considerations. The diffusion rate constant is defined as

$$k_d = k_c \exp[-c^*(x - x_c)] \tag{5.53}$$

where c^* and x_c are adjustable parameters. However, the application of this equation to fit experimental kinetic data of epoxy–amine systems showed that both c^* and x_c must be written as a function of temperature (Cole *et al.*, 1991), a fact that makes it less useful for practical purposes.

The influence of vitrification on kinetics may be operative when T_g is still below T, either when k_c is intrinsically very high or k_d very low. This situation is observed when the reaction involves the diffusion of three functional groups to form the activated complex, as in the case of cyclotrimerization reactions. A particular example was reported by Simon and Gillham

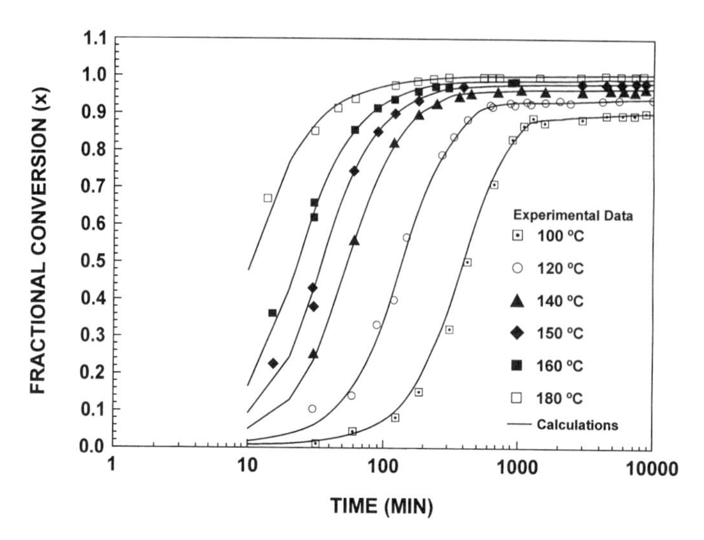

FIGURE 5.8 Comparison of experimental and predicted conversion vs ln (t) curves, using an overall rate constant including chemical and diffusional resistances. (Wisanrakkit and Gillham, 1990 – Copyright 2001 – Reprinted by permission of John Wiley & Sons, Inc.)

(1993), for the formation of a polycyanurate by the cyclotrimerization of a dicyanate ester, catalyzed by copper naphthenate and nonylphenol.

Figure 5.9 shows the evolution of T_g as a function of ln (time) at different cure temperatures. Diffusional restrictions show their influence by a decrease in the slope of the experimental curves at T_g values below the corresponding cure temperatures.

The following kinetic model was used to fit the curves in the chemically controlled regime:

$$dx/dt = K_1(1-x)^2 + K_2x(1-x)^2 \tag{5.54}$$

where

$$1/K_i = 1/k_{ci} + 1/k_d \qquad (i = 1, 2) \tag{5.55}$$

$$k_{ci} = A_i \exp(-E_i/RT) \qquad (i = 1, 2) \tag{5.56}$$

$$k_d = A_d \exp(-E_d/RT)\exp(-c_1/[4.8 \times 10^{-4}(T - T_g) + 0.025]) \tag{5.57}$$

Equation (5.57) is based on the equilibrium free volume of the material calculated with the WLF equation. This may be considered a consistent assumption for $T > T_g$ (most of the experimental data are included in

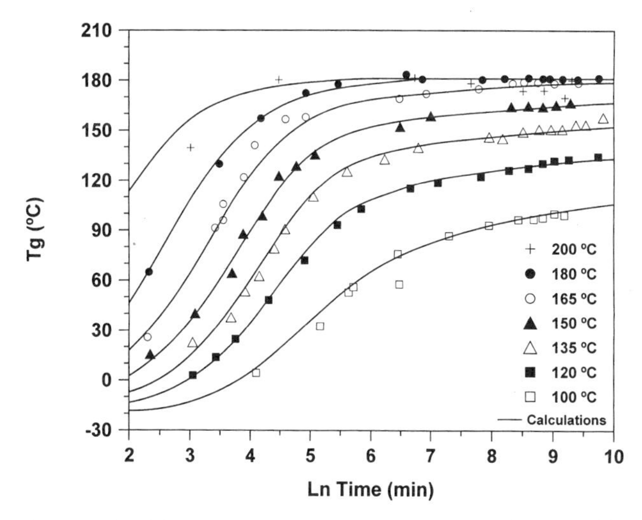

FIGURE 5.9 Glass transition temperature (T_g) vs ln (time) for the catalyzed cyclotrimerization of a dicyanate ester at different cure temperatures. Full lines represent the fitting with a kinetic model that includes diffusional restrictions. (Simon and Gillham, 1993 – Copyright 2001 – Reprinted by permission of John Wiley & Sons, Inc.)

this range). However, extrapolating this hypothesis for $T < T_g$ will produce an underestimation of the total free volume, because it will not account for the nonequilibrium contribution; A_d, E_d, and c_1 were used as adjustable parameters. Full lines in Fig. 5.9 represent the fitting of experimental data with the proposed model.

Deng and Martin (1996) also showed the necessity of including a diffusional resistance in the rate equation for the cyclotrimerization of dicyanates, well before vitrification. They observed a significant decrease in the diffusion coefficient from conversions of about 0.40, using dynamic dielectric analysis. They could fit experimental kinetic data in the whole conversion range using Eq. (5.50). Experimental values of the decrease in the diffusion coefficient with conversion were used to estimate k_d for different cure temperatures.

5.4 EXPERIMENTAL TECHNIQUES

In this section, some useful experimental techniques for the analysis of the reaction kinetics of thermosetting polymers are discussed.

5.4.1 Use of Model Compounds

a. Stepwise Polymerizations

The use of model compounds is a convenient starting point to determine the reaction path, particularly for stepwise polymerizations. For epoxy–amine systems, a monofunctional epoxide such as phenyl glycidyl ether (PGE) is often used for these studies (Verchère *et al.*, 1990; Mijovic and Wijaya, 1994). Figure 5.10 shows the reaction scheme for the curing of a monoepoxide with a diamine.

A set of differential equations for the generation and consumption of the different species may be stated. For example, by assuming that the catalytic mechanism predominates over the noncatalytic mechanism in most of the conversion range, the following kinetic scheme may be written:

$$-d[A_0]/dt = 4k_1[OH][A_0][E] \tag{5.58}$$

$$-d[A_1]/dt = \{(2k_1 + k_2)[A_1] - 4k_1[A_0]\}[OH][E] \tag{5.59}$$

$$-d[A_2]/dt = (2k_2[A_2] - 2k_1[A_1])[OH][E] \tag{5.60}$$

$$-d[A'_2]/dt = (2k_1[A'_2] - k_2[A_1])[OH][E] \tag{5.61}$$

$$-d[A_3]/dt = (k_2[A_3] - 2k_1[A'_2] - 2k_2[A_2])[OH][E] \tag{5.62}$$

$$d[A_4]/dt = k_2[A_3][OH][E] \tag{5.63}$$

where [E] and $[A_i]$ are the concentrations of monoepoxide and diamine with i reacted amine hydrogens, respectively. The concentration of hydroxy groups is given by

$$[OH] = c_0 + x[E]_0 \tag{5.64}$$

where c_0 represents the concentration of hydrogen–donor impurities present in the initial formulation.

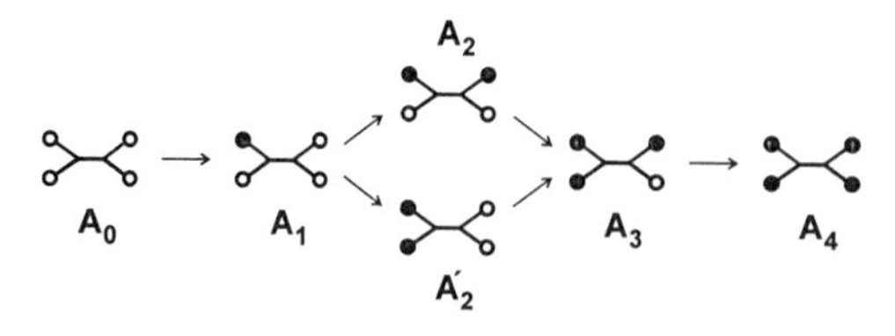

Figure 5.10 Reaction scheme for the curing of a monoepoxide with a diamine. A_i represents a diamine with i reacted amine hydrogens.

The evolution of the concentration of different species may be followed by high–performance liquid chromatography (HPLC) or size exclusion chromatography (SEC). Figure 5.11 shows SEC chromatograms for different reaction times obtained during the reaction of PGE with a cycloaliphatic diamine at 50°C (Verchère *et al.*, 1990). Only concentrations $[A_3]$ and $[A_4]$ could be estimated with enough accuracy from the resulting chromatograms.

The set of differential equations was analytically solved by assuming $R_{21} = k_2 / k_1 =$ constant. A very good fit with experimental results was obtained for $R_{21} = 0.4$, as shown in Fig. 5.12.

b. Chainwise Polymerizations

For chainwise polymerizations, the analysis of model systems implies consideration of the homopolymerization or copolymerization of bifunctional monomers. Kinetic results cannot be directly extrapolated to the case of networks, because very important features such as intramolecular cyclization reactions are not present in the case of linear polymers. However, the nature of initiation and termination reactions may be assessed. For example, using electron spin resonance (ESR), Brown and Sandreczki (1990) identified different types of radicals produced during the homopolymerization of a monomaleimide (a model compound of bismaleimides).

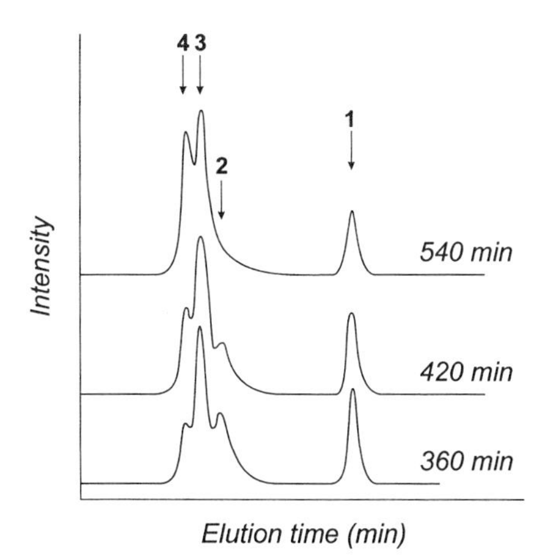

Figure 5.11 SEC chromatograms for different reaction times obtained during the reaction of PGE with a cycloaliphatic diamine at 50°C. 1, PGE; 2, $A_2 + A'_2$; 3, A_3; 4, A_4. (Reprinted with permission from Verchère *et al.*, 1990. Copyright 2001 American Chemical Society)

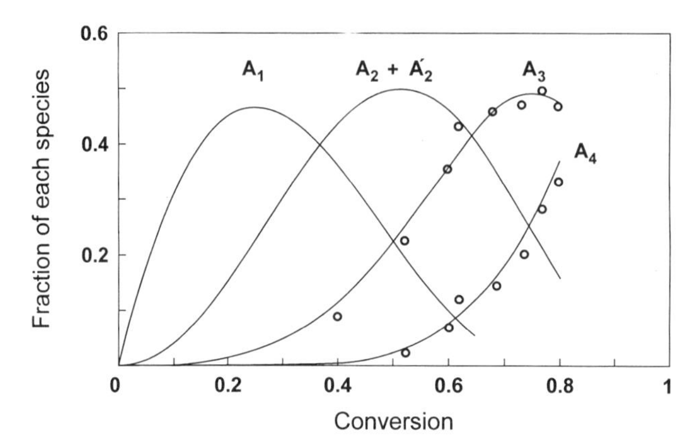

FIGURE 5.12 Comparison between experimental values of the concentration of A_3 and A_4 and theoretical predictions for $R_{21} = 0.4$. (Reprinted with permission from Verchère *et al.*, 1990. Copyright 2001 American Chemical Society)

5.4.2 Determination of the Evolution of Different Relevant Species

The best way to elucidate the reaction path is to follow the evolution of as many independent species and functional groups as possible. For example, analysis of the epoxy–amine reaction following the simultaneous evolution of epoxy and primary amine groups by near infrared spectroscopy (NIR); simultaneous determination of the conversion of double bonds belonging to unsaturated polyester (UP) and styrene (S) using FTIR, as shown in Fig. 5.13 (Yang and Lee, 1988); determination of the evolution of the concentration of free radicals using ESR, as shown in Fig. 5.14 (Tollens and Lee, 1993).

The determination of the evolution of concentrations of different species and functional groups enables one to discern different paths present in the reaction mechanism. For example, Fig. 5.13 shows that as the molar ratio of styrene to polyester C=C double bonds (MR) increases from 1/1 to 4/1, the curves tend to shift downward. For MR = 4/1 there is a very low styrene consumption until the polyester double bonds are converted to 40%. On the other hand, SEM (scanning electron microscopy) shows phase separation of a UP-rich phase in the early stages of the polymerization. Most radicals are probably trapped in this phase, which explains the higher initial conversion of the UP double bonds than styrene double bonds. A kinetic model would have to take this observation into account.

From Fig. 5.14 interesting concepts about the progress of a UP–S copolymerization may be inferred. Let us first analyze the radical spectra.

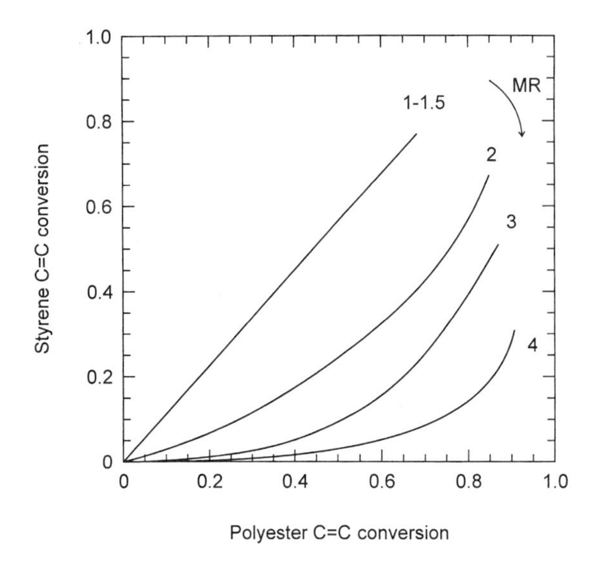

Figure 5.13 Styrene conversion vs polyester C=C conversion for different molar ratios (MR) of styrene to polyester C=C double bonds, followed by FTIR. (Reprinted from Yang and Lee, 1988, Copyright 2001, with permission from Elsevier Science)

For radicals in a liquid environment the excitation and relaxation processes are very fast, which results in a narrow line shape. However, in the solid state these processes are slower, which results in a much broader line shape. Spectra A and B are characteristic of very mobile radicals. These radicals are present in a concentration of about 10^{-6} mol l^{-1}, during the first 3 min of reaction. Spectrum E is characteristic of less-mobile radicals present in a solid environment. Similar spectra were obtained after this time. Then, it is inferred that at 6 min (macro)gelation takes place, which was confirmed by experiments performed with dynamic mechanical analysis.

Prior to gelation, two types of radicals with solid- and liquid-like mobility are present: they are possibly located in microgels (solid-like mobility) and in monomers (liquid-like mobility). The concentration of free radicals increases continuously, so that the pseudo steady-state assumption cannot be applied to model the reaction kinetics.

Also important is that the induction period, which was about 3 min in the ESR experiment, was found to increase to about 30 min in isothermal DSC scans performed at the same cure temperature (Tollens and Lee, 1993). This is possibly due to the presence of dissolved oxygen (coming from air) in the DSC samples. Oxygen is a known inhibitor of the UP–S free-radical polymerization. This is a very important fact: rate equations determined

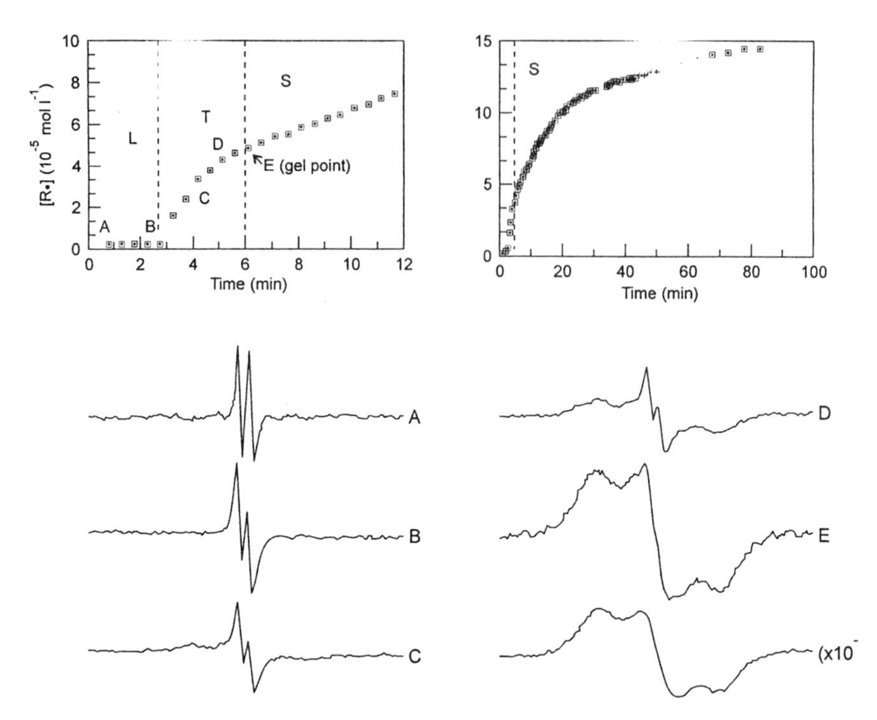

Figure 5.14 Evolution of the concentration of free radicals and radical spectra, followed by ESR, for a UP–S polymerization. (Reprinted with permission from Tollens and Lee, 1993, Copyright 2001, with permission from Elsevier Science)

from DSC in the presence of air (present when encapsulating the sample) cannot be applied to the cure of samples in bulk. The initial period in which inhibitors are being consumed by the generated radicals may be substantially different in both cases.

5.4.3 The Case of Very Fast Reactions

On occasion, it is necessary to establish the kinetics of thermosetting polymers that react at a very fast rate from the time of mixing the initial components. This is the case for thermosetting polymers used in reaction injection molding (RIM) or in foaming processes (Chapter 9). A characteristic of very fast reactions is that the heat of polymerization cannot be dissipated, so that the system may be considered quasi-adiabatic. The reaction rate may be followed using a very thin thermocouple (to avoid heat losses through the wires), immersed in a relatively large volume of sample.

The following thermal energy balance may be written for an arbitrary volume surrounding the measuring point:

$$c_p \, dT/dt = (-\Delta H) \, dx/dt \tag{5.65}$$

where c_p is the specific heat which usually is a function of x and T, and $(-\Delta H)$ is the reaction heat per unit mass, which is a function of temperature. Integration of Eq. (5.65) needs the knowledge of the $c_p(x,T)$ function and of $(-\Delta H)(T)$.

Taking average values of both parameters, Eq. (5.65) may be integrated to give:

$$x = c_p(T - T_0)/(-\Delta H) \tag{5.66}$$

For $x = 1$, $T = T_{max}$, and

$$(T_{max} - T_0) = \Delta T_{ad} = (-\Delta H)/c_p \tag{5.67}$$

Therefore, the reaction rate may be obtained from

$$dx/dt = (dT/dt)/\Delta T_{ad} \tag{5.68}$$

$$x = (T - T_0)/\Delta T_{ad} \tag{5.69}$$

Let us now analyze the application of this technique to the formation of a polyurethane synthesized from toluene diisocyanate (TDI, a mixture of 80% toluene 2,4- diisocyanate and 20% toluene 2,6-diisocyanate, as shown in Fig. 5.15) and a stoichiometric amount of a polyfunctional polyol based on sorbitol, using triethylamine (TEA) as a catalyst (Aranguren and Williams, 1986).

The formulation was intensively mixed for 15 s in a cylindrical vessel of 9.5 cm diameter and 10 cm height. A copper–constantan thermocouple was centered, and the signal continuously monitored. Figure 5.16 shows adiabatic temperature rise curves for different catalyst concentrations. The adiabatic temperature rise was estimated as 155°C.

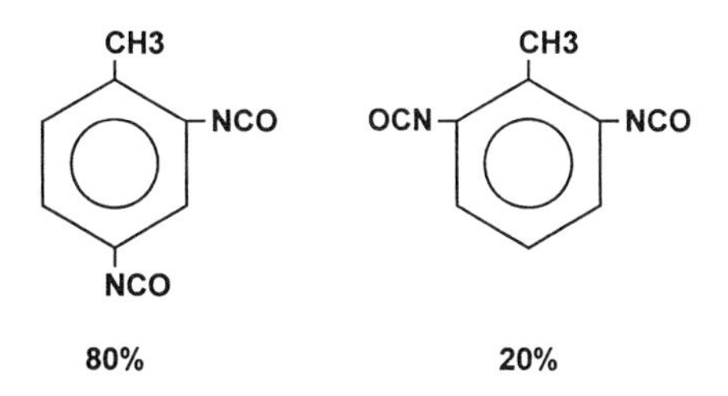

FIGURE 5.15 Isomers of toluene diisocyanate (TDI).

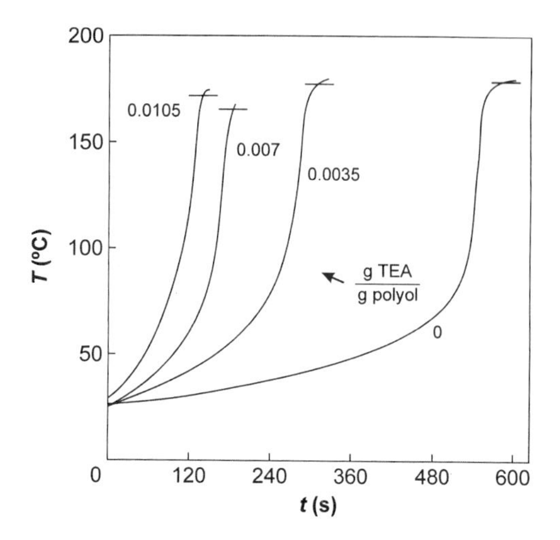

FIGURE 5.16 Adiabatic temperature rise for the TDI–polyol system using different catalyst concentrations. Horizontal bars indicate the maximum temperature used in the regression analysis. (Reprinted from Aranguren and Williams, 1986, Copyright 2001, with permission from Elsevier Science)

To fit experimental curves, a kinetic model has to be proposed. The simplest model is to assume that o-NCO groups react at a different rate than p-NCO groups, but there are no substitution effects (the reactivity of any NCO group does not depend on the reaction of the other NCO group attached to the same ring). Although this last hypothesis is certainly not correct, it is interesting to test the validity of the simplest model to fit experimental results.

As the TDI contains 60% of o-NCO and 40% of p-NCO groups, the overall conversion may be written as

$$x = 0.6x_1 + 0.4x_2 \tag{5.70}$$

where x_1 and x_2 are the conversions of o-NCO and p-NCO groups, respectively. Then,

$$dx/dt = 0.6\ dx_1/dt + 0.4\ dx_2/dt \tag{5.71}$$

The following rate equations represent the proposed kinetic model:

$$dx_1/dt = A_1[NCO]_0 \exp(-E_1/RT)(1 - x_1)(1 - x) \tag{5.72}$$

$$dx_2/dt = A_2[NCO]_0 \exp(-E_2/RT)(1 - x_2)(1 - x) \tag{5.73}$$

with $[NCO]_0 = 4.9 \times 10^{-3}$ Eq cm^{-3}.

Sets of (A_1, E_1, A_2, E_2) values that provided the best fit with experimental results when integrating Eqs (5.71)–(5.73) are shown in Table 5.1. Figure 5.17 shows the fit for the formulation containing the highest amount of catalyst.

At the same catalyst concentration, the activation energy is lower for the p-NCO than for the o-NCO. The ratio of specific rate constants, k_2/k_1, depends on temperature and catalyst concentration. At 25°C, the ratio k_2/k_1 is close to 100 for uncatalyzed formulations, and equal to 32.5 for the highest catalyst amount. However, both ratios are close to 1 at 125°C. Clearly, the rate of NCO consumption at an overall conversion x and temperature T depends on the previous thermal history, which determines the particular concentration of o-NCO and p-NCO groups at the particular values of T and x.

To test the reliability of the rate equation, isothermal determinations were carried out by placing a small amount (about 3 g) of the uncatalyzed formulation between thin aluminum foils. Several of these "sandwiches" were placed between two heated metal plates held at temperatures in the 20–60°C range. The constancy of temperature during each run was controlled by inserting a thin copper–constantan thermocouple inside one of the "sandwiches." Samples were periodically withdrawn and quenched in a solution of dibutylamine in toluene. The amount of unreacted NCO groups was determined by back-titrating with chlorhydric acid (ClH), using bromocresol green as indicator. The reaction could be followed up to overall conversions of about 0.60.

Figure 5.18 shows the overall conversion of NCO groups as a function of time for uncatalyzed samples cured at three different temperatures. Points are experimental values, while full curves are predicted results using the kinetic parameters derived in the adiabatic analysis. The predictive capability of the rate equation is very good, in spite of the strong hypothesis regarding the absence of substitution effects.

Table 5.1 Kinetic parameters for the TDI–polyol polymerization

g TEA/g polyol	0	0.0035	0.0070	0.0105
A_1 (cm^3 Eq^{-1} s^{-1})	4.75×10^{15}	3.87×10^{12}	6.34×10^{11}	1.66×10^{10}
E_1 (kJ mol^{-1})	106.6	85.3	77.7	67.5
A_2 (cm^3 Eq^{-1} s^{-1})	1.10×10^{10}	1.81×10^{8}	1.46×10^{7}	1.82×10^{6}
E_2 (kJ mol^{-1})	63.0	49.9	42.4	36.2

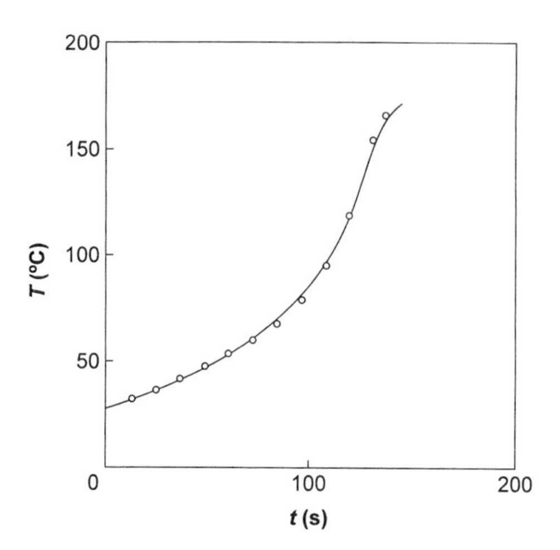

FIGURE 5.17 Fit of the adiabatic temperature rise curve for the formulation containing the highest catalyst amount. Points are experimental values while the full curve represents the best regression. (Reprinted from Aranguren and Williams, 1986, Copyright 2001, with permission from Elsevier Science)

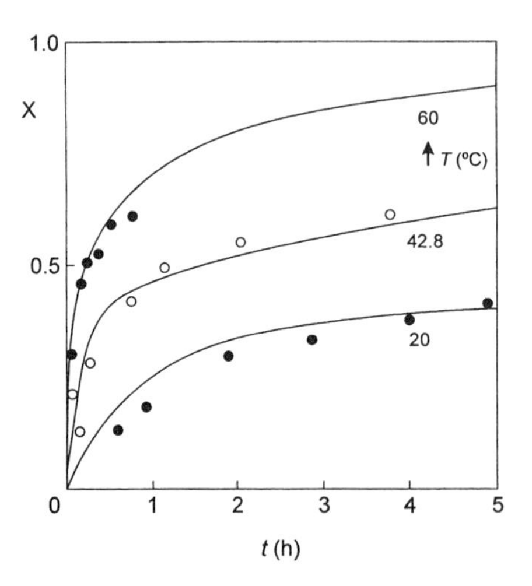

FIGURE 5.18 Overall conversion of NCO groups as a function of time for uncatalyzed samples cured at three different temperatures. Points are experimental values, while full curves are predicted values using the kinetic parameters derived from the adiabatic analysis. (Reprinted from Aranguren and Williams, 1986, Copyright 2001, with permission from Elsevier Science)

The adiabatic temperature rise method may also be used if the reaction is not very fast, by taking into account corrections for heat losses. The thermal energy balance may be written as (Rojas *et al.*, 1981):

$$c_p dT_{exp}/dt = (-\Delta H)dx/dt - U(T_{exp} - T_0) \tag{5.74}$$

where U is a global heat transfer coefficient per unit mass, T_{exp} is the measured temperature, and T_0 is room temperature.

Integration of Eq. (5.74), for sufficiently long times when $dx/dt \rightarrow 0$, leads to

$$\ln(T_{exp} - T_0) = \ln(T_{exp1} - T_0) - U'(t - t_1) \tag{5.75}$$

where (T_{exp1}, t_1) is the particular set of values at the beginning of the integration and $U' = U/c_p$ is assumed to be constant.

From Eqs (5.65) and (5.74) one can derive the relationship between T_{exp} and T, the latter being the hypothetical value of temperature under adiabatic conditions:

$$dT/dt = dT_{exp}/dt + U'(T_{exp} - T_0) \tag{5.76}$$

Integrating Eq. (5.76), leads to

$$T = T_{exp} + \int_0^t U'(T_{exp} - T_0)dt \tag{5.77}$$

which enables the calculation of the adiabatic curve from experimental information.

As an example, the temperature rise for the formation of a polyurethane by reaction of a polymethylenpolyphenyl isocyanate (average functionality = 2.7), with a polyfunctional polyol based on sorbitol, using dibutyltin dilaurate (DBTDL) as a catalyst, is shown in Fig. 5.19 for two different catalyst concentrations (Marciano *et al.*, 1982).

The comparison of time scales of Figs 5.19 and 5.16, shows that the reaction proceeds at a much lower rate in the example reported in Fig. 5.19. This makes it necessary to correct the experimental curves for heat losses.

Figure 5.20 shows a plot of the experimental temperature decay for run 1, after t = 60 min. An excellent linear regression was obtained, which means that U′ can be regarded as a constant value. The adiabatic temperature rise curves were calculated using Eq. (5.77) (plots are shown in Fig. 5.19). The adiabatic curves are now ready for a kinetic analysis.

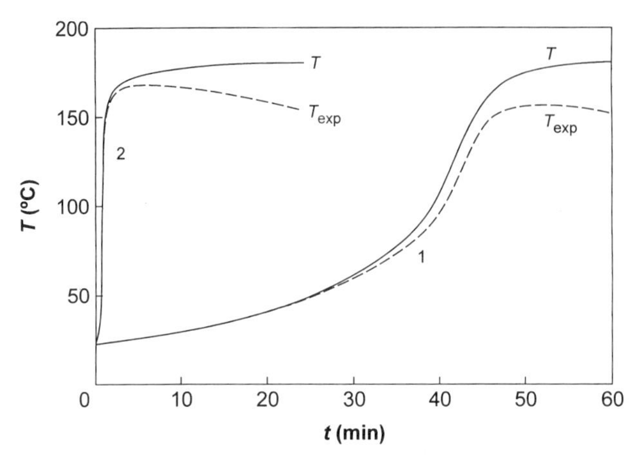

FIGURE 5.19 Experimental and adiabatic temperature rise curves for the synthesis of a polyurethane with two different catalyst amounts. Run 1, dibutyltin dilaurate (DBTDL) = 2.17 mol m^{-3}; run 2, DBTDL = 6.22 mol m^{-3}. (Reprinted from Marciano *et al.*, 1982, Copyright 2001, with permission from Elsevier Science)

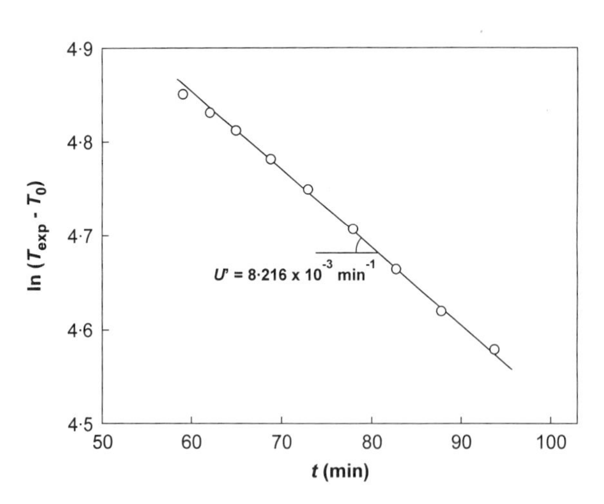

FIGURE 5.20 Experimental temperature decay for run 1 of Fig. 5.19. (Reprinted from Marciano *et al.*, 1982, Copyright 2001, with permission from Elsevier Science)

5.5 KINETIC EQUATIONS AND MASS BALANCES

To obtain the evolution of the concentration of functional groups and temperature, the constitutive equations for the reaction rates must be introduced into mass balances (in fact, balances of functional groups) and in a thermal energy balance. The form of these balances depends on the characteristics of the reactor in which the thermosetting polymerization is taking place (the reactor may be a DSC pan, a heated mold, a flow reactor, etc.).

Let us assume a polymerization that may be described by the evolution of the concentration of three functional groups or active species: $\gamma_i (i = 1\text{–}3)$. If the reaction takes place in an isothermal DSC pan (a batch reactor with uniform values of composition and temperature), the following mass balances may be stated:

$$-d[\gamma_1]/dt = r_1([\gamma_1], [\gamma_2], [\gamma_3], T) \tag{5.78}$$

$$-d[\gamma_2]/dt = r_2([\gamma_1], [\gamma_2], [\gamma_3], T) \tag{5.79}$$

$$-d[\gamma_3]/dt = r_3([\gamma_1], [\gamma_2], [\gamma_3], T) \tag{5.80}$$

where r_1, r_2, and r_3 are the constitutive equations for the reaction kinetics (they may be positive or negative depending on whether $\gamma_1, \gamma_2, \gamma_3$ are consumed or produced in the reaction).

If, instead of isothermal conditions, the cure is performed at a constant heating rate, the constitutive equations for the reaction kinetics are not altered, but a thermal energy balance must be included to describe the system. In this case, this balance may be reduced to

$$dT/dt = q \tag{5.81}$$

If the same polymerization is carried out in a heated mold, concentrations and temperature vary with both position and time. For a one-dimensional system (for example, when variations along the thickness are only relevant), mass and thermal energy balances are written as

$$-\partial[\gamma_1]/\partial t = r_1([\gamma_1], [\gamma_2], [\gamma_3], T) \tag{5.82}$$

$$-\partial[\gamma_2]/\partial t = r_2([\gamma_1], [\gamma_2], [\gamma_3], T) \tag{5.83}$$

$$-\partial[\gamma_3]/\partial t = r_3([\gamma_1], [\gamma_2], [\gamma_3], T) \tag{5.84}$$

$$\rho c_p \partial T/\partial t = \partial/\partial z(k_T \partial T/\partial z) + \Sigma\ r_i(-\Delta H_M)_i \tag{5.85}$$

where ρ is the mass density, k_T the thermal conductivity, and $(-\Delta H_M)_i$ the total reaction heat per mol in reaction i (i = 1, 2, 3).

The system of Eqs (5.82)–(5.85), with the necessary initial and boundary conditions, may be numerically solved (Chapter 9).

NOTATION

a_1, a_2, a_3	= primary, secondary and tertiary amine groups
A, A_1, A_2	= pre-exponential factor of the specific rate constant, s^{-1} or $(m^3)^{n-1}$ $(kmol)^{1-n}$ s^{-1}
A_d	= pre-exponential factor for the diffusionally controlled polymerization rate, s^{-1}
$[A_i]$	= concentration of a diamine with i reacted amine hydrogens, kmol m^{-3}
A_p	= pre-exponential factor of the specific rate constant for the propagation step, m^3 $kmol^{-1}$ s^{-1}
a_T	= shift factor
c^*	= adjustable constant in the equation proposed by Chern and Poehlein (1987)
c_0	= initial concentration of a particular species or a functional group, kmol m^{-3}
c_1	= constant in the WLF equation
c_2	= constant in the WLF equation, K
c_p	= heat capacity per unit mass, kJ kg^{-1} K^{-1}
e	= epoxy group
[E]	= concentration of a monoepoxide, kmol m^{-3}
E, E_1, E_2, E_{-1}	= activation energy, kJ mol^{-1}
E_d	= activation energy for the diffusionally controlled polymerization rate, kJ mol^{-1}
E_p	= activation energy of the propagation step, kJ mol^{-1}
f_n	= efficiency of the decomposition of initiator I_n
I	= inactive state of an initiator
I*	= active state of an initiator
I_n	= generic initiator of a free-radical polymerization
$k, k', k_1, k_2, k'_1, k'_2, k_{-1}$	= specific rate constant, $(m^3)^{n-1}$ $(kmol)^{1-n}$ s^{-1}
K	= equilibrium constant
K_1, K_2, K'_1, K'_2	= specific rate constant, s^{-1}
k_c	= specific rate constant for the chemically controlled polymerization rate, s^{-1}
k_d	= specific rate constant for the diffusionally controlled polymerization rate, s^{-1}
k_{dn}	= specific rate constant for the decomposition of initiator I_n, s^{-1}
k_o	= overall rate constant for the polymerization rate, s^{-1}

k_p = specific rate constant for the propagation step, $m^3\ kmol^{-1}\ s^{-1}$
k_T = thermal conductivity, $kJ\ m^{-1}\ s^{-1}\ K^{-1}$
m = reaction order
M = monomer
MR = molar ratio of styrene to polyester C=C double bonds
n = reaction order
q = rate of temperature variation, $K\ s^{-1}$
r_i = rate of consumption (production) of a generic species i, $kmol\ m^{-3}\ s^{-1}$
R = gas constant = 8.314 $kJ\ kmol^{-1}\ K^{-1}$
R. = free radical
R_{21} = reactivity ratio of secondary to primary amine hydrogens
t = time, s
T = temperature, K
T_g = glass transition temperature, K
T_{max} = maximum temperature attained in an adiabatic process, K
T_r = reference temperature, K
U = global heat transfer coefficient per unit mass, $kJ\ kg^{-1}\ s^{-1}$
U′ = global heat transfer coefficient, U/cp, s^{-1}
x = conversion
x_1, x_2 = conversion of particular functional groups
x_c = arbitrary conversion defined in the equation proposed by Chern and Poehlein (1987)
x_f = maximum attainable conversion under particular cure conditions
z = coordinate in space, m

Greek Letters

α_1 = dimensionless concentration of primary amine groups, $[a_1]/[e]_0$
α_2 = dimensionless concentration of secondary amine groups, $[a_2]/[e]_0$
β = relative concentration of the active state of an initiator, I^*/I_0
$(-\Delta H)$ = total reaction heat per unit mass, $kJ\ kg^{-1}$
$(-\Delta H_M)$ = total reaction heat per mol, $kJ\ mol^{-1}$
ΔT_{ad} = adiabatic temperature rise, $(-\Delta H)/c_p$, K
γ_i = generic functional group
ρ = mass density, $kg\ m^{-3}$
τ = characteristic relaxation time of the polymer structure, s

Subscripts

0 = initial or external value
eq = equilibrium value
exp = experimental value

REFERENCES

Anseth KS, Bowman CN, *Polym. React. Eng.*, 1, 499–520 (1992/93).
Anseth KS, Decker C, Bowman CN, *Macromolecules*, 28, 4040–4043 (1995).
Aranguren MI, Williams RJJ, *Polymer*, 27, 425–430 (1986).
Bowman CN, Peppas NA, *Macromolecules*, 24, 1914–1920 (1991).
Brown IM, Sandreczki TC, *Macromolecules*, 23, 94–100 (1990).
Chern CS, Poehlein GW, *Polym. Eng. Sci.*, 27, 788–795 (1987).
Chiao L, *Macromolecules*, 23, 1286–1290 (1990).
Cole KC, *Macromolecules*, 24, 3093–3097 (1991).
Cole KC, Hechler JJ, Noël D, *Macromolecules*, 24, 3098–3110 (1991).
Deng Y, Martin GC, *Polymer*, 37, 3593–3601 (1996).
Flammersheim HJ, *Thermochim. Acta*, 296, 155–159 (1997).
Flammersheim HJ, *Thermochim. Acta*, 310, 153–159 (1998).
Galante MJ, Oyanguren PA, Andromaque K, Frontini PM, Williams RJJ, *Polym. Int.*, 48, 642–648 (1999).
Horie K, Hiura H, Sawada M, Mita I, Kambe H, *J. Polym. Sci.* A1, 8, 1357–1372 (1970).
Hsu CP, Lee LJ, *Polymer*, 32, 2263–2271 (1991).
Kamal MR, *Polym. Eng. Sci.*, 14, 231–239 (1974).
Kurdikar DL, Peppas NA, *Macromolecules*, 27, 4084–4092 (1994).
Leukel J, Burchard W, Krüger RP, Much H, Schulz G, *Macromol. Rapid Commun.*, 17, 359–366 (1996).
López-Quintela A, Prendes P, Pazos-Pellín M, Paz M, Paz-Abuín S, *Macromolecules*, 31, 4770–4776 (1998).
McCallum JP, Tanner J, *Nature*, 225, 1127–1128 (1970).
Marciano JH, Rojas AJ, Williams RJJ, *Polymer*, 23, 1489–1492 (1982).
Matejka L, Lövy J, Pokorny S, Bonchal K, Dušek, K, *J. Polym. Sci. Polym. Chem.Ed.*, 21, 2873–2885 (1983).
Mauri AN, Galego N, Riccardi CC, Williams RJJ, *Macromolecules*, 30, 1616–1620 (1997).
Mijovic J. Wijaya J, *Macromolecules*, 27, 7589–7600 (1994).
Min BG, Stachurski ZH, Hodgkin JH, Heath GR, *Polymer*, 34, 3620–3627 (1993).
Montserrat S, Málek J, *Thermochim. Acta*, 228, 47–60 (1993).
Ng H, Manas-Zloczower I, *Polym. Eng. Sci.*, 29, 1097–1102 (1989).
O'Neil GA, Torkelson, JM, *TRIP*, 5, 349–355 (1997).
Paz-Abuín S, López-Quintela A, Varela M, Pazos-Pellín M, Prendes P, *Polymer*, 38, 3117–3120 (1997).

Paz-Abuín S, López-Quintela A, Pazos-Pellín M, Varela M, Prendes P, *J. Polym. Sci. A: Polym. Chem.*, 36, 1001–1016 (1998).
Peyser P, Bascom WD, *J. Appl. Polym. Sci.*, 21, 2359–2373 (1977).
Rabinowitch E, *Trans. Faraday. Soc.*, 33, 1225–1233 (1937).
Riccardi CC, Williams RJJ, *J. Appl. Polym. Sci.*, 32, 3445–3456 (1986).
Riccardi CC, Dupuy J, Williams RJJ, *J. Polym. Sci. B: Polym. Phys.*, 37, 2799–2805 (1999).
Rojas AJ, Borrajo J, Williams RJJ, *Polym. Eng. Sci.*, 21, 1122–1127 (1981).
Russell GT, Gilbert RG, Napper DH, *Macromolecules*, 21, 2133–2140 (1988a).
Russell GT, Gilbert RG, Napper DH, *Macromolecules*, 21, 2141–2148 (1988b).
Scott RA, Peppas NA, *Macromolecules*, 32, 6149–6158 (1999).
Sesták J, J. *Thermal Anal.*, 16, 503–520 (1979).
Simon SL, Gillham JK, *J. Appl. Polym. Sci.*, 47, 461–485 (1993).
Smith IT, *Polymer*, 2, 95–108 (1961).
Sourour S, Kamal MR, *Thermochim. Acta*, 14, 41–59 (1976).
Sun X, Chiu YY, Lee LJ, *Ind. Eng. Chem. Res.*, 36, 1343–1351 (1997).
Tollens FR, Lee LJ, *Polymer*, 34, 29–37 (1993).
Verchère D, Sautereau H, Pascault JP, Riccardi CC, Moschiar SM, Williams RJJ, *Macromolecules*, 23, 725–731 (1990).
Williams ML, Landel RF, Ferry JD, *J. Am. Chem. Soc.*, 77, 3701–3707 (1955).
Wisanrakkit G, Gillham JK, *J. Appl. Polym. Sci.*, 41, 2885–2929 (1990).
Yang YS, Lee LJ, *Polymer*, 29, 1793–1800 (1988).

FURTHER READING

Dušek K, *Polym. Gels Networks*, 4, 383–404 (1996).
Prime RB, In *Thermal Characterization of Polymeric Materials*, 2nd edn, Turi EA (ed.), Academic Press, New York, 1997, Vol. 2, pp. 1380–1766.

6

Rheological and Dielectric Monitoring of Network Formation

6.1 INTRODUCTION

Two main transformations can take place during the formation of a polymer network: gelation and vitrification. Gelation corresponds to the incipient formation of an infinite network, while vitrification involves the transformation from a liquid or rubbery state to a glassy state (Chapter 4).

Gelation is characterized by the divergence of the mass-average molar mass, M_w, and the radius of gyration, and by the formation of an insoluble gel. Vitrification occurs when the increasing glass transition temperature, T_g, becomes equal to the reaction temperature. Below $T_{g,gel}$, the temperature at which the time to gel is the same as the time to vitrify, the reactive system vitrifies in an ungelled (liquid) state. Although conversion may increase at a very slow rate during storage, the polymer can be processed as long as it remains in the ungelled state. The thermosetting polymer will not vitrify during an isothermal cure at a reaction temperature higher than T_{g_∞}, the glass transition temperature of the fully cured material (Chapter 4).

The experimental determination of gelation and vitrification is very important for the design of cure cycles (Chapters 4 and 9) and to control the morphology of inhomogeneous polymer networks (Chapter 7) and modified-thermosetting polymers that undergo a phase-separation process during cure (Chapter 8). The rheological and dielectric monitoring of these transformations is analyzed in this chapter.

6.2 EQUILIBRIUM MECHANICAL MEASUREMENTS

A typical evolution of equilibrium mechanical properties during reaction is shown in Fig. 6.1. The initial reactive system has a steady shear viscosity that grows with reaction time as the mass-average molar mass, M_w, increases and it reaches to infinity at the gel point. Elastic properties, characterized by nonzero values of the equilibrium modulus, appear beyond the gel point. These quantities describe only either the liquid (pregel) or the solid (postgel) state of the material. Determination of the gel point requires extrapolation of viscosity to infinity or of the equilibrium modulus to zero.

Accurate measurement of the equilibrium modulus is extremely difficult because its value remains below the detection limit for a considerable time and, theoretically, an infinite time is required to perform the measurement.

Steady shear viscosity measurements are very simple and are often used in practice. Very often a viscosity of 10^3 Pa s is arbitrarily identified with the gel point. But the determined gelation time, t_{gel}, depends on the shear rate, and extrapolation to zero shear rate meets the following difficulties:

1. t_{gel} may depend on the shear rate due to shear thinning at high rates.
2. The network structure near the gel point is very fragile and can be broken by the shear flow experiment.

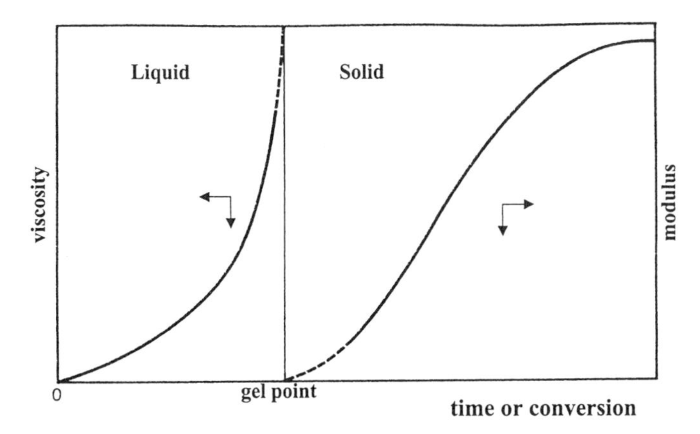

FIGURE 6.1 Schematic evolution of steady-state mechanical properties of a thermoset as a function of reaction time or conversion. Representative properties are the steady shear viscosity for the liquid state and the equilibrium modulus for the solid state.

3. Infinite viscosity is not an unambiguous indicator of gelation: it can equally be caused by vitrification. This means that complementary methods, such as sol fraction measurements, are necessary to distinguish between both phenomena.

For all these reasons, steady-state mechanical measurements, even if they are very simple and very often used in practice, lead to an apparent gel point.

6.3 DYNAMIC MECHANICAL MEASUREMENTS

6.3.1 Search for an Experimental Criterion for Gelation

Dynamic mechanical measurements describe both the liquid and solid states. A freely oscillating torsion pendulum can be used to provide shear moduli data of solid specimens versus temperature or time. A composite specimen made by impregnating a glass-fiber braid with liquid thermoset precursors is used to study liquid systems that change to solids (Babayevsky and Gillham, 1973). Two maxima of mechanical damping are observed during reaction and assigned to gelation and vitrification of the material (Fig. 6.2, Enns and Gillham, 1983). As the measurement is performed at the resonant frequency of the pendulum, this technique is very sensitive. But one big disadvantage is that experiments are not performed at controlled frequencies.

The evolution of the dynamic viscosity η^* (ω, x) or of the dynamic shear complex modulus G^* (ω,x) as a function of conversion, x, can be followed by dynamic mechanical measurements using oscillatory shear deformation between two parallel plates at constant angular frequency, $\omega = 2\pi f$ (f = frequency in Hz). In addition, the frequency sweep at certain time intervals during a slow reaction (x $\sim$ constant) allows determination of the frequency dependence of elastic quantities at the particular conversion. During such experiments, storage $G'(\omega)$, and loss $G''(\omega)$ shear moduli and their ratio, the loss factor $\tan\delta(\omega)$, are obtained:

$$G^*(\omega) = G'(\omega) + iG''(\omega) \tag{6.1}$$

$$\tan\delta = G''/G' \tag{6.2}$$

Typical rheological curves obtained during a diepoxy–diamine reaction are shown in Fig. 6.3. The cure temperature ($T_i = 90°C$) is well above the glass transition temperature of the fully cured network ($T_{g_\infty} \sim 35°C$), which means that only gelation occurs. Three typical regions are observed during cure (Matejka, 1991).

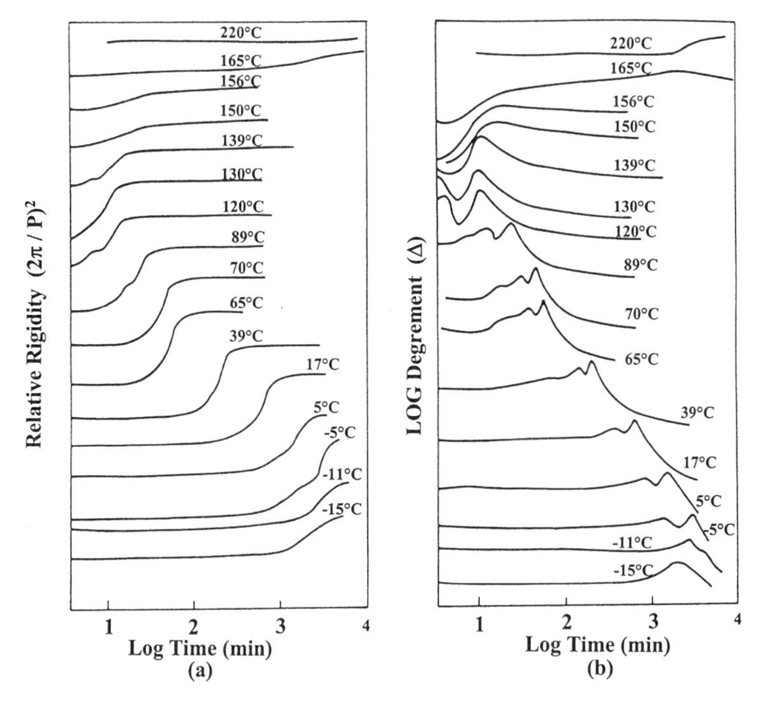

FIGURE 6.2 Torsional braid analysis (TBA) during isothermal reaction of a diepoxy, DGEBA, with a diamine, bis(p-aminocyclohexyl) methane, from −15°C to 220°C: (a) relative rigidity; (b) logarithmic decrement (T_{g0} = −19°C, $T_{g\infty}$ = 170°C, and $T_{g,gel}$ = 49°C). (Enns and Gillham, 1983 – Copyright 2001 – Reprinted by permission of John Wiley & Sons, Inc.)

(a) The pregel region is characterized by an increase in the loss modulus, G″, corresponding to the increase of the real part of dynamic viscosity $\eta'(= G''\omega)$, due to the increasing molar mass of the thermosetting polymer. The storage modulus, G′, is very low and tends to zero at low frequencies. In this region the loss modulus (G″) is higher than the elastic modulus (G′), and the loss factor, $\tan\delta > 1$.

(b) The "critical region" begins with a sudden increase in the storage modulus, G′, by several orders of magnitude. At the intersection of the G′(t) and G″(t) curves, $\tan\delta = 1$. After the intersection point, G′ becomes higher and tan δ becomes less than 1.

The viscous properties are dominant in the liquid state, i.e., $G'' > G'$ and $\tan\delta > 1$, while the elastic properties predominate in the solid state, where $G' > G''$ and $\tan\delta < 1$. For this reason, the G′–G″ crossover ($\tan\delta = 1$) was firstly identified as the gel point (Tung and Dynes, 1982). The

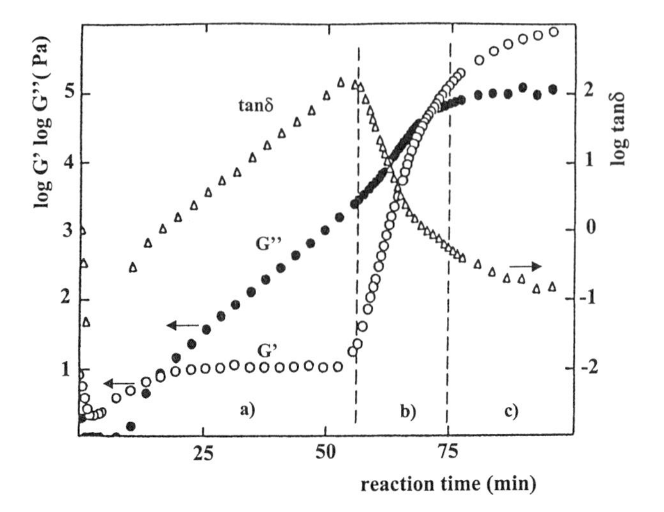

FIGURE 6.3 Evolution of the storage modulus, G′(○), loss modulus, G″(●), and loss factor, tan δ(Δ), during the reaction of pure diglycydyl ether of bisphenol A (DGEBA) with poly(oxypropylene) diamine (PPO). $M_n = 370$ g/mol at $T_i = 90°C$, [NH] / [epoxy] = 2, and f = 10 Hz. a) Pregel region, b) "critical region", c) postgel region. (Matejka, 1991 – Copyright 2001 – Reprinted by permission of Springer-Verlag)

problem is that when a reaction like the one represented in Fig. 6.3 is followed at different angular frequencies, it is found that the reaction time to reach tan $\delta = 1$ increases with angular frequency. As the gel point is a material constant and should not depend on experimental conditions, the crossover point between $G'(t)$ and $G''(t)$ cannot correspond to gelation.

The value of tan δ decreases in the (b) region and the rate of this decrease depends on the angular frequency, ω (Fig. 6.4). As a rough approximation:

1. As $G' \sim \omega^2$ and $G'' \sim \omega$, tan $\delta \sim 1/\omega$ in the Newtonian liquid state.
2. In the solid state $G' \sim$ constant and $G'' \sim \omega$, thus, tan $\delta \sim \omega$ (Ferry, 1980).

Therefore, the drop of tan δ during the reaction is steeper at low angular frequencies. Figure 6.4 reveals that for a particular tan δ value higher than 1 (tan $\delta \sim 2$ in the case of Fig. 6.4), the time to reach the value is independent of frequencies in the 0.1–50 Hz range. This crossover of the tan δ curves at various frequencies can be used as a criterium for the identification of the gel point.

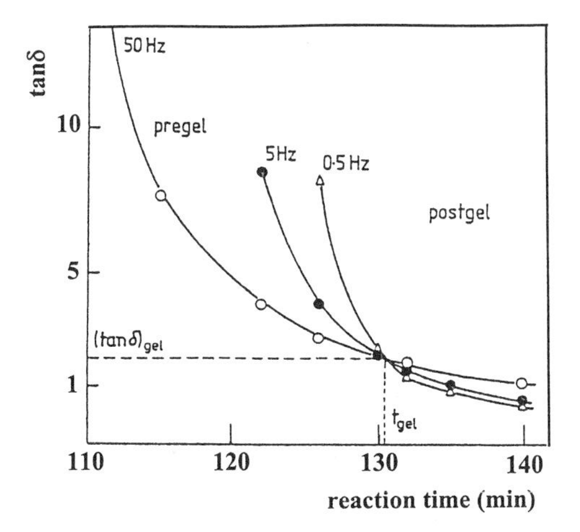

FIGURE 6.4 Decrease in the loss factor (tan δ) during cure, for the same epoxy–diamine system as that represented in Fig. 6.3, at different frequencies of the dynamic measurements. $T_i = 70°C$. (Matejka, 1991 – Copyright 2001 – Reprinted by permission of Springer-Verlag)

(c) Finally, in the postgel region, a slow increase in G', that levels off in the final stages of the reaction is observed; $\tan\delta < 1$ for the fully cured rubbery network ($T_i > T_{g\infty}$). Additional experiments show that the stoichiometric mixture has the highest final modulus and the lowest final loss factor (tan δ) because it forms the most perfect network, with the highest crosslink density.

6.3.2 Experimental Evidence of Singular Power Laws at the Gel Point

The frequency dependence of dynamic mechanical results is of primary importance for the interpretation of data. The dependence of G' and G'' versus ω can be evaluated from experiments. Chambon and Winter (1985) first revealed that the $G'(\omega)$ and $G''(\omega)$ curves, in logarithmic scales, were parallel over a wide range of angular frequencies at the gel point. The validity of a power law

$$G'(\omega) \sim G''(\omega) \sim \omega^{\Delta} \tag{6.3}$$

over the entire frequency range was assumed to be an inherent property of the gel state. An example is given in Fig. 6.5. The critical exponent in this case is $\Delta = 0.72 \pm 0.02$. Before and after the gel time, the storage modulus $G'(\omega)$ decreases rapidly to zero or shows a rubbery plateau at a low angular frequency, respectively.

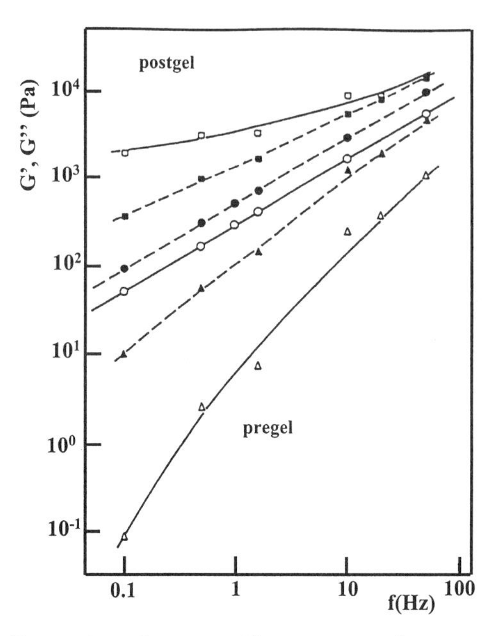

Figure 6.5 Storage G' and loss G'' moduli as a function of frequency, at different cure times, for the same epoxy–diamine system as that represented in Figs 6.3 and 6.4. $T_i = 90°C$. The parameter is the reaction distance from the gel point. ■□ $t = t_{gel} + 5$ min; ○● $t = t_{gel}$; △ ▲ $t = t_{gel} - 5$ min; —— G'; - - - - G''. (Matejka, 1991 – Copyright 2001 – Reprinted by permission of Springer-Verlag)

Similar evolution of the frequency dependence of tan δ and the complex viscosity, η*, during the reaction near the gel point can also be obtained.

6.3.3 Scaling Laws and Gel Point Determination: a Search for Exponents

Many authors have studied the rheological behavior of chemical gels theoretically and experimentally. From the original mean-field theories of Flory and Stockmayer (Flory, 1953) describing the gelation phenomena (Chapter 3), current emphasis has shifted to the utilization of the fractal geometry concepts and the connectivity transition model of percolation (De Gennes, 1979; Stauffer et al., 1982). Under the percolation model, predictions for numerous physical properties are made by assimilating the thermosetting polymer as a polydisperse self-similar distribution of clusters (fractals), which grow from the beginning of the reaction through the gel point. An infinite cluster, with a size and a mass that diverges, appears at the gel point. Above gelation, clusters connect to the infinite cluster and, as a consequence, a strengthening of the network is produced. As the reaction proceeds, finite clusters with a decreasing average size are joined to the infinite cluster, and the process continues to the end of the reaction.

Both the Flory–Stockmayer mean-field theory and the percolation model provide scaling relations for the divergence of static properties of the polymer species at the gelation threshold.

The correlation length in the thermosetting polymer is defined by the z-average cluster radius, which scales with conversion x as

$$R \sim \left[\frac{|x - x_{gel}|}{x_{gel}}\right]^{-\nu} \qquad \text{for } x < x_{gel} \tag{6.4}$$

The mass-average molar mass of clusters in solution diverges as

$$M_w \sim \left[\frac{|x - x_{gel}|}{x_{gel}}\right]^{-\gamma} \qquad \text{for } x < x_{gel} \tag{6.5}$$

For the Flory–Stockmayer theory the predicted exponents are $\gamma = 1$ and $\nu = \frac{1}{2}$, while for percolation they are predicted as 1.76 and 0.88, respectively (Stauffer *et al.*, 1982). Percolation theory has been quite successful in predicting the static properties. For example, R_z and M_w can be determined by static light scattering, and reported values for a polyurethane are $\gamma = 1.65 \pm 0.1$ and $\nu = 0.86 \pm 0.1$ (Adam *et al.*, 1987).

Scaling predictions for the steady-state viscoelastic properties have also been developed. The growth of the equilibrium modulus after the gel point can be described as a function of x by

$$G_0 \sim \left[\frac{|x - x_{gel}|}{x_{gel}}\right]^{u} \qquad \text{for } x > x_{gel} \tag{6.6}$$

and, similarly, the zero shear-rate viscosity as

$$\eta_0 \sim \left[\frac{|x - x_{gel}|}{x_{gel}}\right]^{-s} \qquad \text{for } x < x_{gel} \tag{6.7}$$

But it is difficult to apply zero shear predictions to measurements that have been performed at low, but nonzero, shear rates. Neither s nor u can be decisively determined at a fixed frequency near the gel point: the only way to truly obtain s and u exponents from experiments is through the use of a theory capable of predicting the entire frequency dependence of the viscoelastic response.

Consideration of the dynamics near the gel point led to predictions for the frequency dependence of the $G'(\omega)$ and $G''(\omega)$ moduli. At the gel point, the behavior given by Eq. 6.3 – $G' \sim G'' \sim \omega^{\Delta}$ – is predicted (Hess et al., 1988; Martin et al., 1989; Rubinstein et al., 1989; Hodgson and Amis, 1990).

Calculations based on a Rouse-like dynamics (Ferry, 1980), applied to the percolation clusters, give

$$\Delta = \frac{u}{s+u} \tag{6.8}$$

This model leads to $\Delta = 0.67$ at the gel point, using the zero-frequency values for s and u. Use of the values for s and u calculated by treating the gelation phenomena as a three-dimensional percolation model of a supraconductor/resistor network (electrical analogy), gives $\Delta = 0.72 \pm 0.02$.

The use of the Rouse model is questionable (Martin *et al.*, 1989). A variety of predictions have been made for the critical exponents, but such a discussion is beyond the scope of this book.

It can be concluded that when thermosetting polymers are cured at a temperature T_i above $T_{g\infty} + 50°C$ (as the results presented in Figs 6.4 and 6.5), the polymeric material can be described by the percolation theory, with chains obeying the Rouse model (Eloundou *et al.*, 1996a).

An explanation for the lower experimental critical exponent values is that the experimental gel times, t_{gel}, may be affected by the viscoelastic behavior of the polymer. The largest relaxation time of the polymer and the width of the distribution of relaxation times increase with increasing conversion and diverge at the gel point (Chambon and Winter, 1987; Winter, 1987). According to the width of the relaxation time spectrum, the relaxation times should be of the same order of magnitude as the observation times. In conclusion, the relative magnitude of the polymer relaxation time τ and the characteristic time of the mechanical test, $t_{exp} \sim 1/\omega$, are the crucial factors governing the experimental dynamic results. It seems that this molecular mobility effect is more important in the case of polyurethane chemistry than in the case of epoxy-diamine systems (Izuka *et al.*, 1994; Prochazka *et al.*, 1996; Nicolai *et al.*, 1997). An example of results obtained during polyurethane synthesis is given in Fig. 6.6.

6.3.4 Determination of the Second Phenomenon: Vitrification

Thermosetting polymers represented in Figs 6.4, 6.5, and also 6.6 were cured at $T_i > T_{g\infty}$. In many cases and particularly for high-T_g networks, a precure step is performed at $T_i < T_{g\infty}$. In this case the two transformations, gelation and vitrification, occur and the rheology in the vicinity of the critical gel point could be affected by vitrification.

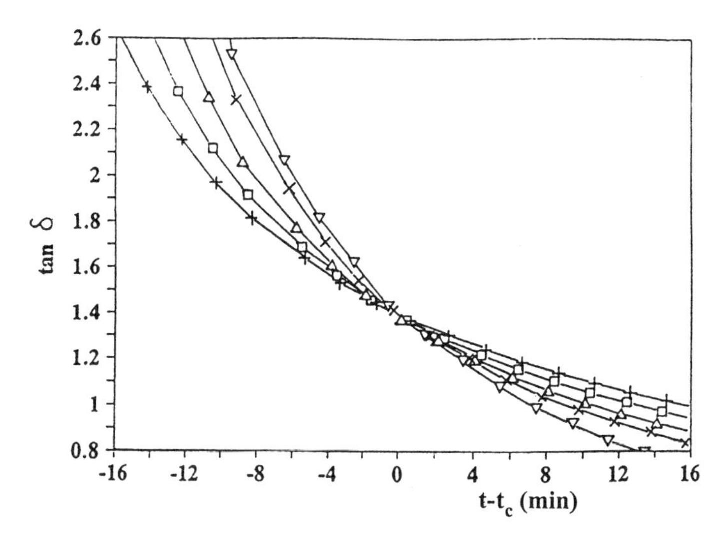

FIGURE 6.6 Evolution of tan δ during polyurethane synthesis at 110°C, at different angular frequencies, ω (s^{-1}) = 1 (▽), 3.162 (X), 10(△), 31.62 (□), and 100 (+). A polycaprolactone diol, M_n = 700 g mol^{-1} was stoichiometrically reacted in bulk with a triisocyanate (the trimer of isophorone diisocyanate). The time t_c at which tan δ is independent of frequency determines the gel point. The critical gel exhibits values of tan $\delta = 1.4$ and $\Delta = 0.61$. (Reprinted with permission from Izuka *et al.*, 1994. Copyright 2001. American Chemical Society)

Figure 6.7 gives the experimental curves of the loss factor (tan δ) at different angular frequencies, as a function of reaction time at a cure temperature, $T_i = 150$°C, for a reactive epoxy system exhibiting a $T_{g_\infty} = 177$°C (Eloundou *et al.*, 1996b). In this case, $T_i \sim T_{g_\infty} -30$°C. The vitrification phenomenon is revealed by a peak at t ~ 350 min, which is frequency dependent. In these conditions, gelation occurs at about 200 min, as revealed by the crossover of tan δ versus time curves, recorded at various frequencies. A value of $\Delta = 0.69$ was found.

As T_i decreases, the viscoelastic characterization of the gelation process becomes more and more disturbed by the vitrification phenomenon, and the scaling laws are no longer well verified. At $T_i = 80$°C, which is close to $T_{g_\infty} - 100$°C or ~ $T_{g,gel} + 30$°C, vitrification and gelation are in the same range of reaction times and the highest frequency curve does not participate in the crossover (Fig. 6.8).

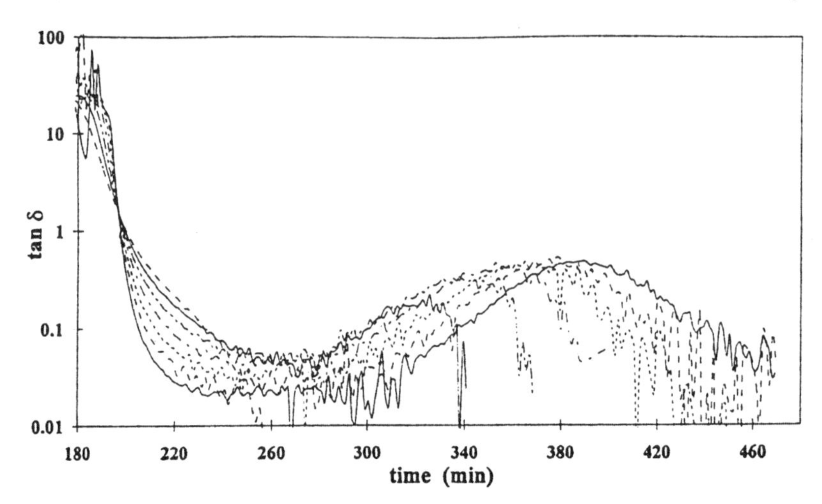

Figure 6.7 Evolution of tan δ during an epoxy network synthesis at 150°C, for various angular frequencies, ω (from 1 to 100 s^{-1}). The diepoxy, DGEBA, was stoichiometrically reacted with 4,4′-methylene bis [3-chloro-2,6-diethylaniline], MCDEA. The characteristics of this system are $T_{g,gel} = 50°C$ and $T_{g\infty} = 177°C$. (Reprinted with permission from Eloundou *et al.*, 1996b. Copyright 2001. American Chemical Society)

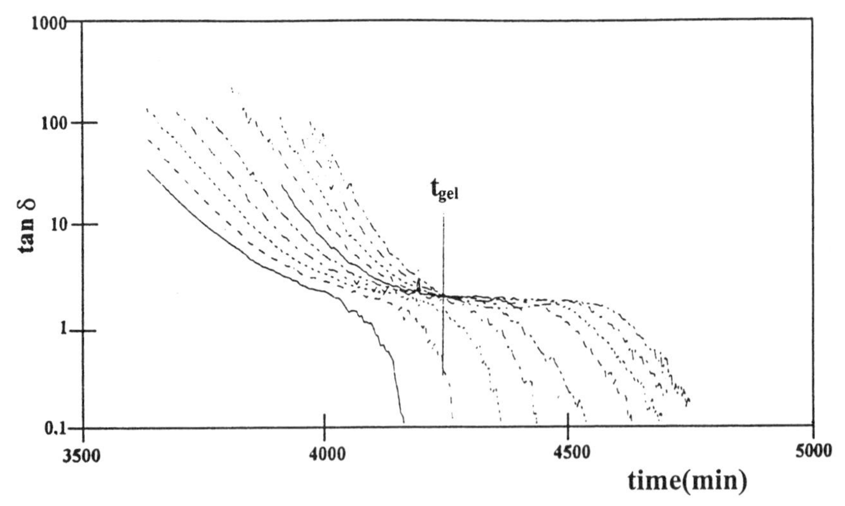

Figure 6.8 Evolution of tan δ during an epoxy network synthesis at 80°C, for various angular frequencies ω (from 1 to 100 s^{-1}). The epoxy system is the same as that in Fig. 6.7). (Reprinted with permission from Eloundou *et al.*, 1996b. Copyright 2001. American Chemical Society)

6.4 DIELECTRIC MONITORING

6.4.1 Dielectric Parameters and Reaction

When applying an alternating electric field to a polymer placed between two electrodes, the response is generally attenuated and the output current is out of phase compared with the input voltage. This response stems from the polymer's capacitive component and its conductive or loss component, as represented by a complex dielectric permittivity measured frequencies f, and temperatures T:

$$\varepsilon^* = \varepsilon' - i\varepsilon'' \tag{6.9}$$

where $\varepsilon'(\omega)$ is the equivalent capacitance of the material and $\varepsilon''(\omega)$ represents its equivalent conductance, both defined as relative values with respect to values of the air replaced by the sensor ($\omega = 2\pi f$).

The applied electric field perturbs the orientational distribution function of the dipolar molecules. Dielectric relaxation due to classical molecular reorientational motions is a form of pure absorption spectroscopy whose frequency range of interest for materials, including polymers, is between 10^{-6} and 10^{11} Hz.

Values of ε' and ε'' are calculated using equations in which the contributions of dipolar effects, ionic displacements, and electrode polarization effects are additive. If conduction effects predominate, i.e., when neither interfacial effects nor dipolar effects are significant, the loss factor is given by (Kranbuehl *et al.*, 1986):

$$\varepsilon'' = \frac{\sigma}{\varepsilon_0 \omega} \tag{6.10}$$

where σ is the ionic conductivity and $\varepsilon_0 = 8.854 \times 10^{-14} C^2 J^{-1} cm^{-1}$. Generally, the magnitude of the low-frequency overlapping values of $\omega\varepsilon''(\omega)$ can be used to measure the change with time of the ionic mobility through the parameter σ.

For a thermosetting polymer, ε' and ε'' can be recorded during reaction at frequencies between 1 and 10^5 Hz. The behavior can be very complex because of the changes in the chemical structure, including ions and polar groups.

Many dielectric studies of epoxy–amine systems have been reported. It appears that the change in dielectric properties during cure may be due to three main contributions:

1. Decrease of σ if ionic conductivity is controlled by the diffusion constant of the impurities already present, which decreases with an increase of the viscosity of the medium.

2. Orientation and oscillation of permanent dipoles in the electric field.
3. Decrease of mean dipolar moment per unit volume due to the increase of the molecular sizes. The term "mean dipolar moment" reminds us that the nature of dipoles is modified by chemical reactions.

Figure 6.9 represents the variations of the product $\varepsilon''\varepsilon_0\omega$ versus reaction time at several frequencies for two epoxy–amine systems. The $\varepsilon''\varepsilon_0\omega$

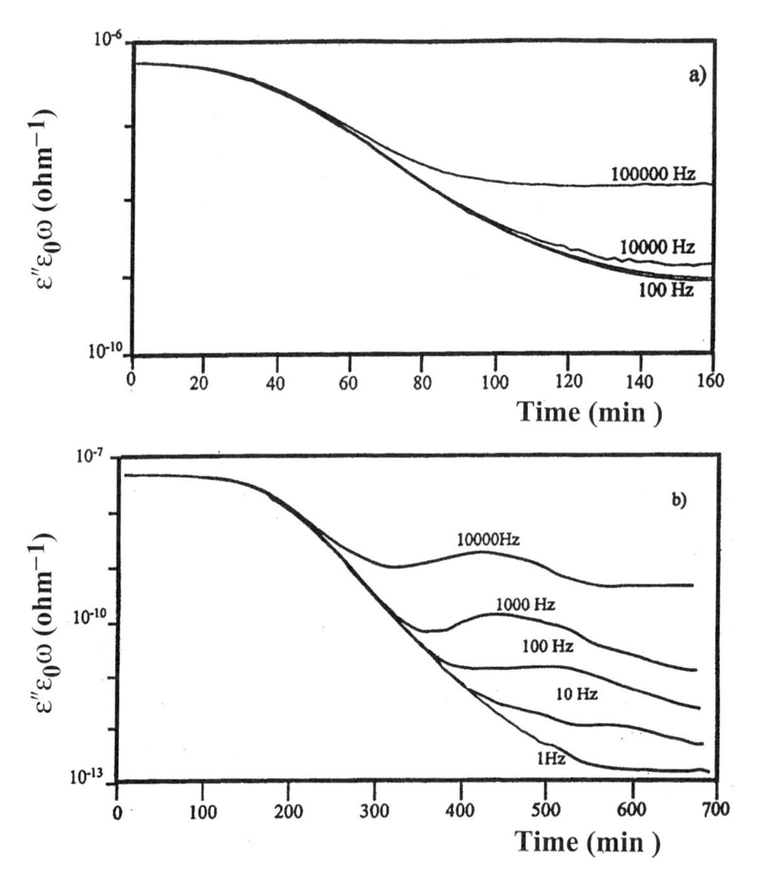

FIGURE 6.9 Experimental curves of the product $\varepsilon''\varepsilon_0\,\omega$ as a function of reaction time, t, for two epoxy systems: (a) the first system is a stoichiometric solution of diglycidyl ether of butanediol with 4,9-dioxadecane-1,12 diamine, at $T_i = 60°C \sim T_{g\infty} + 70°C$; (b) the second system is the same as that for Fig. 6.7, at $T_i = 135°C \sim T_{g\infty} - 40°C$. (Eloundou *et al.*, 1998a. Copyright 2001. Reprinted by permission of Wiley-VCH)

data for all frequencies lie on a single curve at the beginning of the cure. This confirms that the ionic conductivity governs the dielectric behavior during the initial reaction period.

For the reactive system cured at a temperature, T_i higher than its maximum glass transition temperature T_{g_∞} ($= -12^\circ$C), $\varepsilon''\varepsilon_0\omega$ decreases to reach plateau values that are frequency dependent (Fig. 6.9a). As there is no vitrification, the plateau observed corresponds to the behavior of a network in a rubbery state.

For the other reactive system for which T_i is lower than its T_{g_∞} ($= 177^\circ$C), the behavior is initially dominated by ionic conductivity and, as the reaction progresses, curves exhibit peaks due to vitrification (Fig. 6.9b). The higher the frequency, the lower the time at which this relaxation peak appears.

From experiments made at several frequencies, the conductivity curve is built up by superimposing the curves $\varepsilon''\varepsilon_0\omega$ versus time in the regions where they are independent of frequency (Kranbuehl *et al*, 1986; Wasylyshyn and Johari, 1997). Figure 6.10 displays the obtained conductivity, σ, versus reaction time for both epoxy systems (a) and (b). Cure temperatures, T_i, are higher than T_{g_∞} for case (a) and lower than T_{g_∞} for case (b). The initial conductivity, σ_0, increases with temperature. Conductivities are higher for system (a) than for system (b) and vary over 3 and 6 decades, respectively, during reaction.

6.4.2 Can the Gelation Phenomenon be Observed from Dielectric Measurements?

The determination of the critical gel point is not as evident as the determination of vitrification. It is still a subject of controversy.

Experimental $\varepsilon''(t)$ curves exhibit an inflection point that corresponds to the maximum rate of change of conductivity, associated to the maximum decrease of mobility. For this reason this inflection point has been attributed to gelation (Wasylyshyn and Johari, 1997). In some cases good agreement between the inflection time and the gel time obtained by rheological methods was observed. A very complex explanation taking into account intermolecular H bonds, protonic conduction, and a bond percolation law has been proposed to explain this mobility change (Johari and Wasylyshyn, 2000).

When comparing different experiments performed at the same isothermal temperature, one major difficulty is to be able to repeat the same thermal history. One guarantee is to use the same oven and the same sample size for all the experiments. When this precaution is respected, the conclusions may be quite different (Eloundou, 1998b):

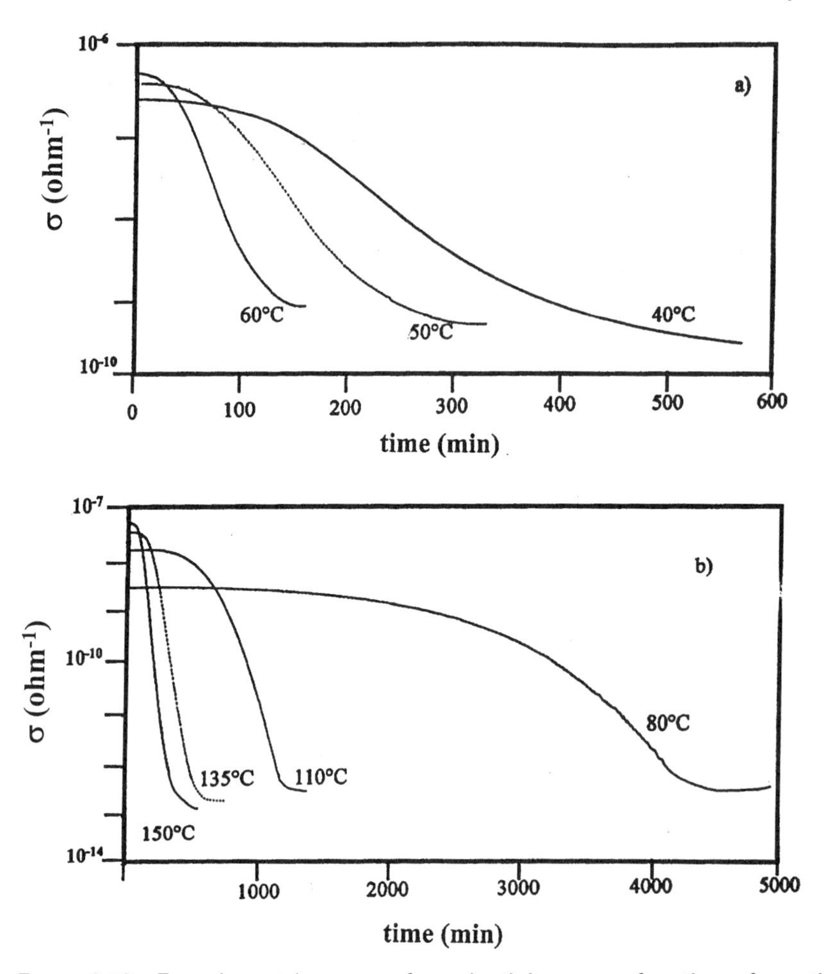

FIGURE 6.10 Experimental curves of conductivity, σ, as function of reaction time, t, at different cure temperatures, T_i: (a) same epoxy system as that in Fig. 6.9a and (b) same epoxy system as that in Fig. 6.9b. (Eloundou *et al.*, 1998a. Copyright 2001. Reprinted with permission of Wiley-VCH)

1. The time and conversion corresponding to the inflection point of log $\sigma(t)$ curves are lower than those of the gel point (e.g., the inflection point is found at $x \sim 0.4$ for epoxy–diamine systems while the gel conversion is found at $x \sim 0.6$).
2. Furthermore, when conductivity is expressed in terms of conversion using the corresponding kinetic information, there is no inflection point on log $\sigma(x)$ curves, as can be seen in Fig. 6.11 for the four epoxy–diamine systems (Eloundou *et al.*, 1998a,

1998b). In particular, a smooth behavior is observed around the gel point, $x_{gel} \sim 0.6$.

3. A correlation exists between conductivity and viscosity but it is limited to the range $x \sim 0$–0.4. The percolation approach leads to similar conclusions.

In conclusion, dielectric measurements seem to be insensitive to gelation. This can be explained by the fact that the dielectric probes, i.e., the ions which are practically always present in the reactive medium, give information only about local viscosities and not the macroscopic changes such as the viscosity near gelation. Nevertheless, dielectric measurements can be used to monitor processing in the pregel region.

6.5 MAIN CONCLUSIONS

1. Dynamic mechanical measurements describe both the liquid and solid states and are the best methods for following the physical changes occurring during the polymerization in the whole conversion range. The main observations are

   - The crossover of tan δ curves measured at various angular frequencies constitutes the best criterium for the gel point determination.
   - At the gel point, $G'(\omega)$ and $G''(\omega)$ curves, plotted on a logarithmic scale, are parallel over a wide range of frequencies. A power law is valid over the entire frequency range, although the value of the exponent Δ is still under discussion. The most accepted value is $\Delta = 0.72 \pm 0.02$.
   - The behavior near the gel point can be disturbed by viscoelastic effects and, more precisely, by the proximity of the vitrification phenomenon. This is defined by a frequency-dependent peak on $G''(t)$ or tan δ (t) curves.

2. Dielectric measurements are also able to monitor the polymerization from the initial liquid state to the final solid state (full cure). The main observations are

   - Vitrification is also defined by a frequency-dependent peak on the $\varepsilon''(t)$ curve.
   - Dielectric measurements are insensitive to gelation. This important point is mainly based on experiments with epoxy–amine reactions for which the dielectric parameters are controlled by ionic conductivity. More experiments with other chemistries are needed to reach a more universal conclusion.

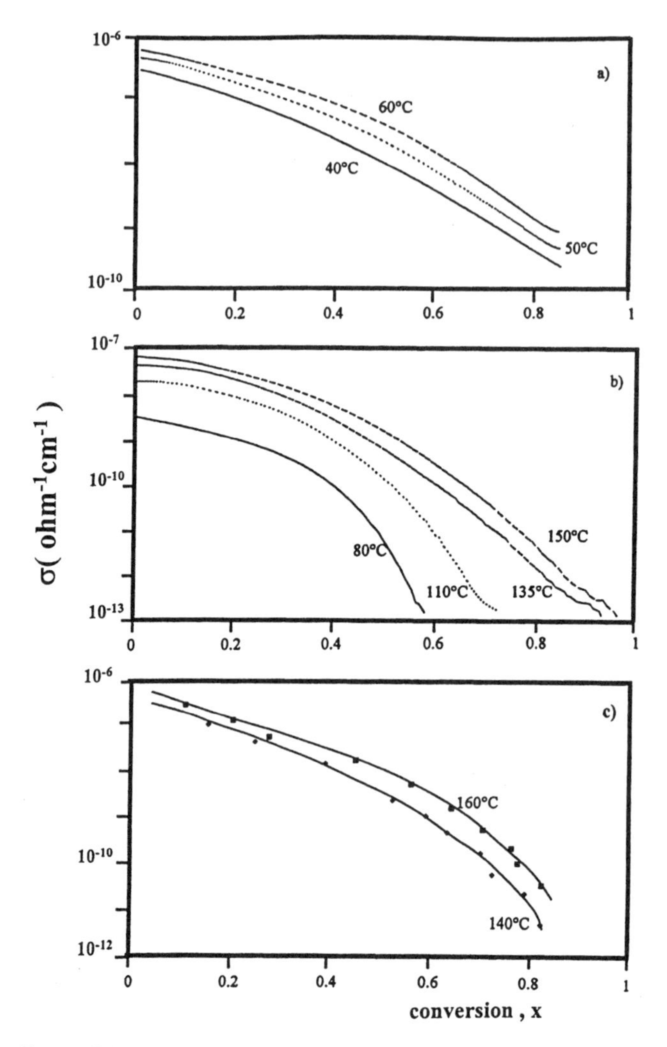

FIGURE 6.11

Nevertheless, with the development of new sensors, microdielectrometry constitutes a nondestructive and convenient method for monitoring *in situ* and in real time the cure of thermosets (Senturia *et al.*, 1982; Kranbuehl *et al.*, 1986). Great efforts have been made and research is still going on to establish relationships between (i) "cure index" and actual conversion of reactive groups, (ii) conductivity and viscosity, and iii) conductivity and glass transition temperature, in order to obtain models for "smart" processing.

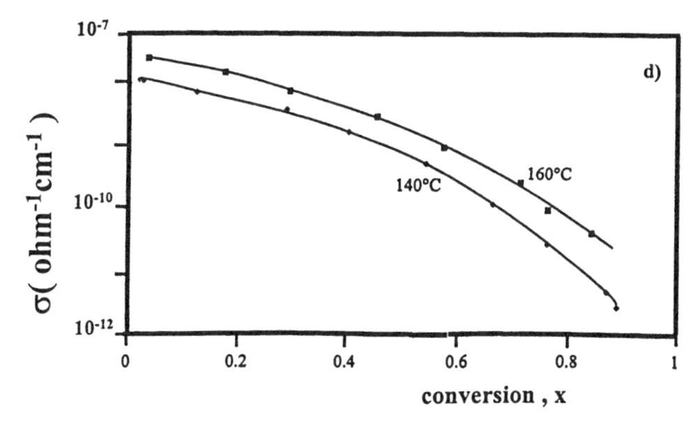

FIGURE 6.11 Curves of conductivity σ as a function of conversion, x ; (a) and (b) are the same epoxy systems as those in Figs 6.8 and 6.9, for (a) $T_{g\infty} = -12^\circ C$; (b) $T_{g\infty} = 177^\circ C$; (c) DGEBA reacting with diaminodiphenyl sulfone, DDS, $T_{g\infty} = 210^\circ C$; (d) DGEBA reacting with 4,4′-methylene bis (2,6-diethylaniline), MDEA, $T_{g\infty} = 170^\circ C$. For the four systems $x_{gel} \sim 0.6$. (Eloundou *et al.*, 1998a. Copyright 2001. Reprinted by permission of Wiley-VCH. Eloundou *et al.*, 1998b – Copyright 2001 – Reprinted by permission of John-Wiley & Sons Inc.)

NOTATION

f	= frequency, Hz
G'	= storage shear modulus, Pa
G''	= loss shear modulus, Pa
G^*	= complex shear modulus, Pa
G_0	= equilibrium shear modulus, Pa
M_n	= number-average molar mass, kg K mol^{-1}
M_w	= mass-average molar mass, kg kmol^{-1}
R_z	= z-average cluster radius, nm
s	= scaling exponent for $\eta_0(x)$
t	= time, s
t_{exp}	= characteristic time of a mechanical test, s
t_{gel}	= gelation time, s
tan δ	= loss factor, tan $\delta = G''/G'$
T_i	= cure temperature, K
T_g	= glass transition temperature, K
$T_{g,gel}$	= cure temperature at which gelation and vitrification occur simultaneously, K
$T_{g\infty}$	= glass transition temperature of the fully cured material, K
u	= scaling exponent for $G_0(x)$
x	= conversion

Greek Letters

$\Delta = u/(s+u)$ = ratio between scaling exponents
γ = scaling exponent for $M_w(x)$
ε' = relative equivalent capacitance of the material with respect to air
ε'' = relative equivalent conductance of the material with respect to air
ε^* = complex dielectric permittivity
η' = real part of the complex viscosity, $\eta' = G''/\omega$, Pa s
η^* = complex viscosity, Pa s
η_0 = zero shear-rate viscosity, Pa s
ν = scaling exponent for $R_z(x)$
σ = ionic conductivity, $ohm^{-1}\ cm^{-1}$
τ = characteristic time of the polymer relaxation, s^{-1}
ω = angular frequency, $\omega = 2\pi f$, s^{-1}

REFERENCES

Adam M, Delsanti M, Munch JP, Durand D, *J. Phys. (Paris)*, 48, 1809–1818 (1987).
Babayevsky PJ, Gillham JK, *J. Appl. Polym. Sci.*, 17, 2067–2088 (1973).
Chambon F, Winter HH, *Polym. Bull.*, 13, 499–504 (1985).
Chambon F, Winter HH, *J. Rheol.*, 31, 683–697 (1987).
De Gennes PG, *Scaling Concepts in Polymer Physics*, Cornell University Press, Ithaca, New York, 1979.
Eloundou JP, Feve M, Gérard JF, Harran D, Pascault JP, *Macromolecules*, 29, 6907–6916 (1996a).
Eloundou JP, Gérard JF, Harran D, Pascault JP, *Macromolecules*, 29, 6917–6927 (1996b).
Eloundou JP, Gérard JF, Pascault JP, Boiteux G, Seytre G, *D. Angew. Makromol. Chem.*, 263, 57–70 (1998a).
Eloundou JP, Ayina O, Ntede Nga H, Gérard JF, Pascault JP, Boiteux G, Seytre G, *J. Polym. Sci., Phys.*, 36, 2911–2921 (1998b).
Enns JB, Gillham JK, *J. Appl. Polym. Sci.*, 28, 2567–2591 (1983).
Ferry JD, *Viscoelastic Properties of Polymers*, 3rd ed, John Wiley & Sons, New York, 1980.
Flory PJ, *Principles of Polymer Chemistry*, Cornell University Press, Ithaca, New York, 1953.
Hess W, Vilgis TA, Winter HH, *Macromolecules*, 21, 2536–2542 (1988).
Hodgson DF, Amis EJ, *Macromolecules*, 23, 2512–2519 (1990).
Izuka A, Winter HH, Hashimoto T, *Macromolecules*, 27, 6883–6888 (1994).
Johari GP, Wasylyshyn DA, *J. Polym. Sci., Polym. Phys.*, 38, 122–126 (2000).
Kranbuehl DE, Delos S, Yi E, Jarvie T, Winfree W, Han T, *Polym. Eng. Sci.*, 26, 338–345 (1986).
Martin JE, Adolf D, Wilcoxon JP, *Phys. Rev.* A., 39, 1325–1332 (1989).
Matejka L, *Polym. Bull.*, 26, 109–116 (1991).

Nicolai T, Randrianantoandro H, Prochazka F, Durand D, *Macromolecules*, 30, 5897–5904 (1997).
Prochazka F, Nicolai T, Durand D, *Macromolecules*, 29, 2260–2264 (1996).
Rubinstein M, Colby RH, Gillmor JR, In *Chemical Physics*, Springer, Berlin, 1989, 51, 66–74.
Senturia SD, Sheppard NF Jr, Lee HI, Day DR, *J. Adhesion*, 15, 69–90 (1982).
Stauffer D, Coniglio A, Adam M, *Adv. Polym. Sci.*, 44, 103–158 (1982).
Tung C-Y M, Dynes PJ, *J. Appl. Polym. Sci.*, 27, 569–574 (1982).
Wasylyshyn DA, Johari GP, *J. Polym. Sci., Polym. Phys.*, 35, 437–456 (1997).
Winter HH, *Polym. Eng. Sci.*, 27 (22), 1698–1702 (1987).

7

Are Cured Thermosets Inhomogeneous?

7.1 INTRODUCTION

7.1.1 Why This Question?

Compared with thermoplastics (TP), the morphology of thermosets (TS) has not been studied thoroughly. Most TS are amorphous and composed of highly crosslinked molecular networks. But the concept of homogeneous infinite networks represented by one giant molecule (Chapter 3), has long been questioned. For example, epoxy networks (the most studied networks) are very often reported in the literature as inhomogeneous. Historically, this claim was supported by electron microscopic observations of free and fracture surfaces of epoxy networks, revealing the presence of a nodular morphology in the range of 10–100 nm. Nodules were suggested to be sites of higher crosslink density, resulting from intramolecular crosslinking and cyclization reactions. This conclusion was supported by the observation that, upon etching, the internodular material was preferentially attacked (Racich and Koutsky, 1976).

However, similar structures were observed with etched surfaces of amorphous linear thermoplastics, such as polystyrene and poly(methyl methacrylate). Moreover, the small-angle X-ray scattering (SAXS) spectra of simple epoxy networks based on diepoxy and diamine monomers were not essentially different from those of common amorphous polymers. Therefore, it was concluded that the inhomogeneous structure was not an inherent property of epoxy networks (Dusek *et al.*, 1978). A strong argument that supported the view of homogeneous polymer networks was the excellent agreement between experimental values of statistical parameters (gel conversion; evolution of the mass-average molar mass, M_w, in the pregel stage and the sol fraction and the crosslink concentration in the postgel stage) with theoretical values arising from mean-field models (Chapter 3).

However, this subject continued to be a controversial matter. Some authors used atomic force microscopy (AFM) in tapping mode to "prove" the existence of a two-phase structure in epoxy networks; this structures comprises a hard microgel phase and a dispersed phase of soft partially reacted material (Vanlandingham *et al.*, 1999). However, the interpretation of this kind of experimental result seems still to depend very much on what one wants to find.

There have been attempts to relate the assumed nodular morphology with the physical properties of networks (Labana *et al.*, 1971). This point is important, because if crosslinked polymers are considered as homogeneous three-dimensional structures, their ultimate properties can be related to the properties of such a continuum. On the other hand, if they are inhomogeneous, the supramolecular structure shown in Fig. 7.1 provides a more fruitful approach to interpreting of macroscopic properties.

The presence of inhomogeneities in some polymer networks, particularly those formed by a free-radical chainwise polymerization, has long been recognized (Labana *et al.*, 1971; Dusek, 1971); however, this does not mean that all thermosets must be inhomogeneous.

7.1.2 What Are Inhomogeneities?

It is not an easy task to define inhomogeneities in the structure of a polymer network. Every system will exhibit the presence of defects and fluctuations of composition in space when the scale of observation becomes smaller and smaller. A hierarchy of structures exists, from atomic dimensions to the macroscopic material. A scheme of different scale levels used to describe linear and crosslinked polymer structures is shown in Fig. 7.2. Inhomogeneities described in the literature for polymer networks are ascribed to permanent fluctuations of crosslink density and composition, with sizes varying from 10 nm up to 200 nm. This means that their size lies in the range of the macromolecular scale.

Another problem for describing a heterogeneous structure is to define boundaries between inhomogeneities and the rest of the structure. In some

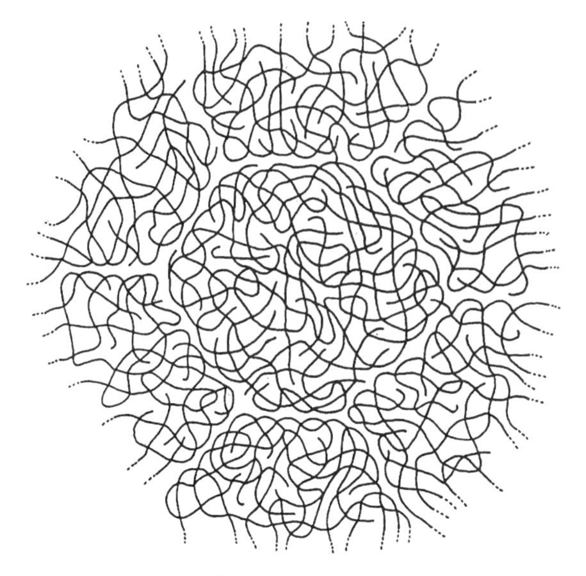

Figure 7.1 Two-dimensional representation of crosslink density inhomogeneities. A "gel ball" at the center is loosely tied to six similar surrounding regions. Crosslinking in bulk requires that branch points and chain ends are evenly distributed independently of morphology. (Labana *et al.*, 1971 with permission from Kluwer Academic)

cases boundaries are not defined or are fractals. Small inhomogeneities that are well defined and well dispersed in a matrix can be described as an ordered structure on a larger scale.

7.1.3 How to Characterize Inhomogeneities?

The presence of inhomogeneities can be characterized by various physical methods. Study of the scattering of electromagnetic radiation from polymers embraces a wide variety of techniques and yields information on diverse intramolecular and intermolecular properties. Methods include X-ray, neutron, and light scattering; some features of the scattering from these diverse sources can be treated with a common formalism. The energy vs wavelength dependence of different types of radiation is shown in Fig. 7.3. X-ray and visible light energy ranges are limited, whereas neutrons obtained from a pulsed source span a vast energy range (Gabrys and Tomlins, 1989). Light, small-angle X-ray or neutron scattering methods can characterize the size, internal structure, and spatial arrangement of inhomogeneities. The detection limit increases as the wavelength, λ, decreases. The average

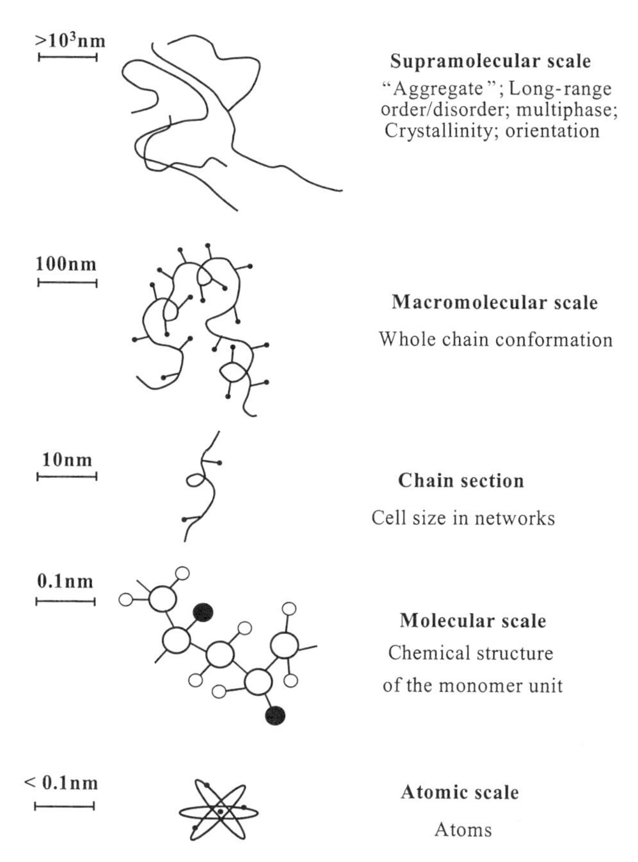

FIGURE 7.2 The different scale levels used to describe linear and crosslinked polymers.

diameter of an inhomogeneity has to be in the order of magnitude of $\lambda/4$ to be detected.

Microscopies offer a more integral response. Other techniques such as thermal and thermomechanical analysis, and methods sensitive to local mobility such as nuclear magnetic resonance (NMR), can also be used.

One major problem of all these techniques is the sensitivity in the parameter selected to detect the presence of inhomogeneities. With visible light for example, inhomogeneous samples can appear transparent if the difference in the refractive index between the phases is less than 0.01. Staining (in the case of transmission electron microscopy, TEM), or chemical etching (in the case of scanning electron microscopy, SEM), can be helpful in revealing the structure.

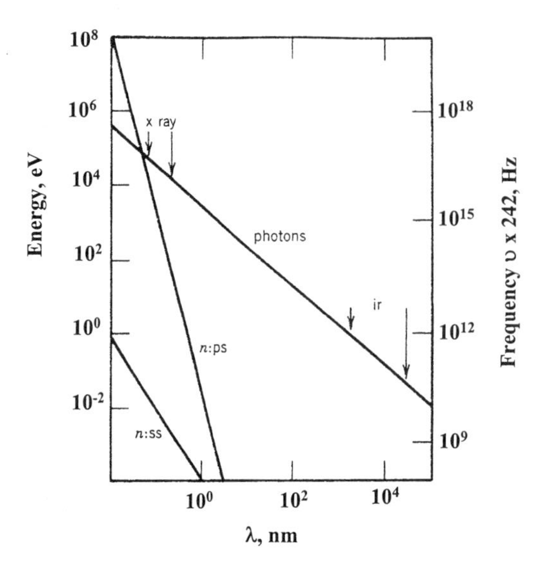

FIGURE 7.3 The energy-wavelength dependence for different types of radiation: photons, infrared, and neutrons. For neutrons (*n*), ps denotes the pulsed source and ss the steady source. (Gabrys and Tomlins, 1989 – Copyright 2001 – Reprinted by permission of John Wiley & Sons, Inc.)

Valuable information on whether the structure is homogeneous or inhomogeneous can also be obtained by analyzing the network formation process. A shift of experimental and estimated statistical parameters (M_w, gel point conversion, sol fraction, etc.) will be observed if inhomogeneities are formed as a result of the crosslinking process.

7.1.4 Aim of This Chapter

The structure of a polymer network at a particular conversion is determined by the structure of the starting components, the initial composition of the system, the reaction paths (mechanism and kinetics), and the network formation history (staging). The aim of this chapter is to try to answer the question 'Are cured thermosets inhomogenous' by giving some examples of both homogeneous and inhomogeneous systems and by analyzing different types of inhomogeneities and the reasons for their formation. Stepwise (polyaddition and polycondensation) and chainwise polymerizations are considered separately, to analyze the way in which the different mechanisms of network formation can fix eventual inhomogeneities produced along the reaction. The possibility of controlling and taking advantage of the formation of a heterogeneous structure is also analyzed.

7.2 THERMOSETS FROM STEP-POLYMERIZATION MECHANISM

7.2.1 Ideal Networks from Stepwise Polymerizations

a. Characterizations

As explained in Sec. 7.1, epoxy networks have been and are still the subject of controversy. This is mainly based on the particular interpretation of results obtained using microscopy techniques. On the contrary, results obtained with small-angle neutron scattering (SANS) proved that typical diepoxy–diamine networks were homogeneous (Wu and Bauer, 1985).

In addition, from thermal and thermomechanical measurements, it is found that typical epoxy–amine networks exhibit one glass transition temperature, T_g, and one sharp well-defined relaxation peak. The same techniques were used for crosslinked polyurethanes based on triol and diisocyanate or diol and triisocyanate (Andrady and Sefcik, 1983). Similar conclusions to those found for epoxy–amine networks were attained.

b. Network Buildup

If inhomogeneities are formed as a result of the crosslinking process, the evolution of the i-mer distribution, the gel point conversion, the sol fraction, etc., would be affected (Sec. 3.2.2, $A_4 + B_2$ stepwise polymerization). For diepoxy–diamine systems, most of the experimental gel point conversions reported in the literature are equal to the theoretical ones within experimental error. They are only determined by the initial composition of the system and the relative reactivities of primary and secondary amino groups (Dusek *et al.*, 1977, Dusek, 1986). One example of the good fit of experimental data to the values calculated assuming homogeneity of the system (Chapter 3) is given in Fig. 7.4. The sol fraction is particularly sensitive to the variations of composition and conversion of reactive groups (Ilavsky *et al.*, 1984).

c. First Conclusion Concerning Stepwise Polymerizations

It can be stated that networks based on a simple formulation (one monomer reacting with a comonomer), obtained from the step-polymerization process will exhibit a homogeneous structure. This is the case for epoxy–amine networks (the most studied) and polyurethane networks that have been used very often as ideal networks for structure–property correlations.

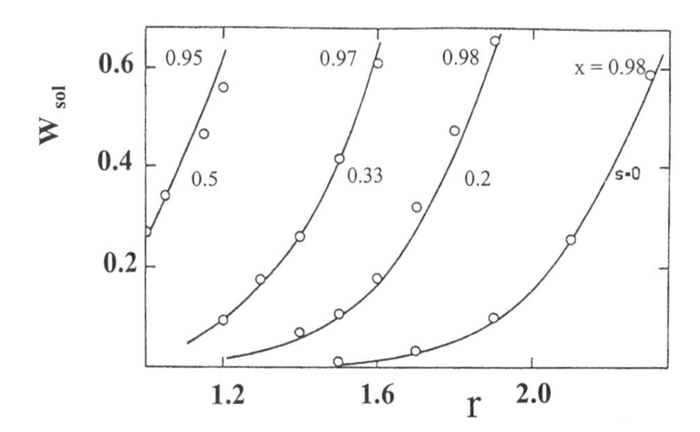

FIGURE 7.4 Mass fraction of sol as a function of the mole ratio of amine and epoxy groups; (○) experimental data, (—) calculated dependences taking epoxy conversion, x, into account.

$$s = \frac{\text{PGE}}{\text{PGE} + 2\,\text{DGEBA}} \quad \text{and} \quad r = \frac{4\,\text{MDA}}{\text{PGE} + 2\,\text{DGEBA}}$$

where PGE = phenyl glycidyl ether (monoepoxy), DGEBA = diglycidyl ether of bisphenol A (diepoxy), and MDA = methylene dianiline. (a) For all the series with constant s, critical molar ratios for gelation (r_c), were determined. The gel point was characterized by the first occurrence of an insoluble gel in dimethyl formamide, DMF. (b) The residual concentration of unreacted epoxy groups was determined by infrared spectroscopy and used to determine the final conversion, x. (c) The mass fraction of sol, W_{sol}, was determined by multiple extraction of networks in DMF, at 130°C. (Ilavsky *et al.*, 1984 – Copyright 2001 – Reprinted by permission of John Wiley & Sons, Inc.)

7.2.2 But the Step-Polymerization Process Can Also Induce Inhomogeneities

a. Thermodynamically Induced Inhomogeneities

If the monomer and the comonomer are initially miscible, reaction products will usually contribute to increasing the miscibility, because of the similarity of the chemical structures of the different i-mers, as their molar masses increase along conversion. Although this is the general rule, some exceptions have been reported. This is the case of a diepoxy (DGEBA, diglycidyl ether of bisphenol A) reacted with a diamine based on a long polypropylene oxide chain (typically, number-average molar mass, $M_n = 2000\ \text{g mol}^{-1}$). The two monomers are initially miscible but, as conversion progresses, a phase separation of domains (possibly) rich in the polypropylene oxide chains takes place (Wu and Bauer, 1985). This process is driven by thermodynamics (change of the interaction energies along polymerization) and leads to inhomogeneous networks.

b. Inhomogeneities Induced by Chemical Reaction

Inhomogeneities can also be formed during reaction without any help of physical interaction; i.e., just by chemistry. For silane condensation reactions in solution (the so-called sol–gel chemistry) but also in bulk, the experimental gel point conversions are very much higher than the theoretical values due to the formation of cagelike structures (Chapter 3). In some cases, using trifunctional silanes with large organic substituents, it is possible to obtain only cage structures without gelation (Fasce *et al.*, 1999). Figure 7.5 shows typical cagelike structures. Such a high percentage of intramolecular reactions is very specific to silane condensation and quite unusual for most step polymerizations in bulk.

Inhomogeneities can be formed in step-polymerization processes when more than two monomers are reacted together. Figure 7.6a represents a crosslinked mixture of a long diol (polyether or polyester, $M_n \sim 1000$–$5000\ g\ mol^{-1}$), a low-molar-mass triol (trimethylol propane, TMP), and a diisocyanate. Chemical or topological clusters are assemblies of covalently bonded units of one kind differing in some properties from the surrounding matrix. The polymer network consists of a dispersion of hard clusters in a soft matrix (Nabeth *et al.*, 1996). The presence of these inhomogeneities can be clearly recorded by SAXS.

In this case of three-monomer polyurethane synthesis, there is no thermodynamic driving force for phase separation. The formation of clusters is fully controlled by the initial composition of the system, the reactivity of functional groups, and the network formation history (one or two stages, macrodiol or triol reacted with diisocyanate first, etc.).

The presence of hard clusters affects mechanical properties. The major problem is the way to define elastically active network chains (EANC) and crosslinks (Chapter 3, Fig. 3.3). It has been demonstrated that hard clusters must be considered as multifunctional crosslinks ($f_c = 6$ in Fig. 7.6a) while macrodiol chains behave as EANC.

Chemical clusters can be obtained also with two monomers, when two reaction mechanisms are in competition, favoring formation of regions of higher and lower crosslink densities. This situation is more complex and more difficult to control. It is certainly the case for dicyanodiamide (Dicy)-cured epoxies: with this hardener an accelerator is always used and a competition between step (epoxy–amine addition) and chain (epoxy homopolymerization) occurs (Chapter 2), leading to inhomogeneous networks.

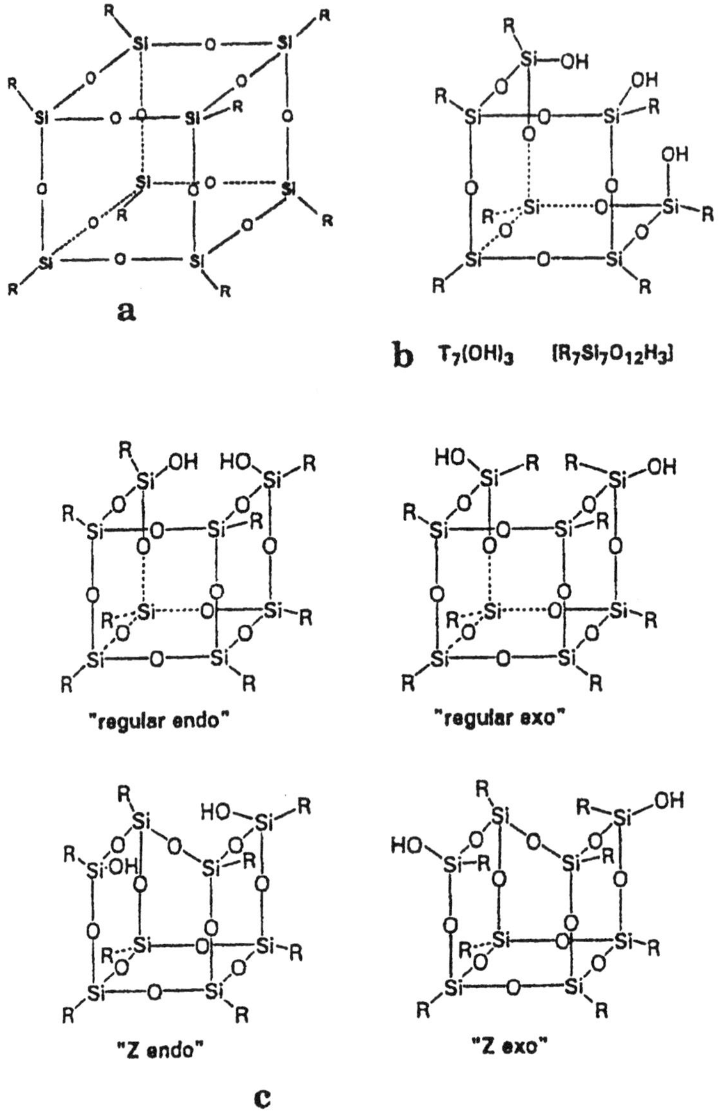

a

b $T_7(OH)_3$ $[R_7Si_7O_{12}H_3]$

"regular endo"

"regular exo"

"Z endo"

"Z exo"

c

FIGURE 7.5 (a) Schematic representation of an octosilsesquinoxane, $(RSiO_{1.5})_8$; (b) typical precursor of a SSQO; (c) incompletely condensed "T8 $(OH)_2$" isomers. (Reprinted with permission from Fasce *et al.*, 1999. Copyright 2001. American Chemical Society)

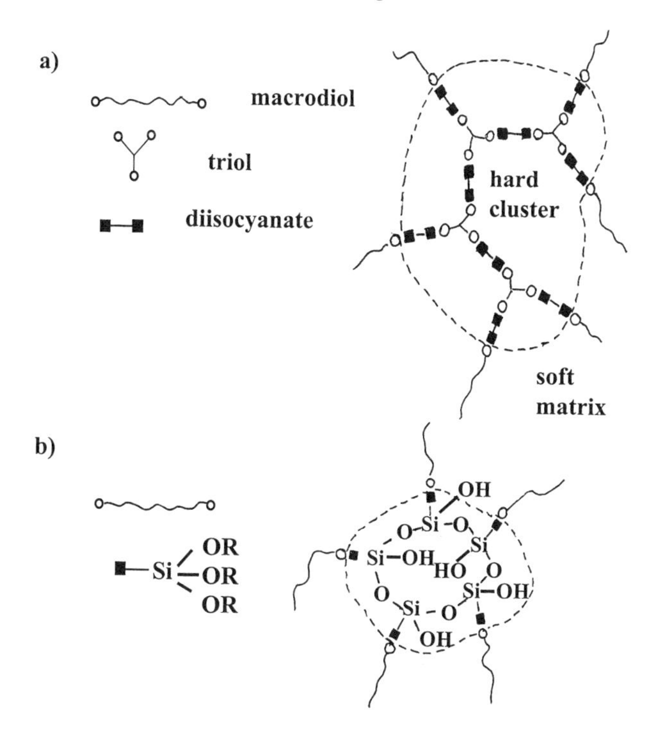

Figure 7.6 Schematic representation of hard clusters in: (a) a polyurethane network composed of a long diol, a triol, and a diisocyanate (polyaddition reactions, Chapter 2). (b) a hybrid inorganic–organic network composed of a silane end-capped long diol (polycondensation reactions, Chapter 2).

c. Both Chemical Clustering and Microphase Segregation

In the case of Fig. 7.6a the cluster formation and the size distribution can be influenced not only by chemical reactions but also by partial miscibility of the substructures during reaction. Polyurethane networks prepared from polyolefin instead of polyester or polyether as macrodiol, can serve as an example. In this particular case an agglomeration of hard domains takes place in the pregel stage, produced by a thermodynamic driving force.

The "sol–gel" chemistry has also been used to prepare inorganic inhomogeneities in an organic matrix. Silane end-capped macrodiols can be used. Hydrolysis and condensation of alkoxy silane groups lead to inorganic hard clusters (Fig. 7.6b). Intramolecular reactions and the miscibility of the soft-segment chains with the relatively polar crosslinks determine the size distribution of the clusters (nanofillers).

It has been demonstrated by SAXS analysis (Fig. 7.7, Cuney *et al.*, 1997), that the control of the intermolecular reactions by chemical reactivities (polycaprolactone, PCL20 case) only gives loose, more or less randomly branched structures of fractal types, whereas thermodynamic microphase segregation (hydrogenated polybutadiene, HPBD case) yields compact particles, possibly with a fractal surface. A loose structure of clusters can interact more strongly with the matrix than the compact clusters, which has a strong effect on the macroscopic properties such as elongation at break.

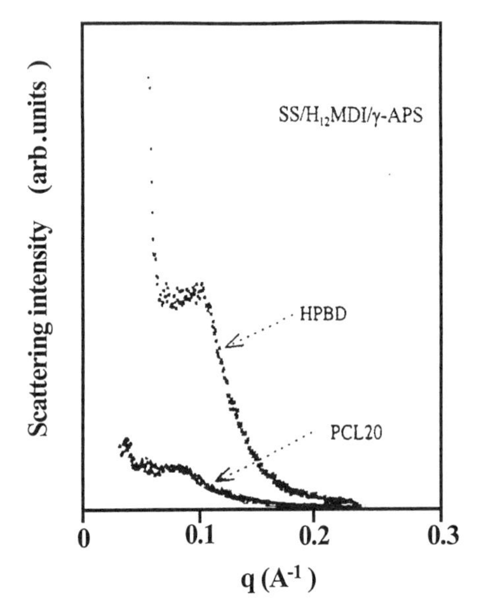

FIGURE 7.7 SAXS profiles for two hydroxyl-terminated oligomers crosslinked by alkoxysilane sol–gel chemistry. First, 1 mole of macrodiol, SS (hydrogenated polybutadiene, HPBD or polycaprolactone, PCL, $M_n = 2$ kg mol^{-1}), was reacted at 80°C with 2 mole of dicyclohexylmethane diisocyanate, H_{12} MDI. After complete reaction, the prepolymer was dissolved in tetrahydrofuran and the γ-aminosilane, γAPS was added dropwise at room temperature. After 1 h of reaction, the solvent was removed under pressure. The final network was obtained in the absence of a solvent by hydrolysis and condensation of the ethoxysilane groups by the addition of 0.1 mol% TFA, trifluoroacetic acid. After stirring at room temperature, the mixture was cast into a mold and cured for 24 h at 100°C under pressure, and then postcured at 150°C for 12 h. (Cuney *et al.*, 1997 – Copyright 2001, Reprinted by permission of John Wiley & Sons, Inc.)

7.3 THERMOSETS FROM CHAIN-POLYMERIZATION MECHANISM

7.3.1 Chainwise Polymerization is Intrinsically an Inhomogeneous Process

As explained in Chapters 2 and 3, the step and chain mechanisms are very different. While for stepwise reactions, the molar masses of the reaction products increase gradually, chain reactions produce long chains dissolved in the monomers from the beginning of the polymerization.

The elementary reaction steps of chain-crosslinking polymerization have been discussed in Chapter 2. After an initiation step, the polymerization proceeds by propagation steps in which monomers are added to the active end of growing chains, accompanied by chain transfer and deactivation of chain ends by termination reactions. The process is controlled by the q parameter (Chapter 3), which gives the probability that the active end of the propagating chain adds another monomer to the growing chain. In the case of free-radical polymerization of a monounsaturated monomer, q tends to 1 due to the high rate of free-radical propagation (Sec. 3.3.1). The increase of viscosity leads to the Trommsdorff effect (or "gel" effect), caused by the onset of the diffusion control for recombination of two macroradicals (Chapters 2 and 5).

For polyfunctional monomers, the most studied system is the free-radical copolymerization of mono- and diunsaturated monomers. The network formation by free-radical chain polymerization is an intrinsic inhomogeneous process, as described by numerical simulations (Boots, 1987). Figure 7.8 illustrates the distribution of reaction products in a two-dimensional lattice, for the homopolymerization of a divinyl monomer at a conversion of double bonds equal to 0.25. A growth step implies the transfer of an active center site to an unreacted neighbor. Termination may either occur due to the encounter of two active chain ends or to the trapping of the active center in a fully reacted region. The lowest possible initiation rate (allowing an initiation event only when no radical remains active on the lattice) corresponds to the largest extent of inhomogeneity (Fig. 7.8a). The initiation rate was artificially increased by limiting the chain length to at most 20 steps (Fig. 7.8b) and 2 steps (Fig. 7.8c). There is a continuous increase in the uniformity of the polymer distribution in space with the increase in the ratio between initiation and propagation rates. For the limiting case plotted in Fig. 7.8c, the chain process has been practically converted into a step-reaction process. As conversion increases, the polymer distribution becomes more uniform for each one of the cases.

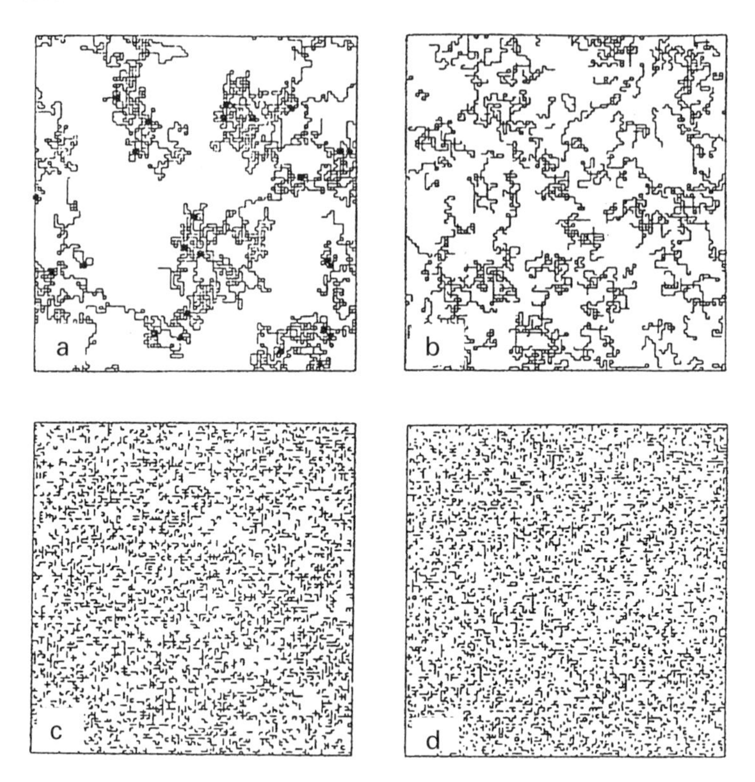

FIGURE 7.8 Snapshots of the polymerization of a tetrafunctional monomer on a 100 × 100 square lattice at a bond conversion of 25%. Bonds between units are indicated, but units are not. Boundary conditions are periodic. (a) A chain reaction of one active radical at a time; when it is trapped (dark squares) a new one starts. The increase in homogeneity by increasing the initiation rate is illustrated in (b) and (c). The initiation rate is artificially increased by limiting the chain length to at most 20 steps (b) and 2 steps (c): note that (c) is as homogeneous as (d), which is obtained for a step polymerization after a conversion of 25% of the reactive groups. (Reprinted from Boots, 1987, Copyright 2001 with permission from Elsevier)

7.3.2 Network Formation in the Case of Free-Radical Polymerization

a. Spatial Correlation and Apparent Reactivity of Double Bonds

As described qualitatively in Fig. 7.8a, the fast chain propagation with respect to initiation is both a source of inhomogeneities and of changes in double-bond reactivities during reaction. This process leads to permanent inhomogeneities (such as those shown in Fig. 7.1).

Abundant experimental evidence is reported in the literature as regards the change of double-bond reactivity along polymerization (Dusek, 1993). Figure 7.9a gives a scheme of the structure of a branched primary chain obtained at the beginning of the process (conversion less than 0.01) and shows some unreacted pendant double bonds and one reactive center. The primary chain can propagate by reacting with one of the monomers, or it can attack pendant double bonds of another macromolecule (intermolecular reaction), leading to more branching and crosslinking, or of the same molecule (intramolecular reaction), leading to cyclization.

At the very beginning of polymerization, there is a very small concentration of macromolecules dissolved in the monomers. For this reason, the probability of the active center meeting a pendant double bond of another primary chain is close to zero and most of the pendant double bonds participate in intramolecular reactions. Spatial correlations increase the effective reactivity of pendant double bonds and explain the occurrence of intramolecular cyclization at the beginning of the reaction.

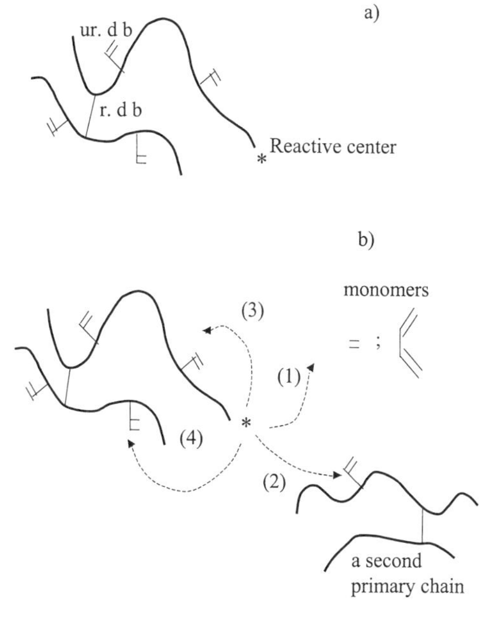

Figure 7.9 Example of a reacting primary chain (a) and possible reactions of a macro reactive center (b); (1) propagation; (2) crosslinking; (3) cyclization; ur.db = unreacted double bond; r.db = reacted double bond.

But a double bond located inside a large branched macromolecule also has less probability of reacting than one located at the periphery: this is called the shielding effect. For this reason, spatial correlations due to excluded volume can also decrease the apparent reactivity of pendant double bonds.

The competition between these opposite effects (increase and decrease of pendant double bond reactivity) depends on the concentration of multifunctional monomers, the length and flexibility of the primary chains, and the quality of thermodynamic interactions between monomers and macromolecules. As a rule, cyclization is more effective at the beginning of polymerization, whereas the steric excluded-volume effects are more effective at the later stages.

In any case, the macromolecules formed at the beginning of polymerization exhibit a large number of intramolecular cycles and are very compact. Cyclization reactions are favored by dilution with a nonreactive solvent.

b. Network Buildup

Figure 7.10 shows a qualitative plot of the increase in the steady-state viscosity as a function of reaction time for the free-radical copolymerization of

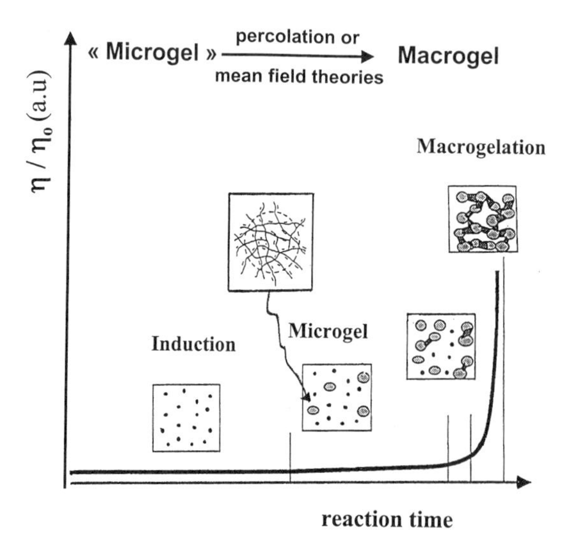

Figure 7.10 Typical (steady-state) viscosity rise curve for the free-radical copolymerization of mono- and multiunsaturated monomers, correlated with the steps of the reaction mechanism.

mono- and multiunsaturated monomers. Due to cyclization, internally crosslinked compact molecules, very often called "microgels," in the literature, but better defined as crosslinked microparticles (CMP), are formed rapidly at the beginning. Intramolecular reactions consume double bonds but do not contribute to the macro-network formation. Peripheral pendant double bonds are more accessible (more reactive) than the ones trapped at the interior of the CMP. Macroscopic gelation, manifested by the increase in the steady-state shear viscosity to infinity, occurs when the structure resulting from covalently bonding the crosslinked microparticles percolates through the material.

Macrogelation proceeds by a mechanism completely different to that described by mean-field models (Chapter 3). In fact, it resembles the chemical aggregation of colloidal particles. At higher conversions the "void" space containing monomers is filled in by polymerization, so that the high-conversion structure appears to be less inhomogeneous than the structure at the gel point; however, permanent inhomogeneities still persist.

7.3.3 Conclusions on the Chain-Polymerization Mechanism

The main conclusions are listed below.

(a) It is a special feature of the free-radical crosslinking copolymerization of multifunctional monomers that compact, internally crosslinked macromolecules are formed from the beginning of the polymerization. Network-formation models that do not take spatial correlations into account cannot be used to obtain a correct description of these systems.

(b) Inhomogeneities are induced by random processes that are difficult to control. The most striking deviation from ideal statistical theories is the shift of the gel-point conversion to higher values. Nonetheless, for most systems the gel-point conversion is relatively low, 5–10%.

(c) A distribution of crosslink densities may be expected to persist in the final structure (qualitatively represented by Fig. 7.1).

(d) Thermodynamic effects may also be present. For example, if CMP are not soluble in the mixture of monomers, they will agglomerate (phase-separate) in the first stages of the polymerization, leading to a broad distribution of particle sizes (including large particles). A well-known example is the cure of unsaturated polyesters, UP resins (Hsu and Lee, 1993). Styrene is generally a poor solvent of the UP prepolymer (Chapter 2). The reaction-induced phase separation of the primary chains occurs very rapidly (after the induction period), due to a thermodynamic driving force. CMP with sizes in the range of 10–20 nm could be isolated at the beginning of reaction.

After phase separation, the sizes increase up to 100 nm. At the same time, trapped radicals were observed by electron spin resonance (esr) spectroscopy.

Reaction-induced phase separation is certainly also the reason for which an inhomogeneous structure is observed for photocured polyurethane acrylate networks based on polypropylene oxide (Barbeau *et al.*, 1999). TEM analysis demonstrates the presence of inhomogeneities on the length scale of 10–200 nm, mostly constituted by clusters of small hard units (the diacrylated diisocyanate) connected by polyacrylate chains. In addition, a suborganization of the reacted diisocyanate hard segments inside the polyurethane acrylate matrix is revealed by SAXS measurements. Post-reaction increases the crosslink density inside the hard domains. The bimodal shape of the dynamic mechanical relaxation spectra corroborates the presence of a two-phase structure.

(e) For low concentrations of the multifunctional monomer (less than 1%), or for short primary chains ($q \leq 0.8$, such as for anionic or cationic polymerization of epoxies, Chapter 2), the formation of CMP will not take place. The same occurs when the tendency toward cyclization decreases by increasing the distance between double bonds, e.g., using a mixture of a monounsaturated monomer with an α, ω-diunsaturated oligomer. These conditions lead to homogeneous polymer networks whose formation may be described by models that do not take spatial correlations into account.

7.4 CONCLUSION

7.4.1 Are Cured Thermosets Inhomogeneous?

The structure of precursors, the number of functional groups per precursor molecule, and the reaction path leading to the final network all play important roles in the final structure of the polymer network. Some thermosets can be considered homogeneous ideal networks relative to a reference state. It is usually the case when networks are prepared by step copolymerization of two monomers (epoxy–diamine or triol–diisocyanate reactions) at the stoichiometric ratio and at full conversion.

Many thermosets are inhomogeneous. This may be the result of the polymerization mechanism (chemically induced inhomogeneities, such as those produced by the free-radical crosslinking polymerization of multifunctional monomers or by the step polymerization of three different monomers), or of the decrease of solubility of reaction products (thermodynamic driving force); or of both factors acting simultaneously, such as the case of several UP formulations. Inhomogeneities formed in the course of polymerization are fixed by the crosslinking reactions.

7.4.2 Is There a Way to Control Inhomogeneities?

The possibility of synthesizing a polymer network containing chemical (topological) clusters by using three monomers of different sizes during a step polymerization was described. In the absence of thermodynamic effects, cluster formation is fully controlled by the initial composition of the system, the relative reactivities of functional groups, and the network-formation history (Nabeth *et al*, 1996; Cuney *et al.*, 1997).

But, in the general case, inhomogeneities are induced by fast chemical reactions and are very difficult to control. For this reason, precursors of controlled architectures have been developed for different purposes, such as to decrease the viscosity in high-solids systems, to lower shrinkage, to improve material properties and, for some electronic and nonlinear optical applications, to build nanosized ordered regions in the networks.

Different architectures, such as block copolymers, crosslinked microparticles, hyperbranched polymers and dendrimers, have emerged (Fig. 7.11). Crosslinked microparticles ("microgels") can be described as polymer particles with sizes in the submicrometer range and with particular characteristics, such as permanent shape, surface area, and solubility. The use of dispersion/emulsion aqueous or nonaqueous copolymerizations of formulations containing adequate concentrations of multifunctional monomers is the most practical and controllable way of manufacturing microgel-based systems (Funke *et al.*, 1998). The sizes of CMP prepared in this way vary between 50 and 300 nm. Functional groups are either distributed in the whole CMP or are grafted onto the surface (core-shell, CS particles).

Dendrimers (Newkome *et al.*, 1996) and hyperbranched polymers, HBP, look like functional microgels in their compactness but they differ in two aspects: they do not contain cyclic structures and, more importantly, they are much smaller, in the range of a few nanometers in size. They are prepared stepwise in successive generations (dendrimers) or they are obtained by the polyaddition/polycondensation of AB_f monomers, where only the A + B reaction is possible (HBP; Voit, 2000). Both molecules have tree-like structures, but a large distribution of molar masses exists in the case of HBP.

Network buildup studies using these preformed molecules are still lacking. Architecture-dependent effects of interactions between reactive groups must certainly influence the reaction kinetics and the network buildup. They should be understood before any reasonable modeling of structure–properties development can be made.

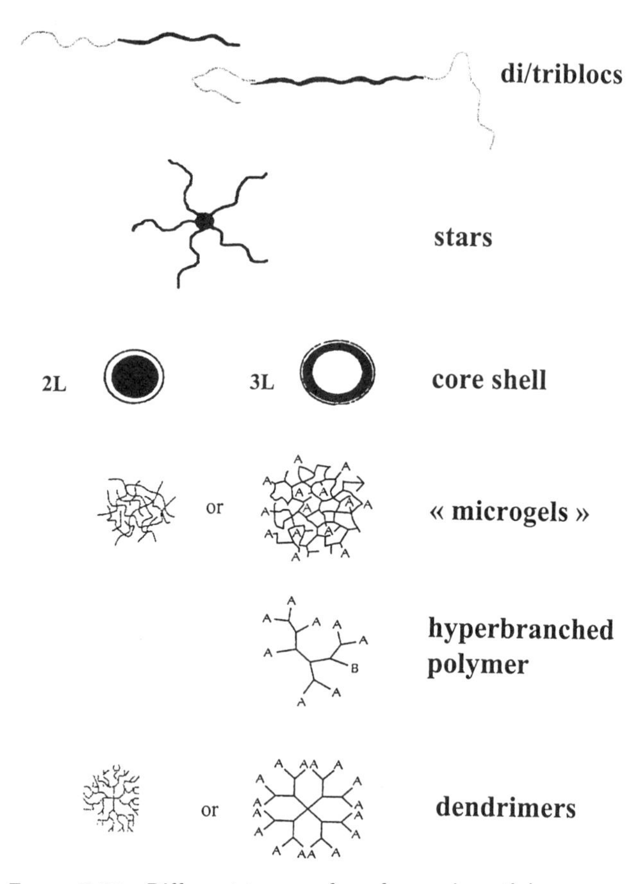

FIGURE 7.11 Different types of preformed particles.

REFERENCES

Andrady AL, Sefcik MD, *J. Polym. Sci. Phys.*, 21, 2453–2463 (1983).

Barbeau P, Gérard JF, Magny B, Pascault JP, Vigier G, *J. Polym. Sci. Phys.*, 37, 919–937 (1999).

Boots HMJ, *Physica* 147A, 90–98 (1987).

Cuney S, Gérard JF, Dumon M, Pascault JP, Vigier G and Dusek K, *J. Appl. Polym. Sci.*, 65, 2373–2386 (1997).

Dusek K, In *Polymer Networks*, Chompff AJ and Newman S (eds), Plenum Press, New York, 1971, pp. 245–260.

Dusek K, In *Epoxy Resins and Composites IV*, Dusek K (ed.), *Adv. Polym. Sci.*, 80, 173–188 (1986).

Dusek K, *Collect. Czech. Chem. Commun.*, 58, 2245–2265 (1993).

Dusek K, Bleha M, Lunak S, *J. Polym. Sci. Chem.*, 15, 2393–2402 (1977).
Dusek K, Plestil J, Lednicky F, Lunak S, *Polymer* 19, 393–397 (1978).
Fasce DP, Williams RJJ, Méchin F, Pascault JP, Llauro MF, Petiaud R, *Macromolecules*, 32, 4757–4763 (1999).
Funke W, Okay O, Müller BJ, *Adv. Polym. Sci.*, 136, 139–234 (1998).
Gabrys B, Tomlins PE. In *Encyclopedia of Polymer Science and Engineering*, 2nd edn, Mark HF, Bikalès NM, Overberger CG, Menges G (eds), *Scattering*, 15, 1–68 (1989).
Hsu CP, Lee LJ, *Polymer*, 34, No. 21 Part 1, 4496–2505; Part 2, 4506–4515; and Part 3, 4516–4523 (1993).
Ilavsky M, Bogdanova L, Dusek K, *J. Polym. Sci. Phys.*, 22, 265–278 (1984).
Labana SS, Newman S and Chompff AJ. In *Polymer Networks*, Chompff AJ, Newman S (eds), Plenum Press, New York 1971, pp 453–487.
Nabeth B, Pascault JP, Dusek K, *J. Polym. Sci., Polym. Phys.* Ed., 34, 1031–1054 (1996).
Newkome GR, Moorefield CN, Vogtle P, *Dendritic Molecules: Concepts, Syntheses, Perspectives*, Wiley-VCH, Weinheim, 1996.
Racich JL, Koutsky JA, *J. Appl. Polym. Sci.*, 20, 2111–2129 (1976).
Vanlandingham MR, Eduljee RF, Gillepsie JW, *J. Appl. Polym. Sci.*, 71, 699–712 (1999).
Voit B, *J. Polym. Sci., Part. A, Chem.*, 38, 2505–2525 (2000).
Wu WL, Bauer BJ, *Polym. Comm.*, 26, 39–42 (1985).

8

Preparation of Modified Thermosets

8.1 INTRODUCTION

Thermoset precursors are frequently formulated by adding other components than monomers: these components, which are generically called modifiers, include small molecules, oils, low- or high-molar mass rubbers or thermoplastics, etc. Depending on the required applications and desired properties, the amount of modifier may vary in a broad range, from about 2 to 50 wt% of the total mixture with monomers.

When rubbery domains in the micrometer range are randomly dispersed in a rigid thermoset matrix, the fracture energy and the toughness can be greatly enhanced (Chapter 13). The use of modifiers is required not only to improve toughness but also for other specific needs such as the dimensional stability of a molded part or the increase of the elastic modulus during the cure of a preimpregnated composite (prepreg), or the generation of microporous structures for thermal or electrical insulation, etc.

Again, as in many other fields covered in the book, modified epoxies are the most studied systems (toughened epoxies for adhesive coatings and composites). But also rubber-modified phenolics and low-profile unsaturated polyesters for sheet and bulk molding compounds have been extensively studied.

There are two main procedures used to generate a second phase in a modified thermoset:

1. The initial mixture is a dispersion of preformed (crosslinked or not) particles in the thermoset precursors and remains heterogeneous during the cure.
2. The initial mixture is homogeneous, and phase separation takes place during the cure of the thermoset. This second technique is called reaction-induced phase separation (Williams *et al.*, 1997) and may lead to several types of morphologies: a dispersion of modifier-rich particles in a thermoset matrix; a dispersion of thermoset-rich particles in a modifier matrix (phase-inverted morphology), or two bicontinuous phases.

8.2 MODIFIERS INITIALLY MISCIBLE IN THERMOSET PRECURSORS

8.2.1 Solubility of Main Modifiers Used

a. Requirements

Modifier miscibility plays an important role in this preparation. On the one hand, modifiers must be miscible with the reactive system; on the other hand, they must phase-separate during cure. Final morphologies are influenced by the phase-separation conditions.

b. Rubbers

A family of carboxy-terminated poly(butadiene co-acrylonitrile) liquid rubbers, abbreviated as CTBN (Table 8.1), was pioneered by BF Goodrich in the late 1960s and early 1970s and introduced commercially as versatile epoxy-toughening agents. They are also called reactive liquid polymers, RLP. Due to their low glass transition temperatures and low molar masses, less than 4 kg mol^{-1}, they are viscous liquids. Miscibility between CTBN and epoxy monomers such as diglycidyl ether of bisphenol A, DGEBA, is matched by incorporating polar acrylonitrile units into a nonpolar immiscible polybutadiene backbone, and also by modifying end groups, e.g., by converting carboxy into epoxy end groups via adduct formation (Riew and Gillham, 1984; Riew, 1989; Chen *et al.*, 1994).

Examples of experimental phase diagrams are given in Fig. 8.1, for different liquid rubbers and the same liquid epoxy prepolymer (Verchère *et al.*, 1989). Cloud-point measurements using light transmission or light scattering are generally used to plot the phase diagram and the cloud-point curve. Liquid rubber modifiers exhibit an upper critical solution tempera-

TABLE 8.1 Examples of some modifiers for thermosetting polymers

Acronym	Name	Formula	Some characteristics
CTBN	Poly(butadiene-co-acrylonitrile)	$HOOC-(CH_2)_2-C(CH_3)_2-[R]-C(CH_3)_2-(CH_2)_2-COOH$ with $R = -(CH_2-CH=CH-CH_2)_a-(CH_2-CH(CH=CH_2))_b-(CH_2-CH(CN))_c-$	$M_n \sim 3500$ g mol^{-1} $T_g \sim -60°C$ $f_{COOH} \sim 1.8–2.0$ wt % AN: 26, 18, 10, 0
PES	Polyether sulfone	$-[C_6H_4-SO_2-C_6H_4-O]-$	$T_g = 220°C$
PEI	Polyether imide	$-[N(CO)_2C_6H_3-O-C_6H_4-C(CH_3)_2-C_6H_4-O-C_6H_3(CO)_2N-C_6H_4]-$	$T_g = 210°C$
PPE	Poly(2,6 dimethyl 1,4-phenylene ether)	$H-[C_6H_2(CH_3)_2-O]-H$	$T_g = 210°C$

PMMA	Poly(methyl methacrylate)	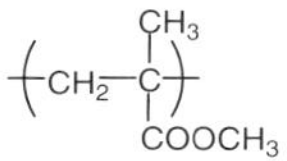 $-(CH_2-C(CH_3)(COOCH_3))-$	$T_g = 110°C$
SAN	Poly(styrene-co-acrylonitrile)	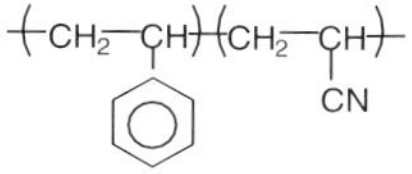 $-(CH_2-CH(C_6H_5))-(CH_2-CH(CN))-$	$T_g = 100°C$
PVAc	Poly(vinyl acetate)	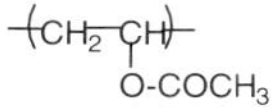 $-(CH_2-CH(O\text{-}COCH_3))-$	$T_g = 40°C$

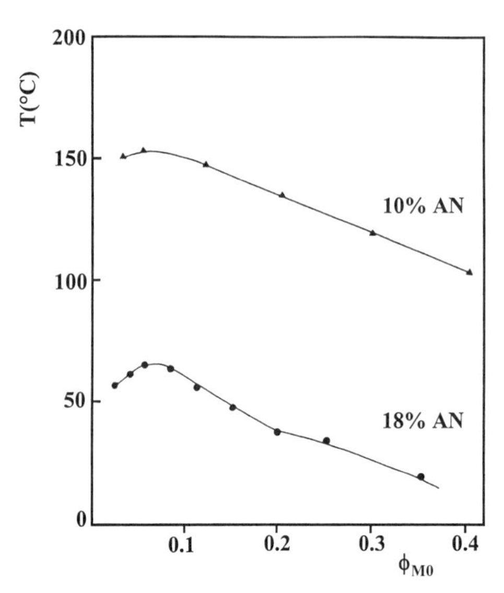

FIGURE 8.1 Cloud-point temperatures versus volume fraction of modifier, for mixtures of diglycidyl ether of bisphenol A, DGEBA ($\bar{n} = 0.15$) with two CTBN copolymers with different acrylonitrile content: 18 and 10 wt%. (Reprinted from Verchère *et al.*, 1989, Copyright 2001, with permission from Elsevier Science)

ture behavior (UCST), i.e., the miscibility increases with increasing temperature. As shown in Fig. 8.1, formulations containing an initial volume fraction of modifier, ϕ_{M0} in the range 10–20 wt% are miscible at $T_i > 65°C$ in the case of CTBN (18 wt% AN), and $T_i > 155°C$ in the case of CTBN (10 wt% AN), while CTBN (26 wt% AN) is totally miscible with DGEBA.

The state of miscibility of any mixture is governed by the Gibbs free energy of mixing, ΔG, which may be described by the lattice theory of Flory–Huggins (Flory, 1953), as follows:

$$\frac{\Delta G}{RT} = \frac{\phi_{TS}}{V_{TS}} \ln \phi_{TS} + \left(\frac{1}{V_M}\right) \Sigma \left(\frac{\phi_{M,i}}{i}\right) \ln \phi_{M,i} + \frac{\chi}{V_{ref}} \phi_M \phi_{TS} \tag{8.1}$$

A unit cell is defined with a molar volume V_{ref}. This reference volume may be selected as the molar volume of the thermoset precursor(s), V_{TS}, or as the molar volume of the constitutive repeating unit of the modifier, V_M. R is the gas constant and ϕ_{TS} and $\phi_{M,i}$ are the volume fractions of the thermoset precursor(s) and of the modifier i-mer, respectively; $\phi_M = \Sigma\phi_{M,i}$.

In principle, the interaction parameter χ must be considered as a function of temperature, T, composition, and the average degree of polymerization of the modifier component. If χ decreases with temperature, then phase diagrams show a UCST behavior. When χ increases with temperature, they show a lower critical solution temperature, (LCST) behavior. The experimental cloud-point curves are fitted by selecting an adequate function for χ.

The interaction parameter χ can also be estimated by the use of solubility parameters (δ); χ is proportional to $(\delta_M - \delta_{TS})^2$. But this approach has a considerable error ($\Delta\delta$ in the range of $\pm$ 0.4 $MPa^{1/2}$), and considers only the excess free enthalpy. For these reasons it is better to determine the miscibility window experimentally.

Although CTBN and derivatives still constitute the most important group of modifiers used in rubber-modified epoxies, several other types of modifiers, such as vegetal oils (castor oil), have been proposed as well.

CTBN and derivatives are also used in unsaturated polyester and vinylester resins but with less success. Studies have shown that the elastomer additive alone is very often immiscible in a polyester resin.

c. Thermoplastic Modifiers

In the 1980s a novel approach to toughen epoxies, consisting of replacing the rubber by an engineering thermoplastic, was developed. Table 8.1 gives the formula of some of the thermoplastics used and Fig. 8.2 gives examples of the phase diagrams. In this case in addition to the cloud-point curve, the

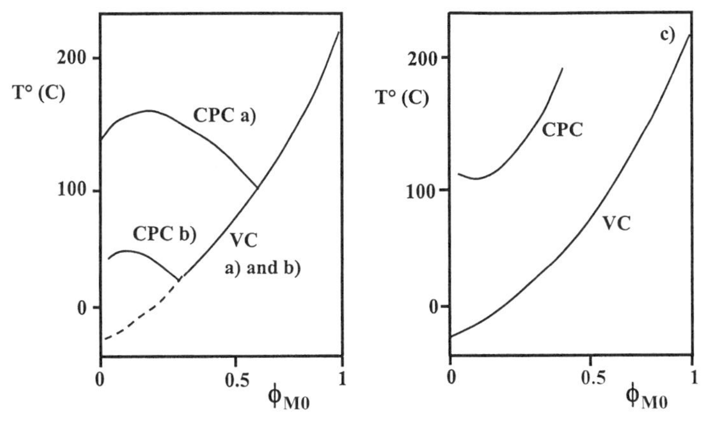

Figure 8.2 Schematic phase diagrams for thermoplastic-epoxy monomer (diglycidyl ether of bisphenol A) blends, (CPC = cloud point curve, and VC = vitrification curve). (a) and (b) UCST (upper critical solution temperature) behaviour for PPE and PEI (respectively) – DGEBA $\bar{n} = 0.15$; (c) LCST (lower critical solution temperature) behaviour for PES-DGEBA $\gamma\ \bar{n} = 0.15$. (Pascault and Williams, 2000 – Copyright 2001. Reprinted by permission of John Wiley & Sons Inc.)

transformation diagrams have to be completed by plotting the vitrification curve. The vitrification curve for a one-phase system can be estimated by the Fox equation:

$$\frac{1}{T_g} = \frac{W_{TS}}{T_{g,TS}} + \frac{W_M}{T_{g,M}} \tag{8.2}$$

Figures 8.2a and b describe a UCST behavior, while Fig. 8.2c represents an LCST behavior. It is interesting to note that in the case of an UCST behavior the cloud-point curve will usually intersect the vitrification curve, while this may not be the case for an LCST behavior.

Experimental cloud-point curves are fitted by Eq. (8.1), selecting an adequate function for χ. Depending on their structures, thermoplastics are more or less soluble in epoxy monomers: poly(methyl methacrylate), PMMA and poly(styrene-co-acrylonitrile), SAN are quite soluble in liquid DGEBA, but the other thermoplastics shown in Table 8.1 are only partially miscible (Pascault and Williams, 2000).

In some cases semicrystalline thermoplastics have been used as modifiers (polyamides, polyesters). With these thermoplastics it is necessary to increase the temperature beyond the melting temperature to obtain a solution.

Another important application is the introduction of a thermoplastic such as poly(vinyl acetate), PVAc, into an unsaturated polyester resin to improve dimensional stability of the mold part. Keeping in mind (sec. 2.3.2) that an unsaturated polyester is dissolved in styrene, the modifier is the third component and the behavior is described with the help of a ternary-phase diagram (Fig. 8.3, Suspène *et al.*, 1991).

8.2.2 Description of the Reaction-Induced Phase Separation Process

a) Case of Stepwise Polymerization

During a step-growth reaction the molar mass of the product increases gradually and the molar mass distribution becomes continuously wider (Sec. 2.2.1). Figure 8.4 gives a scheme of the reaction-induced phase-separation process in a transformation diagram temperature, T, versus composition, ϕ_M, for a UCST behavior. The initial formulation has a volume fraction of modifier equal to ϕ_{M0} and is kept at the cure temperature, T_i, where it is a homogeneous solution. As conversion increases at T_i, the solubility of the modifier decreases, due to the increase in the average size of the distribution of the thermosetting species and the corresponding decrease in the entropic contribution to the free energy of mixing.

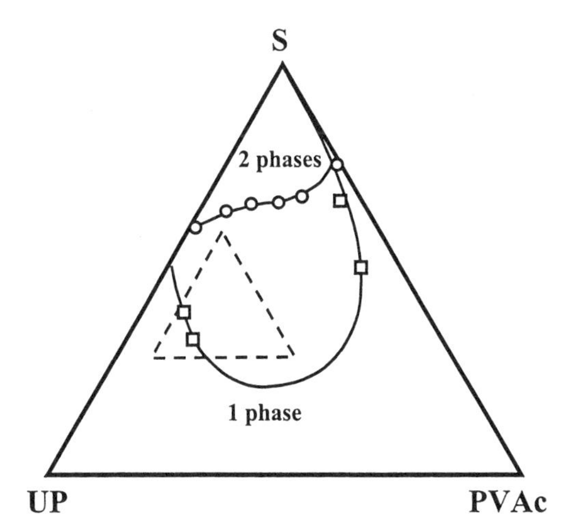

FIGURE 8.3 Phase diagram for styrene (S)–unsaturated polyester (UP) prepolymer–PVAc ternary blends at T = 23°C,–□– UP: M_n = 1690 g.mol^{-1}and M_w / M_n = 7.5; –o– UP: M_n = 1480 g.mol^{-1} and M_w / M_n = 3.1. The dashed triangle represents the formulation range of typical industrial UP resins. (Reprinted from Suspène *et al.*, 1991, Copyright 2001, with permission from Elsevier Science)

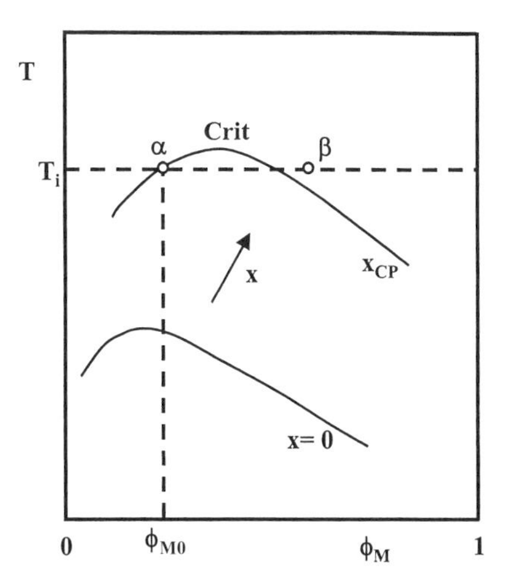

FIGURE 8.4 Temperature vs. composition transformation diagram for a modified thermoset with an upper critical solution temperature (UCST) behavior (Crit = critical point; for α and β see text).

Looking back to Eq. (8.1), the absolute value of the first term decreases with increasing conversion.

The interaction parameter χ may also vary along the reaction, due to the continuous change in the chemical structure of the thermoset. This secondary effect may favor mixing or demixing, depending on whether χ decreases or increases, respectively, with conversion.

Phase separation begins when the cloud-point curve reaches the point α at (ϕ_{M0},T_i). This is defined as the cloud-point conversion, x_{CP}, which is usually lower than the gel conversion, x_{gel}.

As a result of polydispersity effects, the composition of the incipient β-phase segregated at the cloud point is located on a shadow curve, outside the cloud-point curve (point β in Fig. 8.4). The effects of polydispersity on phase diagrams and phase compositions may be found in specialized reviews (Tompa, 1956; Kamide, 1990; Williams *et al.*, 1997). Because $\phi_{M0} <$ ($\phi_{M,crit}(x_{CP})$, the incipient β-phase, which is richer in the modifier, will be dispersed in the α-phase, which is richer in the growing thermosetting polymer. The opposite occurs when $\phi_{M0} > \phi_{M,crit}(x_{CP})$. It has been shown both theoretically (Riccardi *et al.*, 1994 and 1996; Williams *et al.*, 1997), and experimentally (Bonnet *et al.*, 1999) that

1. The β-phase rich in modifier, contains the higher-molar mass species of the modifier as well as a significant amount of thermosetting species, mainly monomers and low-molar-mass species. Therefore the conversion in the β-phase is lower than in the α-phase. On the contrary, the α-phase is richer in low-molar-mass species of the modifier and the high-molar mass species of the thermosetting polymer.
2. As conversion increases beyond x_{CP}, the α and β-phases become richer in their main components.
3. Most of the primary morphology development is arrested at gelation of the α-phase.
4. A secondary phase separation may take place inside the β-phase, and due to diffusion effects it leads to a sub-structure of thermoset-rich domains. Secondary phase separation may be arrested by gelation or vitrification of the β-phase.
5. When the modifier is a high-T_g thermoplastic, a postcure step at $T > T_{g,M}$ is necessary to complete the reaction in the β-phase and to develop the final morphology of the blend.

In Fig. 8.4 an UCST behavior is represented. The corresponding situation for an LCST behavior is a shift of the cloud-point curve to lower temperatures as conversion increases. A similar description of the phase-separation process is valid also for the LCST case.

The reaction-induced phase separation may also be described using conversion, x, versus composition, ϕ_M, transformation diagrams at a constant cure temperature (Fig. 8.5). Cloud point and spinodal curves bound stable, metastable, and unstable regions. Experimental studies of phase separation (Chen *et al.*, 1993; Girard-Reydet *et al.*, 1998), revealed that compositions located close to the critical point (e.g., trajectory 2 in Fig. 8.5) undergo spinodal demixing, while off-critical compositions (e.g., trajectories 1 and 3) exhibit phase separation by a nucleation-growth mechanism.

Independently of the mechanism by which phase separation is produced, final morphologies depend primarily on the location of the trajectory. Trajectory 1 leads to a random dispersion of modifier-rich particles in the thermoset-rich matrix, and trajectory 3 leads to the opposite situation. In a composition region located close to the critical point (trajectory 2), bicontinuous structures may be obtained.

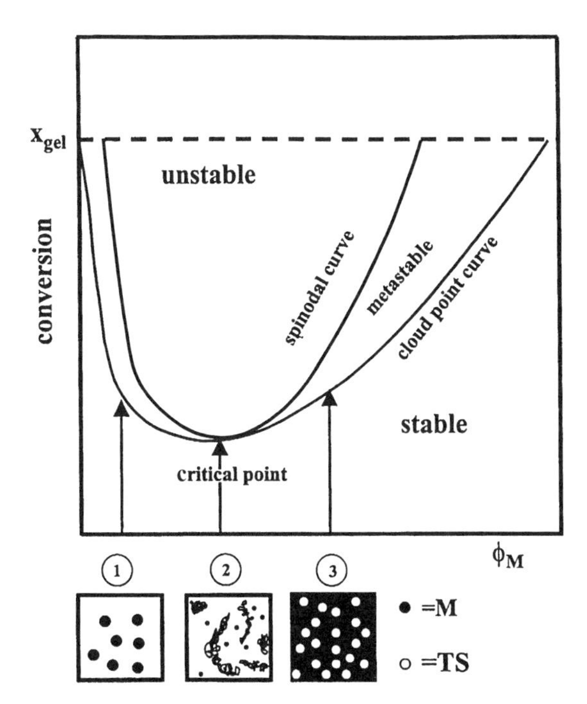

FIGURE 8.5 Conversion vs composition transformation diagram at a constant cure temperature, showing cloud-point curves and spinodal curves that bound stable, metastable, and unstable regions; ①, ②, and ③ represent the three trajectories, starting from different initial thermoplastic concentrations and leading to different morphologies. (Pascault and Williams, 2000 – Copyright 2001. Reprinted by permission of John Wiley & Sons Inc.)

b. Case of Chainwise Polymerization

During chain reactions, when the propagation step is fast compared with the initiation step (such as for free-radical polymerizations) long polymer chains are already formed at the beginning of the reaction (Sec. 2.2.1). For this reason, in the case of chain polymerization a triangular diagram is more useful for describing the miscibility curves at a constant cure temperature. Figure 8.6 shows that for a particular conversion the system attains the cloud-point curve. At the beginning of the phase-separation process the main phase is the one rich in modifier dissolved in the monomer(s). But, as the polymer fraction increases (higher conversions), a phase inversion may take place. The location of phase inversion depends on the initial composition ϕ_{M0}.

In the case of modified UP resins, the situation is complicated by the fact that an inmiscibility region develops between the monomer and polymer also. Complex morphologies may result, such as the bicontinuous structures obtained for PVAc – UP resin blends when $\phi_{M0} \geq 7$ wt%.

There are few results in the literature on the evolution of transformation diagrams during chain reactions of thermosets. From a thermodynamic point of view, before the gel point the behavior is similar to the well-studied synthesis of high-impact polystyrene (Bucknall, 1989). But after the gel point, which arrives at low conversions, the contribution of elastic forces to the free energy of mixing has to be added in Eq. (8.1) (De Gennes, 1979).

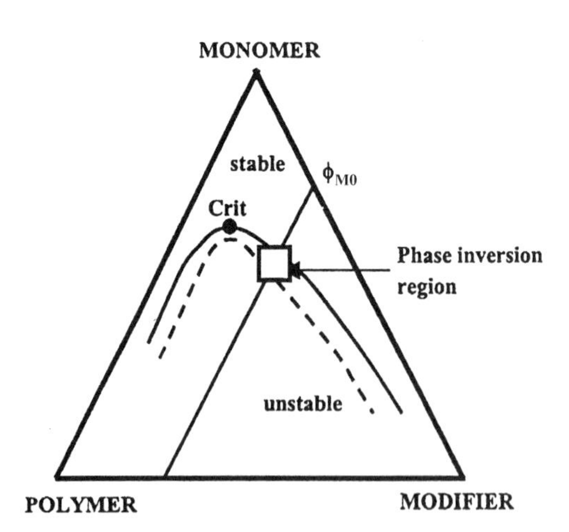

FIGURE 8.6 Ternary phase diagram for a typical chain polymerization (Crit. = critical point).

8.2.3 Control of Generated Morphologies

Morphologies are generally characterized by size and concentration of dispersed-phase particles. Figures 8.7–8.10 give some examples of morphologies obtained in the case of modified epoxies. Morphologies are controlled by many factors, such as miscibility, modifier concentration, temperature, reaction rate, and presence of an emulsifier, etc. (Williams *et al.*, 1997).

The most important factor controlling the morphologies generated is the location of the composition of the initial blend, ϕ_{M0} with respect to the critical composition, $\phi_{M,crit}$ (Figs 8.5 and 8.6). The latter may be calculated from the Flory–Huggins model as applied to a binary blend (step reactions) or a ternary blend (chain reactions), taking into account polydispersity (Kamide, 1990). The size of particles increases with the concentration of the component that forms the dispersed phase. Typically, for $\phi_{M0} < \phi_{M,crit}$, an increase in ϕ_{M0} will lead to an increase in both the volume fraction and the average size of dispersed phase modifier-rich particles.

The initial miscibility also influences the resulting morphologies. For a modifier exhibiting an initial poor miscibility with thermoset precursors,

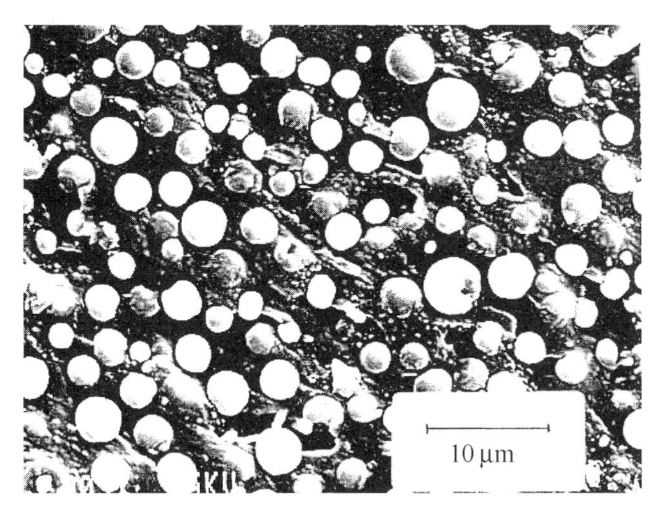

FIGURE 8.7 SEM photograph of a fully cured rubber-modified epoxy network. The rubber CTBN (26 wt% AN) is first pre-reacted with a large excess of diglycidyl ether of bisphenol A (DGEBA) to obtain an epoxy-terminated rubber. Then an equivalent of 15 wt% initial CTBN is introduced in DGEBA-4,4′-diamino diphenyl sulfone, DDS, system precured at 135°C (time > t_{gel}) and then postcured at 230°C. Rubber-rich particles are spherical, D $\sim$ 2.8 $\pm$ 0.5 μm, and well dispersed. (From LMM Library.)

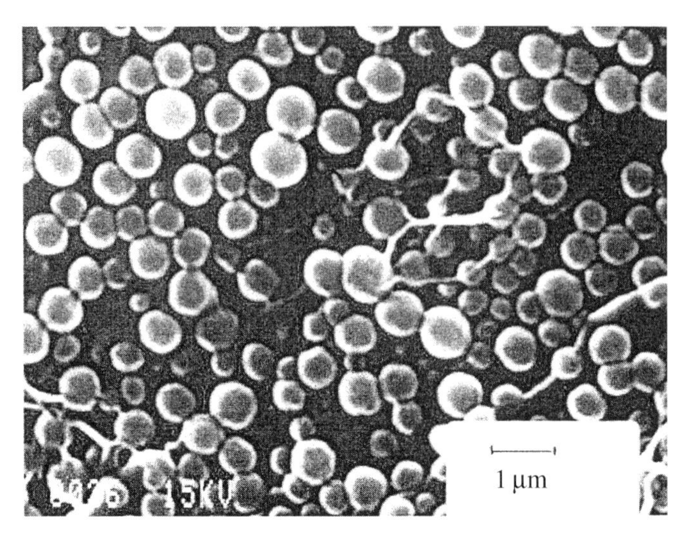

FIGURE 8.8 SEM photograph of a fully-cured rubber-modified epoxy network. The rubber CTBN (18 wt% AN) was also pre-reacted with a large excess of DGEBA and then introduced (15 wt% initial CTBN) in DGEBA – 4,4′ diamino – 3,3′-dimethyldicyclohexylmethane, 3 DCM, precured at 50°C (time > t_{gel}) and postcured at 190°C. Rubber particle sizes are smaller than in Fig. 8.7. (From LMM Library).

phase separation occurs at low conversions when the viscosity is relatively low. This leads to the presence of large particles, because mass diffusion is favored by the low viscosity of the reaction medium. The opposite situation is observed when the modifier presents a high initial miscibility. When conversion x_{CP} approaches x_{gel}, a significant amount of modifier remains dissolved in the thermoset-rich matrix. If a rubber is used as modifier, a significant decrease of the T_g compared with the one of the neat matrix, is observed.

The effect of cure temperature is more difficult to analyze. An increase of cure temperature produces three different effects: an increase of the reaction rate, a decrease of the viscosity, and an increase (UCST) or a decrease (LCST) of the initial miscibility. It has been observed that as the viscosity at the cloud point, η_{CP}, decreases there is an increase in the average size of dispersed phase particles and a corresponding decrease in their concentration.

Increasing the polymerization rate by increasing the concentration of a catalyst or an initiator will decrease the time period in which phase separation may take place. A continuous decrease of both the volume fraction of dispersed phase and the average size of particles is expected.

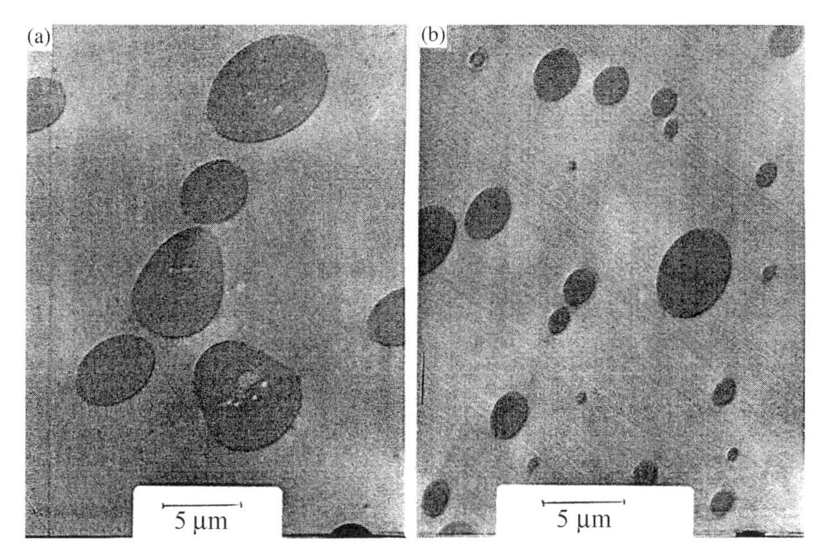

Figure 8.9 TEM photographs of thermoplastic-modified epoxy networks. The system is DGEBA – 4,4′-methylene bis [3-chloro-2,6 diethylaniline], MCDEA, with 10 wt% polyphenylene ether, PPE; (a) is cured at 135°C ($T < T_{g,PPE}$) and (b) at 240°C ($T > T_{g,PPE}$). (a) Epoxy-rich substructures are visible inside PPE-rich particles; (b) the particle sizes decrease with an increase in cure temperature and a binodal distribution is observed. (From LMM Library.)

In some cases, due to coalescence mechanisms (Oswald ripening), bimodal particle-size distributions are observed.

The interfacial adhesion between dispersed particles and the matrix can be improved by functionalizing the modifier with any chemical group that can react with the thermosetting polymer in the selected cure cycle. But when the end groups of the modifier are changed, its miscibility with the thermoset precursors can also be modified. In addition, the covalent bonds introduced between the modifier and the thermosetting polymer can also influence the phase-separation process. Therefore, the particle size and the volume fraction of the modifier-rich phase are changed, and it is not possible to vary just the adhesion and keep exactly the same morphologies (Williams *et al.*, 1997).

Emulsifiers, such as block copolymers, also modify both particle size and interfacial adhesion (Girard-Reydet *et al.*, 1999).

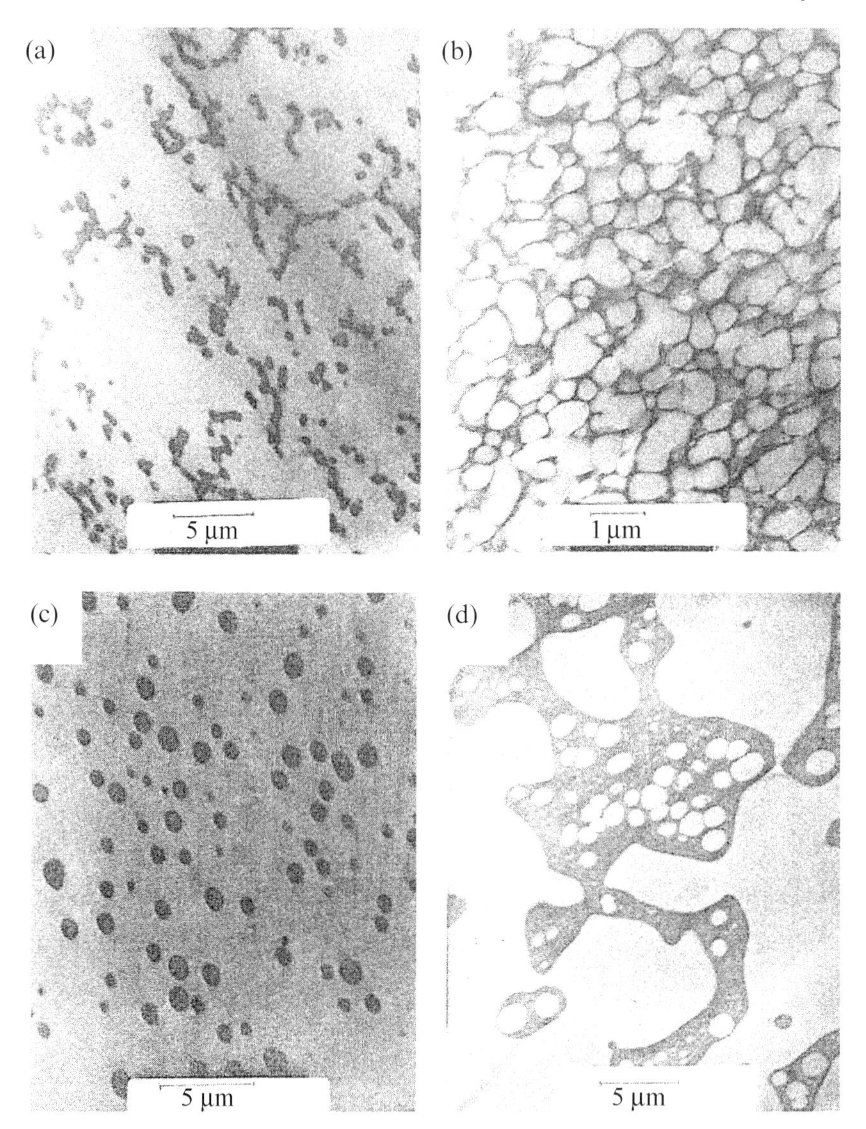

Figure 8.10 TEM photographs of thermoplastic-modified epoxy networks. The epoxy system is DGEBA–MCDEA (same as in Fig. 8.9), and the thermoplastic is a polyetherimide, PEI. The figure illustrates both the influence of the cure schedule and the PEI concentration. (a) 10 wt% PEI precured at 80°C and postcured at 190°C ; (b) 20 wt% PEI, same cure schedule; (c) 10 wt% PEI precured at 160°C and postcured at 190°C; (d) 20 wt% PEI same cure schedule as (c). Phase inversion is around 20 wt%. (From LMM Library.)

8.3 DISPERSION OF AN ORGANIC SECOND PHASE IN THE THERMOSET PRECURSORS

8.3.1 Introduction

Another procedure for the preparation of modified thermosets consists of introducing preformed particles in the initial formulation. This technique is also well documented for modified thermoplastics (Paul and Bucknall, 2000). In Chapter 7 different macromolecular architectures such block copolymers, crosslinked microparticles, hyperbranched polymers, and dendrimers, were presented (Fig. 7.11). All these compact molecules can be used as thermoset modifiers. Thermoplastic powders and core-shell polymers are the more accessible preformed molecules. Some examples are given below.

8.3.2 Thermoplastic Powders

Semicrystalline polyamide fine powders have been used as toughening agents for epoxy networks. The powders can be obtained by grinding granules, or directly by anionic polymerization of lactams, 6 or 12, in an organic solvent from which the formed semicrystalline polymer precipitates. Microporous powders with an average particle size in the range of 10 μm and a narrow particle-size distribution, are obtained.

At their melting temperature these powders can be dissolved in the epoxy monomers and they are able to react and participate in the crosslinking reaction through the amide groups (Lennon *et al.*, 2000). For this reason, the cure cycle must be selected in order to keep the polyamide particles below their melting point (in the range 170°C or 220°C, depending on the type of polyamide used), and thus keep their initial shape and size. But in some cases a partial dissolution of the powder surface can improve the particle–polymer network interactions.

Amorphous polyimide powders prepared by dissolution/precipitation processes, can be used to toughen thermosetting polymers. Polyethylene powders are frequently used in low-shrink unsaturated polyester formulations.

8.3.3 Core-Shell Particles

Core-shell polymers were commercially introduced as impact modifiers for poly(vinyl chloride) PVC, in the 1960s. They are produced by a two-stage latex emulsion polymerization technique (Cruz-Ramos, 2000). The core is a graftable elastomeric material, usually crosslinked, that is insoluble in the thermoset precursors. Typical elastomers used for these purposes are crosslinked poly(butadiene), random copolymers of styrene and butadiene, poly(butyl acrylate), and copolymers. The cores keep their size during the thermoset cure. Core diameters can range from about 50 nm to 5 μm. For toughening applications, a high amount of elastomeric core relative to the total particle is desirable; e.g., from about 50 to 90 wt% of the particles.

The shell component is grafted to the core to effectively stabilize the dispersion of particles along polymerization. Even if the cores are insoluble in the thermoset precursors they require a protective shell for colloidal stability in a nonaqueous environment, due to the low value of its T_g. The shell typically includes chemically grafted polymer chains and nongrafted chains. The number of nongrafted chains is generally unknown, but it should be kept at a minimum to decrease the viscosity of the dispersion. Molar masses of nongrafted polymer chains should be lower than 100 kg mol^{-1}. The shell composition must be miscible with the thermoset precursors to achieve the colloid stability necessary to survive mixing operations at elevated temperatures. Methyl methacrylate is often used as shell monomer. It can be copolymerized with other methacrylates, styrene, or acrylonitrile. Poly(methyl methacrylate) is usually miscible with epoxy monomers but may phase-separate during reaction. If this happens, a partial aggregation of particles may take place.

The interfacial adhesion between dispersed particles and the matrix can be improved by functionalizing the core-shell particles with any chemical group that can react with the thermosetting polymer. For example, glycidyl methacrylate can be introduced in the shell composition to incorporate functional groups that can react with epoxy formulations.

It is not easy to obtain a good dispersion of core-shell particles in an organic medium. The initial mixing operation is very important and can be made by extrusion. But to keep a good dispersion after cure it is necessary to control the miscibility of the shell, its reactivity with the thermoset precursors, and a possible phase-separation process of previously dissolved shell components, induced by the polymerization. An example of the quality of the resulting dispersions is illustrated in Fig. 8.11a and b.

8.3.4 Miscellaneous Modifiers

Other chemistries have been used to prepare preformed reactive modifiers for epoxy and unsaturated polyester resins. Particles are generally composed of

1. A core matching the required size and properties.
2. A shell for stabilizing the colloidal dispersion.

In the case of block copolymers ABA or ABC for example, the mid block B is insoluble and the other blocks are miscible with thermoset pre-

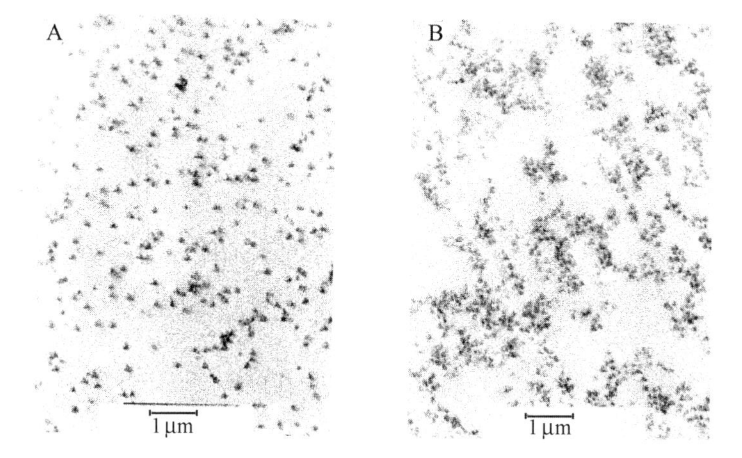

Figure 8.11a TEM photographs of core-shell particles dispersed in an epoxy network, DGEBA-Isophorone diamine, IPD. (A) 10 wt% CS; (B) 15 wt% CS. (From LMM Library.)

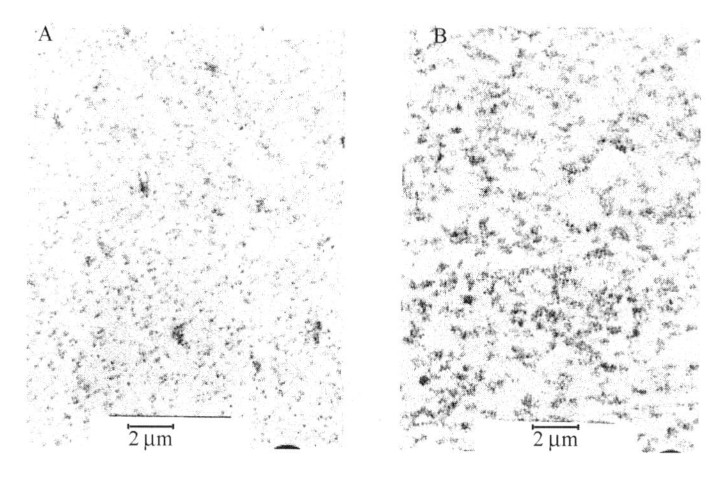

Figure 8.11b Same samples as in Figure 8.11a but observed at a lower magnification (From LMM Library.)

cursors. Typical mid blocks are poly(dimethyl siloxane) or polybutadiene and derivatives. The miscible blocks can be polycaprolactone, poly(methyl methacrylate), or polyethylene oxide. As for core-shell particles, the quality of the dispersion depends on the degree of miscibility of the stabilizing blocks (Fig. 8.12). For high block copolymer concentrations, nanostructured networks can be prepared (Lipic *et al.*, 1998).

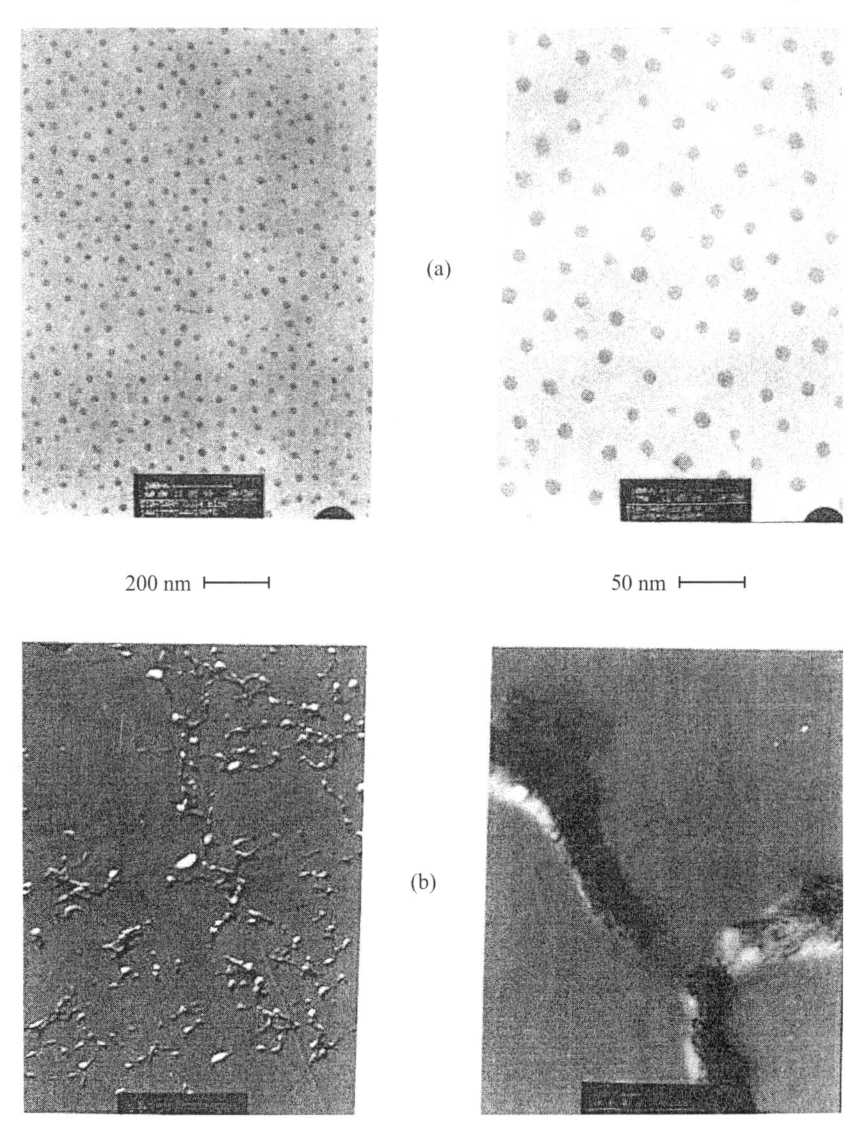

Figure 8.12 TEM photographs of triblock copolymers dispersed in a DGEBA–diamine epoxy network. The triblock copolymer is polystyrene-b-polybutadiene-b-poly(methyl methacrylate), and the epoxy hardener is (a) 4,4′-methylene bis [3-chloro–2,6 diethylaniline], MCDEA, and (b) 4,4′-diamino diphenyl sulfone, DDS. In the case of the epoxy system based on MCDEA, the PMMA block is miscible up to the end of the epoxy reaction. In the case of the epoxy system based on DDS, the PMMA block phase-separates during reaction. (From LMM Library.)

Hyperbranched polymers (Boogh *et al.*, 1999), and crosslinked microparticles based on acrylates and prepared in organic media (Pascault *et al.*, 2000), give size particles in the range of 20–50 nm. Crosslinked epoxy particles made from a latex can be either small, 30-600 nm (Landfester *et al.*, 2000) or very large 10–100 μm (Jansen *et al.*, 1999). In every case the chemistry of the shell has to be controlled.

8.3.5 Concluding Remarks

Depending on applications, both procedures used to generate two-phase morphologies in modified-thermosets – (a) reaction-induced phase separation (RIPS) and (b) the use of preformed particles – have their advantages and disadvantages.

1. Advantages of RIPS are the stability and the low viscosity of the starting homogeneous solution. The possibility of generating a variety of structures in the final material is the most important factor that cannot be produced by the use of preformed particles.
2. A disadvantage of RIPS is the necessity of establishing relationships between cure cycles, generated morphologies, and phase compositions.
3. Another disadvantage of RIPS is that, usually, part of the modifier remains dissolved in the thermoset matrix, even at full conversion. If the modifier is a low-T_g rubber, this leads to a decrease of the T_g of the fully cured material and losses in thermal and load-bearing properties.
4. Regarding the use of preformed particles, the main advantages are the initial control of particle size, volume fraction, and composition of the disperse phase. Also important is the possibility of extending the use of these particles for thermoplastic applications, which means a large expansion of the market.
5. Disadvantages of the use of preformed particles are the relatively high initial viscosity of the dispersions, and the possibility of producing particle agglomeration or even macroscopic segregation during storage or curing.

NOTATION

i = generic number of mers (i-mers) in a molecule of the modifier
M_n = number-average molar mass, kg $kmol^{-1}$
M_w = mass-average molar mass, kg $kmol^{-1}$
R = gas constant, J $K^{-1}mol^{-1}$
T = temperature, K

T_g = glass transition temperature, K
T_i = cure temperature, K
V = molar volume
W = mass fraction
x = conversion

Greek Letters

α = indicates one of the phases in a two-phase system
β = indicates one of the phases in a two-phase system
δ = solubility parameter, $MPa^{1/2}$
ΔG = Gibbs free energy of mixing per unit volume, J/m^3
η = viscosity, Pa s
ϕ = volume fraction
χ = Flory–Huggins interaction parameter

Subscripts

0 = initial value
CP = cloud point
crit = critical point
gel = gel point
i = indicates an i-mer of the modifier
M = modifier
ref = reference
TS = thermoset

REFERENCES

Bonnet A, Sautereau H, Pascault JP, *Macromolecules*, 32, 8517–8523 (1999).
Boogh L, Petterson B, Manson JAE, *Polymer*, 40, 2249–2259 (1999).
Bucknall CB, In *Comprehensive Polymer Science. The Synthesis, Characterization, Reactions and Applications of Polymers, 7,* Allen G and Bevington JC (eds), 1989, pp.27–50, New York: Pergamon Press.
Chen D, Pascault JP, Sautereau H, Vigier G, *Polym. Int.*, 32, 369–379 (1993).
Chen D, Pascault JP, Bertsch RJ, Drake RS, Siebert AR, *J. Appl. Polym. Sci.*, 51, 1959–1970 (1994).
Cruz-Ramos CA. In *Polymer Blends, Volume 2: Performance*, Paul DR, Bucknall CB. John Wiley & Sons, New York, 2000, pp. 137–175.
De Gennes PG. In *Scaling Concepts in Polymer Physics*, Cornell University Press, Ithaca, 1979.
Flory PJ, In *Principles of Polymer Chemistry*, Cornell University Press, Ithaca, New-York, 1953.
Girard-Reydet E, Sautereau H, Pascault JP, Keates P, Navard P, Thollet G, Vigier G, *Polymer*, 39, 11, 2269–2280 (1998).

Girard-Reydet E, Sautereau H, Pascault JP, *Polymer*, 40, 1677–1687 (1999).
Jansen BJP, Tamminga KY, Meijer HEM, Lemstra PJ, *Polymer*, 40, 5601–5607 (1999).
Kamide K, In *Thermodynamics of Polymer Solutions. Phase Equilibria and Critical Phenomena*, Jenkins AD (ed.), Elsevier, Amsterdam, 1990.
Landfester K, Tiarks F, Hentze HP, Antonietti M, *Macromol. Chem. Phys.*, 201, 1–5 (2000).
Lennon P, Espuche E, Sautereau H, Valot E, *J. Appl. Polym. Sci.*, 77, 857–865 (2000).
Lipic PM, Bates FS, Hillmyer MA, *J. Am. Chem. Soc.*, 120, 8963–8970 (1998).
Pascault JP, Williams RJJ. In *Polymer Blends, Volume 1: Formulation*, Paul DR, Bucknall CB (eds), John Wiley & Sons, New York, 2000, pp. 379–415.
Pascault JP, Valette L, Magny B, Barbeau P, French Patent 99/04042 (2000).
Paul DR, Bucknall CB (eds), *Polymer Blends, Volume 1: Formulation*, John Wiley & Sons, New York, 2000.
Riccardi CC, Borrajo J, Williams RJJ, *Polymer*, 35, 5541- 5550 (1994).
Riccardi CC, Borrajo J, Williams RJJ, Girard-Reydet E, Sautereau H, Pascault JP, *J. Polym. Sci. Part B, Polym. Phys.*, 34, 349–360 (1996).
Riew CK, Gillham JK (eds), *Rubber-Modified Thermoset Resins, Adv. Chem. Sci.*, 208, ACS, Washington DC, 1984.
Riew CK. (ed.), *Rubber Toughened Plastics, Adv. Chem. Sci.*, 222, ACS, Washington, DC, 1989.
Suspène L, Fourquier D, Yang YS, *Polymer*, 32, 1593–1604 (1991).
Tompa H. In *Polymer Solutions*, Butterworths, London, 1956.
Verchère D, Sautereau H, Pascault JP, Moschiar SM, Riccardi CC, Williams RJJ, Polymer, 30, 107–115 (1989).
Williams RJJ, Rozenberg BA, Pascault JP, *Adv. Polym. Sci.*, 128, 97–156 (1997)

9

Temperature and Conversion Profiles During Processing

9.1 INTRODUCTION

The precursors of thermosetting polymers are usually one of the ingredients of complex formulations. They may be present in very small amounts, as in the manufacture of abrasive disks where the thermoset acts as an agglutinant; in medium amounts, as in the case of filler-reinforced thermosets; or as the only components, in formulations used for encapsulation purposes. Apart from fillers, fibers, pigments, etc., some formulations contain rubber or thermoplastic modifiers that phase-separate upon the polymerization reaction (cure).

The cure cycle is the temperature vs time schedule used to polymerize the thermoset precursors. The selection of an adequate cure cycle has several purposes. What is desired is to obtain the final part without strains exceeding design tolerances, with a uniform conversion (usually close to the maximum possible conversion), without degradation produced by the high temperatures attained during the cure, with convenient morphologies (in the case of heterogeneous materials), and all this, must be achieved in the minimum possible time for economic reasons.

In this chapter, the evolution of conversion and temperature profiles during typical cure processes is discussed. This is useful for analyzing the possibility of attaining the maximum conversion, avoiding undesired high temperatures, and keeping the cycle time at practical values. These calculations are also necessary for estimating the distribution of stresses generated during the cure (Adolf *et al.*, 1998) or the distribution of morphologies generated in a rubber-modified thermoset (Williams *et al.*, 1987; Fang *et al.*, 1995). These two important problems will not be addressed here – the references mentioned will enable the reader to get acquainted with the complexities involved in the detailed analysis of these subjects.

A survey of typical processing technologies is first presented. Then, some general criteria for an adequate selection of the cure conditions (initial temperature, control of the temperature rise, influence of gelation, and vitrification), are analyzed. The remaining sections are devoted to discussing the influence of selected cure conditions on temperature and conversion profiles generated in several types of processing technologies: cure in heated molds, autoclave molding of graphite/epoxy composites, foaming, and shell molding. A range of adequate cure conditions are discussed for each one of the selected examples. A final design should consider the distribution of stresses generated during the cure as well as the corresponding strains.

9.2 EXAMPLES OF PROCESSING TECHNOLOGIES

The processing of formulations containing thermosetting polymers involves the simultaneous development of the network structure together with the morphology and shaping of the final material. Examples of processing technologies are casting, coating, foaming, molding, pultrusion, filament winding, etc. Some of these technologies (e.g., autoclave molding, pultrusion, filament winding, etc.), are particularly suitable for processing composites made of continuous fibers (glass, carbon, etc.), impregnated with the precursors of the thermosetting polymer.

Casting may be used for encapsulation purposes or for small-volume productions of shaped parts. Monomer casting is possibly the simplest processing technology. It may be used both with bifunctional monomers, such as methyl methacrylate and styrene, leading to linear polymers, as well as with polyfunctional monomers such as those used in epoxy formulations. Parts made from bifunctional monomers made be reshaped by heating because of their thermoplastic nature, while parts made from polyfunctional monomers reach the final shape during the polymerization. Processing errors in the case of thermoplastics may be repaired by reheating and recycling the material, but in the case of thermosets this is not possible. Therefore, cure cycles have to be analyzed in great detail to avoid processing problems.

Coating includes spread, roller, and spray processes for flat substrates, as well as the coating of complex parts using liquid dipping and fluidized-bed equipment.

Foaming requires thermosetting monomers that react very fast at the mixing temperature. The formulation is divided into two or more streams that circulate independently in the foaming machine and react at a fast rate when mixed together. Usually monomers and comonomers are separated, and both a catalyst and a blowing agent are added to one of the streams. The blowing agent evaporates when the heat evolved in the polymerization reaction increases the temperature to its boiling point in the mixture. From this time on, foaming takes place. The end of the rise time is determined by gelation (in open molds) or by the filling time (in closed molds).

Molding processes include compression molding, resin transfer molding (RTM), injection molding and reaction injection molding (RIM), autoclave molding, and several types of specific processes such as shell molding, which is used in foundries.

Compression molding involves the use of preforms (bulk molding compounds, sheets of glass fibers impregnated with the thermoset precursors, etc.) that are placed in a heated mold. When the mold is closed, the plastified preform flows to fill the mold and cure takes place in about 1–2 min. Then, the mold is opened and the final part ejected.

RTM involves the pumping of the thermoset precursors into a heated mold cavity containing preplaced fiber mats. Mold filling and fiber impregnation may be assisted by partial evacuation of the mold.

Conventional injection molding has been adapted for the production of thermosets. In this case, the volume of the ducts between the exit die and the mold cavity is minimized to reduce the scrap as much as possible. Capital investment and operational costs are much less in the case of RIM. In this process, two or more low-viscosity streams are accurately metered, mixed by impingement at high pressures (20–30 MPa) in a hydraulically operated mixhead, and injected into the mold cavity. Mold-fill times are in the order of 1 s and cycle times in the order of 1 min. Although the reaction is activated by mixing, the mold is heated to increase the cure rate for the material located close to the wall. Polyurethanes and polyurethanes–copolyureas are typically processed by RIM. Short glass fibers may be introduced in one of the streams, in which case the process is known as reinforced-reaction injection molding (RRIM). Difficulties associated with this process are the control of both the stability of the suspension of glass fibers and the abrasion that it generates during its flow in the molding machine.

Autoclave molding has been particularly adapted for the cure of laminates of preimpregnated plies of continuous fibers (prepregs). A lay-up of preimpregnated plies with fibers oriented according to the mechanical design of the final part is covered by a porous release film, bleeder cloths, a nonporous release film, and a pressure plate. The ensemble is enclosed in a vacuum bag, which is placed in an autoclave. While a vacuum is being made inside the bag, the temperature is increased up to a particular value, where it is held constant for a predetermined period. At this particular temperature, which must be high enough to lower the viscosity of the thermoset precursors but no so high as to begin the cure, pressure is applied to consolidate the part. This step provokes the elimination of resin, which is absorbed by the bleeder plies, and the increase in the fiber volume fraction from about 50% to about 65%. Then, the temperature is increased again to another plateau value, leading to the cure of the laminated panel.

In the pultrusion process, continuous fibers are impregnated by the thermoset precursors and are pulled through a heated die where the cure takes place: this produces continuous profiles of different shapes at a rate of the order of 1 m min^{-1}.

Filament winding also involves the impregnation of continuous fibers by the thermosetting formulation, and the winding of fibers onto a mandrel with angles that are previously determined in the mechanical design of the part. When the desired thickness is obtained, cure of the composite is performed by heating the ensemble using different procedures. This process is useful for producing tubes and tanks, but it may also be adapted to produce more complex shapes with a computer controlling the winding process using several axes.

Epoxies, unsaturated polyesters, and vinyl esters are typical thermosetting polymers used in pultrusion and filament winding applications.

9.3 SELECTION OF CURE CONDITIONS

9.3.1 Selection of the Initial Temperature

Usually it is desired to start with a low-viscosity formulation to permit the shaping of the part (for example, the filling of a mold). This puts a constraint on the initial temperature, T_0 , which must be high enough to obtain a low initial viscosity but not so high as to advance the cure during the shaping stage and cause premature gelation (in the case of mold filling, this will cause a "short shot").

Figure 9.1 shows a qualitative plot of the viscosity variation produced during the heating of a thermosetting polymer. Initially, viscosity decreases with the increase in temperature, but as cure progresses, an abrupt increase in viscosity is observed. The processability window is the range of T_0 values

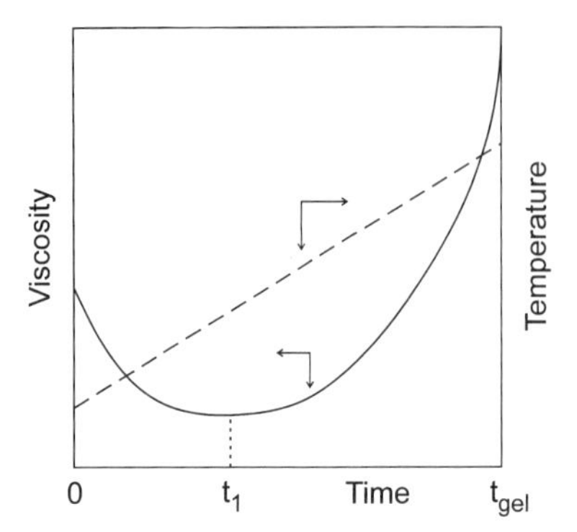

FIGURE 9.1 Variation of viscosity during the heating of a thermosetting polymer.

where both the viscosity and the polymerization rate are low enough to facilitate shaping of the part.

9.3.2 Control of the Temperature Rise

One of the main problems in the selection of a cure cycle is to achieve control of the exothermic polymerization reaction, particularly for the case of large parts. The exothermic character of the polymerization reaction arises from the evolution of the Gibbs free energy:

$$\Delta G = \Delta H - TS \qquad (9.1)$$

For the polymerization to proceed spontaneously, $\Delta G < 0$. But $\Delta S < 0$, because the system evolves to a more ordered state (the number of configurations in which free monomers may be placed in space decreases by the introduction of covalent bonds among themselves); thus, the entropy change does not favor polymerization. Then, the only possibility of getting $\Delta G < 0$ is to have a significantly exothermic reaction ($\Delta H < 0$) to counterbalance the unfavorable entropy change.

The actual temperature variation at a particular location of the part depends on the ratio of the heat dissipation rate to the heat generation rate. This ratio must be kept high enough to control the temperature increase and therefore avoid degradation reactions. As most polymers exhibit a very low thermal conductivity, heat dissipation can only be increased by using large surface areas per unit volume (thin parts) or fillers with high thermal conductivities (aluminum powder, graphite fibers, etc.).

It is also possible to act over the heat generation factor. This can be decreased by diluting the formulation with fillers or fibers. But this depends on the desired mechanical properties, which will be significantly modified by their presence. One method of decreasing the heat generation is to dilute the monomers with partially cured polymer, a procedure used in the manufacture of organic glasses based on diethylene glycol bis (allyl carbonate) (DADC or CR-39) (Fig. 9.2). The polymerization of this monomer can be stopped at the soluble, fusible stage. The dried polymers may be ground into powder, mixed with monomer and peroxide initiator, and molded by heating to give a glasslike hard, clear, thermoset plastic.

9.3.3 Influence of Gelation and Vitrification

Both gelation and vitrification have to be taken into account in the analysis of a cure cycle. As already mentioned, gelation must be avoided during mold filling. In some processing technologies such as in free-rise foaming, gelation determines the maximum height and the apparent density of the final foam (Sec. 9.6).

The influence of vitrification on the thermoset cure is very important, because once the material enters the glassy region the polymerization kinetics is severely retarded. On occasions one can take advantage of this situation, as in the once-in-the-life cure of large structures. By operating close to the vitrification curve (Chapter 4), any thermal excursion following an adiabatic trajectory is arrested by vitrification. Figure 9.3 shows a possible trajectory where periodic increases in the external temperature are followed by adiabatic heatings, ending in the vitrification curve. The cure proceeds until T reaches T_{g_∞} and full cure of the part is achieved.

But vitrification may be a problem when the cure is started at room temperature and no external heat source is provided (the only source of heat generation is the polymerization reaction). This is the case of UV (ultraviolet radiation), EB (electron beam), or X-ray curing processes.

In the cure with high-energy EB irradiation, polymerization proceeds via a free-radical mechanism, where the initiating species are formed by bond cleavage of monomers or other components of the formulation. EB curing of epoxy monomers via a cationic mechanism using onium salts as initiators (Chapter 2) is also possible. In every case, the irradiation is performed at room temperature, but a fast temperature increase usually occurs due to the very high polymerization rate (Glauser *et al.*, 1999).

$$CH_2=CH-CH_2O-\underset{\underset{O}{\|}}{C}-O-CH_2-CH_2-O-CH_2-CH_2-O-\underset{\underset{O}{\|}}{C}-OCH_2-CH=CH_2$$

FIGURE 9.2 Structure of diethylene glycol bis (allyl carbonate).

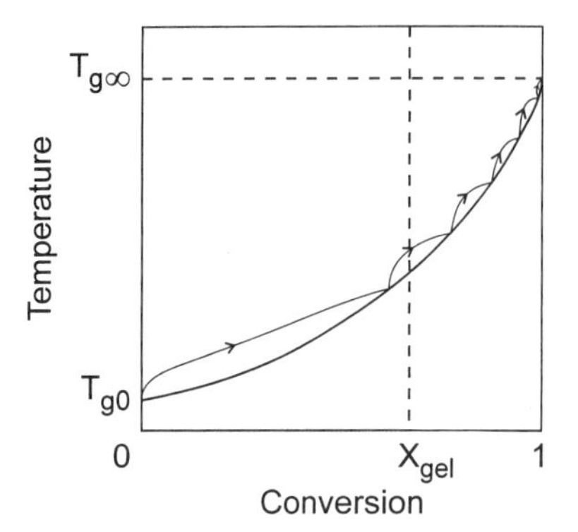

Figure 9.3 Temperature vs conversion transformation diagram representing a cure cycle along the vitrification curve.

The EB cure of vinyl ester resins (VE) based on acrylic and methacrylic end groups was reported by Glauser *et al.*, (1999). Both thin (2 mm thickness) and thick (20 mm diameter) specimens were cured using one to four sweeps of 2.5 Mrad each. After the first dose, the maximum temperature recorded in thick specimens was $T_{max} = 150°C$ for the acrylate-VE and $T_{max} = 110°C$ for the methacrylate-VE. Subsequent doses did not increase T_{max} beyond these values. The higher value obtained for the acrylate resin is due to the larger values of both the heat of polymerization and the propagation rate. Glass transition temperatures after the first dose, defined as the maximum in tan δ measured by dynamic-mechanical thermal analysis, were $T_\alpha = 150°C$ for the acrylate-VE and $T_\alpha = 156°C$ for the methacrylate-VE. Subsequent doses increase these values to 159°C and 177°C, respectively. But complete cure could not be attained. After the four doses, 7% and 15% of residual unsaturations remained in the acrylate-VE and methacrylate-VE, respectively. The temperature increase in thin specimens was significantly less important, as indirectly shown by the resulting T_α values. After four doses, T_α was equal to 117°C for the acrylate-VE and $T_\alpha = 141°C$ for the methacrylate-VE. Thin samples dissipated the heat generated in the polymerization in a more efficient way, so that vitrification took place at lower temperatures (and conversions).

It is surprising to realize how often the undercure produced by vitrification is completely ignored when performing the thermosetting polymerization by irradiation (UV, EB, X-ray) at room temperature. As there is no external heat source, once vitrification sets in conversion may only increase through the continuation of reaction in the glassy state. However, as we have discussed in Chapter 5, polymerization in the glassy state is a self-retarded and very slow process.

Vitrification also has a bearing on the microwave cure of thermosetting polymers. The use of microwave radiation has the potential advantage of significantly reducing cure times, because the polymerization begins at the same time in all the specimen (it is not necessary to wait for thermal energy to diffuse inside the sample). Energy transfer in microwave heating occurs by electrical dipolar coupling of the radiation to permanent dipole moments in the polymer. The rate of conversion of electrical energy into thermal energy is primarily determined by the dielectric loss factor of the material.

Srinivasan *et al.* (1997) analyzed the feasibility for microwave cure of cyanate ester resins (Chapter 2). They observed that when the temperature approached 160°C, polymerization took place at a very fast rate, and the consequent exothermic heat resulted in a tremendous acceleration of the heating rate. The sharp temperature rise was very hard to control by either detuning the cavity or by lowering the input power. Such an uncontrolled process resulted in a charred product. If microwave radiation was turned off at an early stage, reactions were arrested by vitrification, resulting in a partially cured material. As the magnitude of the dielectric loss peaks was very small, it was not possible to reheat the material by microwave radiation. Therefore, vitrification has a significant influence on microwave processing of thermosetting polymers.

An interesting method of eliminating the undercure caused by vitrification when using microwave radiation is to modify the formulation, including the use of polar thermoplastic that phase-separates during cure (Chapter 8). The thermoplastic material can convert microwave energy into heat, which enables the thermosetting polymer to devitrify and reach full cure.

An aspect that has not received enough attention is the influence of pressure on the vitrification curve (Chapter 10). For some processes that operate at very high pressures there is a significant shift of the vitrification curve to lower temperatures: for example, in the processing of phenolic molding compounds, where the polymerization may be arrested by vitrification at much lower temperatures than those predicted using T_g vs conversion values determined at ambient pressure.

9.4 CURE IN HEATED MOLDS

Most thermosetting materials are polymerized in heated molds. Figure 9.4 shows a schematic diagram of the mold; L is the part thickness, which is assumed to be much less than the other two dimensions. Therefore, the system may be modeled as a case of unidimensional heat transfer with simultaneous heat generation.

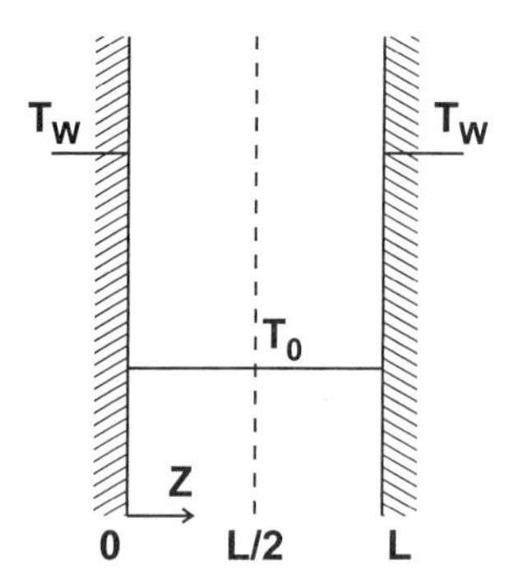

FIGURE 9.4 Schematic diagram of the heated mold (T_w = wall temperature, T_0 = initial temperature, L = part thickness).

The thermal energy balance for this case may be written as

$$\rho c_p \partial T/\partial t = \partial/\partial z(k_T \partial T/\partial z) + \rho(-\Delta H)R_c \tag{9.2}$$

where ρ, c_p, and k_T, are, respectively, density, specific heat and thermal conductivity (in the z-direction if the material is anisotropic). These parameters may vary with temperature (T) and conversion (x). R_c is the polymerization rate in time^{-1} units, which may be expressed by a set of kinetic equations or by a simple rate equation, $R_c(x,T)$, for single-path reactions (Chapter 5). The factor $(-\Delta H)$ is the reaction heat evolved at full conversion, which is expressed per unit mass of the formulation.

The rate at which conversion increases at any point is given by

$$\partial x/\partial t = R_c \tag{9.3}$$

Equation (9.2) simply states that the rate of heat accumulation in a differential volume (first term) is the difference between the heat flow that enters and leaves the volume element by thermal conduction (second term) plus the rate of heat generation by the polymerization reaction (third term).

To illustrate the system's behavior it will be assumed that ρ, c_p, k_T, and $(-\Delta H)$ are constant, and that the polymerization rate may be described by a simple second-order equation:

$$R_c = A(1 - x)^2 \exp(-E/RT) \tag{9.4}$$

where A (time^{-1}) is the pre-exponential factor, E is the activation energy, and R is the gas constant.

To complete the mathematical description it is necessary to state initial and boundary conditions. For illustration purposes, it will be assumed that the initial temperature (T_0) is uniform in the mold, the wall temperature

(T_w) remains constant during the cure, and no reaction occurs during the mold-filling stage. Therefore, the following conditions are stated for T(t,z) and x(t,z):

$$T(0, z) = T_0 \tag{9.5}$$

$$x(0, z) = 0 \tag{9.6}$$

$$T(t, 0) = T_w \tag{9.7}$$

$$(\partial T/\partial z)(t, L/2) = 0 \tag{9.8}$$

Equation (9.8) is a symmetry condition that enables us to state and solve the differential equations in half of the mold (from $z = 0$ to $z = L/2$).

Equation (9.2) may be rewritten in terms of the thermal diffusivity, $\alpha_T = k_T/\rho c_p$, and the adiabatic temperature rise (Chapters 4 and 5), $\Delta T_{ad} = (-\Delta H)/c_p$,

$$\partial T/\partial t = \alpha_T \partial^2 T/\partial z^2 + \Delta T_{ad} R_c \tag{9.9}$$

For generalization purposes, the system of differential equations with the initial and boundary conditions may be conveniently rewritten in terms of dimensionless variables: $z^* = z/L$, $t^* = A\exp(-E/RT_0)t$, $T^* = (T - T_0)/\Delta T_{ad}$. Substituting in Eqs. (9.3) to (9.9) leads to

$$\partial T^*/\partial t^* = W_1 \partial^2 T^*/\partial z^{*2} + (1 - x)^2 \exp[W_2 T^*/(W_3 + T^*)] \tag{9.10}$$

$$\partial x/\partial t^* = (1 - x)^2 \exp[W_2 T^*/(W_3 + T^*)] \tag{9.11}$$

$$T^*(0, z^*) = x(0, z^*) = 0 \tag{9.12}$$

$$T^*(t^*, 0) = W_4; \qquad (\partial T^*/\partial z^*)(t^*, 1/2) = 0 \tag{9.13}$$

The system's behavior depends on the values of four dimensionless groups:

$$W_1 = (\alpha_T/L^2)/[A\exp(-E/RT_0)] \tag{9.14}$$

$$W_2 = E/RT_0 \tag{9.15}$$

$$W_3 = T_0/\Delta T_{ad} \tag{9.16}$$

$$W_4 = (T_w - T_0)/\Delta T_{ad} \tag{9.17}$$

W_1 represents the ratio between the heat diffusion rate and the initial heat generation rate. When $W_1 \to \infty$, heat diffuses at a much higher rate than it is generated. This leads to an isothermal cure at $T = T_w$ (or $T^* = W_4$). When $W_1 \to 0$, the material behaves as a thermal insulator and an adiabatic cure takes place. Dividing Eq. (9.10) with $W_1 = 0$ by Eq. (9.11), and integrating, leads to

$$T^* = x \qquad (9.18)$$

The maximum temperature for the adiabatic cure is $T^* = 1$ or

$$T = T_0 + \Delta T_{ad} \qquad (9.19)$$

For a given formulation, α_T, A, and E are fixed; therefore W_1 may be only varied through T_0 and L. But the initial temperature (T_0) may be selected in a relatively narrow range, because of the necessity of having a low viscosity together with a low polymerization rate during the filling stage. Thus, the specimen thickness (L) is the parameter that plays the most important role in determining the evolution of temperature and conversion profiles in the part: a "small" thickness will lead to an isothermal cure and a "large" thickness to an adiabatic behavior. But how small is "small" and how large is "large" depends on the values of the four dimensionless groups that determine the system's behavior.

The parameter W_2 is called the Arrhenius number. It varies from values of about 15 for low-activation energy systems (e.g., thermosetting polyurethanes), to values of about 40 for high-activation energy systems (e.g., phenolic molding compounds).

The parameter W_3 depends strongly on the presence of fillers, fibers, or any inert modifier in the formulation. For unfilled thermosetting polymers with high values of ΔT_{ad}, W_3 may take values of about 1–2. For formulations including a large fraction of fillers and moderate values of ΔT_{ad}, W_3 may increase to between 4 and 8.

Finally, W_4 depends on the value selected for the wall temperature, which is usually determined by the need to reduce the molding time as much as possible without leading to thermal degradation of the part. Values of W_4 are usually in the range of 0–0.7. Low values are used for systems that react very fast at the mixing temperature (T_0), such as formulations used in RIM applications. High values correspond to the usual case, where the polymerization is activated by heat transfer from the wall.

Depending on the shape of temperature vs conversion trajectories, vitrification may take place at particular locations in the part. If this happens, the constitutive equations describing the kinetics must include the diffusional resistance that characterizes the sharp decrease in the polymerization rate when entering the vitrification region. In particular, vitrification can occur at the wall if $T_w < Tg_\infty$. In the examples that illustrate this section, it will be assumed that vitrification does not take place, but in the following section the influence of vitrification on the cure in a heated mold will be discussed.

Differential Eqs (9.10) and (9.11), with initial and boundary conditions (9.12) and (9.13), may be numerically solved for different sets of values of the four dimensionless parameters, W_1–W_4 (Williams *et al.*, 1985). To illustrate the evolution of temperature and conversion profiles during the cure, values of W_2–W_4 will be kept constant and W_1 will be varied to simulate the influence of the part thickness. The particular case of $W_2 = 40$, $W_3 = 1.5$, and $W_4 = 0.125$ will be analyzed. This represents a process characterized by high values of both the activation energy and the adiabatic temperature rise.

Figure 9.5 shows the evolution of temperature and conversion profiles for $W_1 = 0.01$ (a relatively thick part). As a result of the heat flow from the wall and the high value of the activation energy, the cure takes place first in the proximity of the wall, followed by the material located at the core. But as the material close to the wall is kept at T_w, it cures at a relatively slow rate and determines the molding time (cycle time). The maximum temperature is attained at an intermediate position and is slightly higher than the adiabatic value, $T_{max} > T_0 + \Delta T_{ad}$. This results from the extra heating arising from the presence of a boundary at T_w. The material located at the core undergoes a strictly adiabatic cure, leading to a maximum temperature of $T_0 + \Delta T_{ad}$.

Figure 9.6 shows the evolution of temperature and conversion profiles when the specimen thickness is reduced; i.e., for $W_1 = 0.1$. Again, the material located close to the wall polymerizes at a fast rate, originating thermal and conversion fronts that travel to the core and to the wall. The maximum temperature is obtained at an intermediate location and is higher than that attained in the previous case: $T_0 + \Delta T_{ad} < T_{max} < T_w + \Delta T_{ad}$.

When the thickness is further reduced, giving $W_1 = 1$, a similar situation is observed (Fig. 9.7). But now the maximum temperature takes place at the core and its value is $T_{max} > T_w + \Delta T_{ad}$. This is because the cure of this material is not directly affected by the wall at T_w but by the high-temperature moving front that travels at a fast rate to the core of the part.

Decreasing the part thickness to obtain $W_1 = 10$ produces a dramatic change in the way in which the part cures, as observed in Fig. 9.8. The material located at the core cures first, generating conversion and temperature fronts advancing at a very fast rate to the wall, which acts as a heat sink (notice the short period of time in which most of the material is cured). The maximum temperature is observed at an intermediate position and is lower than in the previous case: $T_0 + \Delta T_{ad} < T_{max} < T_w + \Delta T_{ad}$.

For very thin parts ($W_1 = 100$), the cure proceeds almost isothermally at the wall temperature (Fig. 9.9). A slight temperature increase is rapidly dissipated to the wall, and conversion profiles are almost uniform in the whole specimen.

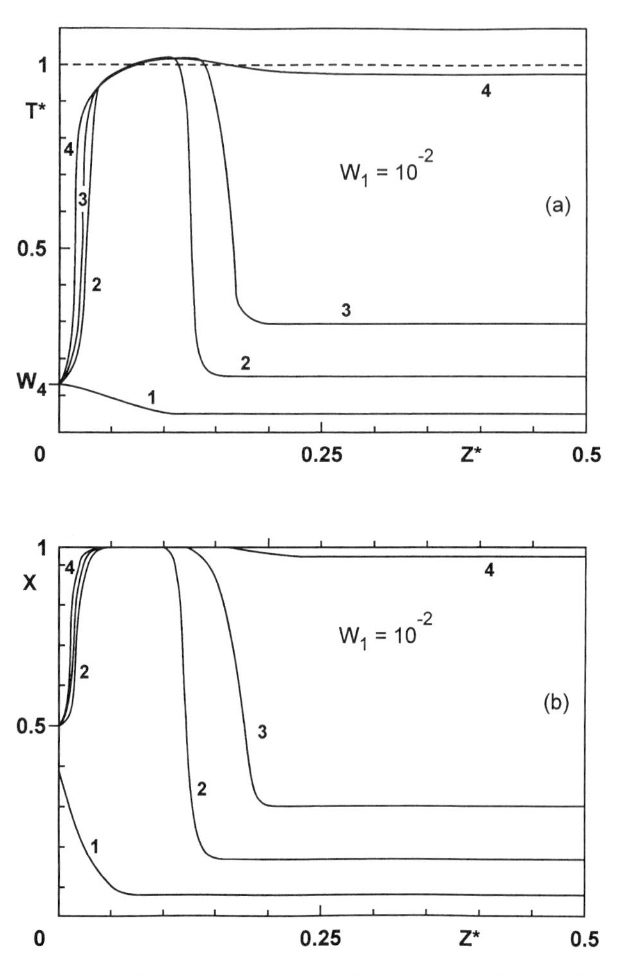

Figure 9.5 Temperature (a) and conversion (b) profiles for $W_1 = 0.01$; (1) $t^* = 0.0312$, (2) $t^* = 0.0417$, (3) $t^* = 0.0432$, and (4) $t^* = 0.0434$. (Reprinted from Williams *et al.*, 1985 by courtesy of Marcel Dekker, Inc.)

Now, the effect of varying W_2, W_3 to W_4 will be discussed. Figure 9.10 shows the maximum dimensionless temperature that is attained during the cure as a function of W_1, and different sets of W_2, W_3, and W_4. The thickness increases to the left ($W_1 \sim L^{-2}$). Cases 4 and 5 correspond to polymerization kinetics exhibiting high values of the activation energy (as in the example illustrated by Figs 9.5–9.9). For these cases, T_{max} exceeds $T_0 + \Delta T_{ad}$ significantly. However, for systems characterized by a low value of the activation energy (cases 1 and 2), the maximum temperature rise varies

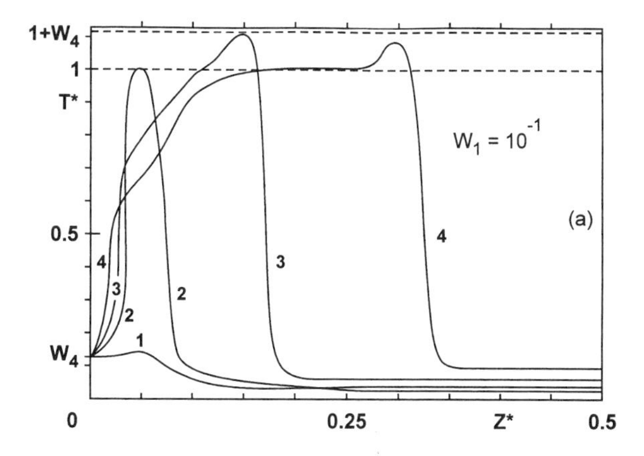

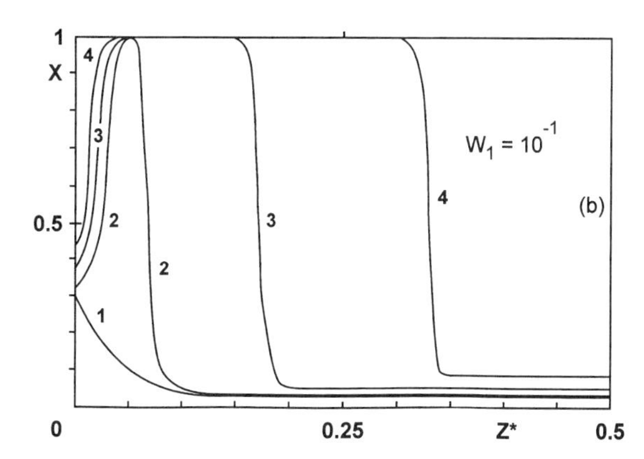

FIGURE 9.6 Temperature (a) and conversion (b) profiles for $W_1 = 0.1$; (1) $t^* = 0.0198$, (2) $t^* = 0.0228$, (3) $t^* = 0.0284$, and (4) $t^* = 0.0365$. (Reprinted from Williams *et al.*, 1985 by courtesy of Marcel Dekker Inc.)

smoothly from the adiabatic limit (for a large thickness) to the isothermal limit (for thin specimens). This is a typical behavior for the cure of polyurethane formulations.

When applying the mathematical model to analyze the cure of a particular system, one has to be aware of the following issues:

1. The quality of the available kinetic equation to fit experimental data in a wide range of isothermal and scanning rate conditions (the numerical solution exhibits a very high parametric sensitivity on the values of the activation energies).

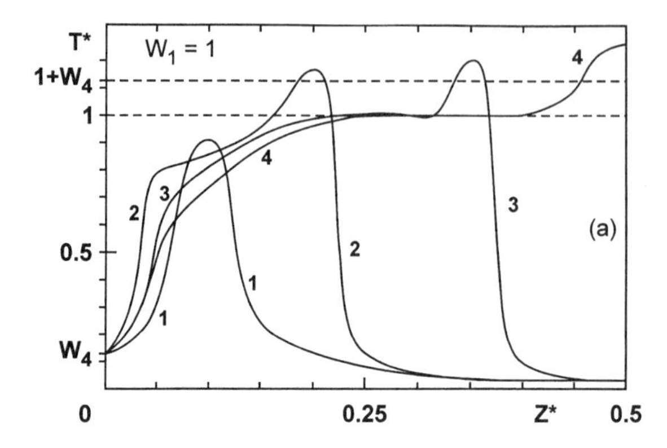

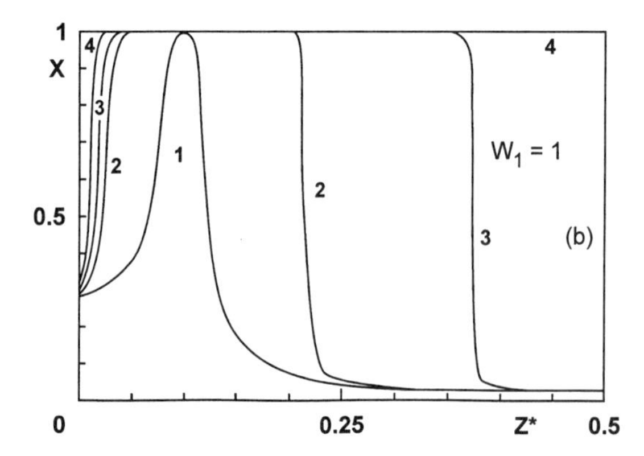

FIGURE 9.7 Temperature (a) and conversion (b) profiles for $W_1 = 1$; (1) $t^* = 0.0179$, (2) $t^* = 0.0186$, (3) $t^* = 0.0202$, and (4) $t^* = 0.0216$. (Reprinted from Williams *et al.*, 1985 by courtesy of Marcel Dekker Inc.)

2. The actual boundary condition at the wall (it is very difficult to get a true isothermal condition at the wall; a better boundary condition arises by stating the continuity of the heat flux at the wall) (Gorovaya and Korotkov, 1996).
3. The possibility of vitrifying part of the material during the cure cycle and the availability of the vitrification curve and of constitutive equations to represent the kinetics in the glassy state.

With these concepts in mind, good predictions of cure cycles may be achieved.

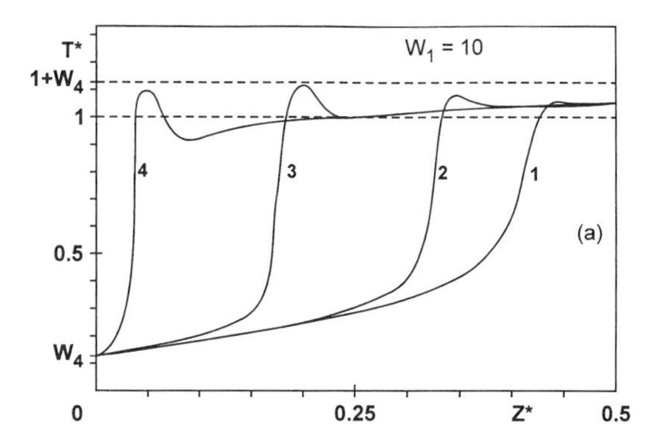

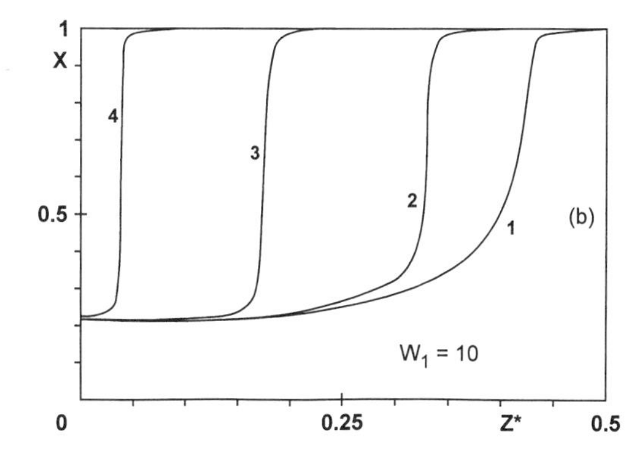

FIGURE 9.8 Temperature (a) and conversion (b) profiles for $W_1 = 10$; (1) $t^* = 0.01251$, (2) $t^* = 0.01255$, (3) $t^* = 0.01268$, and (4) $t^* = 0.01291$. (Reprinted from Williams *et al.*, 1985 by courtesy of Marcel Dekker Inc.)

9.5 AUTOCLAVE MOLDING OF GRAPHITE/EPOXY COMPOSITES

The autoclave molding of graphite/epoxy composites based on tetraglycidyl 4,4′-diaminodiphenylmethane (TGDDM), crosslinked with 4,4′-diaminodiphenylsulfone (DDS), is analyzed in this section (Williams *et al.*, 1990). Figure 9.11 shows a scheme of the mold configuration with the lay-up of preimpregnated plies (prepregs) placed between a heated plaque and the bleeder cloth.

A typical cure cycle is shown in Fig. 9.12. Somewhere in the plateau at 135°C the autoclave pressure is applied. This step provokes the elimination of resin, which is absorbed by the bleeder cloth, and the increase in the fiber

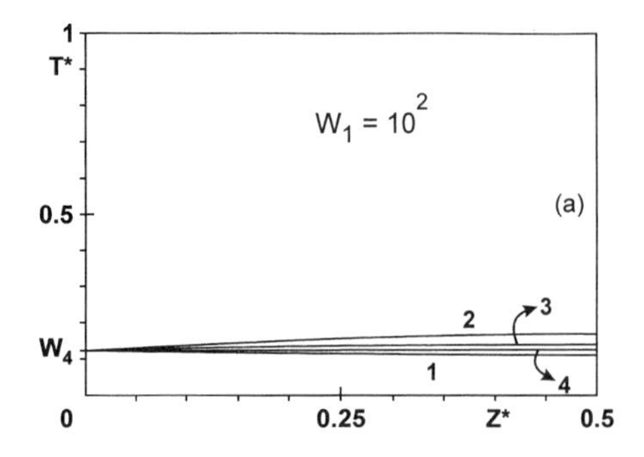

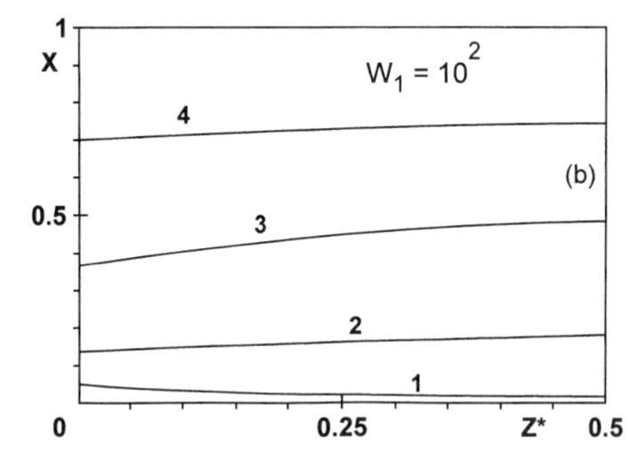

FIGURE 9.9 Temperature (a) and conversion (b) profiles for $W_1 = 100$; (1) $t^* = 0.0019$, (2) $t^* = 0.0069$, (3) $t^* = 0.0257$, and (4) $t^* = 0.1043$. (Reprinted from Williams *et al.*, 1985 by courtesy of Marcel Dekker Inc.)

volume fraction from its initial value in the prepreg (close to 50%) to its final value in the composite (close to 65%). The slow increase of the wall temperature after the plateau favors the partial dissipation of the reaction heat. The slow cooling at the end of cure enables the partial relaxation of residual stresses. In this case, the quality of the molded part is by far more important than the length of the cure cycle.

The evolution of temperature and conversion profiles in the part may be simulated by solving Eqs (9.2) and (9.3), with the following initial and boundary conditions:

$$t = 0, \qquad x = 0, \qquad T = 298\text{K}, \qquad V_f = 0.50$$

$$t = 45 \text{ min}, \qquad V_f = 0.65 \tag{9.20}$$

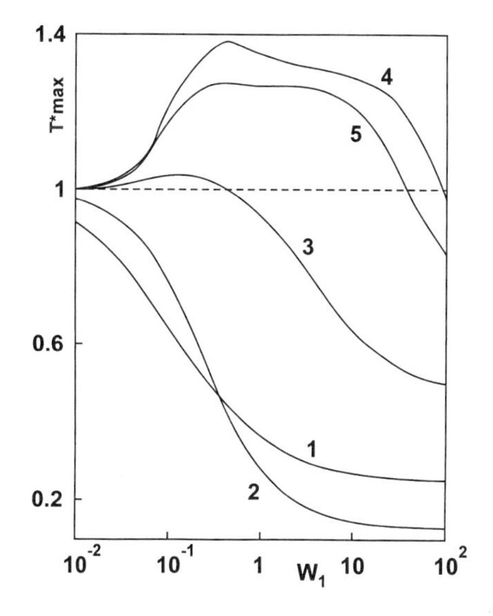

FIGURE 9.10 Maximum temperature attained during the cure as a function of W_1, for different sets of W_2, W_3, and W_4; (1) 17, 7, and 0.25; (2) 17, 3.5, and 0.125; (3) 25, 4, and 0.5; and (4) 30, 3, and 0.7; and (5) 40, 5 and 0.7. (Reprinted from Williams *et al.*, 1985 by courtesy of Marcel Dekker Inc.)

$$z = L, \qquad \partial T/\partial z = 0 \text{ (adiabatic boundary)}$$

$$z = 0, \qquad T = T_w(t)$$

V_f is the volume fraction of fibers, L is taken as 2 cm to simulate the cure of a thick part, and $T_w(t)$ is the cure cycle described in Fig. 9.12. It is (arbitrarily) assumed that the boundary between the laminate and the bleeder cloth behaves adiabatically. A better simulation would have to consider the heat transfer to the bleeder cloth.

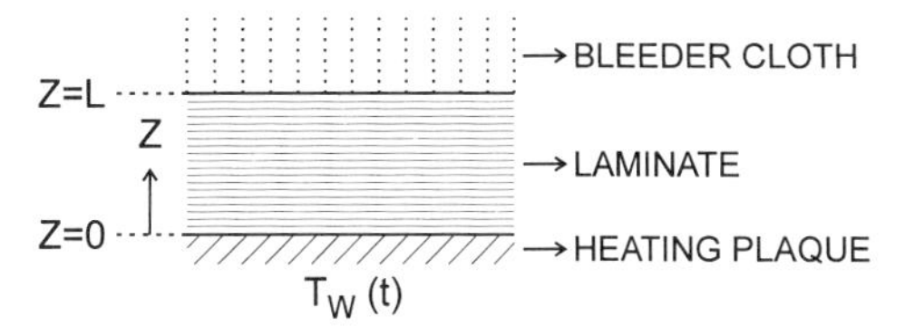

FIGURE 9.11 Lay-up of preimpregnated plies placed between a heated plaque and the bleeder cloth.

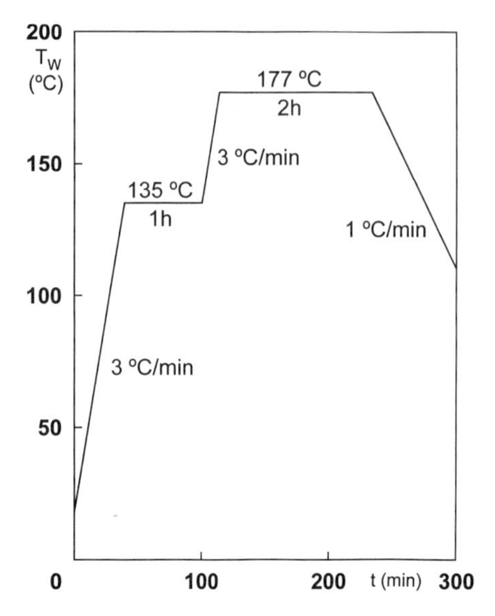

FIGURE 9.12 Typical autoclave cure cycle for a graphite/epoxy composite (TGDDM–DDS).

The constitutive equations for the cure kinetics and the values of the different parameters that are necessary to solve the differential equations were taken from the experimental work performed by Mijovic and Wang (1988):

$$R_c = (k_1 + k_2 x^{1.53})(1 - x)^{0.47} \tag{9.21}$$

$$k_1 = A_1 \exp(-E_1/RT) \tag{9.22}$$

$$k_2 = A_2 \exp(-E_2/RT) \tag{9.23}$$

with $A_1 = 9.43 \times 10^6\ \mathrm{min^{-1}}$, $E_1 = 77.2\ \mathrm{kJ\,mol^{-1}}$, $A_2 = 1.47 \times 10^5\ \mathrm{min^{-1}}$, $E_2 = 58.2\ \mathrm{kJ\ mol^{-1}}$.

The relationship between glass transition temperature and conversion is given by (Williams *et al.*, 1990):

$$(T_g - T_{g0})/(T_{g\infty} - T_{g0}) = \lambda x/[1 - (1 - \lambda)x] \tag{9.24}$$

with $T_{g0} = 258$ K, $T_{g\infty} = 505$ K and $\lambda = 0.51$.

The cure kinetics is calculated from Eq. (9.21) outside the vitrification region ($T > T_g$). For $T \leq T_g$, the reaction rate is (arbitrarily) neglected.

The density of the composite is defined as

$$\rho = \rho_m(1 - V_f) + \rho_f V_f \tag{9.25}$$

where densities of matrix (ρ_m) and fibers (ρ_f) are given by

$$\rho_m(\text{kg m}^{-3}) = 1261.3 - 0.1T(K) + 90\ x \qquad (x < 0.45)$$

$$\rho_m(\text{kg m}^{-3}) = 1301.3 - 0.1T(K) \qquad (x \geq 0.45) \tag{9.26}$$

$$\rho_f(\text{kg m}^{-3}) = 1790$$

The mass fraction of fibers is calculated by

$$w_f = V_f \rho_f / \rho \tag{9.27}$$

The heat of reaction at full conversion has the value

$$(-\Delta H)(\text{J kg}^{-1}) = 6.1 \times 10^5 (1 - w_f) \tag{9.28}$$

The specific heat of the composite is defined as

$$c_p = (1 - w_f)c_{pm} + w_f c_{pf} \tag{9.29}$$

where

$$c_{pm}(\text{J kg}^{-1}\text{K}^{-1}) = 1277.3 + 2.503T(K) - 590.7\ x$$

$$c_{pf}(\text{J kg}^{-1}\text{K}^{-1}) = 933.4 + 0.913\ T(K) - 4.081 \times 10^7 / T^2(K^2) \tag{9.30}$$

The thermal conductivity of the composite may be calculated by combining the thermal conductivities of the matrix, k_{Tm}, and fibers, k_{Tf}, using the parallel model at the corresponding volume fraction of fibers, V_f :

$$k_T = k_{Tf} V_f + k_{Tm}(1 - V_f) \tag{9.31}$$

where

$$\begin{aligned} k_{Tm}(\text{J s}^{-1}\text{m}^{-1}\ \text{K}^{-1}) &= 0.161 + [0.00147\ T(K) - 0.417]\ x \\ k_{Tf}(\text{J s}^{-1}\ \text{m}^{-1}\ \text{K}^{-1}) &= 2.514 \end{aligned} \tag{9.32}$$

Figure 9.13 shows the evolution of conversion and temperature profiles in the part. An almost uniform cure takes place up to the time at which T_w reaches the plateau at 177°C. At 132 min, both conversion and temperature attain the maximum values at the adiabatic boundary ($z = L$). At this time, T_{max} exceeds T_w by about 90°C, which may produce an incipient thermal degradation. At 168 min, the final conversion profile is attained (T is almost uniform in the part and equal to T_w). The maximum conversion of the material located close to the metallic plaque is $x_m = 0.872$ due to the effect of vitrification (and the assumption of $R_c = 0$ when $T \leq T_g$). Therefore, vitrification produces a conversion profile in the cured part. A postcure step would be necessary to completely cure the composite material.

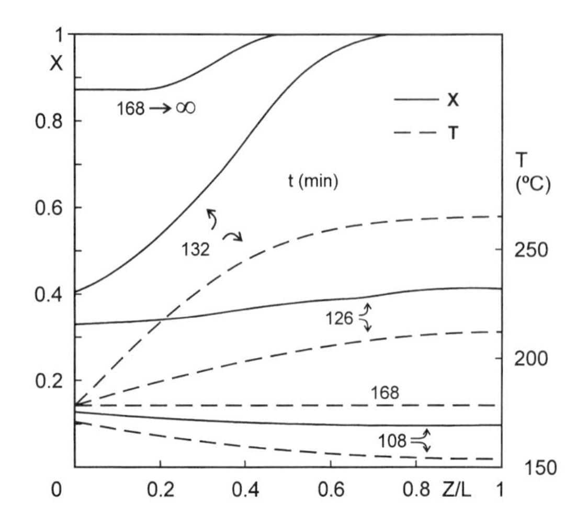

Figure 9.13 Evolution of conversion and temperature profiles for a graphite / epoxy composite of thickness L = 2 cm, cured in an autoclave. (Williams *et al.*, 1990. Reprinted with permission of SPE)

9.6 FOAMING

Thermosetting polymers may be foamed by adding a blowing agent to the formulation as well as a convenient concentration of a catalyst to make the system highly reactive at the mixing temperature (T_0). As the exothermic reaction takes place, the boiling point of the foaming agent is reached (cream time, t_c). Its evaporation produces bubbles that become trapped in the viscous reaction mass, producing its expansion. A relatively high viscosity and the presence of a surfactant are necessary conditions for obtaining a uniform dispersion of bubbles in the thermosetting polymer. The foam rises until the mold is filled (closed mold) or until gelation takes place (free rise).

Here, the case of free rise in an open mold is analyzed. Because of the very low thermal conductivity of a plastic foam, an adiabatic process will be considered. The nonadiabatic case in a closed mold leads to integral-skin foams. These foams exhibit a mass density gradient, with unfoamed skins in contact with the mold walls (Marciano *et al.*, 1986).

The main assumptions of the present analysis are

1. The system is adiabatic.
2. Only the blowing agent is vaporized.
3. Vaporization takes place at equilibrium and is not limited by the rate of mass transfer to the growing bubbles.

4. Bubbles are retained in the foam, effectively contributing to its expansion.
5. Foaming is arrested by gelation of the thermosetting polymer (evaporation of the blowing agent after this point only contributes to cell rupture, leading to the opening of the cellular structure).

The thermal energy balance, expressed per unit polymer mass, may be written as:

$$(c_p + L_{B0}c_{pB})dT/dt = (-\Delta H)R_c - \Delta H_v \, dG_B/dt \tag{9.33}$$

where c_p is the specific heat of the formulation devoid of blowing agent, c_{pB} is the specific heat of blowing agent (average of values in liquid and gaseous states), L_{B0} is the initial mass of blowing agent per unit polymer mass, ΔH_v is the heat of vaporization of blowing agent per unit mass, and G_B = mass of blowing agent in the gas phase per unit polymer mass.

The balance of reactive functionalities is given by

$$dx/dt = R_c \tag{9.34}$$

The cream temperature (T_c) is the temperature at which the foam starts to rise. It decreases by increasing the initial blowing agent concentration in solution. An experimental solubility curve plotted as L_B (mass of blowing agent in the liquid phase per unit polymer mass) vs T_c may be experimentally obtained.

The vaporization rate of blowing agent may be calculated as

$$dG_B/dt = 0, T < T_c \tag{9.35}$$

$$dG_B/dt = -dL_B/dt = -(dL_B/dT)(dT/dt), \qquad T \geq T_c \tag{9.36}$$

T_c is obtained from the solubility curve at the particular value of L_{B0} used in the formulation. The factor (dL_B/dT) is the slope of the solubility curve at any temperature (arising from the hypothesis of vaporization at equilibrium).

The foam density at any time during the expansion period results from

$$\rho = (1 + L_{B0})/(\rho_p^{-1} + G_B/\rho_G) \tag{9.37}$$

where ρ_p is the polymer density (assumed constant) and ρ_G is the gas density, calculated from the ideal gas law:

$$\rho_G = pM_B/RT \tag{9.38}$$

where p is the atmospheric pressure.

The final density results from Eq. (9.37) by substituting the values of G_B and ρ_G at $x = x_{gel}$.

The foam expansion may be characterized by the expansion ratio, defined as the ratio of the height H at any time, with respect to the initial height, H_0:

$$H/H_0 = \rho_0/\rho \tag{9.39}$$

which implicitly assumes that no mass is lost from the system up to the gel conversion.

A comparison between theoretical predictions and experimental results is illustrated for the foaming of a phenolic resin of the resol type (Vázquez and Williams, 1986).

An NaOH-catalyzed resol (molar ratio NaOH/phenol = 0.03) was synthesized using a formaldehyde/phenol molar ratio of 1.5. After 3 h reaction at 70°C, oxalic acid was added, carrying the pH to 6–6.5, and water distilled under vacuum until the viscosity (at 20°C) increased to 1.5–2 kg m^{-1} s^{-1}. The condensation of the resol takes place through the free CH_2OH groups located in the o, o′, and p positions of the phenolic ring (Chapter 2).

The fast condensation required by the foaming process may be achieved by using strong acids. In this case, the acid catalyst consisted of 12.5 parts H_3PO_4 (85%) and 1–15 parts H_2SO_4 (49%) in 100 parts of resol by weight. Condensation proceeds through the following reactions:

$$R - CH_2OH + R'H \rightarrow R - CH_2 - R' + H_2O \tag{9.40}$$

$$R - CH_2OH + R'CH_2OH \rightarrow R - CH_2 - O - CH_2 - R' + H_2O \tag{9.41}$$

$$R - CH_2 - O - CH_2 - R' \rightarrow R - CH_2 - R' + CH_2O \tag{9.42}$$

where $R'H$ represents any free position located in an o, o′, or p position of a phenolic ring.

The polymerization kinetics were determined using the adiabatic temperature rise method (Chapter 5), in a pressurized vessel, to avoid the vaporization of volatiles (and the consequent heat losses). A second-order kinetics was obtained:

$$R_c = A(1 - x)^2 \exp(-E/RT) \tag{9.43}$$

with $A = 8.86 \times 10^5\ s^{-1}$ and $E = 48.4\ kJ\ mol^{-1}$.

Phenolic foam formulations consisted of the following components, in parts by weight: resol, 100; sulfuric acid (49%), 10; phosphoric acid, 12.5; 1,1,2-trichloro-1,2,2-trifluoromethane (R113, blowing agent), 5; polyoxyethylene sorbitan monostearate (surfactant), 0.6.

Mixing was carried out with a stirrer at 2200 rpm and the formulation was rapidly poured into a large mold provided with thermocouples, located at various heights, to monitor the foam-rising process.

Figure 9.14 shows the influence of the blowing agent concentration on the experimental cream temperature, T_c. For mass ratios of R113 higher than 0.03, T_c is less than the normal boiling point of R113. This means that the vapor pressure of other volatiles (water, formaldehyde) is also contributing to the total pressure. The curve may be described by the following equation:

$$L_B = 4.13 \times 10^5[T(^\circ C)]^{-4.233} \tag{9.44}$$

For $L_B = L_{B0} = 5/123.1 = 4.062 \times 10^{-2}$, the cream temperature is $T_c = 45.2^\circ C = 318.4$ K.

Parameters appearing in the equations describing the system's behavior had the following values: $c_p = 1950$ J kg^{-1}K^{-1}, $c_{pB} = 962$ J kg^{-1}K^{-1}, $\Delta H_v = 1.52 \times 10^5$ J kg^{-1}, $(\Delta H) = -1.52 \times 10^5$ J kg^{-1} (by coincidence it is the same absolute value as the vaporization heat of the blowing agent), $\rho_0 = 1240$ kg m^{-3}, $(pM_B/R) = 2285$ K kg m^{-3}, $T_0 = 288.5$ K, and $H_0 = 1.54$ cm.

Figure 9.15 shows predicted and experimental values of temperatures and relative heights achieved during the foaming process. Predicted values were obtained by solving the differential equations. A reasonable description of the foaming process is obtained, although the foam expands more than theoretically expected from the blowing agent evaporation.

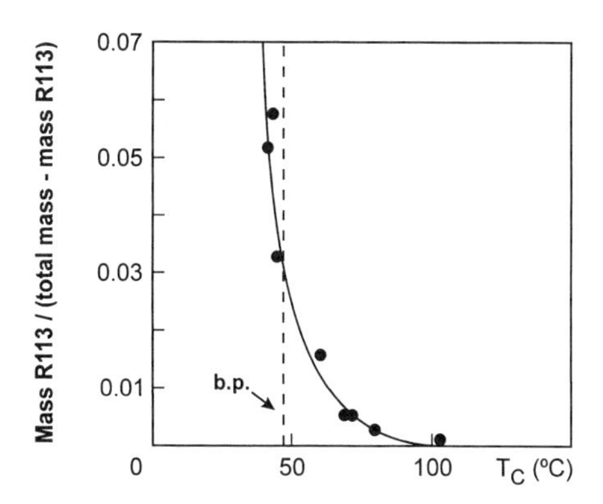

FIGURE 9.14 Influence of the blowing agent concentration on the observed cream temperature; the dashed line represents the normal boiling point of R113. (Vázquez and Williams, 1986. Reprinted with permission from RAPRA Tech. Ltd)

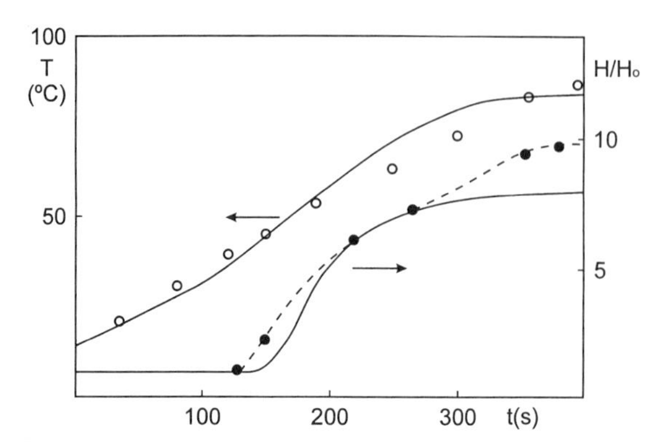

FIGURE 9.15 Temperature and relative height evolution during the foaming process. Full curves are the theoretical values predicted by solving the differential equations describing the system's behavior. (Vàzquez and Williams, 1986. Reprinted with permission from RAPRA Tech. Ltd)

The final density was predicted from Eq. (9.37) assuming that all the initial blowing agent evaporates and expands up to T $=355$ K (experimental temperature at which the foam stops rising). The predicted value was $\rho_{theor} = 145$ kg m^{-3}, while the experimental value was $\rho_{exp} = 128$ kg m^{-3}.

The shape of the experimental H/H_0 vs T curve shows an overexpansion at T close to 55°C. A good hint to explain this overexpansion is the characteristic formaldehyde smell, easily noticed when temperature reaches values close to 70°C. The fraction of the initial formaldehyde amount (F) which must evaporate to account for the observed overexpansion may be calculated from:

$$\rho_{exp} = (1 + L_{B0})/(\rho_p^{-1} + G_B/\rho_G + F/\rho_F) \tag{9.45}$$

A value of $F = 1.0 \times 10^{-3}$ is necessary to account for the experimental density of the final foam. This represents 0.4% of the formaldehyde initially added to synthesize the resol, and is in the order of its free formaldehyde content.

9.7 SHELL MOLDING (CRONING PROCESS)

As a final example of modeling thermoset processing, the process called the shell molding or Croning process will be analyzed. This is just one of many processes found in a variety of industries (automobile, electronics, sports goods, furniture, etc.) that involve the cure of thermosetting polymers. The selected example is a typical process used in foundries.

Among the foundry processes, the shell molding or Croning process has evolved into an accepted production method using both shell molds and shell cores. Invented in Germany in 1943 by J. Croning, this process has been industrially adapted because it offers high production rates. A scheme of the process is described in Fig. 9.16. Its principles are as follows:

1. A set of metal pattern plates is heated to 200°C–280°C and coated with a silicone release agent.
2. Dry sand, precoated with a phenolic novolac–hexamethylenetetramine (hexa) formulation, is dropped upon the pattern. After about 15 s–30 s, the mold is turned over (180°C), leaving a shell of several millimeters of sand adhered to the metal pattern. The rest of the sand that falls upon mold inversion is supplemented with fresh coated sand and used in the following cycle.
3. The cured shell is ejected from the mold and postcured for several minutes in a continuous oven, at temperatures close to 300°C.
4. Shell parts are assembled and positioned in the final mold where casting of the liquid metal takes place.
5. The phenolic binder is carbonized in the casting process but the resistance of the shell part is enough to enable the formation of a solid metal skin before the shell part disintegrates. The metallic part is removed from the mold and the remaining sand with the burnt resin is brushed aside.

Modeling of this process may be performed by assuming the validity of the following hypotheses (Aranguren *et al.*, 1984):

- unidimensional heat transport
- semi-infinite medium (short contact times)
- constant density, ρ, and specific heat, c_p, but variable thermal conductivity, $k_T(T)$

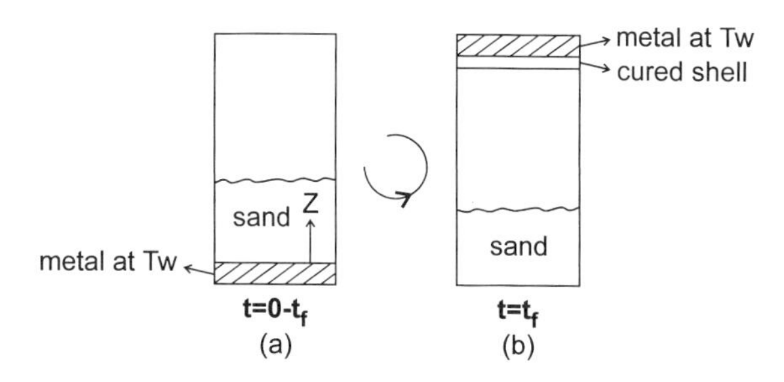

FIGURE 9.16 Scheme of the shell molding process: (a) at $t = 0 - t_f$, (b) at $t = t_f$ (inversion time).

- negligible contribution of the heat of reaction, due to the small mass fraction of thermosetting resin in the coated sand (about 4%)
- the thickness of the shell remaining adhered to the metallic plate at the inversion time, t_f, is defined as the section where the conversion of the thermosetting polymer is $x \geq x_{gel}$.

With the previous hypotheses, balances of thermal energy and reactive functionalities may be written as

$$\rho c_p \partial T/\partial t = \partial/\partial z(k_T(T)\partial T/\partial z) \tag{9.46}$$

$$\partial x/\partial t = R_c \tag{9.47}$$

Initial and boundary conditions are

$$\begin{aligned} &t = 0, \quad x = 0, \quad T = T_0 \\ &z = 0, \quad T = T_w \\ &z \to \infty, \quad T = T_0 \end{aligned} \tag{9.48}$$

Since the contribution of the reaction heat was neglected, Eqs (9.46) and (9.47) are uncoupled.

An approximate analytical solution of Eq. (9.46) may be obtained by assuming that heat transport is confined to a thickness $\delta(t)$, verifying

$$T = T_0 \text{ and } \partial T/\partial z = 0 \text{ for } z = \delta(t) \tag{9.49}$$

A possible solution is

$$(T - T_0)/(T_w - T_0) = (1 - z/\delta(t))^m, \qquad m > 1 \tag{9.50}$$

Integrating both sides of Eq. (9.46) on z, between 0 and $\delta(t)$, we obtain

$$\begin{aligned} \rho c_p \int_0^{\delta} (\partial T/\partial t)dz &= (k_T \partial T/\partial z)(z = \delta) - (k_T \partial T/\partial z)(z = 0) \\ &= -k_T(T_w)(\partial T/\partial z)(z = 0) \end{aligned} \tag{9.51}$$

From Eq. (9.50),

$$\partial T/\partial t = (T_w - T_0)m(1 - z/\delta)^{m-1}(z/\delta^2)d\delta/dt \tag{9.52}$$

Then, integrating with respect to z leads to

$$\int_0^{\delta} (\partial T/\partial t)dz = [(T_w - T_0)/(m + 1)]d\delta/dt \tag{9.53}$$

Substituting Eq. (9.53) and the derivative of T with respect to z, obtained from Eq. (9.50), in Eq. (9.51), gives

$$\delta\, d\delta/dt = \alpha_{Tw}\, m\,(m+1) \tag{9.54}$$

where $\alpha_{Tw} = k_T(T_w)/\rho c_p$ is the thermal diffusivity evaluated at the wall.

Integrating Eq. (9.54) leads to

$$\delta = [2m(m+1)\,\alpha_{Tw} t]^{1/2} \tag{9.55}$$

To evaluate the adjustable parameter m, results valid for a system with a constant thermal diffusivity, α_T, will be used. Solving Eq. (9.46) for $\alpha_T = k_T/\rho c_p$ = constant, and using initial and boundary conditions given by Eq. (9.48), leads to (Bird *et al.*, 1960):

$$(T - T_0)/(T_w - T_0) = 0.01 \tag{9.56}$$

for

$$z = \delta = 4(\alpha_T t)^{1/2} \tag{9.57}$$

We will use the δ value defined by Eq (9.57) with an average value of the thermal diffusivity,

$$\alpha_{Tav} = k_{Tav}/\rho c_p \tag{9.58}$$

with

$$k_{Tav} = \int_{T_0}^{T_w} k_T(T)\, dT/(T_w - T_0) \tag{9.59}$$

Equating Eq. (9.55) with Eq. (9.57) (using α_{Tav}), and solving for m, leads to

$$m = 0.5[(1 + 32\alpha_{Tav}/\alpha_{Tw})^{1/2} - 1] \tag{9.60}$$

It must be verified that $\alpha_{Tav}/\alpha_{Tw} > 0.25$ to obtain $m > 1$, as required.

Then, an analytical expression that approximates temperature profiles in the semi-infinite medium with variable thermal diffusivity is obtained by inserting Eq. (9.57) with $\alpha_T = \alpha_{Tav}$ into Eq. (9.50):

$$(T - T_0)/(T_w - T_0) = (1 - z/[4(\alpha_{Tav} t)^{1/2}])^m \tag{9.61}$$

where m is given by Eq. (9.60).

Replacing the T(t,z) function in the constitutive equations for the cure kinetics, R_c, and integrating Eq. (9.47), leads to the evolution of conversion for any z-position. The thickness L of the cured shell is the value of z for which $x = x_{gel}$ at $t = t_f$. For $z < L$, $x > x_{gel}$ at t_f.

We now compare theoretical predictions with experimental results obtained using silica sand coated with a 4% resin consisting of 100 parts by weight of a phenol–formaldehyde novolac and 5, 10, or 15 parts of hexamethylenetetramine (hexa) as hardener (Aranguren *et al.*, 1984).

The cure kinetics were determined using differential scanning calorimetry (DSC). The following second-order rate equation could adjust the experimental results obtained for formulations containing 5–15 parts of hexa:

$$R_c = A(1 - x)^2 \exp(-E/RT) \tag{9.62}$$

where $A = 1.75 \times 10^{27}\ s^{-1}$ and $E = 235\ kJ\ mol^{-1}$. The very high value of the activation energy leads to the development of sharp conversion profiles. This is the basis of the success of the Croning process, because only a very small fraction of the sand that falls upon mold inversion exhibits a partial cure of the resin; thus, it is possible to reuse the sand in the following cycle.

Figure 9.17 shows the gel conversion, estimated from a statistical analysis of the complex cure reaction, as a function of the amount of hexa used in the formulation. As a result of the sharp conversion profile in the part, the influence of the gel conversion value on the predicted thickness is rather small.

The apparent density of the coated sand was $\rho = 1394\ kg\ m^{-3}$, the specific heat was $c_p = 695\ J\ kg^{-1}\ K^{-1}$, and the thermal conductivity exhibited the following variation with temperature:

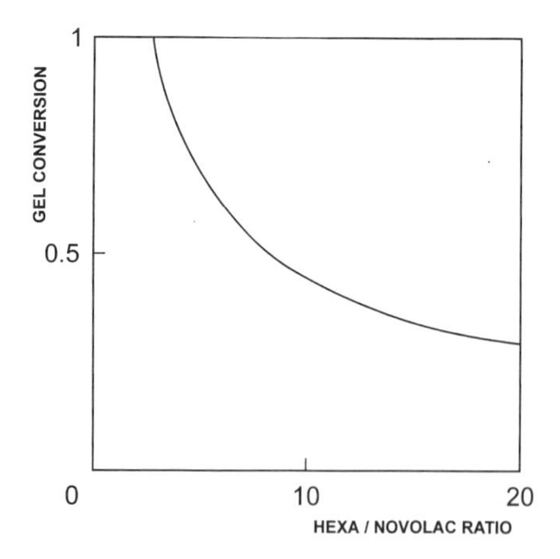

Figure 9.17 Gel conversion as a function of the amount of hexa in the formulation. (Aranguren *et al.*, 1984. Reprinted with permission from SAMPE)

$$k_T(Wm^{-1}K^{-1}) = 0.387, \qquad T = 298\ K$$
$$k_T(Wm^{-1}K^{-1}) = 7.70410^{-3}\ T - 1.909, \qquad 298K < T \leq 386K$$
$$k_T(W\ m^{-1}K^{-1}) = 1.068, \qquad T > 386K$$

The uncoated sand showed a k_T value close to 0.387 W m^{-1} K^{-1}, while the cured coated sand exhibited a k_T value equal to 1.068 W m^{-1} K^{-1}. The increase in the thermal conductivity with temperature, in the range 298–386 K, is assigned to the melting of the novolac, producing an agglomeration of sand particles that gives a continuous path for the heat flux through the solid material.

Figure 9.18 shows experimental values of the shell thickness plotted as a function of the thickness predicted by the theoretical model. A very good agreement was obtained for large variations of the wall temperature and the amount of hexa in the formulation. Visual inspection of the shell sand revealed the presence of a distinct hard layer with a characteristic yellow color. The rest of the sand falls easily upon mold inversion, which means that the cure proceeds through the development of sharp conversion profiles.

Predicted conversion profiles for a particular case are plotted in Fig. 9.19. For contact times such as those used in industrial practice (up to 30 s), they are extremely sharp. There is a moving conversion boundary, generated by the high value of the activation energy of the novolac–hexa reaction.

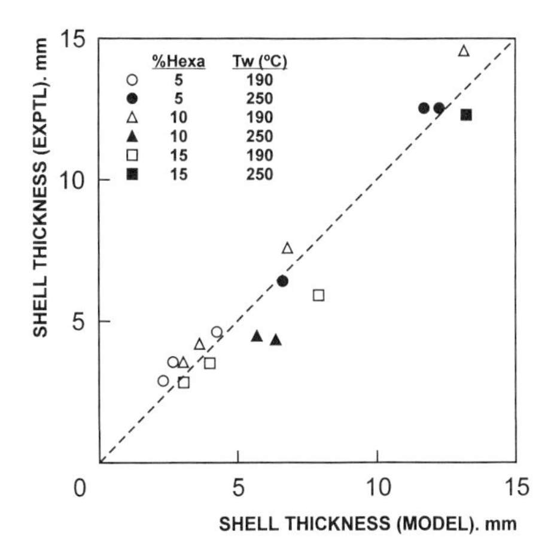

FIGURE 9.18 Experimental thickness of the shell sand vs thickness predicted by the model for several experimental conditions. (Aranguren *et al.*, 1984. Reprinted with permission from SAMPE)

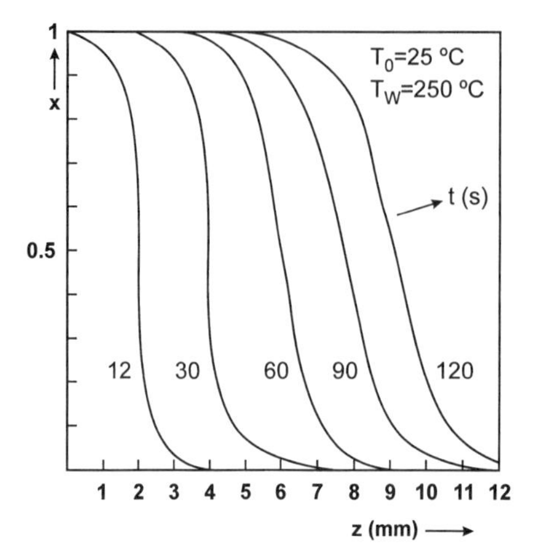

FIGURE 9.19 Conversion profiles for $T_w = 250°C$, $T_0 = 25°C$, and hexa–novolac mass ratios in the range of 5–15 parts per 100 parts of novolac. (Aranguren *et al.*, 1984. Reprinted with permission from SAMPE)

The theoretical model may be used to optimize the cure cycle for a given formulation. Wall temperature and the required shell thickness are the most important parameters that determine the process time.

9.8 CONCLUSIONS

Through the analysis of the particular selected examples it was shown that it is possible to get a good description of temperature and conversion profiles generated during the cure of a thermosetting polymer. Thermal and mass balances, with adequate initial and boundary conditions, may always be stated for a particular process. These balances, together with constitutive equations for the cure kinetics and reliable values of the necessary parameters, can be solved numerically to simulate the cure process.

The use of empirical rules is very often misleading; for example, it is frequently stated that "the cycle time is directly proportional to the specimen thickness" or that "the maximum temperature in the part is limited by the adiabatic temperature rise." Both statements have no general validity, as discussed in the analysis of the cure in heated molds.

When the filling stage of the process is also relevant in the analysis, as in RIM, the momentum balance must be added to the analysis. This, in turn, requires the knowledge of constitutive equations describing the variation of the apparent viscosity with temperature and conversion. Gelation becomes a relevant transition in this analysis.

Important concepts in any analysis of the cure of thermosetting polymers are the possibility of reaching vitrification at particular locations in the part and the need to provide heat from an external source to devitrify the material and continue the cure. The analysis of vitrification cannot be avoided when the cure is performed at room temperature using different types of radiation (e.g., UV, EB, microwave) to initiate the polymerization. A temperature vs conversion transformation diagram (Chapter 5) may be useful in the design of cure cycles.

NOTATION

A, A_1, A_2 = pre-exponential factor of the specific rate constant, s^{-1}
c_p = heat capacity per unit mass, kJ kg^{-1} K^{-1}
c_{pB} = specific heat of a blowing agent, kJ kg^{-1} K^{-1}
E, E_1, E_2 = activation energy, kJ mol^{-1}
F = mass of formaldehyde evaporated per unit polymer mass
G_B = mass of blowing agent in the gas phase per unit polymer mass
H = height, m
k_1, k_2 = specific rate constants, s^{-1}
k_T = thermal conductivity, kJ m^{-1} s^{-1} K^{-1}
L = part thickness, m
L_B = mass of blowing agent per unit polymer mass
m = adjustable parameter
M_B = molar mass of blowing agent, kg $kmol^{-1}$
p = pressure, N m^{-2}
R = gas constant = 8.314 kJ $kmol^{-1}$ K^{-1}
R_c = polymerization rate, s^{-1}
t = time, s
t^* = dimensionless time, $A \exp(-E / RT_0)\, t$
t_c = cream time in a foaming process, s
t_f = inversion time of the mold in the Croning process, s
T = temperature, K
T^* = dimensionless temperature, $(T - T_0)/\Delta T_{ad}$
T_α = glass transition temperature determined using dynamic mechanical analysis, K
$\tan \delta$ = loss tangent
T_c = cream temperature in a foaming process, K
T_g = glass transition temperature, K
T_w = wall temperature, K
V_f = volume fraction of fibers in a composite material
w_f = mass fraction of fibers
W_1 = dimensionless group, $(\alpha_T/L^2)/[A \exp(-E/RT_0)]$

W_2 = dimensionless group, E/RT_0
W_3 = dimensionless group, $T_0/\Delta T_{ad}$
W_4 = dimensionless group, $(T_w - T_0)/\Delta T_{ad}$
x = conversion
x_{gel} = gel conversion
z = coordinate in space, m
z* = dimensionless coordinate, z/L

Greek Letters

α_T = thermal diffusivity, $k_T/\rho c_p$, $m^2\ s^{-1}$
$\delta(t)$ = thickness of a part affected by heat transport from the wall, m
ΔG = change in the Gibbs free energy, kJ
ΔH = enthalpy change, kJ
$(-\Delta H)$ = reaction heat per unit mass, $kJ\ kg^{-1}$
ΔH_v = heat of vaporization of blowing agent per unit mass, $kJ\ kg^{-1}$
ΔS = entropy change, $kJ\ K^{-1}$
ΔT_{ad} = adiabatic temperature rise, $(-\Delta H)/c_p$, K
λ = adjustable parameter in the T_g vs x relationship
ρ = mass density, $kg\ m^{-3}$
ρ_F = mass density of formaldehyde in the gas phase, $kg\ m^{-3}$
ρ_G = mass density of the gas phase in a foam, $kg\ m^{-3}$
ρ_p = mass density of the polymer phase in a foam, $kg\ m^{-3}$

Subscripts

0 = initial value
∞ = value at full conversion
av = average value
exp = experimental value
f = fibers
m = polymeric matrix
max = maximum value
theor = predicted value

REFERENCES

Adolf DB, Martin JE, Chambers RS, Burchett SN, Guess TR, *J. Mater. Res.*, 13, 530–550 (1998).
Aranguren MI, Borrajo J, Williams RJJ, *SAMPE J.*, 3, 18–23 (1984).
Bird RB, Stewart WE, Lightfoot EN, *Transport Phenomena*, Wiley, New York, 1960, p. 354.
Fang DP, Frontini PM, Riccardi CC, Williams RJJ, *Polym. Eng. Sci.*, 35, 1359–1368 (1995).
Glauser T, Johansson M, Hult A, *Polymer*, 40, 5297–5302 (1999).

Gorovaya TA, Korotkov VN, *Composites Part A*, 27, 953–960 (1996).
Marciano JH, Reboredo MM, Rojas AJ, Williams RJJ, *Polym. Eng. Sci.*, 26, 717–724 (1986).
Mijovic J, Wang HT, *SAMPE J.*, 2, 42–55, 191 (1988).
Srinivasan SA, Joardar SS, Kranbeuhl D, Ward TC, McGrath JE, *J. Appl. Polym. Sci.*, 64, 179–190 (1997).
Vázquez A, Williams RJJ, *Cell. Polym.*, 5, 123–140 (1986).
Williams RJJ, Rojas AJ, Marciano JH, Ruzzo MM, Hack HG, *Polym.-Plast. Technol. Eng.*, 24, 243–266 (1985).
Williams RJJ, Adabbo HE, Rojas AJ, Borrajo J. In *New Polymeric Materials: Reactive Processes and Physical Properties*, Martuscelli E, Marchetta C (eds), VNU Science Press, Utrecht, Netherlands, 1987, pp. 1–17.
Williams RJJ, Benavente MA, Ruseckaite RA, Churio MS, Hack HG, *Polym. Eng. Sci.*, 30, 1140–1145 (1990).

FURTHER READING

Kresta JE (ed.), *Reaction Injection Moulding and Fast Polymerization Reactions*, Plenum, New York, 1982.
Macosko CW, *RIM: Fundamentals of Reaction Injection Moulding*, Hanser, Munich, 1989.
Ryan AJ, Stanford JL. In *Comprehensive Polymer Science*, Vol. 5, Eastmond GC, Ledwith A, Russo S, Sigwalt P (eds), Pergamon, Oxford, 1989, Chapter 25, pp 427–454.
Stanford JL, Elwell MJA, Ryan AJ. In *Processing of Polymers*, Meijer HEH (ed.), VCH, Weinheim, 1997, Chap 9, pp. 465–512.

10

Basic Physical Properties of Networks

10.1 INTRODUCTION

To understand polymer properties, especially thermomechanical behavior, one needs to know their basic physical properties such as packing density, free volume, cohesive energy density, chain mobility, glass transition temperature and, indeed, crosslink density in the case of networks. These properties are more or less directly linked to the chemical structure at the nanometric (molecular) and large (macromolecular) scale. If the material is heterogeneous, one also needs information describing this heterogeneity (supramolecular or morphological scale). The conceptual and experimental investigation tools differ strongly from one scale to another (Table 10.1); it is thus important to recognize that, to be fully efficient, an investigation on structure–property relationships in the field of networks must be multidisciplinary.

Time and energy can be saved if one recognizes that there is only one qualitative difference between a linear and a tridimensional polymer: the existence in the former and the absence in the latter of a liquid state (at a macroscopic scale). For the rest, both families display the same type of boundaries in a time–temperature map (Fig. 10.1). Three domains are characterized by (I) a glassy/brittle behavior (I), (II), a glassy/ductile behavior, and (III) a rubbery behavior. The properties in domain I are practically

TABLE **10.1** The three structural scales of polymer structure. Conceptual and experimental tools

Structural scale	Typical size	Typical entities	Academic domain	Main tools
Molecular	0.1–1 nm	Chemical groups, monomer units, CRU	Organic chemistry	IR, NMR
Macromolecular	10–100 nm	Network chains/strands, crosslinks network defects (dangling chains)	Macro-molecular science	Rubber elasticity, solvent swelling
Supramolecular	10–10^5 nm	Packing defects, nodular/globular morphology, multiphase structure	Materials science	Microscopies, scattering methods, thermal analysis

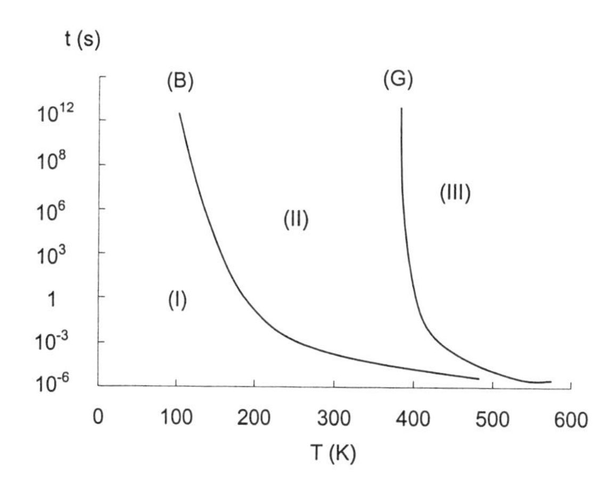

FIGURE **10.1** Time–temperature map. Shape of main boundaries for linear or network polymers. (I) Glassy brittle domain; B, ductile–brittle transition. (II) Glassy ductile domain; G, glass transition. (III) Rubbery domain. The location of the boundaries depends on the polymer structure but their shape is always the same. Typical limits for coordinates are 0–700 K for temperature and 10^{-3} s. (fast impact) to 10^{10} s; e.g., 30 years static loading in civil engineering or building structures. Fpr dynamic loading, t would be the reciprocal of frequency. For monotone loading, it could be the reciprocal of strain rate $\dot{\varepsilon} = dl/ldt$.

insensitive to the large-scale structure; they depend essentially on the cohesive energy density, which is directly linked to the molecular scale structure. In contrast, the properties in domain (III) are almost independent of the molecular scale structure; they depend essentially on the crosslink density (entanglement density in linear polymers). The intermediary domain II is especially interesting for users because a relatively high stiffness (typically $E \approx 1$–3 GPa) can be combined with a high ductility/toughness/impact resistance. Since there is a very abundant literature on structure–property relationships of linear polymers (Van Krevelen, 1990; Porter, 1995; Mark, 1996), we have chosen to recall very briefly the aspects of the physical behavior common to linear and tridimensional polymers and to insist on the differences between both families. Volumetric properties will be especially developed because they are, often abusively in our opinion, used for the interpretation of thermomechanical behavior.

10.2 PROPERTIES IN THE GLASSY STATE

10.2.1 Volumetric Properties

a. Density at Ambient Temperature

Density, ρ, depends on atomic composition, which can be represented by a single quantity – the average atomic mass M_a – defined by

$$M_a = \frac{\text{Molar mass of the CRU}}{\text{Number of atoms of the CRU}} \quad (10.1)$$

The CRU (constitutive repeating unit) can be considered as the "monomer unit" of the network. It has been defined in Chapter 2.

The density of many linear, as well as tridimensional, polymers has been plotted against M_a in Fig. 10.2. One can estimate ρ from a power law:

$$\rho = kM_a^{2/3} \quad (10.2)$$

where $k \sim 320$, ρ is in kg m^{-3} and $\overline{M}_a$ in g mol^{-1}

This empirical relationship remains valid for light inorganic materials such as calcium carbonate, silica, aluminium, etc., and allows the density to be predicted with a maximum error of 10% in the range of most usual organic network densities ($1100 \leq \rho \leq 1400$ kg m^{-3}). In this range, density can be approximated by a linear relationship:

$$\rho = 350 + 120\,M_a \qquad (\text{kg m}^{-3}) \quad (10.3)$$

Typical density values for networks are 1150–1330 kg m^{-3} for amine-crosslinked epoxies ($M_a \approx 7 \pm 1$); 1120–1180 kg m^{-3} for styrene-crosslinked vinyl esters ($M_a \approx 6.3$–7.1); and 1170–1220 kg m^{-3} for styrene-crosslinked

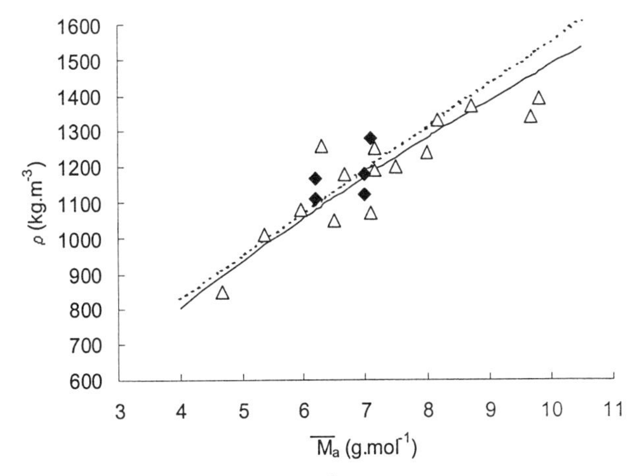

FIGURE 10.2 Density (kg m^{-3}) against average atomic mass M_a (g mol^{-1}) for linear polymers (△) and networks (◆). Curve (full line), $\rho = 320\ \overline{M}_a^{2/3}$; straight line (dashed line), $\rho = 350 + 120\overline{M}_a$.

polyesters ($M_a \approx 8$ for a maleate/phthalate (1/1) of propylene glycol cross-linked by 35% of styrene). Many points corresponding to these systems have been plotted in Fig. 10.2. They reveal no systematic difference (within the observed scatter) between networks and linear polymers.

In all structural series, the density is effectively an increasing function of M_a, e.g., of the content of "heavy" atoms such as O or S.

Let us consider the variation of the composition in a two-component system (R + H), e.g., unsaturated polyester–styrene or epoxide–amine. Are density measurements capable of detecting such a variation?

If y is the molar ratio H/R and the CRU contains 1 mole of R and y moles of H, the molar mass M and the number of atoms N are

$$M = M_R + yM_H$$
$$N = N_R + yN_H$$

The average atomic mass is therefore

$$M_a = \frac{M}{N} = \frac{M_R + yM_H}{N_R + yN_H} \tag{10.4}$$

By derivation, one obtains

$$\frac{dM_a}{dy} = \frac{N_R M_H - N_H M_R}{(N_R + yN_H)^2} \tag{10.5}$$

But

$$\frac{dM_a}{dy} = \frac{DM_a}{d\rho}\frac{d\rho}{dy} \quad \text{so that} \quad \frac{d\rho}{dy} = \frac{N_R M_H - N_H M_R}{(N_R + yN_H)^2}\frac{d\rho}{dMa} \tag{10.6}$$

In a first approximation, one can use Eq. (10.3) for $d\rho/dM_a$, so that

$$\frac{d\rho}{dy} = 120\frac{N_R M_H - N_H M_R}{(N_R + yN_H)^2} \tag{10.7}$$

The sensitivity of the method is an increasing function of the difference of average atomic mass ($M_{aH} - M_{aR}$). Let us consider four cases of industrial networks (Table 10.2).

Using simple laboratory equipment, it is possible to detect variations of the density of the order of 1 kg m^{-3}; therefore, density measurements would be useless for diglycidyl ether of bisphenol A and diamino diphenyl methane (DGEBA–DDM), because both components have close M_a values. They would be moderately sensitive for unsaturated polyesters of maleate/phthalate (1/1) of propylene glycol crosslinked by styrene (36 wt%), and relatively sensitive for DGEBA cured by diamino diphenyl sulphone (DDS) or phthalic anhydride (PA), for which relative variations of the molar ratio y of about 5% could be detected. In certain cases, where some comonomer (PA or styrene) can be lost by evaporation during the cure, density measurements can constitute a simple and efficient method of control way (e.g., for non-filled materials).

b. Packing Density at Ambient Temperature

In Fig. 10.2, the scattering is obviously linked to differences in molecular packing. In Eq. (10.2), one could tentatively assume that the factor k is a function of the packing density. For instance, for the four materials of Table

Table 10.2 Sensitivity of density to variations of the composition. The definitions of M_R, N_R, M_H, N_H, and y are given in the text.

Family	Name	M_R (g mol^{-1})	N_R	M_H (g mol^{-1})	N_H	y	$d\rho/dy$ (kg m^{-3})
Epoxide–amine	DGEBA-DMM	340	49	198	29	0.5	−4.7
Epoxide–amine	DGEBA–DDS	340	49	248	29	0.5	68.2
Epoxide–anhydride	DGEBA–PA	340	49	148	15	2	41.4
Polyester–styrene	UP–S (36% styrene)	362	44	104	16	2	25.3

10.2, we would have the data of Table 10.3. The hierarchy of packing densities would be :

DGEBA–DDS = DGEBA-DDM > DGEBA–PA > UP–S

There is no clear correlation between k and density or M_a values.

The use of k values to represent packing density would be questionable, owing to the empirical character of Eq. (10.2). The following definition is better (Bondi, 1968; Van Krevelen, 1990):

$$\rho^* = \frac{V_w}{V} \left\} \begin{array}{l} V_w = \sum_i V_{wi} \\ V = M/\rho \end{array} \right. \tag{10.8}$$

where V_w is the van der Waals volume obtained by summation of molar increments and V is the molar volume of the CRU. According to Van Krevelen, ρ^* would be almost constant ($\rho^* \approx 0.65$) for linear polymers in the amorphous glassy state. Systematic determinations in amine-crosslinked epoxies (Bellenger *et al.*, 1988), and in vinyl ester networks (Bellenger *et al.*, 1994), showed that ρ^* is almost independent of the crosslink density. But packing density increases with the concentration of hydrogen bonds for both linear polymers ($\rho^* \approx 0.635$) for polystyrene and $\rho^* = 0.717$ for poly (vinyl alcohol)), and for networks ($\rho^* = 0.664$ for a styrene-crosslinked polyester almost free of hydroxyls, $\rho^* = 0.68$ for a DGEBA network containing 4–5 mole kg^{-1} of hydroxyls, and $\rho^* \approx 0.70$ for a triglycidyl amino phenol (TGAP) network containing 6.5–7 mole kg^{-1} of hydroxyls) (Fig. 10.3). It is noteworthy that in most of the structural series of amine-crosslinked epoxies (reactions are described in Chapter 2), crosslink density varies in the same way as hydroxyl concentration with the composition which can lead to erroneous interpretations. Other structural factors also probably have a small influence on packing density; e.g., aromatic networks seem to be systematically less densely packed than comparable aliphatic ones, but this effect seems difficult to quantify owing to data scattering.

Table 10.3 Density, average atomic mass, and prefactor of Eq. (10.2) for the four networks of Table 10.2.

Network	ρ (kg m^{-3})	M_a (g mol^{-1})	k
DGEBA–DDM	1200	6.913	330.6
DGEBA–DDS	1247	7.307	331.2
DGEBA–PA	1273	8.05	316.9
UP–S	1182	7.50	308.5

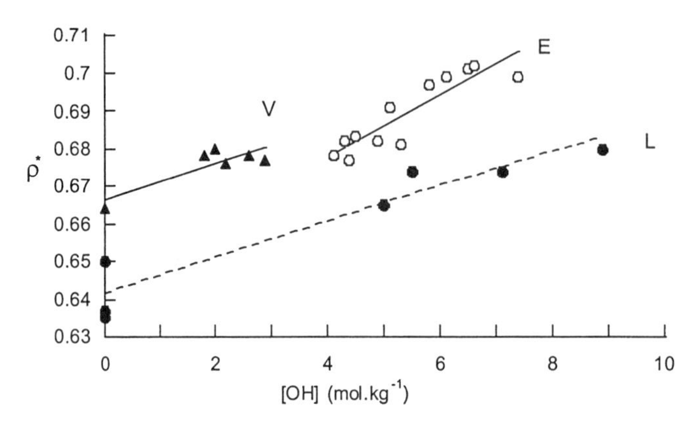

FIGURE 10.3 Packing density, $\rho^* = V_W/V$, versus hydroxyl concentration: V (▲), vinylesters; E (○), amine crosslinked epoxies ; L (●), linear polymers.

To appreciate an eventual effect of crosslink density on mass density, we have to compare networks having the same hydrogen bond concentration. This is possible, for instance, in the series B of Table 10.4 (Morel *et al.*, 1989).

One can see in series B that [OH] is constant within ± 1% and that the packing density is almost constant or increases slightly with the crosslink density. In contrast, in series A, the packing density increases significantly from A1 to A4, e.g., with the crosslink density, but the OH concentration increases at the same time. To summarize: all the results of Table 10.4 are consistent with a relatively strong influence of OH groups, $\Delta\rho \cdot /\Delta[\mathrm{OH}] \approx 6 \times 10^{-3}$ kg mol^{-1}, and a smaller influence of crosslinks, $\Delta\rho \cdot /\Delta n \approx 10^{-3}$ kg mol^{-1}. In fact, one can reasonably assume that all the intermolecular attractive forces participate in the increase of packing density. The effect of these forces can be introduced in structure–property relationships through the concept of cohesive energy density (Sec. 10.2.2).

c. "Anomalous" Density Variations with the Structure in Amine-Cured Epoxies

In many structural series based on a given epoxide–amine pair differing by the amine/epoxide molar ratio (e.g., Won *et al.*, 1991) or by the degree of cure conversion (e.g., Venditti and Gillham, 1995), it has been observed that the density is a decreasing function of the crosslink density, and authors such as Venditti and Gillham (1995) have suggested that both phenomena would be linked to the free volume fraction "trapped" in the polymer. In fact, in series where the atomic composition and the cohesive energy density

TABLE **10.4** Density, hydroxyl concentration, crosslink density, packing density, ultrasonic bulk modulus, average atomic mass (g mol^{-1}, prefactor of Eq. (10.2) calculated from Eq. (10.19), calculated density from Eq. (10.21). Experimental data from Morel *et al.* (1989).

Code	Network	ρ (kg m^{-3})	[OH] (mol kg^{-1})	n (mol kg^{-1}	ρ^*	K (GPa)	M_a (g mol^{-1})	k	$\rho^{(calc)}$ (kg m^{-3})
A0	DGEBA-DDM	1200	4.56	2.28	0.678	6.31	6.913	330.7	1201
A1	DGEBA–TGAP (75-25) DDM	1217	5.26	3.30	0.684	6.75	6.926	334.9	1215
A2	DGEBA–TGAP (50/50) – DDM	1238	5.91	4.23	0.693	7.18	6.948	340.0	1234
A3	DGEBA–TGAP (20/80)–DDM	1257	6.62	5.25	0.700	7.72	6.985	344.0	1239
A4/B1	TGAP–DDM	1269	7.05	5.88	0.703	8.02	7.004	346.6	1270
B2	TGAP–DDM/AN (75–25)	1263	7.09	4.98	0.702	8.17	6.988	345.5	1275
B3	TGAP–DDM/AN (50–50)	1263	7.13	4.10	0.703	8.05	6.973	346	1268
B4	TGAP–DDM. AN (25–75)	1259	7.19	3.24	0.702	7.62	6.957	345.5	1251
B5	GAP–AN 100	1255	7.20	2.40	0.700	–	6.941	344.9	1255

are almost constant, such as series B of Table 10.4, or DGEBA-cycloaliphatic amine systems studied by Won *et al.*, (1991), or polyesters (Shibayama and Suzuki 1965), the density appears to be rather an increasing function of the crosslink density. In some of these series, the modulus tends to be a decreasing function of the crosslink density as found for instance in series B of Table 10.4. The tensile modulus is about 3.25 GPa for TGAP–DDM (crosslink density 5.88 mol kg^{-1}) and 3.91 GPa for TGAP-aniline (AN) (crosslink density 2.40 mol kg^{-1}).

It thus seems that there is no direct link between volumetric and elastic properties in the glassy state and that the "anomalous" density variations cannot be attributed to a crosslink density effect, either direct (on molecular packing) or indirect (through internal antiplasticization as discussed below). It seems reasonable to correlate this behavior with the presence of unreacted epoxides. The density would be (in the systems under consideration) a continuously increasing function of the amine/epoxide ratio, owing to the strong contribution of OH and NH groups to cohesivity. In the presence of unreacted epoxide groups, this effect would be counterbalanced by a specific effect – for instance, a decrease of Van der Waals volume linked to epoxide ring opening (Bellenger *et al.*, 1982) – or a hypothetical effect of dangling chains on packing (only present in networks containing unreacted epoxides for the systems displaying the "anomalous" effect).

d. Thermal Expansion in Glassy State

When studied in a limited temperature interval around the ambient temperature, expansion can be considered linear:

$$V \approx V_{og}(1 + \alpha_g T) \tag{10.9}$$

The expansion coefficient ranges generally between 1.4×10^{-4} and 4×10^{-4} K^{-1}, as for linear polymers, and does not display any significant variation with the crosslink density.

Studies in larger temperature intervals showed that expansion obeys a parabolic law, both for linear polymers (Bongkee, 1985), and epoxy thermosets (Skourlis and McCullough, 1996):

$$V = V_o(1 + AT^2) \tag{10.10}$$

This is practically equivalent to an exponential law, $V = V_0 \exp AT^2$, which can be derived from a thermodynamic approach (Bongkee, 1985).

For a wide family of epoxy networks, A ranges between 2×10^{-7} and 5×10^{-7} K^{-2}; it seems to be essentially influenced by the chain stiffness (Tcharkhtchi *et al.*, 2000):

$$A \approx 60 \times 10^{-8} - 5 \times 10^{-8}\,\mathrm{Ar} \tag{10.11}$$

where Ar is the concentration of aromatic units in mol kg^{-1}.

Indeed, the linear expansion coefficient near α_g temperature T_a, is linked to A by

$$\alpha_g = 2AT_a \tag{10.12}$$

Near the ambient temperature $\alpha_g \approx 600A$.

10.2.2 Cohesive Properties

a. Definitions

In linear polymers, cohesion results from weak (compared with covalent bonds) intermolecular attractive forces (Van der Waals) of various types: London, Debye, Keesom, and hydrogen bonding. In a first approach, they can be considered undistinguishable, and one can define cohesive energy as the whole energy of intermolecular interactions. For small molecules, cohesive energy is easy to determine from calorimetric measurements since vaporization corresponds to the rupture of intermolecular bonds. Then, for 1 mole,

$$E_C = \Delta H_v - RT \qquad (10.13)$$

where E_c is the molar cohesive energy, ΔH_v is the molar vaporization enthalpy, and the term RT corresponds to the gas expansion – (pressure) (volume variation) = RT.

In polymers, it is usual to consider that E_c corresponds to 1 mole of CRU. Two materials can be compared through their cohesive energy density (CED), defined by $CED = E_c/V$, where $V = M/\rho$, is the molar volume of the CRU with molar mass M. CED is expressed in J m^{-3}; typical values for polymers range between 2×10^8 to 8×10^8 J m^{-3}. It is usual to convert these values into MPa (200 MPa < CED < 800 MPa). A practically important quantity is the solubility parameter, δ, defined by $\delta = (CED)^{1/2}$ and expressed in $J^{1/2}\ m^{-3/2}$ or, better, in $MPa^{1/2}$.

In certain circumstances, it may be necessary to distinguish between the different types of interactions. This can be performed in several ways (Barton, 1983; Van Krevelen, 1990). The most usual method is to make a distinction between dispersion (London), dipolar (Debye–Keesom) and hydrogen-bonding components, each one being characterized by its contribution to CED and the corresponding solubility parameter, $\delta_d, \delta_p, \delta_h$, respectively, such that $\delta = (\delta_d^2 + \delta_p^2 + \delta_h^2)^{1/2}$.

b. Methods of Determining Cohesive Properties

The methods of determining δ or CED are discussed below.

(i) Calorimetric Measurement of ΔH_v. Unfortunately, this method does not work in the case of polymers because they undergo thermal degradation long before vaporization.

(ii) Polymer–Solvent Interactions. From a thermodynamic approach (Hildebrand), it can be shown that, for a given polymer of solubility parameter δ_P, the miscibility with a solvent is a decreasing function of $|\delta_p - \delta_S|$ where δ_S is the solvent's solubility parameter. For a thermoset that is totally insoluble in any solvent, the miscibility can be, easily quantified, in principle, by the equilibrium concentration of the solvent determined, for instance, from weight uptake measurements in saturated vapor. There are, however, many strong obstacles to such determinations (Bellenger *et al.*, 1997), and literature data are scarce. This method cannot be reasonably recommended as an efficient tool for determining cohesive properties. Inverse gas chromatography can be also used to determine solubility parameters.

(iii) Bulk Modulus at High Frequency/Low Temperature. The unrelaxed bulk modulus determined below the main secondary transition, e.g. almost free of viscoelastic effects, is almost proportional to CED (Morel *et al.*, 1989; Bellenger *et al.*, 1994):

$$K \approx 11\text{CED} \tag{10.14}$$

This is probably the simplest and most accurate way of determining the cohesive energy density. The measurements are easy to perform if homogeneous, relatively thick (few mm) samples, with two parallel and smooth surfaces, are available.

(iv) Calculation from Molar Increments. There are many ways to calculate the cohesive energy of a given CRU (Van Krevelen, 1990). The simplest method is to consider E_c as an additive molar function:

$$E_c = \sum_{\text{CRU}} E_{ci} \tag{10.15}$$

where E_{ci} is the molar increment of the ith group of the CRU. Indeed, E_{ci} depends on the polarity of the corresponding group, but also on its size.

For very rough estimation, one can consider the cohesive energy/molar mass ratio as

$$\varepsilon_i = E_{ci}/M_i \tag{10.16}$$

Some typical values are

Very low polarity groups such as CF_2, CF_3: $\varepsilon_i \approx 85$ J g^{-1}

Low polarity groups such as hydrocarbon groups, tertiary amines, chlorine, ether: $\varepsilon_i = 200$–400 J g^{-1}

Esters: $\varepsilon_i \approx 400$ J g^{-1}

Amides, primary amines: $\varepsilon_i \approx 600$–$800$ J g^{-1}

Alcohols: $\varepsilon_i \approx 1800$ J g^{-1}

It is clear that hydrogen-bond donor groups, such as alcohols or amides have an especially high contribution to cohesion, whereas hydrocarbon and halogenated groups display the lowest "cohesivity." This method is essentially valid in a comparative way, but it can lead to serious discrepancies with experimental results, despite the fact that E_c is not very sensitive to structural irregularities.

c. Cohesion and Packing Density

From the results of Table 10.4, one can write

$$\rho^* = 15 \times 10^{-3}\,K + 0.584 \tag{10.17}$$

Using the relationship between the bulk modulus and the cohesive energy density leads to

$$\rho^* = 0.17\text{CED} + 0.584 \tag{10.18}$$

where CED is expressed in GPa (GJ m^{-3}).

However, from a practical point of view, it is easier to start from Eq. (10.2) to predict density. In this case, a relationship between the parameter k and a quantity characterizing the cohesivity, for instance the OH concentration or CED, may be found.

According to the data of Table 10.4, k varies almost linearly with these quantities:

$$k = 10K + 267 \tag{10.19}$$

or, from Eq. (10.14),

$$k = 110\text{CED} + 267 \tag{10.20}$$

Finally, both main effects on density of cohesion and atomic mass are taken into account in the following relationship:

$$\rho = (267 + 10K)M_a^{2/3} \tag{10.21}$$

The corresponding calculated values are compared with experimental ones in Table 10.4. The error of the prediction is generally less than 1% (5% including vinyl esters and polyesters, for which K varies between 4.5 and 5.9 GPa).

d. Data onThermosets

Data are available for unsaturated polyesters (Deslandes *et al.*, 1998), cross-linked- PMMA (Bellenger *et al.*, 1997), vinyl esters (Bellenger *et al.*, 1994), and amine-crosslinked epoxies (Bellenger *et al.* 1989). Some results are summarized in Table 10.5.

From the point of view of cohesive properties, polymer networks:

- behave as linear polymers – the influence of crosslinking on cohesion can be considered negligible, at least in a first approximation;
- cannot be accurately characterized from solvent absorption experiments
- can be easily (but only globally) characterized from unrelaxed modulus measurements (especially from measurements of the velocity of ultrasonic waves)
- can be characterized from calculations assuming the molar additivity of group contributions.

Table 10.5 Hildebrand's solubility parameter and cohesive energy density determined from this; cohesive energy density from bulk modulus; Hydroxyl concentration for some networks. (a) Molar ratio dimethacrylate/methacrylate = 5×10^{-4} (500 ppm); (b) aromatic poly(bismaleimide) from BASF.

Family	Network	δ ($MPa^{1/2}$)	CED (MPa)	CED (MPa) from bulk modulus	[OH] ($mol\ kg^{-1}$)
UP	30% styrene	22.2	493	525	0
VE	30% styrene	23.0	529	540	2.86
VE	23% styrene	21.2	449	460	0
EPO	DGEBA–DDM	23.3	543	574	4.56
EPO	TGAP–AN	24.9	620	690	7.20
EPO	TGAP–DDM	24.7	610	729	7.05
EPO	TGAP–DDS	25.6	655	–	6.50
PMMA	500 ppm (a)	23.0	529	522	0
BMI	(b)	24.2	586	–	0

10.2.3 Local Mobility

a. Principles of Measurement

Polymer glasses are characterized by a residual segmental mobility in a more or less wide temperature interval below T_g. This mobility is responsible for the existence of relaxations often called secondary relaxations that are in opposition to the main (cooperative) relaxation occurring at $T \geq T_g$. There are many types of localized motions that involve, for example, rotations of lateral groups, oscillations of aromatic rings, around the chain axis, chair–chair isomerization in cycloaliphatic rings, or crankshaft-like motions on short-chain segments. After a perturbation, the system returns to its initial equilibrium or to a new equilibrium at a rate which depends essentially on the distance from equilibrium and on molecular mobility. Molecular mobility is represented by a characteristic time or relaxation time, or by an average relaxation time, or by a spectrum of relaxation times.

Under a perturbation P_o, the response (Q) rate, for the simplest case of first-order relaxation kinetics, can be written as

$$\frac{dQ}{dt} = \frac{1}{\tau}(Q - Q_0) \tag{10.22}$$

where τ is the characteristic time (relaxation time) of the process and Q_o the equilibrium response,

$$Q = Q_0\left(1 - \exp - \frac{t}{\tau}\right) \tag{10.23}$$

Let us consider now the case of a sinusoidal perturbation: $P^* = P_o \exp i\omega t$ (using complex notation). The same differential equation may be applied, but with an oscillatory equilibrium:

$$\frac{dQ}{dt} = -\frac{1}{\tau}(Q - \Delta S P_0 \exp i\omega t) \tag{10.24}$$

where $S = Q_0/P_0$.

Defining

$$S^* = Q^*/P^* \tag{10.25}$$

leads to a sinusoidal response with the same frequency as the perturbation, so that

$$Q^* = P_0 S^* \exp i\omega t) \tag{10.26}$$

From Eq. (10.24), we obtain

$$S^* = \frac{\Delta S}{1 + i\omega\tau} = \frac{\Delta S(1 - i\omega\tau)}{1 + \omega^2\tau^2} = \frac{\Delta S}{1 + \omega^2\tau^2} - \frac{\Delta S i\omega\tau}{1 + \omega^2\tau^2} \tag{10.27}$$

$$S^* = S' - iS'', \text{ where } S' = \frac{\Delta S}{1 + \omega^2\tau^2} \text{ and } S'' = \frac{\omega\tau\Delta S}{1 + \omega^2\tau^2} \tag{10.28}$$

S' and $S''$are plotted versus $\omega\tau$ in Fig. 10.4. One sees that the system displays a transition from the unrelaxed state ($S' \approx \Delta S$) to the relaxed state ($S' = 0$), at a given frequency such that $\omega\tau = 1$, e.g., $\omega = \tau^{-1}$; τ is the characteristic time of the relaxation process, directly linked to molecular mobility, e.g., to thermal agitation, τ is a decreasing function of temperature, i.e., $\tau = f(T)$; and ω^{-1} can be considered as the characteristic time of the perturbation.

It may be concluded from the above results that the transition between unrelaxed and relaxed states occurs when the characteristic times of the relaxation process and of the experiment used to study the relaxation (here sinusoidal perturbation), are equal. Therefore the following points can be made:

1. A given type of local motion in the glassy state must be responsible for a characteristic transition. This transition will be arbitrarily called β, γ, δ, etc., in the order of decreasing temperatures or increasing frequencies; α is reserved in amorphous polymers to the transition between glass (local motions) and rubbery (cooperative motions) domains.

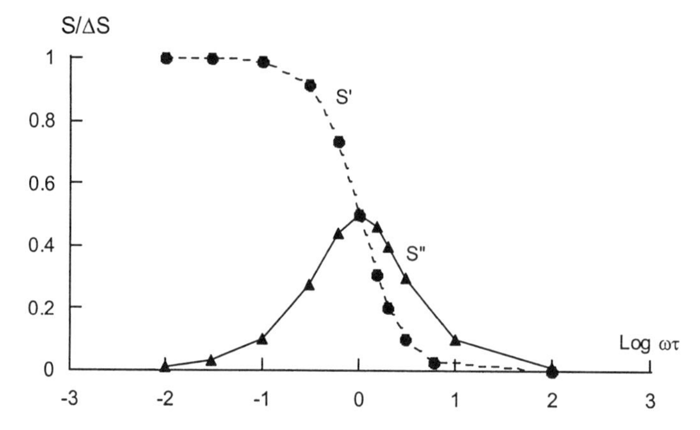

FIGURE 10.4 Real S′ (●) and imaginary S″ (▲) part of S normed by △S (see text) versus log ωτ for a dynamic perturbation in the case of a first order relaxation process.

2. A given transition is characterized by an inflection point in the curve S′ (real part of S*) = f (T) at ω constant, or f(ω) at T constant, and by a maximum on the curve S″ = f (T) or f(ω).
3. A given transition is essentially characterized by two quantities: its amplitude ΔS and its location in the frequency scale at a given temperature, in the temperature scale at a given frequency or, better, in the frequency–temperature map.
4. In the simple case of Fig. 10.4, it can be easily demonstrated that the area under the curve S″ = f(ω) is proportional to ΔS. Both quantities characterize the "activity" of the relaxation in the physical domain under consideration: for instance, ΔS can correspond to the gap between two distinct conformations of the chain.

- When the perturbation is a stress, the response is a strain and S″ is the complex compliance, $J^* = J' - iJ''$ (dynamic mechanical analysis: DMA)
- When the perturbation is a strain, the response is a stress and S″ is the complex modulus, $E^* = E' + iE''$ (dynamic mechanical analysis). The damping factor, $\tan \delta = E''/E'$, is a useful parameter (Chapters 6 and 11).
- When the perturbation is an electric field, the response is the material polarization and one can define a complex dielectric constant, which is the electrical equivalent of a compliance: $\varepsilon^* = \varepsilon' - i\varepsilon''$ (dielectric spectroscopy; Chapter 6).

- When the perturbation is a magnetic field, the response is the material magnetization related to a complex magnetic susceptibility: $\kappa = \kappa' - i\kappa''$ (NMR and ESR spectroscopies).

All these methods can be used to study the glassy-phase transitions in complementary ranges of frequency, typically 10^{-2}–10^{2} Hz for DMA ; 10–10^{6} Hz for dielectric spectroscopy and 10^{6}–10^{10} Hz for NMR and ESR spectroscopies. Photophysical measurements give access, eventually,to very high frequencies/very short times. Static mechanical testing (creep, stress, relaxation), or electrical testing (thermally stimulated depolarization currents) give access to very low frequencies. It must be noted that the "activity" of a given relaxation, as represented by ΔS, depends on the selected property. Certain relaxations display a strong mechanical activity and a low electrical activity, whereas others display the opposite behavior.

Actually, relaxation processes do not follow a first-order kinetics. In a glass there is a relatively wide variety of situations leading to the existence of a spectrum of relaxation times. A "static" relaxation kinetics can be approximated by

$$S = S_0\left[1 - \exp - \left(\frac{t}{\tau}\right)^{\beta}\right] \text{(Kolrausch, Williams, Watt)} \tag{10.29}$$

where the exponent $\beta(< 0 < \beta \leq 1)$ expresses the broadness of the relaxation spectrum. Let us recall that β is temperature dependent. Thus, a given transition is essentially characterized by its amplitude ΔS and its location in ω and T scales, but it can also be characterized by its width, which is generally larger than the theoretical one for the first-order process ($\Delta\omega = 1.2$ decades.

Generally, the average relaxation times and the transition frequencies in the glassy state obey the Arrhenius law:

$$\tau = \tau_0 \exp\frac{H}{RT} \text{ or } \omega = \omega_0 \exp - \frac{H}{RT} \tag{10.30}$$

b. Sub-Glass Transitions in Thermosets

Some thermomechanical spectra of thermosets (at 1 Hz), are shown in Fig. 10.5. Some networks display at least two secondary transitions between 100 K and T_g. Their assignment is often controversial and, to our knowledge, there is no widely accepted way of predicting their position (T_β) and their intensity.

From the practical point of view, it is interesting to make the following distinctions:

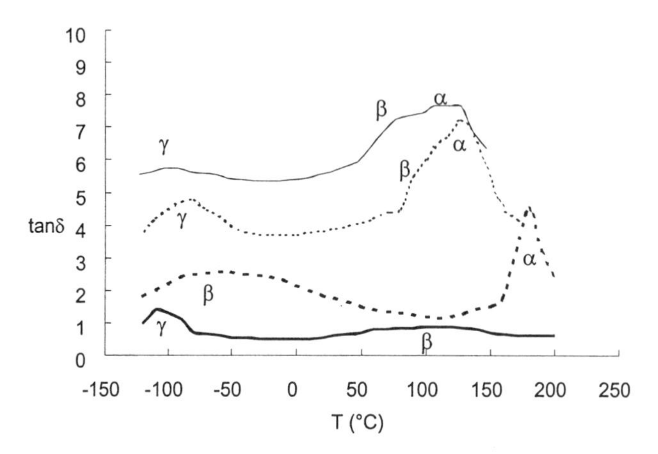

Figure 10.5 Examples of thermomechanical spectra (tan $\delta = f(T)$ at 1 Hz frequency) of some thermosets: PMR15 (polynadimide) (▬▬▬) ; EP (.........) ; VE (- - - - - -); and UP (———). The spectra are arbitrarily vertically shifted.

(i) Polymer networks with their β transition located above the ambient temperature, such as unsaturated polyesters (T_β(1 Hz) ≈ 350 K), or aromatic polyimides (T_β(1 Hz) ≈ 330–400 K, activation energy $H_\beta \approx$ 100–180 kJ mol^{-1}), can be distinguished from networks having their β transition below the ambient temperature, for instance amine-crosslinked epoxies: T_β(1 Hz) ≈ 190–240 K, $H_\beta \approx 70 \pm 30$ kJ mol^{-1}, or phenol–formaldehyde networks: T_β(1 Hz) ≈ 210 K, $H_\beta \approx 140$ kJ mol^{-1}. As most of the mechanical tests are made at ambient temperature, the corresponding β relaxation is inactive for the former, in the most frequent use conditions, but is active for the latter. This has very important practical consequences.

(ii) Networks with a highly active β transition – e.g., stoichiometric–epoxide–amine networks, in which the β relaxation is linked to crankshaft motions of the CH_2-CH(OH)-CH_2 segment, and which have an especially intense β dissipation peak – can be distinguished from networks having a low intensity β dissipation peak – e.g., styrene-crosslinked vinyl esters, polyimides, etc.

The β relaxation can play a very important role in fracture properties. For comparable T_g values, for instance ≈ 100°C, amine-crosslinked epoxies are considerably more ductile and tough than unsaturated polyesters.

(iii) In networks with a highly active β transition, e.g. essentially amine-cured epoxies, structural modifications leading to a decrease of the glass transition temperature (internal or external plasticization) lead to a decrease of the β-transition intensity, which leads to antiplasticization, as observed in linear polymers such as poly(vinyl chloride) (Chapter 11). This phenomenon has received various interpretations (Ngai *et al.*, 1996; Heux *et al.*, 1998).

10.2.4 Conclusion

Most of the physical properties of thermosets in the glassy state can be interpreted on the basis of three groups of basic properties:

- volumetric properties – density, packing density, expansion, free volume
- cohesive properties – CED, solubility parameter(s), nature of intermolecular interactions
- local mobility – β and γ motions, relaxations, secondary transitions.

Generally speaking, these properties are almost independent of the crosslink density in the same way as they are almost independent of the chain length in linear polymers. Thus, there are no fundamental differences between thermosets and amorphous thermoplastics in the glassy state, although certain second-order effects linked to crosslinking can be observed, sometimes, on packing density and local mobility.

The cohesive and volumetric properties (under the dependence of, essentially, atomic composition and CED), can be predicted or at least estimated from the CRU structure.

Parameters related to local mobility are more difficult to predict, although there have been significant advances in this field (Ngai, *et al.* 1986). A complete understanding of these properties, including phenomena such as antiplasticization, will probably require the knowledge of factors such as the spatial free volume fluctuations.

Certain important physical properties of polymer glasses depend also on their nonequilibrium character (creep, yield stress, fracture properties).

10.3 GLASS TRANSITION, STRUCTURE–PROPERTY RELATIONSHIPS

10.3.1 Introduction

The glass transition phenomenon has been presented in Chapter 4. Here, only structure–property relationships will be briefly examined.

There is a relatively abundant literature on the relationships between T_g and the network structure, but it is difficult to make a coherent synthesis of the published data owing to the great diversity of empirical, semiempirical, and physical approaches to the problem.

It must be first remarked that structure–T_g relationships can be reasonably established on the basis of experimental data obtained on networks of well-defined structure. Thus, it seemed interesting to us to distinguish three cases:

- Homogeneous "ideal" networks, also called "closed" networks, result from a single-step polymerization mechanism of a stoichiometric mixture of monomers, reacted to full conversion. Many amine-crosslinked epoxies of T_g < 200°C and polyurethanes obtained using a single isocyanate monomer and a single polyol belong to this family.
- Homogeneous "nonideal," e.g. "open" networks, obtained from the same chemistry as the previous ones. These networks contain dangling chains as a result of incomplete cure, nonstoichiometric composition, or presence of monofunctional monomers.
- Inhomogeneous "open" networks characterized by spatial fluctuations of the crosslink density (nodular/globular morphologies, microgels, see Chapter 7).

10.3.2 "Ideal" Networks

a. Epoxy Networks

Some important trends of the structure–property relationships in this field are well illustrated by a comparison of some stoichiometric, fully cured epoxide–amine networks (Table 10.6).

Three structural parameters have been calculated from the theoretical CRU for these networks:

- The crosslink density, n, expressed in terms of moles of crosslinked mers (network crosslinks);
- The chain stiffness expressed by the flex parameter F which, for a given chain of molar mass M_e, is given by : $F = M_e/N_e$, where N_e is the number of elementary (undeformable) segments. F is essentially an increasing function of the content of aromatic nuclei, Ar.
- The concentration of hydroxyl groups [OH], which has a predominant contribution to the CED. It is clear that this latter variable plays a negligible role in the series under study: the highest cohesive energy density (A) corresponds to the lowest T_g value.

The first two variables play an important role, especially chain stiffness, as shown by the comparison of systems DDM and DA12 based on the same epoxide, which have the same crosslink density and very close OH concentrations, e.g., close CED values. Thus, it appears that any widely

TABLE 10.6 Comparison of some stoichiometric, fully cured epoxide–amine networks: Tg, glass transition temperature; [Ar], aromatic group concentration; F, flex parameter (see text); n, crosslink density; and [OH], hydroxyl concentration. Note DGEBD = diglycidyl ether of butane diol; DGEBA = diglycidyl ether of bisphenol A (i = degree of polymerization); and TGAP = triglycidyl derivative of p-aminophenol; $DA_j = H_2N - (CH_2)_j - NH_2$; DDM = diamino diphenyl methane.

Code	Sample	T_g (K)	[Ar] (mol g^{-1})	F (g mol^{-1})	n (mol kg^{-1})	[OH] (mol kg^{-1})
A	DGEBD–DA2	273	0	14	4.31	8.62
B	DGEBA–DA2	396	4.88	22.5	2.44	4.88
C	DGEBA–DA12	366	4.55	23	2.27	4.54
D	$DGEBA_0$–DDM	480	6.83	33	2.28	4.56
E	$DGEBA_6$–DDM	395	7.10	33	0.473	3.78
F	TGAP–DDM	494	5.88	23	5.88	7.05
G	TGAP–DA12	386	2.34	17	5.85	7.03

applicable T_g–structure relationship must involve at least two variables that express, respectively, the chain stiffness and the crosslink density. To check this, it is convenient, in principle, to start from a structural series based on a linear polymer (glass transition T_{gl}), with variable but slight crosslink densities: in this case, crosslinking does not modify significantly the chain flexibility and one can establish the effect of the crosslink density on T_g.

A simplified version of the free volume theory, considering that T_g is an iso-free volume point and that the free volume "absorbed" by one crosslink is independent of the crosslink density, leads to (Fox *et al.*, 1955):

$$T_g = T_{gl} + K_{FL}\nu_e \tag{10.31}$$

where K_{FL} is a constant for the structural series under consideration and ν_e is the concentration of elastically active network chains (EANC).

One can try to derive such a law from a copolymer law of the Fox type, considering a CRU containing crosslinks (of which the mass is negligible compared with the chain mass). Then,

$$\frac{M}{T_g} = \frac{M_L}{T_{gl}} - N\Theta \tag{10.32}$$

where Θ is the contribution of a crosslink to the $M\,T_g^{-1}$ value, M is the molar mass of the CRU, and M_L is the molar mass of the linear chains. $M_L \sim M$, since the crosslinks have a negligible mass. Then,

$$\frac{1}{T_g} = \frac{1}{T_{gl}} - \frac{N}{M}\Theta = \frac{1}{T_{gl}}\left(1 - \frac{N}{M}\Theta\, T_{gl}\right) \tag{10.33}$$

For low crosslink densities, such that $\frac{N}{M}\Theta T_{gl} \ll 1$,

$$T_g \sim T_{gl} + \frac{N}{M}\Theta T_{gl}^2 = T_{gl} + \frac{2}{\varphi}\Theta T_{gl}^2 \nu_e \tag{10.34}$$

where φ is the crosslink functionality.

This equation is equivalent to the Fox–Loshaek relationship, provided that

$$K_{FL} = \frac{2}{\varphi}\Theta T_{gl}^2 \tag{10.35}$$

These relationships show that (i) if the "copolymer" approach is considered as a correct model, this means that the Fox–Loshaek relationship is only valid at low crosslink densities (as effectively found experimentally); (ii) The Fox–Loshaek constant is an increasing function of T_{gl}, e.g., of the chain stiffness, as effectively found. K_{FL} varies from 6–10 K kg mol^{-1} for fully aliphatic to 50–60 K kg mol^{-1} for highly aromatic systems. Thus, it appears that both chain stiffness and crosslinking effects cannot be considered independent. The effect of crosslinking, represented for instance by (T_g–T_{gl}) in the Fox–Loshaek equation, is an increasing function of the chain stiffness. This dependence is taken into account in the approach based on the theory of conformational entropy (Di Marzio, 1964) and leads to

$$T_g = \frac{T_{gl}}{1 - K_{DM} F \nu_e} \tag{10.36}$$

where K_{DM} is a universal constant and F is the flex parameter. The higher is F, the higher is the inertia of rotatable units and the lower is the whole chain mobility.

Let us remark that Di Marzio's equation reduces to the Fox–Loshaek equation at low crosslink densities:

$$T_g \sim T_{gl} + K_{DM} F\, T_{gl} \nu_e \tag{10.37}$$

so that $K_{FL} = K_{DM}\ F\ T_{gl}$. Thus, it is difficult to distinguish between both approaches at low crosslink densities, typically for $\nu_e \leq 1/10\, K_{DM} F$ (see below for the values of K and F).

Application of this approach to the usual thermosets, such as epoxies, carries many difficulties:

1. These networks are made of various types of chains and, eventually, of various types of crosslink. How do we calculate F?

2. There is no starting linear polymer; T_{gl} corresponds to a hypothetical copolymer containing all the network difunctional units and must be calculated. The following solutions have been proposed (Bellenger *et al.*, 1981):

For a given crosslink connected with φ chains, F is given by

$$F = \frac{1}{\varphi}(F_1 + F_2 + \cdots + F_\varphi) \tag{10.38}$$

where F_1, F_2 ..., F_φ, are the flex parameters of the chains linked to the crosslink. If there are distinct types of crosslinks in the CRU, F will be the number-average of the F_i values: $F = \Sigma N_i F_i / \Sigma N_i$, where N_i is the number of crosslinks of type i for which $F = F_i$. T_{gl} can be calculated using a copolymer law. The Fox approach gave the better results:

$$\frac{M_L}{T_{gl}} = \sum \frac{M_i}{T_{gi}} \tag{10.39}$$

where M_L is the molar mass of the difunctional units of the network, e.g., it is the whole molar mass of the CRU, M, minus the mass of the network crosslinks (in number N in the CRU): $M_L = M - NM_x$, where M_x is the molar mass of a crosslink. (Remark: generally, $NM_x \ll M$.)

Some component values of T_{gi} are listed in Table 10.7. It can be easily checked that these values are widely applicable since they allow us to predict the T_g of linear polymers such as polycarbonate, polysulphones, phenoxy, etc. For example, for the polysulphone

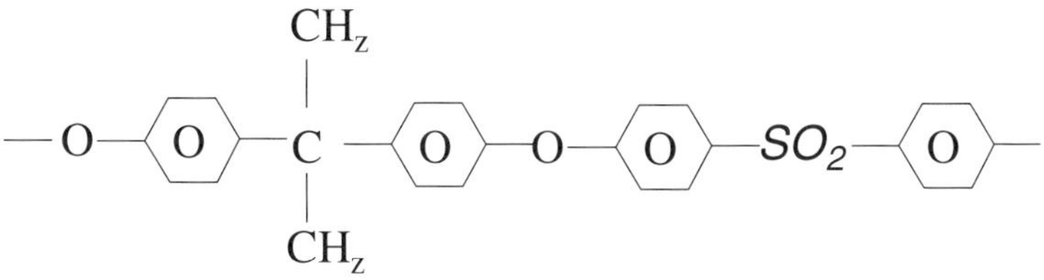

$$-O-\langle O \rangle-\underset{CH_z}{\overset{CH_z}{\underset{|}{\overset{|}{C}}}}-\langle O \rangle-O-\langle O \rangle-SO_2-\langle O \rangle-$$

($M = 442\ \mathrm{gmol}^{-1}$)

$$4 \times \frac{76}{453} + \frac{42}{554} + \frac{64}{782} + 2 \times \frac{16}{254} = 0.9547$$

so that

$$T_g = \frac{442}{0.9547} = 463\,\mathrm{K}$$

against 463 K (experimental value).

TABLE 10.7 Component values of some important groups for the calculation of T_{gi} from the Fox equation. Remark (a): these component values are only applicable to the bridges between two aromatic nuclei.

Unit	T_{gi}(K)
$-C_6H_4-C(CH_3)_2-C_6H_4-O-CH_2-CH(OH)-CH_2-O$	364
$-CH_2-$	233
$-O-$	254
$>C-(CH_3)_2$ (a)	554
$>SO_2$ (a)	782
$-CH_2-CH(OH)-CH_2-$	236
$-C_6H_4-$ (meta)	432
$-C_6H_4-$ (para)	453
$>C=O$	444

The larger the CRU, the higher is the number of elementary contributions to sum and the higher is the resulting incertitude in T_g prediction. The Di Marzio's equation can be rearranged as follows:

$$K_{DM}F = \frac{T_g - T_{gl}}{\nu_e T_{gl}} \tag{10.40}$$

By plotting $(T_g–T_{gl})/\nu_e T_{gl}$ versus F, one must obtain a straight line of slope K_{DM}. This approach was tested on about 100 systems of T_g values ranging from 270 to 520 K, and crosslink densities ranging from 0 to about 10 mol kg^{-1}. We obtained excellent results with $K_{DM} = 2.91$, when the crosslink density was expressed in terms of crosslink concentration (for trifunctional crosslinks). This leads to $K_{DM} \sim 2$, when the crosslink density is expressed in chain concentration.

For epoxy systems, no exception was found to the rule expressed by the above relationships. There is only one eventual ambiguity, illustrated as follows: let us consider the CRU of a system based on a difunctional epoxide (E) and a tetrafunctional amine $D–(NH_2)_2$:

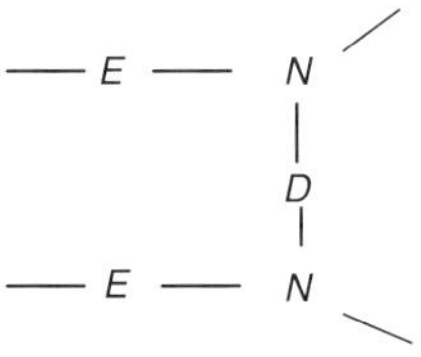

```
                  /
—— E ——   N
          |
          D
          |
—— E ——   N
                  \
```

In the case where D is a small group – e.g., a single rotatable unit such as a methylene or a phenylene – is it pertinent to consider that the CRU is composed of two trifunctional crosslinks

```
\
 N–
/
```

or rather one tetrafunctional crosslink

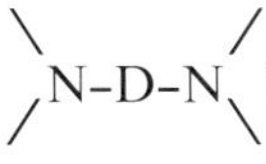

```
\       /
 N-D-N   ?
/       \
```

We lack a rigorous demonstration, but we find empirically that better results were obtained taking the first option (trifunctional crosslinks).

b. Other "Ideal" Networks

With regards to networks other than epoxies, one can encounter problems essentially linked to the nature of crosslinks, e.g., in the following materials which are not necessarily "ideal" networks but for which the hypothesis of ideality can be useful as a first approximation for evaluating the influence of certain structural parameters on T_g.

Styrene–Divinylbenzene Networks. Using ionic polymerization methods, Rietsch *et al.* (1976) prepared polystyrene (PS) networks with a well-controlled length of elastically active chains and crosslinks of variable functionality. In a given series, the glass transition temperature obeys the classical free volume theory:

$$T_g = T_{gl} + \frac{K_R}{M_e} \quad (K_R = \text{constant for a given series})$$

where M_e is the molar mass of the PS precursor and thus of an elastically active chain. But K_R increases with the crosslink functionality φ: from 13 K kg mol^{-1} for $\varphi = 3$, to 78 K kg mol^{-1}, for $\varphi = 10$–12. This effect, unpredicted by the free volume (or Di Marzio) theory, is attributed to steric constraints around crosslinks.

The concept of a polyfunctional hard cluster, in which a crosslinks-rich region constitutes a rigid phase, distinct from the rubbery one, has been developed by Nabeth *et al.* (1996), to explain the properties of certain polyurethane networks but it could be eventually applicable to the networks studied by Rietsch *et al.* (1976).

Polyurethane Networks. Andrady and Sefcik (1983) have applied the same relationship as Rietsch *et al.* (1976), to the glass transition temperature of networks based on poly(propylene oxide) diols with a controlled molar mass distribution, crosslinked by aromatic triisocyanates. They obtained a K_R value of 25 K kg mol^{-1}, about twice that for PS networks. They showed that the length distribution of elastically active chain lengths, directly related to the molar mass distribution of the starting poly(propylene oxide), has practically no effect on T_g.

Phenolics. The crosslink is an aromatic nucleus

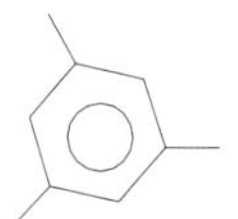

with a non-negligible contribution to the CRU mass. How do we take into account its "copolymer effect"?

Unsaturated Polyesters. The crosslink is the reacted fumerate unit

CH-CH

How do we take into account its eventual internal rotations? How do we distinguish between tetrafunctional (four elastically active chains) and trifunctional (three elastically active chains and one dangling chain) crosslinks?

Polynadimides. The exact structure of the network crosslink is somewhat controversial but it could resemble the structure shown in Fig.10.6. These polymers cumulate all the problems encountered in other polymers:

Is it really pertinent to consider that we are in the presence of hexafunctional crosslinks? In this case, how do we take into account their "copolymer effect"? In fact, if the black junctions in Fig. 10.6 connect one crosslink directly to another, we are in the presence of crosslink lines rather than dispersed individual crosslinks. Does this feature modify the whole glass transition behavior? There is, to our knowledge, no satisfactory answer to these questions, and the research field remains largely open in this domain.

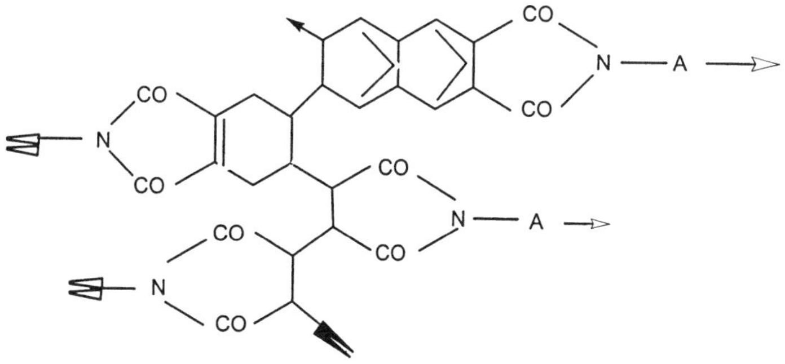

FIGURE 10.6 Tentative representation of the CRU of a polynadimide. A is a linear aromatic polyimide chain of molar mass $\sim$ 1500 g mol^{-1}. The male and female complementary junctions are distinguished.

10.3.3 Homogeneous, Nonideal Networks

The concentration of elastically active chains and dangling chains may be obtained using statistical models (Chapter 3). In this section, however, a simple approach will be used for estimation purposes. Let us call b the concentration of dangling chains. How do we determine the concentration of elastically active network chains? Let us imagine that we link each dangling chain extremity to another one. From b dangling chains, b/2 new elastically active chains will be generated, resulting in an "ideal" network with an average molar mass, M_{eo}, between crosslinks:

$$\nu_e + \frac{b}{2} = \frac{1}{M_{eo}} \text{ so that } \frac{1}{M_e} = \frac{1}{M_{eo}} - \frac{b}{2} \tag{10.41}$$

where M_e is the equivalent molar mass between crosslinks for the network under consideration. Let us apply this relationship to a crosslinked linear polymer in which the dangling chains are the chain ends of the initial linear polymer (b=2/M where M is the initial molar mass of the polymer).

$$\frac{1}{M_e} = \frac{1}{M_{eo}} - \frac{1}{M} \tag{10.42}$$

rather than

$$\frac{1}{M_{eo}} - \frac{2}{M}$$

as frequently found in the literature.

Remark: This approach is only valid for low concentrations of dangling chains, typically when one dangling chain is isolated from any other one by at least one crosslink.

How do we take into account the contribution of dangling chains to T_g? In linear polymers, we know that chain ends carry on a free volume excess and, thus, play a plasticizing effect that can expressed through a copolymer law:

$$\frac{M}{T_{gp}} = \frac{M}{T_{gp\infty}} + 2B \tag{10.43}$$

where T_{gp} is the T_g of the linear polymer of molar mass M and $T_{gp\infty}$ is the T_g of a hypothetical polymer of infinite length (determined by extrapolation). B is the chain end contribution. Let us note that for high molar masses, this equation is equivalent to the Fox–Flory equation:

$$T_g = T_{g\infty} - \frac{K_{FF}}{M} \tag{10.44}$$

Then,

$$\frac{1}{T_{gp}} = \frac{1}{T_{gp\infty}} + \frac{2B}{M} = \frac{1}{T_{gp\infty}}\left(1 + \frac{2BT_{g\infty}}{M}\right) \text{ so that } T_{gp} = T_{gp\infty} - \frac{2BT_{g\infty}^2}{M} \tag{10.45}$$

where $K_{FF} = 2BT_{gp\infty}^2$. (Note the similarity to the Fox–Loshaek equation, in which the crosslinking constant is also proportional to T_{gp^2}; e.g., it depends sharply on chain stiffness.)

Let us now consider a network based on a linear prepolymer having reactive groups in the chain ($T_g = T_{gp}$), reacted with a molecular "hardener" S (mass fraction s_t), and a CRU based on one prepolymer mole. The following relationship may be written:

$$\frac{M}{1 - s_t} \frac{1}{T_{gl}} = \frac{M}{T_{gp\infty}} + \frac{s_t}{1 - s_t} \frac{M}{T_{gs}} + 2B \tag{10.46}$$

where T_{gs} is the component value of the reacted "hardener" S. In styrene-crosslinked UP, for example, $s_t \sim 0.3$–0.4 and $T_{gs} = 380$ K. Then, a study of prepolymers of variable length allows us to determine $T_{gp\infty}$ and B, and T_{gl}, using the above equation.

In practice, it appears that dangling chains have a smaller plasticizing effect in networks than in linear polymers because the free volume effect of the chain end is counterbalanced by some physical crosslinking role of the branching point. Incidentally, we lack the theoretical tools to determine the critical length below which a pendant group can be considered as a simple lateral group, for instance a methyl group, and above which it is a true dangling chain with a specific contribution to the mobility. Here, also, the research field is largely open.

10.3.4 Inhomogeneous, "Nonideal" Networks

Incompletely cured networks constitute an important case of nonideal networks. They can be considered homogeneous if they result from step polycondensation. The following factors are expected to have an influence on T_g (beyond the gel point):

- crosslink density
- chain flexibility
- contribution of free chain ends to mobility
- effect of branching points
- plasticizing effect of the soluble fraction (in relation to its average molar mass).

Most of these structural parameters can be derived from a statistical study (Chapter 3), but what is not obvious is the determination of their contribution to T_g. Surprisingly, very simple relationships are available to estimate the evolution of the glass transition temperature with conversion, $T_g = f(x)$ (Chapter 4).

One possible way of characterizing the network inhomogeneity linked to spatial fluctuations of the crosslink density (Chapter 7) could be to study the width of the α dissipation band in the DMTA or DS spectra. The comparison between tetraglycidyl methylene dianiline (TGMDA)–DDS (bandwidth about 35 K) and poly(bismaleimide) (BMI) (bandwith about 85 K) networks is shown in Fig. 10.7. BMI is presumably more heterogeneous than TGMDA–DDS. Is it possible to appreciate the amplitude of crosslink density fluctuations from these data? We can tentatively use the Di Marzio's equation, which gives by derivation:

$$\frac{dT_g}{d\nu_e} = \frac{K_{DM} F T_g}{1 - K_{DM} F \nu_e} = K_{DM} F \frac{T_g^2}{T_{gl}} \tag{10.47}$$

This relationship was applied to some networks described in Table 10.6, and led to the data of Table 10.8.

One sees clearly that the higher is T_g, the higher is the sensitivity of T_g to small variations of the crosslink density. It is thus not very surprising to find relatively broad glass transitions in high-T_g polymers, which confirms the importance of cure schedule and TTT diagrams in order to obtain well-defined networks.

10.4 PROPERTIES IN THE RUBBERY STATE

10.4.1 Introduction

Above T_g, the network chains have sufficient thermal energy to overtake the potential barriers linked to Van der Waals interactions. They undergo fast conformational changes through cooperative segmental motions, but cross-linking prevents any liquid flow. We are thus in the presence of a peculiar state of matter, which displays at the same time liquid and solid (elastic) properties: the rubbery state.

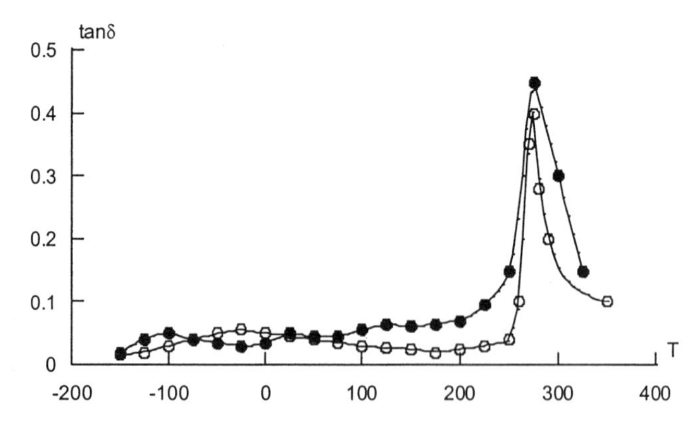

FIGURE 10.7 Comparison of the damping peaks of tetraglycidyl methylene diamine and diamino diphenyl sulphone (TGMDA–DDS epoxy network) (○) and poly(bismaleimide) (BMI network) (●).

TABLE 10.8 Sensitivity of T_g to variations of the crosslink density according to the Di Marzio's equation

Network	T_g (K)	dT_g/dn (K kg mol^{-1})[a]	$\frac{\Delta n}{n}$ (10 K) %[b]
DGEBD–DA2	273	14	17
DGEBA–DA2	396	32	13
DGEBA–DDM	480	60	8
TGAP–DDM	494	57	3

[a] n is the crosslink density expressed in mol kg^{-1}; [b] $\Delta n/n$ is the relative variation of crosslink density to give a T_g variation of 10 K.

Most of the physical properties of networks in the rubbery state can be linked to two groups of quantities that characterize respectively, the equilibrium entropic elasticity and the relaxation kinetics (linked to the segmental mobility).

10.4.2 Entropic Elasticity

a. Basic Theory

The equilibrium (relaxed) elastic properties of polymers in the rubbery state display two very important features:

1. The elastic modulus is an increasing function of temperature. In the ideal case, it is proportional to T.
2. Elasticity is nonlinear; the secant modulus (the stress/strain ratio) decreases rapidly as strain increases. It can be reversible at very high strains, but this property can be rarely checked for in thermosets owing to their brittleness at $T \geq T_g$.

These very original features gave rise to a great amount of research in the 1930s (Guth, Mark, Khun, Meyer, etc.), and the results were synthetized in the 1940s by Flory (1953), in a theory which can be summarized, in its simplest version, as follows:

Rubber deformation occurs at nearly constant volume. The work of deformation dW is equal to the free energy dF at constant volume:

$$dW = dU - TdS = fdl$$

where U is the internal energy, S is the entropy, f is the force, and l is the displacement). Thus, where

$$\frac{dW}{dl} = f = \left(\frac{dU}{dl}\right)_T - T\left(\frac{dS}{dl}\right)_T \tag{10.48}$$

Since f is experimentally found to be proportional to T, one can deduce that $dU/dl = 0$, and

$$f = -T\left(\frac{dS}{dl}\right)_T \tag{10.49}$$

We are in the presence of an "entropic" deformation. Elasticity results from chain conformational changes due to stretching rather than from changes of intermolecular distances as in polymer glasses, metals, and other classic materials.

To determine $(dS/dl)_{T''}$, one can use the Boltzmann law that links the entropy S to the number of complexions of the system:

$$S = k_B \text{Ln}\, \Omega \qquad (k_B = \text{ Boltzmann's constant}) \tag{10.50}$$

For one chain and one conformation of this chain, expressed by the coordinates of its extremities, Ω is the probability of this conformation. One starting point of the theory consists of expressing mathematically this probability by the Gauss function or, better, by the inverse Langevin function (which is equivalent to the Gauss function at low strains). For a "Gaussian" chain, a deformation state characterized by the three components of the extension ratio, $\lambda = l/l_0$, along the three space directions, leads to the following change of free volume energy:

$$\Delta F = -T\Delta S = \frac{kT}{2}(\lambda_x^2 + \lambda_y^2 + \lambda_z^2 - 3). \tag{10.51}$$

For a macroscopic sample (n_e moles of chains), it is assumed that the deformations are affine; e.g., that macroscopic extension ratios Λ_x, Λ_y, and Λ_z are equal to microscopic ones λ_x, λ_y,.and λ_z, Then,

$$\Delta F = n_e \frac{RT}{2}(\Lambda_x^2 + \Lambda_y^2 + \Lambda_z^2 - 3) \tag{10.52}$$

For a unidirectional loading:

$$\frac{f}{s_c} = RT\nu_e(\Lambda - \Lambda^{-2}) = \frac{RT\rho}{M_e}(\Lambda - \Lambda^{-2}) \tag{10.53}$$

where s_o is the initial sample cross section, ν_e is the number of elastically active network chains (EANC) per volume unit, M_e is the molar mass of network chains, and ρ is the density.

The stress, $\sigma = f/s$, is therefore given by

$$\sigma = RT\nu_e(\Lambda^2 - \Lambda^{-1}) \tag{10.54}$$

Since $\Lambda = 1 + \varepsilon$ and $\varepsilon \ll 1$, the tangent modulus is given by

$$E = 3RT\nu_e = \frac{3RT\rho}{M_c}; \qquad G = E/3 \tag{10.55}$$

where E and G are, respectively, the Young's modulus and the shear modulus.

Let us consider a rubbery thermoset at 450 K; $\rho \approx 1300 \pm 300$ kg m^{-3};M_c ranges generally between 0.1 and 2 kg mol^{-1}, so that one expects that 150 MPa $\leq E \leq$ 7 MPa.

The glassy state modulus is of the order of 1 GPa just below T_g. One can imagine the case of a very highly crosslinked polymer in which the rubbery modulus would be equal to the glassy one (no gap at T_g). The corresponding M_e value would be (at 500 K for instance) $M_e \approx 16$ g mol^{-1}, which is an unrealistic value.

Remark: As previously stated, the above theory is the simplest one. There are other physical approaches of entropic elasticity, especially the theory of phantom networks (Queslel and Mark, 1989), in which the cross-links freely fluctuate around their mean position. The above relationship then becomes:

$$E = 3RT(\nu_e - n) \tag{10.56}$$

where n is the concentration of crosslinks. Thus,

$$E = 3RT\nu_e\left(1 - \frac{2}{\varphi}\right) \tag{10.57}$$

where φ is the crosslink functionality.

b. Nonideal Behavior of Rubbers

The above relationships result from an approach based on at least eight hypotheses that define an "ideal" rubber. Thermosets are generally far from the ideal case for the following reasons.

Broad or Polymodal Distribution of Chain Lengths. A given chain is characterized by its maximum (full) extension ratio:

$$\Lambda_{max} = k_c M_e^{1/2} \tag{10.58}$$

where k_c depends on the chain tortuosity in its equilibrium state and other geometrical factors. One sees that the maximum chain elongation is an increasing function of the chain length (M_e). Thus, for a given macroscopic extension ratio, Λ, certain chains will be fully extended whereas others will remain partially coiled. It is clear that, in this case, the above relationships cannot predict the exact behavior.

Variations of the Equilibrium Distance r_o between Crosslinks. Thermodynamical calculations start from the hypothesis that the (gaussian) chain is unperturbed and lead to

$$\Delta F = \frac{3k_BT}{2}\frac{\langle r^2\rangle}{\langle r_0^2\rangle} \tag{10.59}$$

where r and r_o are the values of the average distance between crosslinks in the loaded and unloaded states, respectively. In the basic theory, r_o is considered temperature independent but, in fact, it can vary with T through the conformational state (the trans and gauche conformations can have distinct spatial extensions). Furthermore, crosslinking can perturb, at least locally, the chain equilibrium and this perturbation can be relatively important in the case of short chains (thermosets). Then, if the perturbed distance is r_c,

$$\Delta F = \frac{3k_BT}{2}\left(\frac{\langle r^2\rangle}{\langle r_c^2\rangle}\frac{\langle r_0^2\rangle}{\langle r_0^2\rangle}\right) = \varnothing\frac{3k_BT}{2}\left(\frac{\langle r^2\rangle}{\langle r_0^2\rangle}\right) \tag{10.60}$$

More generally, the basic Flory relationship can be extended as follows:

$$G = \varnothing RT\nu_e$$

where $\varnothing = G$ (experimental)/G (theoretical), takes into account the selected physical approach (for instance $\varnothing = 1 - 2/\varphi$ in the theory of phantom networks), and all the eventual causes of nonideality.

Mooney–Rivlin correction. In the 1940's, Mooney and Rivlin showed that, generally, the basic force–elongation relationship must be corrected by a term proportional to the reciprocal extension ratio:

$$\frac{f}{s_c} = (\Lambda - \Lambda^{-2})(C_1^M + C_2^M\Lambda^{-1}) \tag{10.61}$$

where C_2^M depends on the network structure and takes low values in densely crosslinked systems or in solvent-swollen systems

c. Experimental data on Thermosets

Despite the preceding remarks and the fact that thermosets are far from being "ideal" rubbers, the basic rubber elasticity theory works surprisingly well in most practical cases, as illustrated by the data of Table 10.9.

The rubber elasticity theory in its simplest version always predicts a good order of magnitude for the modulus, with $0.4 \leqslant \varnothing \leqslant 1.6$ for every system under consideration. It seems difficult to go beyond, predicting for instance $\varnothing$ from the network structure, for many reasons:

TABLE 10.9 Literature values of ∅ for thermosets

Family	System	Reference	∅
EP	DGEBA–DGEBD–DCM	Urbaczewski-Espuche *et al.*, 1991	0.98 ± 0.04 for less than 10% DGEBD
EP	DGEBA–DGEBD–DCM	Urbaczewski-Espuche *et al.*, 1991	0.42–0.43 for 50–100% DGEBD
UP	Maleate/phthalate/ styrene	Bellenger *et al.*, 1992	0.88–1.51 for propylene glycol systems
UP	Maleate-phthalate/ styrene	Bellenger *et al.*, 1992	0.53–0.89 for other diols
VE	Epoxy-methacrylate/ styrene	Ganem *et al.*, 1994	0.95–1.50 increases with styrene content

1. Very precise modulus measurements using suitable extensometric methods at high temperatures are not very easy to perform.
2. The rubbery modulus may change during the measurement at high temperatures, because the polymer may participate in both postpolymerization and degradation reactions.
3. The usual methods give not exactly the equilibrium modulus value: viscoelastic effects and internal energy effects (modulus component not proportional to T), are usually not taken into account.
4. There is only an alternative way of determining crosslink density by a non empirical method, using the theory of equilibrium swelling of a network in a solvent (Flory and Rehner, 1943). This method, usual in the domain of rubbers, needs the knowledge of a polymer–solvent interaction coefficient proportional to $(\delta_p - \delta_s)^2$, which is not very easy to determine accurately. Furthermore, damaging by swelling stresses and the need to work at elevated temperatures complicate the analysis seriously for the usual thermosets

Thus, at the present state of our knowledge, rubber elasticity can only be considered as a relatively rough method for crosslink density determinations, but it is practically the unique possible method for thermosets. It is valid for relative determinations in sample series where there are no big structural changes (for instance aliphatic or aromatic) other than crosslink density changes, and no complications linked to chemical reactions occurring in the timescale of the experiments (postpolymerization, degradation, etc.).

10.4.3 Properties Linked to the Segmental Mobility

a. Thermal Expansion in the Rubbery State

Expansion seems to be more linear than in glassy state:

$$V = V_{ol}(1 + \alpha_l T) \tag{10.62}$$

where the index l (liquid) is kept for the rubbery state because there is no difference between both states from the point of view of volumetric properties; α_l. appears generally as a decreasing function of the crosslink density, n, in polyesters (Shibayama and Suzuki, 1965), as well as in epoxies (Won *et al.*, 1991) (Fig. 10.8). The curves are almost parallel; the difference between both families is probably not significant and essentially due to experimental discrepancies. What is significant is the pseudo hyperbolic decrease of α_l with n.

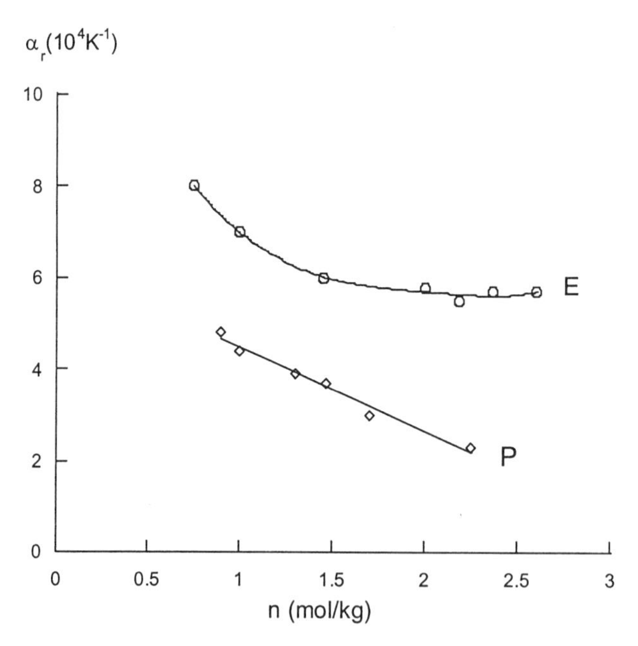

FIGURE 10.8 Expansion coefficient in rubbery state for amine-crosslinked epoxies, E (○), and for crosslinked aliphatic polyesters, P (◇), versus crosslink density (n).

b. Time–Temperature Equivalence Principle

It is well established that between T_g and about T_g + 50 K, the relaxation kinetics obeys the WLF law (Williams *et al.*, 1955). If P_r is a property depending on the macromolecular mobility (relaxation modulus, complex modulus, viscosity, diffusion rate, etc.), the time–temperature equivalence principle may be formulated as

$$P_r(t, T) = P_r\left(\frac{t}{a_T}, T_g\right) \tag{10.63}$$

where a_T is a shift factor, depending only on temperature through the WLF law:

$$\log a_T = \frac{-C_1^g(T - T_g)}{C_2^g + (T - T_g)} \tag{10.64}$$

It was initially stated that C_1^g are C_2^g were universal constants ($C_1^g \approx 17$; $C_2^g \approx 50$ K), but C_1^g can vary between 2 and 50 and C_2^g between 14 and 250 K (Mark, 1996). Epoxy values have been found in the low part of thcsc intervals: $C_1^g \approx 10$, $C_2^g \approx 40 \pm 15$ K (Gerard *et al.*, 1991), whereas unsaturated polyester values can be relatively high: $C_1^g/C_2^g = 15$–55; = 73–267 K (Shibayama and Suzuki, 1965). There is, to our knowledge, no synthetic study on the "ideality" and crosslinking effects on $C_1^g$and C_2^g. The time-temperature equivalence principles will be examined in detail in Chapter 11, which is devoted to elasticity and viscoelasticity.

c. Free Volume

The free volume fraction, f_T, is usually defined from the classical linear model of thermal expansion.

$$f_T = f_g + \alpha(T - T_g) \tag{10.65}$$

where $\alpha = \alpha_l - \alpha_g$ is the difference between rubbery and glassy expansion coefficients and f_g is the free volume fraction "frozen" at T_g.

The WLF relationship can be derived from the Doolittle relationship that links the mobility M to the free volume fraction:

$$M = M_0 \exp -\frac{B_D}{f_T} = M_0 \exp -\frac{B_D}{f_g + \alpha(T - T_g)} \tag{10.66}$$

where B_D is usually considered equal to unity. It results in

$$C_1^g = \frac{B_D}{2.3f_g} \text{ and } C_2^g = \frac{f_g}{\alpha} \tag{10.67}$$

In principle, these relationships open the way to a determination of f_g which is found to decrease with crosslink density as well in "ideal" epoxy networks (Gerard *et al.*, 1991), as in "nonideal" polyesters (Shibayama and Suzuki, 1965). However, it must be recognized that, in both series of data, it is impossible to have consistent values of , C_1^g, C_2^g, α, and f_g except if B_D varies with the structure, which can be considered as a serious argument against the free volume interpretation of WLF parameters.

There is a possible direct way of determining f_g through positron annihilation measurements that allow us to determine the volume of holes in polymer glasses, with typical diameters in the range 0.05–0.25 nm. Using this method, it is possible to determine f_g values of the same order as the ones derived from WLF parameters. For instance, in DGEBA systems cured by oligomeric polysulphone diamines, Jean *et al.* (1995) obtained f_g values ranging from about 0.018 (the highest crosslink density) to 0.037 (the lowest crosslink density).

10.5 CONCLUSIONS

As in linear polymers, the relative influence of the molecular structure (scale of nanometers and monomers), and the macromolecular structure (crosslink density), on network properties, depends on temperature, as shown in Fig. 10.9. In the glassy state, the physical behavior is essentially controlled by cohesion and local molecular mobility, both properties being mainly under the dependence of the molecular scale structure. As expected, there are only second-order differences between linear and network polymers. Here, most of the results of polymer physics, established on linear polymers, can be used to predict the properties of thermosets. Open questions in this domain concern the local mobility (location and amplitude of the β transition).

In the glass transition region, both structural scales play a significant role and their effects cannot be dissociated. The crosslinking effect (macromolecular scale) on T_g, is an increasing function of the chain stiffness, which is under the dependence of molecular scale factors (essentially aromaticity).

The equilibrium properties in the rubbery state are almost exclusively governed by the macromolecular scale structure: crosslinking suppresses liquid flow, decreases the number of available network complexions, and the gap between equilibrium (unstretched) and fully stretched network states. With regard to time-dependent properties (viscoelasticity, time–

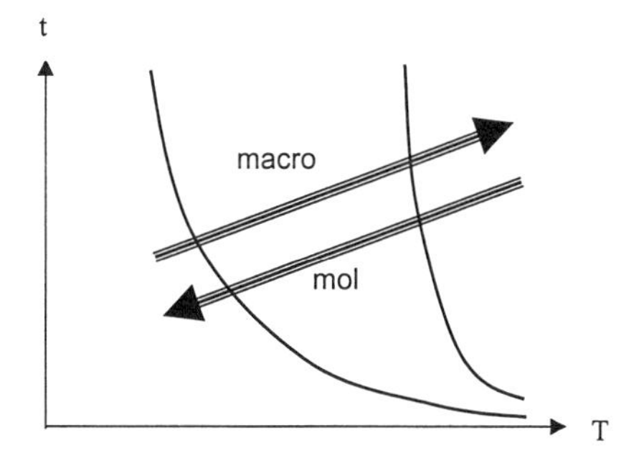

FIGURE 10.9 The relative influence of the molecular (mol) and macromolecular (macro) scale, i.e., essentially crosslinking, on physical properties. The arrows indicate an increasing influence of the structural scale.

temperature equivalence), there are no fundamental differences between network and linear polymers.

NOTATION

A	= "parabolic" expansion coefficient according to Eq. (10.10), K^{-2}
Ar	= concentration of aromatic groups, mol kg^{-1}
a_T	= shift factor (time–temperature equivalence principle)
b	= dangling chain concentration, mol kg^{-1}
B	= contribution of 1 mole of chain end to MTg^{-1} for a linear polymer, mol kg^{-1} K^{-1}
B_D	= Doolittle constant
C_1^M, C_2^M	= Mooney–Rivlin coefficients
C_1^g, C_2^g	= WLF Coefficients for the reference temperature = T_g, C_1^g dimensionless C_2^g in K
CED	= cohesive energy density, J m^{-3} or MPa
dF	= elementary variation of the free energy, J
E, E^*, E',E''	= Young's modulus (experimental, complex, real part, and imaginary part), Pa
E_c	= cohesive energy, J mol^{-1}
F	= flex parameter, g mol^{-1}
f	= force
f_T	= free volume fraction at temperature T
f_g	= free volume fraction at T_g
G, G^*, G',G''	= shear modulus (experimental, complex, real part and imaginary part), Pa
H	= activation enthalpy, J mol^{-1}
ΔH_v	= vaporization enthalpy (molar), J mol^{-1}

J, J*, J', J''	=	compliance (experimental, complex, real part, and imaginary part), Pa
K	=	bulk modulus, Pa
k	=	pre-factor of the density-average atomic mass relationship, kg m^{-3} mol $^{2/3}$ $g^{-2/3}$
k_B	=	Boltzmann's constant, J mol^{-1} K^{-1}
k_C	=	ratio: maximum chain extension ratio/(molar mass)$^{1/2}$, $mol^{1/2}$ $kg^{-1/2}$
K_{FL}	=	Fox and Loshaek constant (K_{FL} = $dT_g/d(_e)$, K kg mol^{-1}
K_{FF}	=	Fox and Flory constant (K_{FF} = $dT_g/d(M^{-1})$ for linear polymers), K kg mol^{-1}
K_{DM}	=	Di Marzio's constant
l	=	displacement, m
M	=	molar mass of the CRU, kg mol^{-1}
M_d	=	Average atomic mass kg mol^{-1}
M_L	=	molar mass of the difunctional units of the CRU, kg mol^{-1}
M_x	=	molar mass of an elastically active network chain, kg mol^{-1}
M_e	=	molar mass of an elastically active network chain, kg mol^{-1}
n	=	crosslink density (moles of crosslinks/mass unit), mol kg^{-1}
N	=	number of crosslinks in the CRU
P, (P*)	=	perturbation (complex perturbation)
P_r	=	property
Q, (Q*)	=	response to the perturbation (complex response)
r	=	distance between chain extremities in loaded state, m
r_o	=	distance between chain extremities in unloaded state, m
R	=	gas constant, J mol^{-1} K^{-1}
s, s_o	=	sample cross section in, respectively, loaded and unloaded state, m^2
s_t	=	styrene mass fraction
S	=	entropy, J mol^{-1} K^{-1}
S*	=	ratio Q*/P*
t	=	time, s
T	=	temperature, K
T_a	=	ambient Temperature, K
T_g	=	glass transition temperature, K
T_α, T_β, T_γ etc., α, β, γ, etc.	=	Transition temperatures, K
T_{gl}	=	glass transition temperature of the linear polymer on which the network is based, K
$T_{g\infty}$	=	glass transition temperature of a hypothetical linear polymer of infinite length, K
U	=	internal energy, J
V	=	volume, m^3
V_W	=	van der Waals volume, m^3

V_{og}, V_{ol} = extrapolated volume at 0 K from glassy state and rubbery state, respectively, m^3
W = work of deformation, $\surd$
y = molar ratio of monomers in a two-component systems

Greek Letters

α_g, α_l = cubic expansion coefficients in glassy and liquid (rubbery) states, K^{-1}
β_δ = exponent of the KWW law
δ = Hildebrand's solubility parameter, MPa $^{1/2}$)
ε = strain
ε_i = cohesive energy/molar mass ratio for a group i, J g^{-1}
λ_x, λ_y, λ_z = microscopic extension ratios in directions x, y, and z
Λ_x, Λ_y, Λ_z = macroscopic extension ratios in directions x, y, and z
ρ = density, kg m^{-3}
ρ^* = packing density (V_w/V)
τ = relaxation time, s
φ = crosslink functionality
$\varnothing$ = ratio E/3RT ν_e
ω = angular frequency, rad.s^{-1}

REFERENCES

Andrady AL, Sefcick MD, *J. Polym. Sci. Polym. Phys. Ed.*, 21, 2453–2463 (1983).

Barton AFM, *Handbook of Solubility Parameters and Other Cohesive Parameters*, CRC Press, Boca Raton Florida, 1983.

Bellenger V, Dahoui W, Verdu J, *J. Appl. Polym. Sci.*, 33, 2647–2650 (1987).

Bellenger V, Morel E, Verdu J, *J. Polym. Sci., B: Polym. Phys.*, 25, 1219–1234 (1987).

Bellenger V, Dahoui W, Morel E, Verdu *J, J. Appl. Polym Sci.*, 15, 563–571 (1988).

Bellenger V, Morel E, Verdu J, *J. Appl. Polym. Sci.*, 37, 2563–2576 (1989).

Bellenger V, Mortaigne B, Grenier MF, Verdu J, *J. Appl. Polym. Sci.*, 44, 643–651 (1992).

Bellenger V, Verdu J, Ganen M, Mortaigne B, *Polym. and Polym. Compos.*, 2, 9–16 (1994).

Bellenger V, Kaltenecker-Commerçon J, Tordjeman P, Verdu J, *Polymer*, 38, 4175–4184 (1997).

Bondi A, *Physical Properties of Molecular Crystals, Liquids and Glasses*, Wiley, New York, 1968.

Bongkee Cho, *Polym. Eng. Sci.*, 25, 1135–1138 (1985).

Deslandes N, Bellenger V, Jaffiol F, Verdu J, *J. Appl. Polym. Sci.*, 69, 2663–2671 (1998).

Di Marzio EA, *J. Res. NBS*, 68A, 611–617 (1964).

Flory PJ, *Principles of Polymer Chemistry*, Cornell University Press, Ithaca, New York, 1953.
Flory PJ, Rehner J, *J. Chem. Phys.*, 11, 521–526 (1943).
Fox TG, Loshaek S, *J. Polym. Sci.*, 15, 371–390 (1955).
Ganem M, Lafontaine M, Mortaigne B, Bellenger V, Verdu J, *J. Macromol. Sci. Macromol. Phys.* B, 33, 155–172 (1994).
Gerard JF, Galy J, Pascault JP, Cukierman S, Halary JL, *Polym. Eng. Sci.*, 31, 615–621 (1991).
Heux L, Lauprtre F, Halary JL, Monnerie L, *Polymer*, 39, 1269–1278 (1998).
Jean YC, Deng Q, Nguyen TT, *Macromolecules*, 28, 8840–8844 (1995).
Mark JE, *Physical Properties of Polymers Handbook*, American Institute of Physics, Woodbury, New York, 1996.
Morel E, Bellenger V, Bocquet M, Verdu J, *J. Mater. Sci.*, 24, 69–75 (1989).
Nabeth B, Pascault JP, Dusek K, *J. Polym. Sci. Polym. Phys. Ed.*, 34, 1031–1054 (1996).
Ngai KL, Rendell RW, Rajagopal AK, Teitler S, *Ann NY Acad. Sci.*, 484, 150 (1986).
Ngai KL, Rendell RW, Yee AF, Plazek DJ, *Macromolecules*, 24, 61–67 (1996).
Porter D, *Group Interaction Modelling of Polymer Properties*, Marcel Dekker, New York, 1995.
Queslel JP, Mark JE, In *Comprehensive Polymer Science*, vol. 2, Pergamon Press, New York, 1989, pp. 271–309.
Rietsch F, Daveloose D, Froelich D, *Polymer*, 17, 859–863 (1976).
Shibayama K, Suzuki J, *J. Polym. Sci.* A, 3, 2637–2651 (1965).
Skourlis TP, Mc Cullough RL, *J. Appl. Polym. Sci.*, 62, 481–490 (1996).
Tcharkhtchi A, Gouin E, Verdu J, *J. Polym. Sci. Polym. Phys. Ed.*. 38, 537–543 (2000).
Urbaczewski-Espuche E, Galy J, Gerard JF, Pascault JP, Sautereau H, *Polym. Eng. Sci.*, 31, 1572–1580 (1991).
Van Krevelen DW, *Properties of Polymers*, 3rd ed, Elsevier, Amsterdam, 1990.
Venditti A, Gillham JK, *J. Appl. Polym. Sci*, 56, 1687–1705 (1995).
Williams M, Landel R, Ferry J, *J. Am. Chem. Soc.*, 77, 3701–3707 (1955).
Won YG, Galy J, Pascault JP, Verdu J, *Polymer*, 32, 79–83 (1991)

11

Effect of Crosslink Density on Elastic and Viscoelastic Properties

11.1 INTRODUCTION

This chapter is devoted to a short description of low-strain mechanical properties of polymers in the solid state and in the glass transition region, with an emphasis on the effect of crosslinking on these properties. There are three degrees of complexity in the description of this behavior, depending on the number of variables taken into account in the constitutive equations under consideration.

Lowest Level. At the lowest level, these equations could involve only two variables, the stress σ and the strain ε:

$$f(\sigma, \varepsilon) = 0$$

This means that the mechanical behavior is regarded in relatively sharp intervals of time and temperature. Then, engineering moduli are generally sufficient to describe the material's behavior at low strains. They are interrelated by the following relationships:

$$E = 3K(1 - 2\nu); \qquad G = \frac{3}{2}\frac{(1 - 2\nu)}{(1 + \nu)}K = \frac{E}{2(1 + \nu)} \tag{11.1}$$

where E, G, and K are the tensile (Young), shear (Coulomb), and bulk modulus, respectively and ν is the Poisson's ratio. For an ideal elastic body, $\nu = 0.33$, leading to $E = K$ and $G = 3E/8$. For an ideal rubber, $\nu \to 0.50$ and $E \to 3\,G$ and $K \to \infty$.

E can be determined from a uniaxial tensile ($E = \sigma/\varepsilon$), or a uniaxial compressive test, or a flexural test (Fig. 11.1a and Chapter 12); G can be determined from a shear test, $G = s/\gamma$, where s is the shear stress and γ is the shear strain (Fig. 11.1c); K can be determined from a compressibility test,

$$K = \left(\frac{1}{V}\frac{dV}{dp}\right)^{-1}$$

where V is the volume and p is the hydrostatic pressure (Fig. 11.1b); and ν can be determined from two independently determined values of modulus, or from a tensile test using a bidimensional extensometer.

In practice, the values of these elastic quantities are needed for rough evaluations in mechanical design of parts working at ambient temperature, as illustrated for instance by Ashby (1992).

Second Level. At the second level, the constitutive equations must involve two (or more) additional variables. For instance:

$$f(\sigma, \varepsilon, \dot{\varepsilon}, T) = 0$$

where $\dot{\varepsilon}$ is the strain rate and T is the temperature.

These new variables are necessary to take into account viscoelastic effects linked to molecular motions. These effects are non-negligible in the glassy domain between boundaries α and β in the map of Fig. 11.2, and they are very important in the glass transition region (around boundary α). Here, we need relationships that express the effects of $\dot{\varepsilon}$, $\dot{\sigma}$ (the stress rate may be used instead of the strain rate), and T on the previously defined elastic properties. Also numerical boundary values of elastic properties are required, characterizing unrelaxed and relaxed states (see Chapter 10).

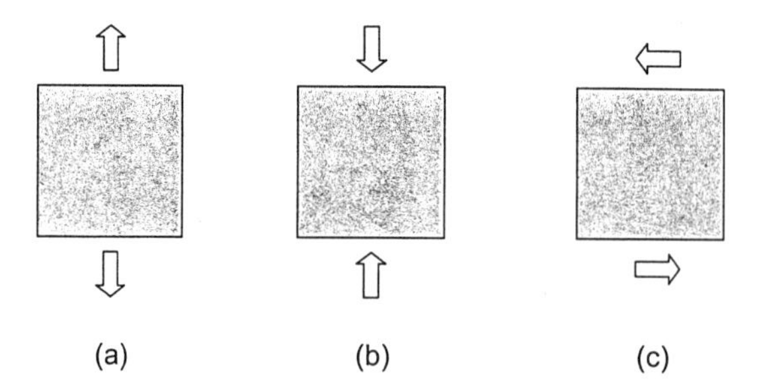

Figure 11.1 Mechanical tests to determine (a) E; (b) K, and (c) G.

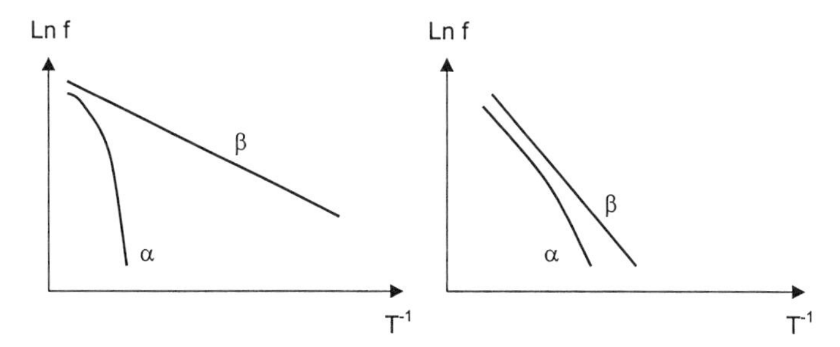

FIGURE 11.2 Shape of relaxation maps (coordinates of transitions α, β, and γ in a graph ln (frequency) – reciprocal temperature). Left: polymers having their α and β transitions well separated (example: polycarbonate, amine-crosslinked epoxy). Right: polymers with close α and β transitions (example : polystyrene, unsaturated polyester).

There are three main experimental approaches for mechanical characterization in this domain. They correspond to particular solutions of the material's state equation:

- static tests: $\varepsilon = \varepsilon_0 =$ constant (relaxation) or $\sigma = \sigma_0 =$ constant (creep)
- monotonous tests with loading rate $\dot{\varepsilon}$ or $\dot{\sigma} =$ constant (for instance tensile tests): $\dot{\varepsilon} = \frac{1}{l}\frac{dl}{dt}$
- dynamic tests: $\varepsilon = \varepsilon_0 \sin \omega t$ or $\sigma = \sigma_0 \sin \omega t$.

Polymers are generally assumed to obey the Boltzmann superposition principle in the domain of small strains. When there are changes of loading conditions, the effects of these changes are additive when the corresponding responses are considered at equivalent times. For instance, if different stresses $\sigma_0, \sigma_1, \sigma_2, \ldots$ are applied at different times $0, t_1, t_2, \ldots$, respectively, the resulting strain is

$$\varepsilon(t) = J(t)\sigma_0 + J(t - t_1)\sigma_1 + \cdots + J(t - t_i)\sigma_i \tag{11.2}$$

where J(t) is the time-dependent creep compliance.

In the same way, if different strains $\varepsilon_0, \varepsilon_1, \varepsilon_2, \ldots$ are applied at times 0, $t_1, t_2, \ldots$, the resulting stress is

$$\sigma(t) = E(t)\varepsilon_0 + E(t - t_1)\varepsilon_1 + \cdots + E(t - t_i)\varepsilon_i \tag{11.3}$$

where E(t) is the time-dependent relaxation modulus.

Thus, what is needed is the knowledge of J(t) or E(t). It is generally convenient to use dynamic tests to determine J(ω) or E(ω), and then apply adequate mathematical transformations to obtain J(t) or E(t).

In general, polymers obey a time–temperature superposition principle:

$$P_r(t, T) = P_r\left(\frac{t}{a_T}, T_R\right) \tag{11.4}$$

where P_r is a property, T_R is a reference temperature, and a_T is a shift factor that depends only on temperature. An interesting characteristic of polymers is that $a_T = f(T)$ takes distinct mathematical forms below and above T_g.

Third Level. At the third level of complexity, the unsteady character of the polymer linked to the fact that it is out of equilibrium in the glassy state must be taken into account. The behavior of the material depends not only on the mechanical stimuli and environmental conditions but also on its thermomechanical history since its processing. In other words, time must be added as a variable to the constitutive equations:

$$f(\sigma, \varepsilon, \dot{\varepsilon}, T, t) = 0$$

From a practical point of view, the main consequence of physical ageing by structural relaxation is embrittlement (decrease in fracture resistance; Chapter 12). For the other aspects of mechanical behavior, ageing has either no effect or a favourable effect (increase of relaxation times, leading to a decrease of creep or relaxation rates). This is the reason why, in most thermoset applications, the knowledge of short-term properties is considered to be sufficient for engineering design, as far as fracture and durability are not concerned.

Thus, the present chapter contains essentially two sections, devoted to the first and second degree of complexity, respectively.

11.2 ENGINEERING ELASTIC PROPERTIES IN GLASSY STATE

11.2.1 Bulk Modulus

a. Short Theoretical Survey

The bulk modulus K can be derived directly from an expression for the intermolecular energy potential u(d), where d is the intermolecular distance (Fig. 11.3). The most usual expression for u(d) is the Lennard–Jones relationship:

$$u(d) = -(\mathrm{CED})\left[\left(\frac{d_0}{d}\right)^{12} - 2\left(\frac{d_0}{d}\right)^{6}\right] \tag{11.4}$$

where CED is the cohesive energy density (Chapter 10).

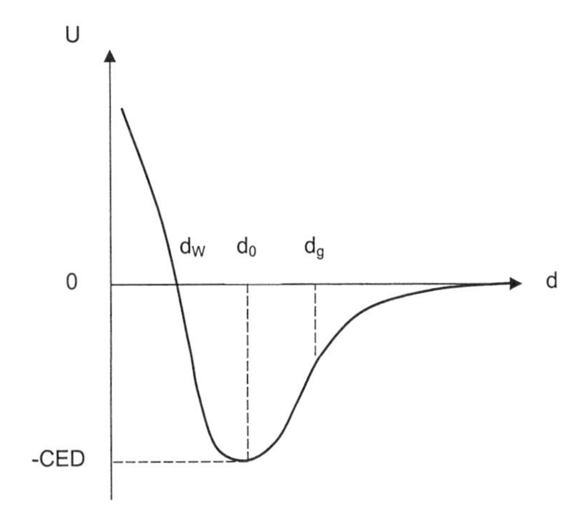

FIGURE 11.3 Variations of the intermolecular potential energy u(d) with intermolecular distance d: d_0 is the equilibrium distance at which $u(d) = -CED$ (CED = cohesive energy density); d_W is the van der Waals distance at which $V = V_W$; and d_g is the abscissa of the inflection point and corresponds to the glass transition.

The bulk modulus can be derived from this equation, since

$$K = \left[\frac{1}{V}\frac{\partial V}{\partial p}\right]^{-1} \text{ and } p = \frac{\partial u}{\partial V} \text{ with } \frac{d_0}{d} = \left(\frac{V_0}{V}\right)^{1/3} \tag{11.5}$$

These relationships lead to the following results:

(i) The bulk modulus is proportional to the cohesive energy density:

$K = a_c(CED)$

where the theoretical value of a_c is 8 (Tobolsky, 1960).

(ii) However, since the intermolecular distances increase with temperature, the intermolecular energy potential and the CED are temperature-dependent. It is then convenient to consider rather a temperature-independent quantity $(CED)_W$, defined by

$$(CED)_W = \frac{CED}{\rho^*} = \frac{E_{coh}}{V_W} \tag{11.6}$$

where V_W is the van der Waals volume and ρ^* the packing density (Chapter 10). It has been shown that K (at T = 0 K) = $11(CED)_W$ and $K(T_g) = 5.7(CED_W)$, just below T_g (Porter, 1995). K at a given temperature $T < T_g$ can be interpolated by the following relationship:

$$K(T) = K(0K)\left(1 - \alpha_K \frac{T}{T_g}\right) \tag{11.7}$$

where $\alpha_K \sim 0.5$.

(iii) There is apparently no discontinuity of the temperature dependence of K at secondary (sub-glass) transitions. In contrast, K varies by a factor of about 2 at T_g: $K = 3(CED)_W$, just above T_g. In thermosets this gap is expected to be reduced as the crosslink density increases.

b. Experimental Data

Ultrasonic measurements (5 MHz) made at ambient temperature on amine-crosslinked epoxies (Morel *et al.*, 1989), or styrene-crosslinked vinyl esters (Bellenger *et al.*, 1994), show that K is effectively proportional to the cohesive energy density:

$$K = (11 \pm 1)(CED) \tag{11.8}$$

CED is determined by calculation, using the Van Krevelen increment values (Van Krevelen, 1990). There is no apparent effect of the crosslink density on K, which seems to depend (as CED), only on the molecular scale structure (polarity, hydrogen bonding). The following results were obtained with quasi-static tensile measurements (10^{-3}–10^{-4} s^{-1} strain rate) in a temperature range 200 K–T_g, using a bidimensional extensometer to determine E and ν (from which K and G could be determined) (Verdu and Tcharkhtchi, 1996):

1. K values are of the same order as ultrasonic ones (typically 5–7 GPa).
2. K values are almost constant in the temperature interval between 200 K and T_g−30, whereas the above theory would predict significant variations (typically 15–25% between 200 and 400 K).
3. The bulk modulus (K) is unaffected by the β transition; it is thus not surprising to find that it is independent of the frequency/strain rate.

To summarize, the bulk modulus of thermosets is proportional to the cohesive energy density and does not depend practically on temperature in the 200 K – (T_g − 30 K) temperature range. There is no significant effect of crosslink density on K, which can be predicted (in the temperature interval under consideration) using K = 11 (CED), with an incertitude of about 10%.

11.2.2 Shear Modulus

The shear modulus can also be derived from an intermolecular energy potential, and its value at 0 K is also proportional to the cohesive energy density. Typically, at very low temperatures, $\nu \approx \frac{1}{3}$, so that $G \approx \frac{3}{8}$ K, and thus

$$G(0\ K) \approx 4\ (CED)_W \tag{11.9}$$

But the big difference with K is that G is directly affected by molecular motions and decreases almost discontinuously at sub-glass (secondary) transitions (Fig. 11.4). Thus, G is more difficult to predict than K, since the prediction must take into account the characteristics of secondary transitions, which are difficult to establish theoretically (Chapter 10).

From the engineering point of view, there are at least two relatively simple approaches to the problem.

For Porter (1995), the important variable is the cumulative loss tangent, $\tan \Delta_\beta$, defined by

$$\tan \Delta_\beta = \tan \delta_b + \sum \tan \delta_\beta \tag{11.10}$$

where $\tan \delta_b$ is the dissipation factor in the unrelaxed state and $\sum \tan \delta_\beta$ is the sum of dissipation factors for all the secondary transitions occurring at temperatures lower than the temperature under consideration and in the selected timescale; $\tan \delta_b$ is generally lower than 0.01, so that $\tan \Delta_\beta$ is essentially representative of the mechanical activity of the secondary transi-

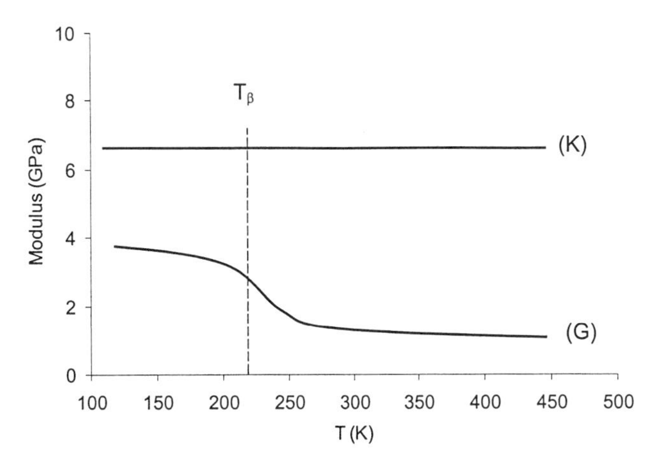

FIGURE 11.4 Typical shape of the temperature variation of the bulk (K) and shear (G) modulus in the glassy state, around a relatively intense β transition.

tions in the range of (T, t) conditions under consideration. Let us remind ourselves (see Chapter 10) that certain secondary transitions, such as T_β in amine-crosslinked epoxies, are very active and lie at a relatively low temperature, for instance 200–240 K at 1 Hz. Other secondary transitions, such as $T\gamma$ in polyesters (180 K at 1 Hz), have a low activity, whereas others, such as T_β in polyesters ($\sim$ 350 K at 1 Hz) or in polyimides (390–420 K at 1 Hz), are obviously inactive at ambient temperature, except eventually in long-term creep or relaxation experiments.

In the frame of this analysis, the Poisson's ratio depends only on the cumulative loss tangent:

$$\nu = 0.5[1 - 0.33(1 - \tan \Delta_\beta^{1/2})^2] \quad (11.11)$$

Then, the knowledge of K and ν allows us to calculate all the other elastic constants. Typically, the Poisson's ratio of glassy thermosets at temperatures close to 20°C ranges between 0.37 (styrene-crosslinked polyesters) and 0.42 (certain amine-crosslinked epoxies), which corresponds to $\tan \Delta_\beta$ values ranging between about 0.01 and 0.1.

The shear modulus is then given by

$$G = \frac{K}{\dfrac{3}{\left(1 - \tan \Delta_\beta^{1/2}\right)^2} - 0.33} \quad (11.12)$$

The second method (in fact the method is applied to Young's modulus but it can be transposed to G) starts from the following equation (Gilbert *et al.*, 1986):

$$G = G_0\left(1 - \alpha_G \frac{T}{T_g}\right) - \sum \Delta G_\beta \quad (11.13)$$

where α_G has values in the order of 0.3–0.5 and $\sum \Delta G_\beta$ is the sum of modulus gaps at the sub-glass transitions located at temperatures below the test temperature; ΔG_β, $\Delta G\gamma$, etc., have to be experimentally determined.

If the hierarchy of low-temperature/high-frequency (unrelaxed) G values corresponds more or less to the hierarchy of cohesive energy densities, it is completely modified at ambient temperature/low frequency – i.e., in the usual mechanical testing conditions – owing to the importance of the modulus gap at T_β (Fig. 11.5).

11.2.3 Tensile Modulus

If one considers that K (cohesive energy density + packing density (expansion)) and G (cohesive energy density + packing density + molecular

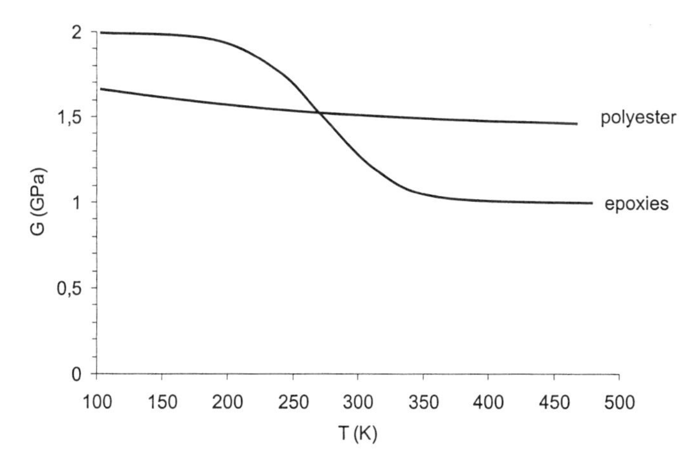

FIGURE 11.5 Temperature variation of G for a network with low local mobility (UP) and a network with high local mobility (EPO); for the latter, $(\Delta G/G) \sim 1/2$.

mobility) are the basic elastic quantities, then, the other elastic quantities, E and ν, are fully determined by K and G values.

By eliminating ν, E may be expressed as a function of G and K:

$$E = 3K(1 - 2\nu) = \frac{3G}{1 + \frac{G}{3K}} \tag{11.14}$$

$G \leq 3K/8$, so that $G/3K < 1/8$ and, roughly, $E \approx 3G$.

It is thus expected that E varies with loading conditions as G, displaying noticeable gaps $\Delta E_\gamma \Delta E_\beta$, etc. at the γ, β, etc., transitions.

As in the case of G, the unrelaxed tensile modulus will be found more or less proportional to the cohesive energy density, whereas the relaxed modulus will depend sharply on the activity of local motions.

Polystyrene (PS) and bisphenol A polycarbonate (PC), are good examples of linear polymers with respectively low (PS) and high (PC) local mobility. Their characteristics are summarized in Table 11.1 (Porter, 1995).

It is observed that E and G are lower for PC than for PS, despite the fact that PC is more cohesive than PS. This is because the β-dissipation peak is considerably more intense for PC than for PS.

Unsaturated polyesters (UP) or vinyl esters (VE) on one side, and amine-crosslinked epoxies on the other side, can be considered as the thermoset counterparts of, respectively, polystyrene and polycarbonate. Tensile moduli determined at relatively low strain rates (10^{-3}–10^{-4} s^{-1}) are typically $3.0 < E < 4.5$ GPa for UP and VE and $2.4 < E < 3.0$ GPa for epoxies,

TABLE 11.1 Comparison of low-frequency elastic properties of linear polymers with low (PS) and high (PC) local mobility, respectively (After Porter, 1995.)

Polymer	K (GPa)	tan Δ_β	G (GPa)	E (GPa)	ν
PS	3.5	0.008	1.2	3.0	0.36
PC	3.8	0.050	0.9	2.3	0.40

despite the fact that amine-crosslinked epoxies are more cohesive than UP or VE.

11.2.4 Poisson's Ratio

As for tensile modulus, ν can be expressed as a function of G and K:

$$\nu = \frac{3K - 2G}{6K + 2G} = \frac{1}{2}\left[1 - \frac{G/K}{1 + \frac{2}{3}G/K}\right] \tag{11.15}$$

Derivation with respect to G gives

$$\frac{d\nu}{dG} = -\frac{1}{2K}\frac{1}{\left(1 + \frac{2}{3}\frac{G}{K}\right)^2} \approx -\frac{1}{2K} \tag{11.16}$$

As ν varies in the opposite way as G, it increases with T (almost discontinuously at secondary transitions), and reaches a value of the order of 0.45 ± 0.01 at $T_g - 20$ K. Then, it undergoes a rapid increase and attains a value very close to 0.50 in the rubbery region. The shape of temperature variations of ν is represented in Fig. 11.6.

The role of local mobility on the temperature or frequency variations of ν appears clearly when one compares amine-crosslinked epoxies to unsaturated polyesters, at ambient temperature under usual tensile testing conditions ($\dot{\varepsilon} \approx 10^{-4}\,s^{-1}$):

ν (UP) ~0.38–0.39, whereas ν (epoxies) ~ 0.41 ± 0.02,

the area of the sub-glass transition

$$A_\beta = \int_{150}^{270} \tan\,\delta\, dT$$

is of the order of 1–2 K for epoxies against 0.1 K for UP.

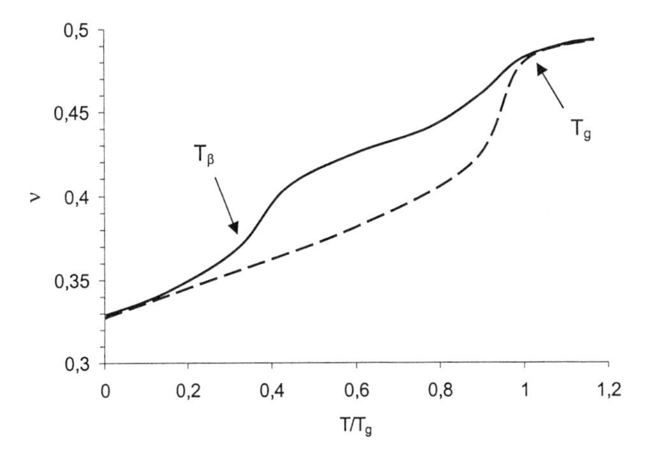

FIGURE 11.6 Shape of the temperature variation of the Poisson's ratio for a polymer with (full line) and without (dashed line) intense β transition.

For an epoxide–amine network displaying a strong β dissipation peak, e.g., the diglycidyl ether of bisphenol A and diaminodiphenylmethane (DGEBA–DDM) system, ν (5 MHz) $\sim 0.37 \pm 0.01$ whereas ν (tensile $\sim 10^{-3}\ s^{-1}$) $\sim 0.41 \pm 0.01$.

The temperature coefficients of the Poisson's ratio

$$\dot{\nu} = \frac{1}{\nu}\frac{d\nu}{dT}$$

for various networks, have been compared at ambient temperature (Verdu and Tcharkhtchi, 1996). For amine-crosslinked epoxies, $\dot{\nu} \approx (0.5 \pm 1.5)10^{-4}\ K^{-1}$, whereas the temperature coefficient of the shear modulus

$$\dot{G} = \frac{1}{G}\frac{dG}{dT}$$

varies almost proportionally, as expected:

$$\dot{\nu}/\dot{G} \approx -1.8 \pm 0.4$$

For styrene-crosslinked unsaturated polyesters and vinyl esters, $\dot{\nu}$ is considerably higher: $\dot{\nu} \approx (20 \pm 5)10^{-4}\ K^{-1}$ and $\dot{\nu}/G \sim -3$. It is noteworthy that in such networks ($T_g \sim 100°C$) the temperature interval between ambient temperature and T_g is a diffuse transition region in which tan δ increases almost continuously with temperature.

Since ν is a viscoelastic quantity, it is expected to increase with the loading time in the case of static loading (Theocaris and Hadjijoseph, 1965).

All the above observations seem to justify Porter's approach (Eq. 11.11)), according to which the Poisson's ratio should depend only on the cumulative loss tangent. It was found that the unrelaxed Poisson's ratio determined from ultrasound (5 MHz) propagation rate, for 12 of amine-crosslinked epoxy stoichiometric networks, displays only small variations ($\Delta\nu \leq 0.01$), in spite of the relatively large variations of the cohesive energy density ($0.59 < \text{CED} < 0.66$ GPa) and the crosslink density (2.0–5.9 mol kg^{-1}).

11.2.5 Antiplasticization

The antiplasticization phenomenon is presumably common to all the polymers exhibiting a relatively strong β transition, well separated from the α transition. It has been observed for both linear (PVC, polycarbonate, polysulphones) and network polymers (amine-crosslinked epoxies). For the case of thermosets, the phenomenon may be a consequence of both "internal" (change of the network structure) and "external" (incorporation of miscible additives) modifications of the structure or the composition; but it always seems to be a consequence of the plasticization, as shown in Fig. 11.7.

Plasticization, whose main manifestation is the decrease of the glass transition temperature (α transition in dynamic mechanical spectra), is generally accompanied by an increase of the glassy modulus in the temperature interval between T_β and T_α, an effect is known as antiplasticization.

A good example of internal antiplasticization is given by the triglycidyl aminophenol–diaminodiphenylmethane–aniline (TGAP–DDM–AN) epoxy system, in which aniline works as a chain extender (internal plasticizer)

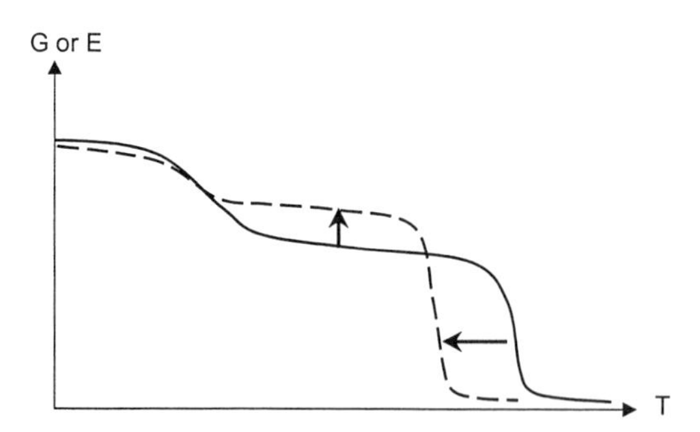

FIGURE 11.7 Shape of the modulus (G or E) versus temperature curves for a given polymer (full line) and its plasticized analog (dashed line).

TABLE **11.2** Crosslink density, glass transition temperature, and Young's modulus (E_T from tensile test at 10^{-3}, s^{-1} strain rate, E_u from ultrasonic propagation at 5 MHz frequency) for triglycidyl aminophenol–diaminodyphenylmethane–aniline (TGAP–DDM–AN) networks. (After Morel *et al.*, 1989.)

Aniline (mol%)	n (mol kg^{-1})	T_g (K)	E_T (GPa)	E_u (GPa)
0	5.88	499	3.25	6.51
25	4.98	470	3.53	6.30
50	4.10	440	3.66	6.10
75	3.24	408	3.81	6.14

(Morel *et al.*, 1989) (Table 11.2). In this series, the tensile modulus increases 17% when 75% of DDM is replaced by aniline, whereas the ultrasonic modulus decreases slightly (~ 6%).

Nonstoichiometric systems or incompletely cured stoichiometric systems behave as if they were internally plasticized systems. The modulus variation with cure conversion (x), for a typical stoichiometric epoxide-amine network, is shown in Fig. 11.8.

Good examples of this behavior have been given, for instance, by Gillham and coworkers (see Venditti and Gillham, 1996). These authors have proposed an explanation in terms of volumetric properties, which has been briefly discussed in Chapter 10. In our opinion, antiplasticization remains an open research field. It appears impossible to suppress totally

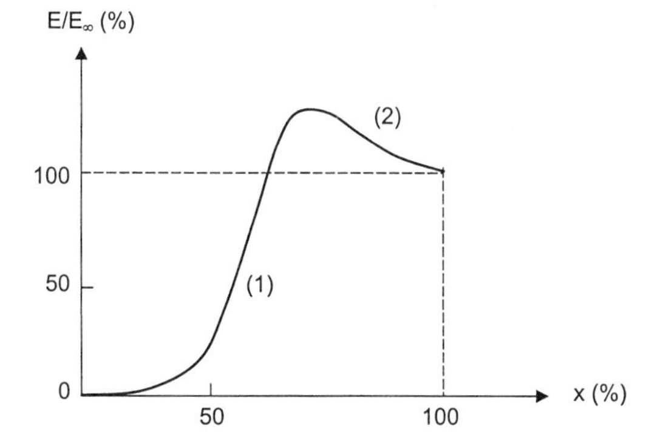

FIGURE **11.8** Shape of isothermal, low-frequency, modulus variations against cure conversion for a stoichiometric epoxide–amine system. (1) vitrification; (2) antiplasticization decreasing with conversion.

viscoelastic effects through antiplastization, but it is relatively easy to obtain a modulus increase of 20–30%. This could be practically interesting in certain cases, provided that direct plasticization effects (decrease of T_g) are acceptable for the considered applications.

11.3 VISCOELASTIC PROPERTIES

11.3.1 Introduction

As shown in Chapter 10, molecular dynamics in polymers is characterized by localised and cooperative motions that are responsible for the existence of different relaxations (α, β, γ). These, in turn, are responsible for energy dissipation, mechanical damping, mechanical transitions and, more generally, of what is called a viscoelastic behavior – intermediary between an elastic solid and a viscous liquid (Ferry, 1961; McCrum *et al.*, 1967).

In the domain of small strains, viscoelasticity is responsible for

- creep or relaxation, in the case of static loading
- deviations from linearity (purely elastic behavior), in the case of continuous loading
- damping, energy loss by dissipation, in the case of dynamic loading.

Let us remind ourselves that for a purely elastic solid $\sigma = E\varepsilon$, where E is independent of ε, whereas for a purely viscous (Newtonian) liquid $s = \eta\dot{\gamma}$, where η (viscosity) is independent of $\dot{\gamma}$ (shear rate).

A viscoelastic solid is characterized by the fact that its modulus E is a function of time. Thus, the response of the material to a loading program, $\dot{\varepsilon}(t)$ or $\dot{\sigma}(t)$ needs the application of the Boltzmann superposition principle (Sec. 11.1). In the case of programmed strain:

$$\varepsilon(t) \rightarrow \sigma(t) = \int_{-\infty}^{t} E(t - t')\dot{\varepsilon}(t)dt' \tag{11.17}$$

or

$$\gamma(t) \rightarrow s(t) = \int_{-\infty}^{t} G(t - t')\dot{\gamma}(t')dt' \tag{11.18}$$

In the case of programmed stress:

$$\sigma(t) \rightarrow \varepsilon(t) = \int_{-\infty}^{t} D(t - t')\dot{\sigma}(t')dt' \tag{11.19}$$

or

$$s(t) \rightarrow s(t) = \int_{-\infty}^{t} J(t - t')\dot{s}(t')dt' \tag{11.20}$$

where E and G are moduli and D and J are compliances. Here, t' is a dummy variable (t is a constant corresponding to the actual time). The boundary ($-\infty$) expresses only that the state of the system at time t depends on the whole loading history.

These relationships show that to model the mechanical behavior of a given viscoelastic material we need to know the function(s) describing the time variation E(t), G(t), D(t), J(t) of the moduli or compliances under consideration.

One of the relationships mostly used in this domain is the Kolrausch–Williams–Watt (KWW) equation:

$$E = (E_\infty - E_0)\exp - \left(\frac{t}{\tau}\right)^{\beta} + E_0 \tag{11.21}$$

where E_0 and E_∞ are the relaxed and unrelaxed modulus, respectively.

Thus, the goal of mechanical characterization is to determine E_∞, E_0, τ, and β. Good examples of the applications of this relationship for linear polymers are given by Matsuoka (1986). The parameter β, which is linked to the width of the relaxation spectrum, varies with temperature in the same way as tan δ; it can take values of the order of 0.01–0.1 in the glassy state, and 0.3–1 at T_g.

Dynamic testing is especially useful in characterizing the viscoelactic behavior in wide time or/and temperature ranges. For the case of a sinusoidal strain:

$$\varepsilon = \varepsilon_0 \sin \omega t \rightarrow \sigma = \sigma_0 \sin(\omega t + \delta) \tag{11.22}$$

A complex modulus is defined as $E^* = E' + iE''$

with

$$E' = \frac{\sigma_0}{\varepsilon_0} \cos \delta \text{ (storage modulus)}$$

and

$$E'' = \frac{\sigma_0}{\varepsilon_0} \sin \delta \text{ (loss modulus);}$$

so that

$$\tan \delta = \frac{E''}{E'} \text{(damping factor)}$$

From a controlled stress experiment, $\sigma = \sigma_0 \sin \omega t$. A complex compliance may be defined:

$$D^* = D' - ID''$$

Dynamic testing leads to two sets of data: e.g.,

$$E' = f'(\omega, T) \text{ and } E'' = f'(\omega, T)$$

which can be used directly (for instance in problems of vibrations, acoustic damping, etc.), or translated into terms of "static" mechanical parameters to solve classical mechanical problems. But dynamic testing is also interesting because it allows us to study polymer transitions and thus gives useful analytical information. Dynamic mechanical thermal analysis (DMTA) can be considered as a basic analytical tool in the study of thermosets and thermoset-based composites.

Whatever the selected method (static, monotonous, or dynamic), it gives access to a limited range of timescales. For example it is almost impossible to perform static experiments in times less than 1 s, or dynamic tests at frequencies lower than 10^{-3} Hz, or tensile tests at strain rates higher than $10^3\ s^{-1}$. These timescales are, however, indirectly accessible because the polymers generally obey a time-temperature superposition principle:

$$P_r(t, T) = P_r\left(\frac{t}{a_T}, T_R\right) \text{ (}P_r\text{ being a polymer property)}$$

where the shift factor a_T depends only on temperature. This dependence reflects the temperature effects on molecular dynamics.

The knowledge of $a_T = f(T)$ is thus doubly important: from the practical viewpoint it can give access to extreme timescales that are very difficult or even impossible to study experimentally; from the theoretical viewpoint, it supplies very useful information on molecular dynamics.

11.3.2 Transitions

a. Homogeneous Materials

Dynamic testing: DMTA, DMA, torsional braid analysis (Enns and Gillham, 1983) is first used as a thermal analysis method to detect the transitions, using dissipation peaks. Certain commercial DMTA instruments have a relatively low accuracy in measuring forces and/or strains. In contrast, they give relatively accurate values of the damping factor tan δ, so that dissipation spectra $\tan \delta = f(\omega, T)$, are very useful analytical tools.

A typical dissipation spectrum is shown in Fig. 11.9. This spectrum displays generally two or three main peaks in the interval ⌊100 K to $T_g + 30$

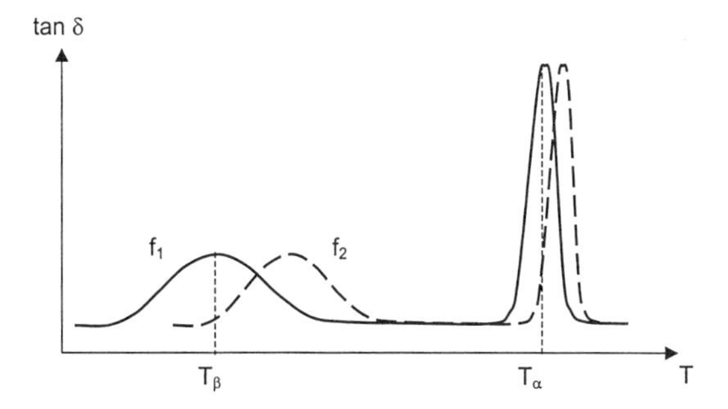

FIGURE 11.9 Dissipation spectrum of an epoxy network at two frequencies f_1 and f_2 so that $f_1 << f_2$.

K]. The peak associated with the glass transition temperature (which can be easily identified by the corresponding modulus gap: more than one decade) is called the α peak. The corresponding temperature T_α , increases slightly when the frequency increases.

The peaks associated with secondary transitions in the glassy state (sub-glass transitions), are in principle, T_β, T_γ, T_δ, ..., in decreasing temperature order. They are wider, lower and more influenced by a frequency change than the α peak. The nomenclature of secondary transitions is not rigorously logical; e.g., in epoxies, there is a wide consensus to call β the peak lying at about 220 ± 20 K for a frequency of 1 Hz. However, these materials often exhibit a wide dissipation peak at about 320–360 K, generally called ω. The name of a given peak reflects its position in the spectrum but also the history of its discovery.

As shown in Chapter 10, every peak corresponds to a type of molecular motion. From the coordinates (ω, T) of the peaks, a relaxation map can be built (see Fig. 11.2). The following observations can be made:

- The apparent activation energies decrease in the order $\alpha > \beta > \gamma >$, etc. Typically, E_{act} (α) is several hundreds of kJ mol^{-1}, whereas E_{act} (β) is generally not higher than 100 kJ mol^{-1} , etc.
- The α-transition does not follow an Arrhenius law; the corresponding curve displays a negative concavity.
- All the straight lines tend to be tangent to the α-transition curve.

It can be recalled that there is no systematic difference between the relaxation spectra of linear and network polymers (they always have the shape of Fig. 11.8). From the practical point of view, the following characteristics are important:

1. *The* T_α *value measured at a low frequency* ($T_\alpha \approx T_g$, see the WLF equation below). It is a key characteristic of the polymer (Chapter 10).
2. *The location of* T_β *with respect to ambient temperature*, T_a , *in the timescale under consideration*. In a first approach, one can distinguish between the cases where $T_\beta < T_a$ (epoxy, phenolics) and $T_\beta > T_a$ (UP, VE, PI). The networks belonging to the first family have relatively low moduli (E < 3 GPa, G < 1 GPa), and can display a ductile or semiductile behavior. The networks belonging to the second family have relatively high moduli (E > 3 GPa, G > 1 GPa), and generally exhibit a brittle behavior.
3. *Unrelaxed and relaxed moduli for each transition*. Let us recall that, generally, the ratio E (unrelaxed)/ E (relaxed) is higher than 10 for the α transition (it is a decreasing function of the crosslink density), whereas it is generally lower than 2 for secondary transitions where it does not depend directly on crosslink density but rather on the mechanical activity of the corresponding molecular motions.
4. *Width of the transition*. The α transitions are generally considerably sharper than secondary ones. The width of the α transition can be related to the degree of inhomogeneity of the spatial distribution of crosslink density (see Sec. 10.3.4). In contrast, secondary transitions can be considered "intrinsically broad;" their width cannot be related to structural inhomogeneities.
5. *Frequency–temperature relationships* (see below).

Some characteristics of the main transitions of industrial thermosets have been described in Chapter 10.

b. Inhomogeneous and Modified Thermosets

DMTA is a very interesting tool for characterizing heterogeneous materials in which domains of distinct T_g values coexist. The most interesting cases involve modified thermosets of different types (see Chapter 8). Examples are the use of rubbers (e.g., liquid polybutadiene and random copolymers), or thermoplastics (e.g., polyethersulphone or polyetherimide in epoxy matrices or poly(vinyl acetate) in unsaturated polyesters), as impact modifier (epoxies), or low-profile additives (polyesters). The modifier-rich phase may be characterized by the presence of a new α peak (Fig. 11.10). But on occasions there may be superposition of peaks and the presence of the modifier cannot be easily detected by these techniques. If part of the added polymer is soluble in the thermoset matrix, its eventual plasticizing effect can be determined from the corresponding matrix T_g depletion, and the

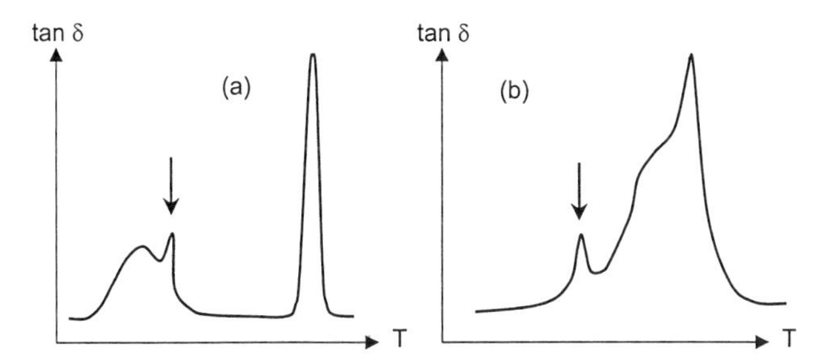

Figure 11.10 Examples of dissipation spectra (1 Hz frequency). (a) amine–epoxy network containing a carboxyl-terminated polybutadiene. The rubber α peak (arrow) is superimposed on the β peak of epoxy. (b) unsaturated polyester containing polyvinylacetate. The α peak (arrow) is at the onset of the β–α dissipation band.

corresponding added polymer concentration can be deduced from the usual plasticization theories.

As discussed in Chapter 7, radical polymerization can induce marked inhomogeneities, easily observable by DMTA. For example, in photocured polyurethane–acrylate networks, two α peaks corresponding to distinct phases, separated by about 40–100 K, depending on composition and cure conditions, can be observed in the DMTA dissipation spectrum at 10 Hz (Barbeau *et al.*, 1999).

11.3.3 Modeling The Viscoelastic Behavior in the Frequency Domain

The complex modulus components E′ and E″ (or G′ and G″ or their compliance counterparts), are functions of the loading frequency or angular frequency ω. To try to identify these functions, it is usual to determine E′ and E″ experimentally in a wide interval of temperature and frequency and to build $E'' = f(E')$ (Cole–Cole) plots.

In the simplest case of a single relaxation time element, it was shown that (see Chapter 10):

$$E' \frac{E\omega^2\tau^2}{1+\omega^2\tau^2} \text{ and } E'' = \frac{E\omega\tau}{1+\omega^2\tau^2} \tag{11.23}$$

The curve $E'' = f(E)'$ is then a half circle intercepting the abscissa at $E' = 0$ and $E' = E$, with a maximum at $E'' = E/2$ for $E' = E/2$ ($\omega\tau = 1$). In the general case where the relaxed modulus takes a nonzero value, the following relationships hold:

$$E' = E_0 + \frac{(E_\infty - E_0)\omega^2\tau^2}{1 + \omega^2\tau^2} \text{ and } E'' = \frac{(E_\infty - E_0)\omega\tau}{1 + \omega^2\tau^2} \tag{11.24}$$

where E_0 and E_∞ are the relaxed and unrelaxed modulus, respectively. Each transition is characterized by a pair of (E_0, E_∞) values. For example, in the case of the glass transition, E_0 would be the equilibrium rubbery modulus (typically 10–100 MPa for common thermosets) and E_∞ would be the modulus at the glassy plateau just above the α transition (typically $E_\infty \approx$ 1–2 GPa). In the case of the β transition, E_0 and E_∞ would be of the same order of magnitude (typically $E_0 \approx$ 1–2 GPa, $E_\infty \approx$ 3–5 GPa for amine-cross-linked epoxies).

Actually, polymer relaxations are complex processes characterized by a relaxation spectrum (coexistence of many relaxation time). Their Cole–Cole plots are generally nonsymmetric (Fig. 11.11).

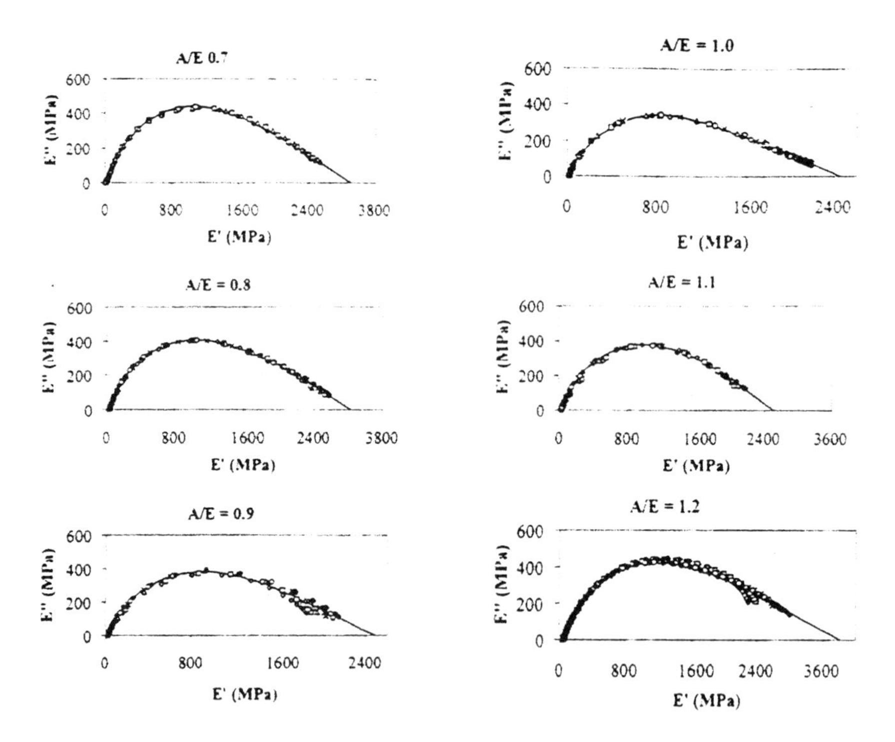

FIGURE 11.11 Cole–Cole plots for networks resulting from the condensation of diglycidyl ether of bisphenol A (DGEBA) and diethyltoluenediamine (ETHA), with various amine/epoxide molar ratios (numbers on the figures). Reprinted from Tcharkhtchi *et al.* 1998. Copyright 2001 with permission from Elsevier Science.

Various models can be proposed for the general case. If a good fit with experimental measurements is found, their parameters can be identified and eventually interpreted in terms of width of the relaxation spectrum, effect of crosslinking, etc.

Models proposed by Havriliak and Negami (1966) and by Perez (1992) usually give a good fit of Cole–Cole plots, provided there is no overlapping between α and β relaxations (Fig. 11.12).

The Havriliak–Negami model was first established for the complex dielectric constant (equivalent to the complex compliance). Its mechanical translation can be written as

$$E^* = E_\infty - \frac{E_\infty - E_0}{[1 + (i\omega\tau)^{1-\alpha'}]^{\gamma'}} \tag{11.25}$$

The Perez model comes from an approach in which the source of mobility is the existence of quasi-punctual defects characterized by positive or negative fluctuations of packing density, whereas classical free volume theories take into account only the domains of low packing density, e.g., the "holes." The model leads to the following equation for the complex modulus:

$$E^* = E_\infty - \frac{E_\infty - E_0}{1 + (i\omega\tau)^{-x} + Q)i\omega\tau)^{-x'}} \tag{11.26}$$

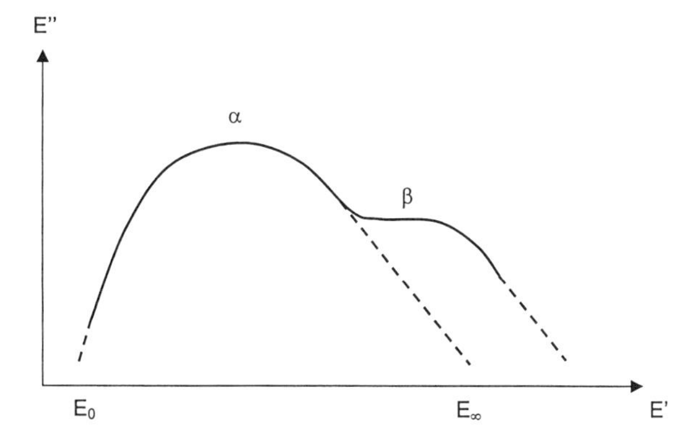

FIGURE 11.12 Shape of the Cole–Cole plots in the case of overlapping of α and β transitions.

11.3.4 Practical Use of Cole–Cole Plots. Characteristics of Thermosets

When Cole–Cole plots are established from the usual DMTA experiments, care must be taken on storage modulus measurements. As a matter of fact, these systems are relatively accurate in phase measurements ($\tan\delta$), but modulus values (E') can be less precise. Since E'' is determined from $E'' = E' \tan\delta$,

$$\frac{\Delta E''}{E''} = \frac{\Delta E'}{E'} + \frac{\Delta \tan\delta}{\tan\delta} \approx \frac{\Delta E'}{E'}$$

The relative errors on both complex modulus components are expected to be almost equal. These errors can be of the order of 50% or more for certain commercial apparatus, and they vary presumably with temperature, so that E' measurements have to be calibrated in both rubbery and glassy domains.

Remark: The model parameters are not very sensitive to variations in the rubbery modulus E_0. E_∞ an be determined only by extrapolation, which is also a source of errors.

The use of Cole–Cole plots is not very developed in practice, despite the fact that they open the way for the modeling of the viscoelastic behavior in dynamic as well as in static loading cases (through Laplace transform). By contrast, these plots could be interesting from the fundamental point of view if certain parameters would reveal a clear dependence with the crosslink density. The effects of crosslinking are difficult to detect on the usual viscoelastic properties, except for the variation of the rubbery modulus E_0.

A comparison of results obtained for several networks based on the same epoxide–amine pair, but with variable amine/epoxide molar ratios, and thus variable crosslink densities, is shown in Table 11.3 (Tcharkhtchi *et al.*, 1998). Havriliak–Negami and Perez models cannot be distinguished from one another by the quality of the fit of experimental curves, within experimental uncertainty. From a mathematical point of view, the Havriliak–Negami model is better than the Perez model because it has less parameters to fit (four parameters against five). In contrast, physical arguments could favor the Perez model, for which the parameters have a physical interpretation.

Crosslink density directly affects E_0 (through rubber elasticity), and has an indirect influence on E_∞ (through the antiplasticization effect). Cole–Cole plots open the way to analyzing the distribution of relaxation times (the exponents α and γ or χ or χ' are linked to the width of the distribution of relaxation times). According to the results of Table 11.3, these exponents seem to depend more on the molecular-scale structure (they vary almost

TABLE 11.3 Havriliak–Negami and Perez (see text) parameters for the glass transition region of DGEBA–ETHA networks differing in the amine/epoxide molar ratio. (After Tcharkhtchi *et al.*, 1998.)

			Havriliak–Negami			Perez	
A/E	E_0(MPa)	E_∞(MPa)	α'	γ'	κ	κ'	Q
0.7	14	2900	0.430	0.355	0.230	0.630	0.70
0.8	30	2820	0.445	0.355	0.240	0.640	0.48
0.9	10	2520	0.405	0.330	0.220	0.620	1.00
1.0	12	2440	0.400	0.280	0.200	0.625	0.60
1.1	6	2520	0.475	0.410	0.235	0.580	0.95
1.2	30	3400	0.410	0.260	0.185	0.560	0.70

monotonically with the amine–epoxide molar ratio) than on the network structure (they would exhibit an extreme value at the stoichiometric ratio). However, different conclusions were obtained from studies of other structural series. The number of investigated systems is again too low to permit us to establish definite conclusions on structure–property relationships.

With regards to the translation of dynamic data in terms of static loading data through appropriate mathematical transformations, both the Havriliak–Negami and Perez models have been claimed to be counterparts of the KWW model:

$$E = (E_\infty - E_0) \exp -(t/\tau)^\beta + E_0 \tag{11.27}$$

However, this subject remains controversial. The KWW model is probably a relatively good approximation of both models, but only in a restricted range of α' and γ' values for the Havriliak–Negami model (Alvarez *et al.*, 1991), or a restricted range of χ and χ' values for the Perez model.

11.3.5 Time–Temperature Relationships

As discussed in Chapter 10, network polymers – as linear polymers – obey the time–temperature equivalence principle in the domain where they are stable, both chemically (no postcure, no thermal degradation), and physically (no orientation relaxation, water desorption, physical aging, etc.).

A typical relaxation map (relaxation time or frequency at the maximum of dissipation peaks against reciprocal absolute temperature) is shown in Fig. 11.2. The general characteristics of relaxation maps are the following:

(a) The slope (apparent activation energy) is always in the order $\alpha > \beta > \gamma$. The order of magnitude of apparent energies is generally few hundreds of kJ mol^{-1} for the α transition and few dozens of kJ mol^{-1} for sub-glass transitions.
(b) The behavior obeys Arrhenius law (straight lines) for subglass ($\beta, \gamma, \ldots$) transitions but not for the α transition (associated with the glass transition) that follows a WLF equation.
(c) All the curves tend to be tangent in the high-temperature region. In other words, the β and γ transitions tend to be overlapped by the α transition at high frequencies (typically $\geq 10^5$ Hz), which can be easily observed in dielectric spectra. In a given structural series, the slope of the α-relaxation curve is an increasing function of T_g (and thus of the crosslink density).

Sub-glass transitions are generally determined by the molecular (local) scale structure. Their location in the (t, T) space undergoes only a second-order influence of the macromolecular (network) structure through internal antiplasticization effects. By contrast, glass transition is directly under the influence of the network structure (Chapter 10), so that it appears interesting to study the influence of crosslinking on the parameters of the time–temperature relationship (WLF equation):

$$\text{lnlog2}_T\, a_T = \frac{-C_1^g(T - T_g)}{C_2^g + (T - T_g)} \qquad (11.28)$$

The apparent activation energy in a region close to T_g can be determined by deriving $d \ln a_T / d\,(T^{-1})$:

$$\frac{E_a(T_g)}{R} = \frac{d(\ln a_T)}{d(T^{-1})} = 2.3 \frac{C_{1_i}^g}{C_2^g} T_g^2, \text{ so that } E_a(T_g) = 2.3\,R \frac{C_1^g}{C_2^g} T_g^2 \qquad (11.29)$$

C_1^g/C_2^g displays relatively small variations with the structure, $C_1^g/C_2^g \approx \frac{1}{3}$ to $\frac{1}{4}$, so that, in a first approximation:

$$E_a(T_g) \approx 2.5 T_g^2 \qquad (11.30)$$

It can be easily checked that this relationship agrees, at least semi-quantitatively, with the available experimental data.

The physical meaning of C_1^g and C_2^g and their variation with the structure has been examined in Chapter 10. It will be only recalled here that:

1. The usual free volume interpretation of $C_1^g$and C_2^g is questionable: $C_1^g = B_D/f_g$ and $C_2^g = f_g/\Delta\alpha$, where B_D is the Doolittle constant, f_g is the free volume fraction at T_g and $\Delta\alpha = \alpha_l - \alpha_g$ is the free volume expansion coefficient. It appears impossible to

have a consistent set of data with a single value of B_D (independent of network structure). In other words, the product $C_1^g C_2^g \Delta\alpha$, which is expected to be constant, varies considerably with the structure.

2. In a given structural series, C_1^g, and C_2^g seem to increase with the crosslink density (Shibayama and Suzuki, 1965; Gerard *et al.*, 1991).
3. The ratio C_2^g/C_1^g varies only slightly and remains on the order of 3.

11.4 CONCLUSIONS

As for linear polymers, the low-strain mechanical properties of thermosets in the glassy state are essentially under the influence of two factors: cohesion and segmental mobility.

The bulk modulus depends practically only on cohesion. It does not exhibit viscoelastic effects and depends only slightly on temperature, except at the glass transition where it varies by a factor of about 2.

The tensile and shear modulus differ from the bulk modulus in fact that they are influenced by the molecular mobility, and therefore display an almost discontinuous decrease at each secondary transition. Their relative variation at T_g is 10–100 times higher than that of the bulk modulus. The Poisson's ratio varies in the opposite sense in the region $T < T_g$. It tends toward 0.5 when the temperature increases beyond T_g.

The main characteristics of viscoelasticity can be summarized as follows:

1. Variable amplitude of modulus variations, depending on the activity of secondary relaxations.
2. Broadness of the relaxation spectrum, as attested, for example, by the low value of the exponent β in the Kolrausch–Williams–Watt expression.
3. Existence of a time–temperature equivalence that takes a different mathematical form in the glassy state (Arrhenius) and in the glass transition and rubbery regions (WLF).

Research on the eventual specific effects of crosslinking on characteristics other than the glass transition temperature or the rubbery modulus remains a widely open domain.

NOTATION

a_c = bulk modulus/cohesive energy density ratio
D = compliance (tensile), Pa^{-1}
d = intermolecular distance, m
d_o = equilibrium intermolecular distance, m
E_o, E_∞ = Young's moduli (relaxed and unrelaxed), Pa
G = shear modules
$\dot{G}$ = temperature coefficient of the shear modulus, K^{-1}
J = compliance (shear), Pa^{-1}
K = bulk modulus
l = length, m
p = pressure, Pa
Q = Perez coefficient
s = shear stress, Pa
$\dot{s}$ = shear stress rate, $Pa.s^{-1}$
$\tan \delta$ = damping, dissipation factor, or loss tangent
$\tan \Delta_\beta$ = cumulative loss tangent

Greek Letters

α' = Havriliak–Negami exponent
α_G = temperature coefficient of shear modulus
α_K = temperature coefficient of bulk modulus
γ = shear strain
ΔG_β = shear modulus gap at T_β, Pa
γ' = Havriliak–Negami exponent
ε = strain
$\dot{\varepsilon}$ = strain rate, s^{-1}
ν = Poisson's ratio
$\dot{\nu}$ = temperature coefficient of Poisson's ratio, K^{-1}
τ = relaxation time, s
χ, χ' = Perez coefficients
ω = angular frequency

REFERENCES

Alvarez F, Alegria A, Colmenero *J, Phys. Rev.* B, 44, 7306–7312 (1991).
Ashby MF, *Materials Selection in Mechanical Design*, Pergamon Press, Oxford, 1992.
Barbeau P, Gerard JF, Magny B, Pascault JP, Vigier G, *J. Polym. Sci.* B, 37, 919–937 (1999).
Bellenger V, Verdu J, Ganem M, Mortaigne, B, *Polym. and Polym. Compos.*, 2, 9–16 (1994).

Bondi A, *Physical Properties of Molecular Liquids, Crystals and Glasses*, Wiley, New York, 1968.

Enns JB, Gillham, JK, *J. Appl. Polym. Sci.*, 28, 2567–2591 (1983).

Ferry JD, *Viscoelastic Properties of Polymers*, Wiley, New York, 1961.

Gerard JF, Galy J, Pascault JP, Cukierman S, Halary JL, *Polym. Eng. Sci.*, 31, 615–621 (1991).

Gilbert DG, Ashby MF, Beaumont PWR, *J. Mater. Sci.*, 21, 3194–3210 (1986).

Havriliak S.J, Negami S, *J. Polym. Sci.* C, 14, 99 (1966).

McCrum NG, Read BE, Williams JG, *Anelastic and Dielectric Effects in Solid Polymers*, Wiley, London, 1967.

Matsuoka S, In: *Failure of Plastics*, Chapter 4, Brostow W, Corneliussen R (eds), Hanser, New York, 1986.

Morel E, Bellenger V, Bocquet M, Verdu J, *J. Mater. Sci.*, 24, 69–75 (1989).

Perez J, *Physique et Mécanique des Polymères Amorphes*, Lavoisier, Paris, 1992.

Porter D, *Group Interaction Modelling of Polymer Properties*, Marcel Dekker, New York, 1995.

Shibayama K, Suzuki Y, *J. Polym. Sci. A*, 3, 2637–2651 (1965).

Tcharkhtchi A, Lucas AS, Trotignon JP, Verdu J, *Polymer*, 39, 1233–1235 (1998).

Theocaris PS, Hadjijoseph C, *Kolloïd. Z.*, 202, 133–139 (1965).

Tobolsky AV, *Properties and Structure of Polymers*, Wiley, New York, 1960.

Van Krevelen DW, *Properties of Polymers. Their Estimation and Correlation with Chemical Structure*, Elsevier, Amsterdam, 1990.

Venditti R, Gillham JK, *J. Appl. Polym. Sci.*, 56, 1687–1705 (1996).

Verdu J, Tcharkhtchi A, *Angew Macromol. Chem.*, 240, 31–38 (1996).

12

Yielding and Fracture of Polymer Networks

12.1 INTRODUCTION

There are two main sets of materials properties involved in any problem of mechanical reliability.

- *Yielding properties.* Yielding or plastic deformation can be considered as the boundary between the domain of reversible and permanent deformation. In normal conditions, the material must be used below the yield point, often called by metallurgists the elastic limit.
- *Fracture properties.* If, accidentally, the material goes beyond its yield point, a new boundary becomes important: the ultimate fracture stress (or strain) where the integrity of the structure (or the part) is lost. Then the problem of stress concentration at defects becomes crucial and we need specific theoretical and experimental tools, e.g., fracture mechanics, to study these phenomena.

For a given material, there are generally several possible mechanisms of yielding and fracture, each characterized by the influence of temperature, loading rate, hydrostatic pressure, time (physical aging). A vast literature deals with the influence of network structure on yielding or on fracture properties, but we have to be very careful with the results obtained because of the different types of networks used in these experiments.

We can distinguish first between homogeneous and inhomogeneous networks (Chapter 7). In homogeneous networks a distinction between ideal and nonideal structures may be performed. This concept is presented in Chapters 10 and 11.

1. An ideal network (e.g., polyurethanes (PU) or epoxies) has no heterogeneity, no dangling chains, no sol fraction, and can be built up with step polymerization using a stoichiometric ratio of reactive groups and fully cured (Crawford and Lesser, 1998). With a good control of the chemistry it is possible to change the molar mass of prepolymers or to introduce chain extenders in order to study the influence of crosslink density or chain flexibility (Urbaczewski-Espuche *et al.*, 1991).
2. A nonideal network may be obtained as in the previous case but using different nonstoichiometric molar ratios or arresting the polymerization at different conversions, to modify the structure. In these cases, the presence of a sol fraction and dangling chains will introduce an additional plasticization effect, surimposed on the new architecture (Vallo *et al.*, 1993).
3. The last case concerns inhomogeneous networks often produced by chain polymerization (acrylate networks, unsaturated polyesters), where a gradient of crosslink densities is the result of the reaction mechanism and, in some cases, thermodynamic effects (Chapter 7).

This chapter is structured into three parts, devoted to (a) the main experimental tools used in this domain, (b) yielding, and (c) fracture of thermosets, respectively.

In particular, we will try to point out the specificity of thermosets compared with thermoplastics. A very good reference for the yielding and fracture of thermoplastics is the book of Kinloch and Young (1983).

12.2 MECHANICAL TESTS USED FOR YIELDING AND FRACTURE

12.2.1 Classical Tests

Most of the mechanical tests mentioned here are well described by Brown (1999).

Very often uniaxial tensile tests are used for polymer characterization. In the case of thermosets, at temperatures below T_g, the stress–strain curve is a straight line (elastic region) and fracture occurs (due to flaws contained in the specimen) well before yielding (Fig. 12.1a). So other tests have to be

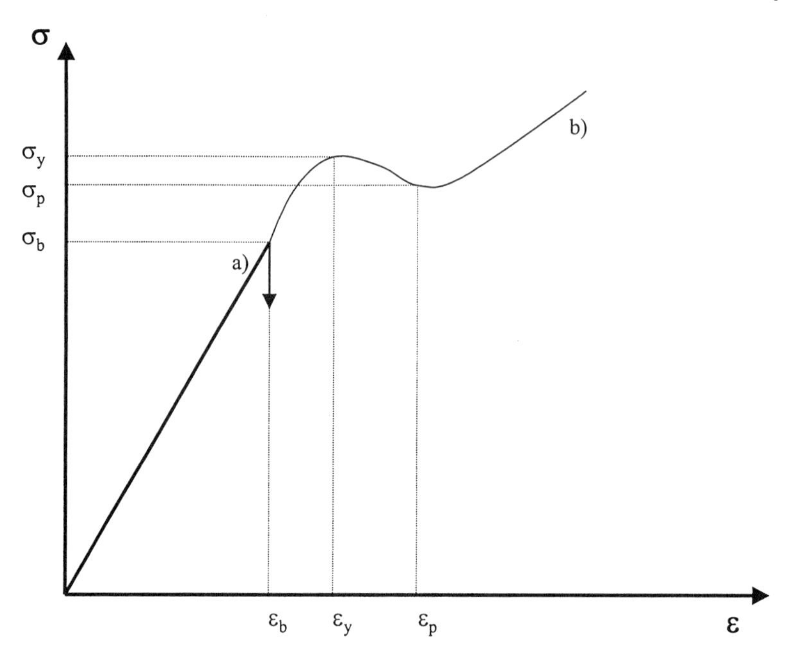

FIGURE 12.1 Typical stress–strain curves for thermosets at a temperature below T_g: (a) uniaxial tensile test; (b) uniaxial compression test.

performed to study yielding – for instance, a uniaxial compression test (Fig. 12.1b). The beginning of the curve is obviously the same. The Young's modulus is the same in tensile or in compression test (the differences often mentioned in the literature are the results of noncareful measurements). This curve exhibits an upper yield stress (σ_y, ϵ_y) and a plateau (σ_p). Very often yielding is characterized by σ_y and ϵ_y, but these values have to be used with caution because they depend on physical aging. This is not the case for σ_p (see below).

After the σ_P region, a strain-hardening phenomenon occurs. Usually the curves use nominal stress, σ_n = (load/initial cross section), and nominal strain, $\epsilon_n = dL/L_0$.

But true stress (load/actual section) and true strain, calculated as the natural logarithm of the relative height, $\ell n(L/L_0)$, must be used to obtain better information on the material. The accuracy of strain measurements may be improved by the use of extensometers or strain gauges and actual cross section determination needs the use of double extensometers or stiffness correction of the machine and rigs (Cook *et al.*, 1998).

Nevertheless, a simple calculation of true stress is possible, assuming constant volume (which is not exactly true because the Poisson's ratio is

$\nu = 0.3$–0.4 and not 0.5) (Chapter 11); it gives a good order of magnitude, $\sigma_{true} = \sigma_n(1 + \epsilon_n)(\epsilon_n > 0$ in tension).

Plane strain compression tests (Fig. 12.2b) may be also used. One principal dimension remains constant: thus $\sigma_1 = \nu\sigma_2$, where σ_1, σ_2 and σ_3 are the principal stresses in the three directions x, y and z. This test has to be performed with care, due to different breadth dies, and friction coefficients, but it is nevertheless used to complete the yielding criteria curves (see below).

Most of the time, the three-point bending test (Fig. 12.2c) induces fracture without exhibiting yielding. The stress calculated here is a maximum stress due to the inhomogeneity of the stress field along the thickness:

$$\sigma_{max} = \frac{3F_{max}D}{2bh^2}$$

Pure shear tests are often performed with torsional bars or flexural tests with very close spans, or using grooved tensile specimens (Fig. 12.2a). They provide the shear yielding stress, τ_y, which can be related to σ_y using yielding criteria (see below).

The fracture properties of thermosets are often very difficult to measure the brittleness of these materials. If a thermoset is tested in uniaxial tensile mode, the stress and strain at break, σ_b and ϵ_b, depend on the temperature and strain rate, $\dot{\epsilon} = d\epsilon/dt$, and also on the sample dimensions (length and cross section). Thus, the parameters σ_b and ϵ_b are not intrinsic values of the materials because they depend on the

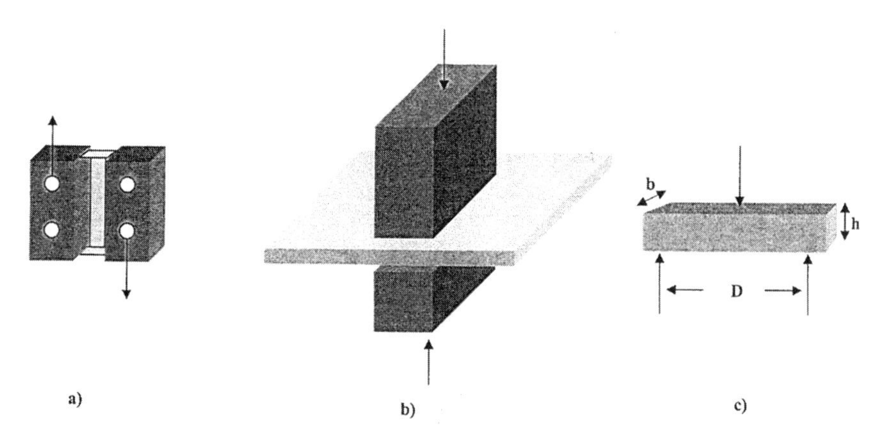

FIGURE 12.2 Schematic representations of (a) pure shear test; (b) plane strain compression test; and (c) three-point bending test.

sample geometry. The smaller the sample, the higher the value of σ_b, due to the lower probability of having defects in the material. To overcome this experimental fact, one can use standards and compare the results obtained (Brown, 1999).

12.2.2 Fracture Mechanics

The best approach, however, consists of controlling the defect size and geometry and taking into account the corresponding stress-field inhomogeneity. This is realized in the frame of linear elastic fracture mechanics (LEFM), which was first applied to metals and ceramics and then adapted with success to polymers (Williams, 1984).

Three modes are clearly defined for crack propagation from a very thin (radius of the order of $10\,\mu$m) notch-machined in the specimen (Fig. 12.3). This notch induces a stress concentration effect, higher than those produced by all the other defects already present in the specimen, which governs the fracture initiation. For isotropic materials, mode I (the most severe) is generally used and gives the lowest value of toughness. In the case of adhesives and laminates, modes II and III are also performed.

LEFM, for fractures occurring in the elastic domain, leads us to characterize the materials using critical characteristics for crack initiation (or arrest in the case of stick-slip propagation).

Griffith used an energy balance approach to predict the crack propagation conditions (see Williams, 1984). The driving force is the elastically stored energy in the notched samples, which can be used to create new surfaces. A parameter G_C, the critical elastic strain energy release rate [G_{Ic} in mode I], can be determined and expressed in $J\,m^{-2}$.

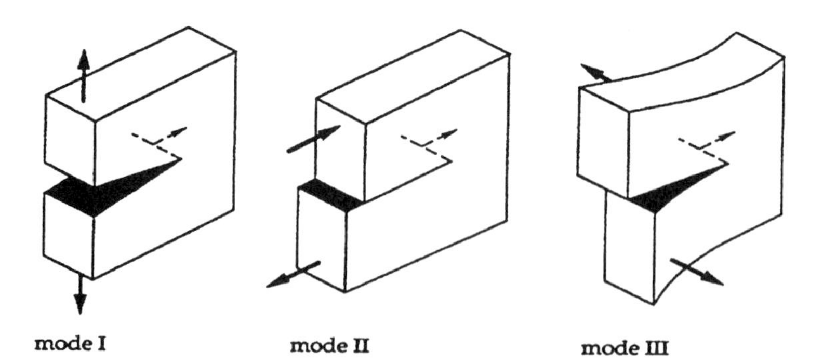

FIGURE 12.3 Different crack propagation modes used in fracture mechanics.

Alternatively to the Griffith's energy criterion, In the 1950s Irwin calculated the stresses in the vicinity of a crack tip based on linear elasticity theory. He introduced a constant, K, the stress intensity factor, to characterize the magnitude of the stress field. If K (K_I in mode I) reaches a critical value K_c (K_{Ic}), then the crack will propagate in the material; K_{Ic} is generally expressed in $MPa\,m^{1/2}$.

At the crack tip, there is a stress concentration effect but, in LEFM, stress is limited to σ_y over a certain domain, the plastic zone. The dimensions of this plastic zone can be calculated using Irwin's theory and is: $\frac{1}{2\pi}\left(\frac{K_{Ic}}{\sigma y}\right)^2$ in plane stress conditions.

K_{Ic} and G_{Ic} are intrinsic parameters that do not depend on crack length and sample size; they depend only on temperature and $\dot{\epsilon}$. Values of K_{Ic} at room temperature, in static conditions, range from $0.5\,MPa\,m^{1/2}$ for very brittle materials up to $2\,MPa\,m^{1/2}$, for toughened thermosets. The values of G_{Ic}, which depend also on Young's modulus (see below), are in the range of $100–200\,J\,m^{-2}$ to $2000\,J\,m^{-2}$. For valid determinations of K_{Ic} and G_{Ic}, samples have to be thick enough to reduce the plastic zone effect and be in plane strain conditions. The edge notch (length a) has to be neither too short nor too long for reducing the plastic-zone size effect (Williams and Cawood, 1990).

Single-edge-notched (SEN) or compact-tension (CT) specimens may be used. Shape factors are tabulated (Williams and Cawood, 1990).

In plane strain conditions, K_{Ic} and G_{Ic} are related through:

$$G_{Ic} = \frac{K_{Ic}^2}{E}(1 - \nu^2) \tag{12.1}$$

Different types of crack propagation may occur (Fig. 12.4). Note, that, even if the crack propagations are different, K_{Ic} values are the same because

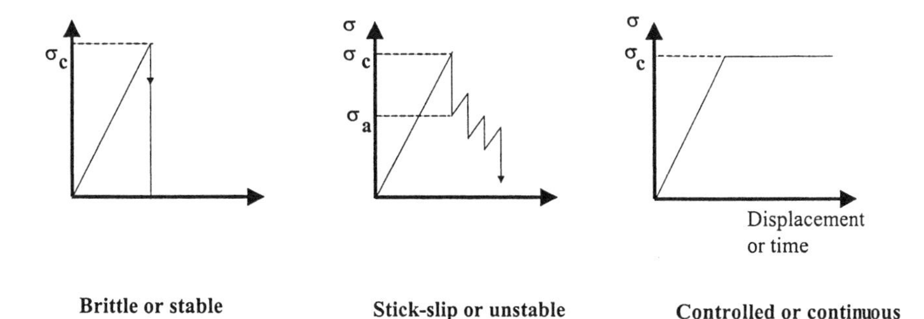

FIGURE 12.4 Different types of crack propagation.

only the critical stress (σ_C) is necessary for K_{Ic} calculations. In the case of stick-slip propagation, one may also calculate arrest values, K_{Ia}.

It is not easy to perform good LEFM experiments because round notch, insufficient thickness etc., tend always *to increase* K_{Ic} and then give wrong information.

This approach using LEFM is a very useful tool for polymer fracture characterization and may be extended to impact characterization (determination of K_{Id}, G_{Id}, in dynamic conditions), and for fatigue crack propagation studies (see Sec. 12.4.4).

When fracture occurs out of the linear domain (fracture with yielding), e.g., for highly toughened thermosets or at high temperatures, it is still possible to apply fracture mechanics, with the energetic theory. For instance, the J-integral may be used for bulk materials (Williams, 1984) or Essential Work for Fracture for thin films (Mai and Powell, 1991; Liu and Nairn, 1998).

12.2.3 Ductile-Brittle Transition

Yielding and fracture are two very important properties of materials, particularly for thermosets. Both aspects can be associated by considering the ductile–brittle transition temperature, T_B (Fig. 12.5).

For a given strain rate, $\dot{\epsilon}$, the temperature domain can be divided into two regions. At low temperatures, the fracture is brittle with a nearly constant stress, σ_b. The temperature is not high enough to induce chain mobility and the fracture occurs from stress concentration at the defects, leading to the formation of holes and then crack propagation.

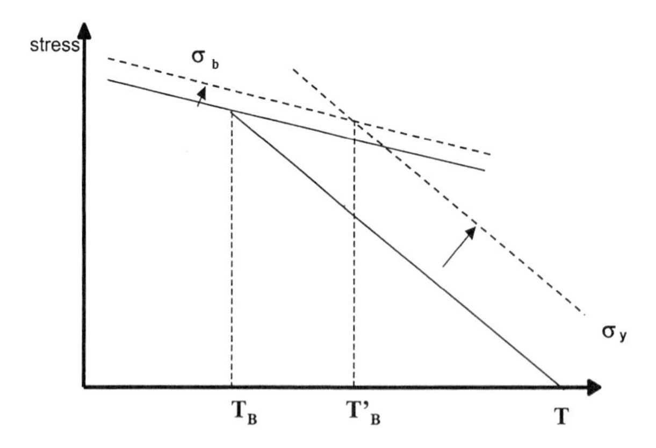

FIGURE 12.5 Effect of strain rate on ductile–brittle transition temperature: — low speed; --- high speed.

At higher temperatures, the failure occurs with yielding, which is now the predominant deformation mechanism. From an experimental point of view, σ_y decreases strongly with temperature. These two domains define what is called the ductile–brittle transition temperature, T_B, which is a very important characteristic for polymers. The ductile–brittle transition is also associated with a stiffness–toughness balance. Note that it is also possible to determine a ductile–brittle strain rate transition varying $\dot{\epsilon}$ at a given temperature.

Thus, the understanding of thermosets fracture needs the complete description of the yielding and the influence of both experimental variables, $(T, \dot{\epsilon})$, on the one hand, and the relationship with structural parameters, on the other hand. Unfortunately, few results are available in the literature dealing with the ductile–brittle transition of thermosets. Very often it is stated that thermosets are more brittle than thermoplastics but this depends only on the location of the test temperature compared with the ductile–brittle transition temperature.

12.3 YIELDING OF NETWORKS

Yielding is a permanent deformation that cannot be completely recovered without increasing temperature. Very often this behavior is ascribed to a region located beyond the elasticity region where the strain is instantaneously recovered. But the situation is not so simple and the concept of anelasticity must be used (Perez, 1998). In the region located beyond but close to the elasticity region, strains can be recovered in times of the order of magnitude of the test time (generally, 3 to 4 times the test time). Numerous studies on linear polymer glasses (Perez, 1988) and thermosets (Girard-Reydet *et al.*, 1997) demonstrated that the upper yield point (σ_y, ϵ_y) (Fig. 12.1) cannot be considered as the actual yield point because most of the deformation can be recovered. In opposition, the lower yield point (σ_p, ϵ_p) where a permanent deformation is evidenced, constitutes a better definition of the yielding process.

The yielding of networks will be described first, beginning with the analysis of deformation mechanisms and the influence of physical aging. The effect of hydrostatic pressure will be treated with yielding criteria. The influence of physical $(T, \dot{\epsilon})$ and structural parameters on yielding will then be described for ideal and nonideal networks.

12.3.1 Deformation Mechanisms

Yielding needs cooperative chain motions and local mobility. The motions involved in the β transition are often regarded as precursors of the yielding process (Perez, 1988). The molecular deformation mechanisms in epoxies below T_g were analyzed using small angle neutron scattering (SANS) and deuterated diglycidyl ether of bisphenol A, DGEBA (Wu and Bauer, 1988).

Local shear induces changes of chain conformations. The more flexible amine chains are preferentially aligned in the stretched direction without a significant increase in the chain length between crosslinks.

No clear experimental evidence of crazing in thermosets has been provided, although claims of their existence have been reported on occasions (Morgan *et al.*, 1982). "Craze-like structures" or dilatation bands are evidenced in toughened or low- crosslinked thermosets (Chapter 13) but, in general, the low molar mass between crosslinks does not permit crazing (Kinloch, 1985). The micromechanism of deformation can be homogeneous or inhomogeneous with the presence of localized shear bands (Morgan *et al.*, 1984).

12.3.2 Influence of Physical Aging on Yielding

Thermosets, like other amorphous polymers, can exhibit physical aging when kept at temperatures below T_g (Chapter 10). The upper yield stress measured at room temperature (below T_g) increases with prolonged aging time (Cook *et al.*, 1999), like the modulus or the specific density.

The aging rate, $\Delta\sigma_y/\Delta t$, is reduced if the crosslink density increases (Cook *et al.*, 1999). This effect could be erased by rejuvenating the sample by heating above T_g or applying large deformations above ϵ_y (Oyanguren *et al.*, 1994).

12.3.3 Influence of the Stress State on Yielding Stress and Yielding Criteria

Yielding occurs not only in the uniaxial stress state but also in more complex stress fields where σ_1, σ_2 and σ_3 can vary. Therefore, it is very important to establish criteria to predict the yielding of thermosets in complex stress states, with a few experimental measurements with simple geometries.

A yielding criterion gives critical conditions (at a given temperature and strain rate) where yielding will occur whatever the stress state. Two main criteria, originally derived by Tresca and von Mises for metals, can be applied to polymers (with some modifications due to the influence of hydrostatic pressure):

(i) *Tresca criterium.* This expresses that yielding occurs if the maximum shear stress reaches a critical value:

$$(\sigma_1 - \sigma_3) = \pm 2k; \qquad (\sigma_1 - \sigma_2) = \pm 2k; \qquad (\sigma_2 - \sigma_3) = \pm 2k$$

In plane stress ($\sigma_3 = 0$) then, $\sigma_1 = \pm 2k$; $\frac{(\sigma_1 - \sigma_2)}{2} = \pm 2k$; $\sigma_2 = \pm 2k$; and $2k = 2\tau_m = \sigma_y$, the uniaxial tensile yield stress. This criterion may be represented as a regular centered, symmetrical hexagon in the plane (σ_1, σ_2) (Fig. 12.6).

(ii) *The von Mises criterion.* This quadratic criterion may be expressed as

$$(\sigma_1 - \sigma_2)^2 + (\sigma_2 - \sigma_3)^2 + (\sigma_3 - \sigma_1)^2 = \text{const} = 9\tau_{oct}^2 = 6\tau_y^2$$

where τ_{oct} is the octahedral shear stress on planes (1,1,1).

For a uniaxial tensile test:$\sigma_2 = \sigma_3 = 0$, $\sigma_1 = \sigma_y$, then $\tau_{oct}^2 = 2/9\sigma_y^2$, and the criterion may be expressed as

$$(\sigma_1 - \sigma_2)^2 + (\sigma_2 - \sigma_3)^2 + (\sigma_3 - \sigma_1)^2 = 2\sigma_y^2 \tag{12.2}$$

In the plane stress state ($\sigma_3 = 0$), Eq. (12.2) gives $\sigma_1^2 + \sigma_2^2 - \sigma_1\sigma_2 = \sigma_y^2$, which is the equation of a symmetrical ellipse, around the Tresca polygon (Fig. 12.6).

One main difference between both criteria is the value of the pure shear stress: $\sigma_y/2$ compared with $\sigma_y/\sqrt{3}$.

Both criteria predict the same value of the yield stress for uniaxial tensile and compressive states, although it was well established that for

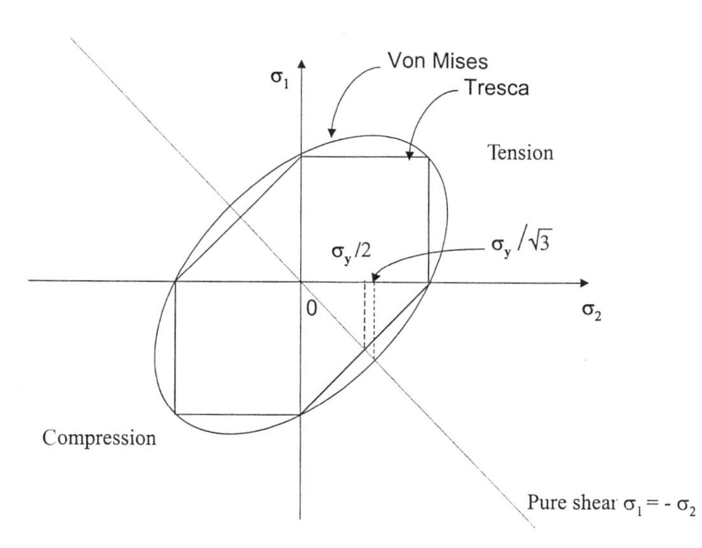

FIGURE 12.6 Unmodified Tresca and von Mises criteria in plane stress conditions ($\sigma_3 = 0$).

polymers $\sigma_{yc} \gg \sigma_{yt}$ due to the effect of hydrostatic pressure, which is non-negligible in such low-density materials.

The hydrostatic pressure $p = (\sigma_1 + \sigma_2 + \sigma_3)/3$ depends on the stress state ($p = 0$ in pure shear $\sigma_3 = 0$, $\sigma_1 = -\sigma_2$).

Therefore, the Tresca criterion has to be modified with the introduction of hydrostatic pressure. The simplest way is a linear dependence:

$$k = k^0 - \mu p$$

and

$$\sigma_y = \sigma_y^0 - 2\mu p = 2\tau_m - 2\mu p$$

In uniaxial tensile test

$$\sigma_y = \sigma_{yt} \qquad p = \frac{\sigma_y}{3} \quad \text{(convention: p > 0 in tension)}$$

so

$$\sigma_{yt} = 2\tau_m - 2\mu\sigma_{yc}/3$$

In an uniaxial compressive test:

$$\sigma_{yc} = 2\tau_m + 2\mu\sigma_{yt}/3$$

which leads to the determination of μ, with two independent experiments

$$\mu = \frac{3}{2}\left[\frac{\sigma_{yc} - \sigma_{yt}}{\sigma_{yc} + \sigma_{yt}}\right]$$

For epoxy–amine networks, at room temperature and $\dot{\epsilon} \approx 10^{-4}\,s^{-1}$, $\mu \approx 0.2$ (Crawford and Lesser, 1999).

To modify the von Mises criterion to take into account the influence of hydrostatic pressure p, we can introduce two constants in Eq. (12.2), $C_0 (C_0 = \tau_y)$ and C_1:

$$(\sigma_1 - \sigma_2)^2 + (\sigma_2 - \sigma_3)^2 + (\sigma_3 - \sigma_1)^2 = 6(C_0 - C_1 p)^2 \qquad (12.3)$$

C_0 and C_1 can be determined with at least two independent measurements. The constant C_1 is obviously different from WLF or Mooney–Rivlin equations (Chapter 11).

For a uniaxial tensile test: $\sigma_1 = \sigma_{yt}$; $\sigma_2 = \sigma_3 = 0$; $p = \sigma_{yt}/3$

$$\begin{aligned} 2\sigma_{yt}^2 &= 6(C_0 - C_1\sigma_{yt}/3)^2 \\ \sigma_{yt} &= \sqrt{3}(C_0 - C_1\sigma_{yt}/3) \end{aligned} \qquad (12.4)$$

For a uniaxial compressive test: $\sigma_1 = \sigma_{yc}$; $\sigma_2 = \sigma_3 = 0$; $p = -\sigma_{yc}/3$

$$\sigma_{yc} = \sqrt{3}\left(C_0 + C_1\sigma_{yc}/3\right) \tag{12.5}$$

By subtracting Eq. (12.4) from Eq. (12.5), we obtain

$$\sigma_{yc} - \sigma_{yc} = \frac{\sqrt{3}}{3} C_1\left(\sigma_{yt} + \sigma_{yc}\right)$$

and

$$\boxed{C_1\sqrt{3}\left(\frac{\sigma_{yc} - \sigma_{yt}}{\sigma_{yc} + \sigma_{yt}}\right)} \tag{12.6}$$

By replacing C_1 in Eq. (12.4), we can obtain

$$\boxed{C_0 = \frac{2}{\sqrt{3}}\left(\frac{\sigma_{yt}\sigma_{yc}}{\sigma_{yt} + \sigma_{yc}}\right)} \tag{12.7}$$

For epoxy networks $C_0 = \tau_y$, and is in the range 30–40 MPa and $C_1 \approx 0.2$–0.4, at 25°C and under static conditions.

Therefore, both modified criteria (hexagon and ellipse) appear now shifted to the low values of σ_1, but are still symmetrical between σ_1 and σ_2. Generally, for thermosets the von Mises criterion gives a better fit with experimental results than the Tresca criterion.

From an experimental point of view, typically $\sigma_{yt}/\sigma_{yc} \approx 0.75$, demonstrating that the influence of hydrostatic pressure is not negligible and must be taken into account for submarine applications (pipelines, submarines, syntactic foams, etc.).

12.3.4 Yielding Theories and Influence of Physical Parameters (T, $\dot{\epsilon}$)

For many years, several authors have tried to explain and predict the yield stress of polymers (crosslinked or not), as a function of the experimental test parameters (T, $\dot{\epsilon}$) and/or structural parameters (chain stiffness, crosslinking density). These models would be very useful to extrapolate yield stress values in different test conditions and to determine the ductile–brittle transition.

Different approaches were used to describe the yielding of polymers quantitatively. Some theories took into account the free volume fraction. Eyring considered thermally activated mechanisms, and Robertson's model was based on changes of chain conformations. Argon's and Bowden's models were based on a metallurgical approach and a dislocation theory. A brief summary of the existing yielding theories is presented.

a. Empirical Brown's Relationship

Thirty years ago, Brown (1971) proposed the following empirical rule: the ratio σ_y/E is a constant value in the range of 1/60 to 1/30.

As demonstrated by many authors (Urbaczewski -Espuche *et al.*, 1991; Vallo *et al.*, 1993), Brown's rule has no general validity, especially when applied to nonstoichiometric networks.

b. Use of Free Volume for Yielding Prediction

The total free volume at a given temperature T is $f_T + f_M$, where f_T is the "thermal" free volume:

$$f_T = f_g + \alpha(T - T_g)$$

where $\alpha = \alpha_\ell - \alpha_g$ and f_M is the "mechanical" free volume (see below).

Let us consider the case of tensile loading of a thermoset sample in the glassy state (typical Poisson's ratio: $\nu \approx 0.40$). Tension in the elastic domain leads to a small swelling that contributes to the free volume. The corresponding free volume fraction is

$$f_M = \frac{\Delta V}{V} = (1 - 2\nu)\epsilon$$

where ϵ is the strain. By applying the same approach as for pressure effects (Chapter 10), one can write

$$f = f_T + f_M = f_g + \alpha(T - T_g) + (1 - 2\nu)\epsilon$$

The glass transition temperature T_{gs} under stress is thus

$$T_{gs} = T_{g0} - \frac{\epsilon(1 - 2\nu)}{\alpha} \approx T_{g0} - 500\epsilon$$

where T_{g0} is the glass transition temperature without stress.

One can consider that at a given temperature T, yielding occurs when the polymer reaches a critical state, at which it becomes suddenly rubbery. Then, the above relationship allows us to determine the elongation at yield, ϵ_y:

$$\epsilon_y \approx \frac{T_{g0} - T}{500} \tag{12.8}$$

Despite the oversimplicity of the starting hypotheses, this relationship gives a good order of magnitude.

c. Eyring's Theory

A general trend for polymers is that σ_y increases when the test temperature decreases and $\dot{\epsilon}$ increases, i.e.,

$$\dot{\epsilon} = \frac{d\epsilon_n}{dt} = \frac{dL/L_0}{dt}$$

suggesting the presence of thermally activated mechanisms.

In Eyring's theory, yielding occurs by stress and temperature-activated jumps of molecular segments (McCrum *et al.*, 1992). The applied stress reduces the activation barrier (ΔH) and segment motions define an "activation volume," V^*.

If the applied strain rate is equal to the rate of molecular jumps, then

$$\frac{\sigma_y}{T} = \frac{2}{V^*}\left(\frac{\Delta H}{T} + 2.3R \log\frac{\dot{\epsilon}}{\dot{\epsilon}_0}\right) \tag{12.9}$$

where $\dot{\epsilon}_0$ is a constant and R is the gas constant.

This model provides a good fit of experimental data for thermosets, on a large range of values of $\dot{\epsilon}$ (Mayr *et al.*, 1998). The temperature dependence is not so clear because the model takes into account only one energy barrier, associated with the main relaxation, and intermolecular interactions are neglected.

The activation energy, ΔH, resulting from the fit is very high (in the order of 300 kJ mol^{-1}) and is not comparable to any activation energy corresponding to molecular motions. The activation volume is near 2–3×10^{-27} m^3 molecule^{-1} for epoxy networks (Urbaczewski–Espuche *et al.*, 1991; Tcharkhtchi *et al.*, 1998) and about 10 times larger than the activation volume determined for yielding (Perez and Lefebvre, 1995).

Eyring's equation may be regarded as a good phenomenological description of yield stress as a function of test parameters ($T, \dot{\epsilon}$), but it cannot be related to physical processes at the molecular scale. The equation can be used at high $\dot{\epsilon}$ for impact properties and for the prediction of the ductile–brittle transition temperature. Eyring's equation can be modified with two sets of parameters if two relaxations are involved in the range of temperatures and strain rates (Bauwens-Crowet *et al.*, 1972).

Eyring's equation can be rewritten as

$$\sigma_y = \frac{2\Delta H}{V^*} - 4.6TR \log\frac{\dot{\epsilon}_0}{\dot{\epsilon}}$$

$$\sigma_y = 4.6R \log\frac{\dot{\epsilon}_0}{\dot{\epsilon}}\left[\frac{2\Delta H}{V^* 4.6R \log\frac{\dot{\epsilon}_0}{\dot{\epsilon}}} - T\right]$$

Thus at a given strain rate ($\dot{\epsilon}$ = constant): $\sigma_y = C(T_1 - T)$, where C and T_1 are constants depending on $\dot{\epsilon}$, $\dot{\epsilon}_0$, ΔH, and V^*.

From an experimental point of view it can be observed that $\sigma_y \to 0$ if the test temperature $T \to T_g$. So it appears that the so-called Kambour's relationship,

$$\sigma_y = C(T_g - T) \tag{12.10}$$

can be inferred from Eyring's equation.

The linear dependence of σ_y with $T_g - T$ is generally well fitted for thermosets (Tcharkhtchi *et al.*, 1996; Cook *et al.*, 1998), and is very useful for the determination of the ductile–brittle transition.

d. Robertson's Theory

The model of Robertson (Robertson, 1966; Escaig and G'Sell, 1982; Cook *et al.*, 1998) is based on a statistical distribution of stiff and flexible bonds (gauche → trans conformation). The applied stress improves the density of flexible bonds up to a critical value where the material becomes equivalent to a liquid.

Using the WLF equation for the viscosity (see Chapter 11), an equation is obtained with only two adjustable parameters: the activation volume ν_a induced by the conformation change and the energy difference between both states.

Robertson's theory fits experimental data for both thermoplastics and thermosets (Cook *et al.*, 1998): values for the activation volume are comparable with the theoretical ones ($\nu_a \sim 0.1\ nm^3$), but exhibit some discrepancy at high temperatures.

As in Eyring's model, only intrasegmental forces are taken into account.

e. Argon's Theory (Argon, 1973; Argon et al., 1977)

In this theory, yielding is produced by local molecular kinks at the nanoscale level. The formation of a kink pair is modeled by a wedge disclination. Permanent yielding is achieved when the surrounding molecules perform a similar process, which can relieve the local stored elastic energy of the initial kink pair.

It can be shown that

$$\left(\frac{\tau_y}{G}\right)^{5/6} = A - B\left(\frac{T}{G}\right)$$

where τ_y is the shear yield stress, G is the shear modulus, T is the absolute temperature,

$$A = \left(\frac{0.077}{1-\nu}\right)^{5/6}$$

and B is a constant depending on the polymer structure.

A good agreement has been reported for epoxies (Yamini and Young, 1980) and bismaleimide / unsaturated polyester blends (Cook *et al.*, 1998). The model gives some molecular basis to the macroscopic yielding process.

f. Bowden's Theory

It is assumed that yielding starts with the nucleation of a thermally activated disk-shaped sheared region, which grows in size as the stress is increased (Bowden and Raha, 1974). The mathematical treatment is based on dislocation theory using the Burger's vector b. Yielding is assumed to occur for a critical value of the energy of the dislocation loops and τ_y is calculated using b as the only fitting parameter.

At low temperatures the theoretical maximum shear yield stress in most glassy polymers is

$$\tau_y \sim \frac{G}{\pi\sqrt{3}} \sim G/5$$

where G is the shear modulus.

The agreement between theory and experiment is very good for epoxy networks over a wide range of temperatures (Yamini and Young, 1980; Cook *et al.*, 1998), especially near T_g, in contrast with Argon's model. Kitagawa (1977) generalized Bowden's theory and assumed that a power-law relationship exists between τ and G:

$$\frac{T_0\tau}{T\tau_0} = \left(\frac{T_0 G}{TG_0}\right)^n$$

where τ_0 and G_0 are taken at the reference temperature T_0 (usually room temperature).

A good agreement was found for thermosets, as demonstrated by Yamini and Young (1980) and Cook et al. (1998). The exponent n for all amorphous thermoplastics was about 1.6 and about 1.9 for epoxy thermosets (Fig. 12.7). As n is significantly higher than 1, Brown's rule cannot be applied.

g. Conclusions about Yielding Theories

1. Eyring's equation is the only relationship describing, with a good agreement, the dependence of yield stress on both temperature and strain rate. Unfortunately, this equation is phenomenological, and the determined constants have no physical meaning.
2. At a given strain rate, the relationship $\sigma_y = C(T_g - T)$ is obeyed for both thermosets and thermoplastics.
3. The ratio σ_y/E is not generally constant, as assumed by Brown, but follows a power-law relationship, as described by Kitagawa.

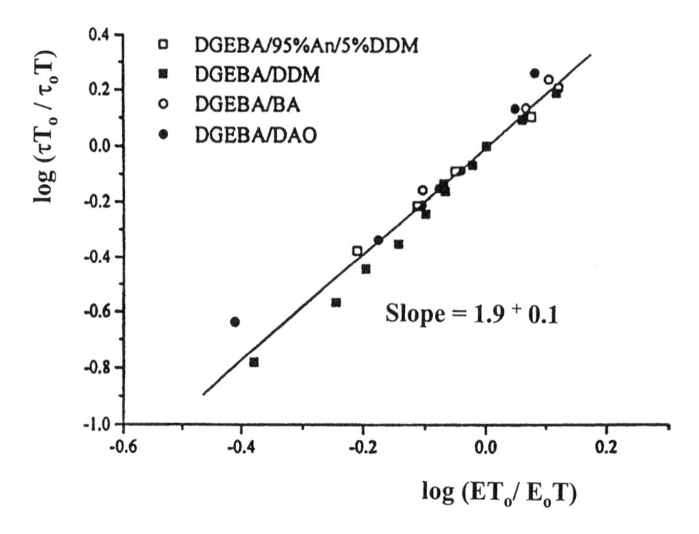

FIGURE 12.7 Plot of yield stress and modulus data according to Kitagawa's equation ($T_0 = 22°C$) for different epoxy formulations (see Table 12.1 for abbreviations). (Reprinted from Cook *et al.*, 1998, Copyright 2001, with permission from Elsevier Science.)

4. Argon and Robertson built models based on local shear yielding of polymer chains. The main drawback of these models is that they include adjustable parameters.

In conclusion, the yield behavior of thermosets is similar to that found for other glassy polymers. The presence of crosslinks does not basically affect the yield behavior of polymer networks.

12.3.5 Influence of Network Structure on Yielding

a. Introduction

There is a vast amount of literature dealing with the influence of network structure (essentially crosslink density) on yielding. Key structural parameters related to yielding will be considered separately for ideal, nonideal, and inhomogeneous networks.

b. Influence of Crosslink Density for Ideal Networks

There are several ways to modify the crosslink density of ideal networks. The first one is the use of monomers with the same structure but with different molar masses. Many workers have reported on epoxy networks with different molar-mass DGEBAs. Lemay *et al.* (1984) showed that the yield stress in epoxy–amine networks depends only on the difference between the temperature of testing and T_g. The previously mentioned Kambour's relation, $\sigma_y = C(T_g - T)$, seems to be valid in most of the cases and can be considered as a general trend (Cook *et al.*, 1998; Tcharkhtchi *et al.*, 1998; Crawford and Lesser, 1999). Therefore, the main concept is that crosslink density affects σ_y through its influence on T_g. If measurements are carried out at the same $(T_g - T)$ for every polymer network, no direct effect of crosslink density on σ_y would be recorded.

The same trend was found for polyurethanes based on a triisocyanate and polypropylene oxide diols with various molar masses (Bos and Nusselder, 1994).

Another way to vary crosslink densities of epoxies is to use mixtures of diamines with different functionalities, $f_e = 2$, 3, and 4.

A monoamine ($f_e = 2$) with a similar structure to a diamine ($f_e = 4$), plays the role of a chain extender without producing a significant change in chain flexibility. Based on this concept, workers have reported using the following mixtures: IPD/TMCA (Won *et al.*, 1990), 3DCM/MCHA (Galy *et al.*, 1994)), (Urbaczewski-Espuche *et al.*, 1991; DDM/An (Cook *et al.*, 1998), and DAO/BA (Crawford and Lesser, 1998, 1999; Mayr *et al.*, 1998) (Table 12.1).

For all these series it was found that σ_{yc} (and σ_p), measured at a constant temperature below T_g, increases in an approximately linear way with the crosslink density (Fig. 12.8). But again, this trend just reflects the influence of crosslink density on T_g.

Table 12.1 Abbreviations used for monoamines and diamines

BA	Butylamine
DAO	Diamino octane
DDM	Diamino diphenyl methane
An	Aniline
3DCM	Diamino-dimethyldicyclohexylmethane
MCHA	Methyl cyclohexylamine
IPD	Isophorone diamine
TMCA	Trimethylcyclohexylamine
HMDA	Hexamethylenediamine
HA	Hexylamine
TGAP	Triglycidylaminophenol

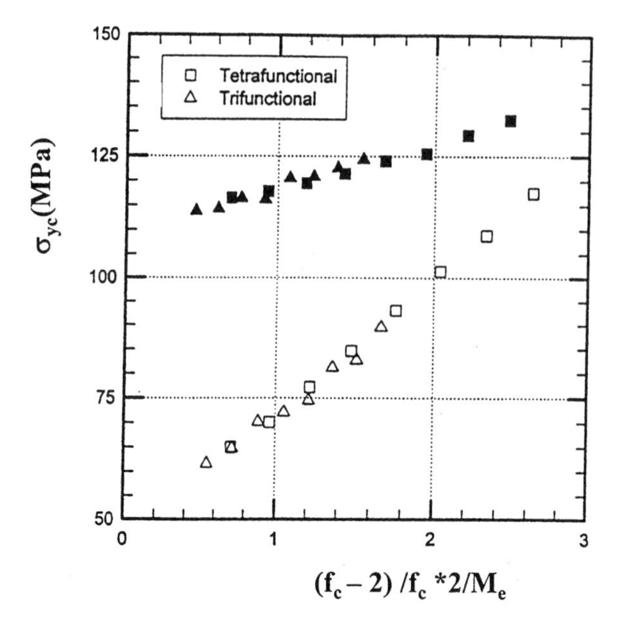

FIGURE 12.8 Variation in the compression yield stress with $(f_c - 2)/f_c * 2/M_e$ for both aliphatic (open symbols) and aromatic (closed symbols) networks. F_c is the crosslink functionality. (Crawford and Lesser, 1998, Copyright 2001, with permission of John Wiley & Sons, Inc.)

c. Influence of Chain Flexibility in the Case of Ideal Networks

The chain flexibility of epoxy networks may be changed easily by replacing the aromatic epoxy monomer, DGEBA, by an aliphatic one, DGEBD (diglycidyl ether of 1–4-butanediol). It was shown that σ_y at room temperature increased linearly with the aromatic group concentration (Urbaczewski-Espuche *et al.*, 1991). Similar trends were found by varying the hardener from an aromatic to an aliphatic one (Cook *et al.*, 1998; Crawford and Lesser, 1998) (Fig. 12.8). But again, this trend arises from the effect of chemical structure on T_g. Unfortunately, the fact that $\sigma_y \sim T_g - T$ was not considered in most of the reported results concerning the influence of network structure on σ_y.

d. Nonideal Networks Obtained by Step Polymerization

When the polymer network is formed by a single type of reaction, conversion of functional groups has a unique meaning. This leads to a direct relationship between glass transition temperature and conversion (Chapter 4). In this case, the increase in conversion (and T_g) induces an increase in σ_y, measured at room temperature.

This may not be the case if during the network formation there are two (or more) possible reaction paths in competition; e.g., epoxy networks obtained using dicyandiamide as hardener (Amdouni *et al.*, 1990).

Series of nonideal networks obtained using off-stoichiometric epoxy–amine formulations have also been employed for yielding studies. As the highest T_g corresponds to the stoichiometric formulation, it is expected that σ_y, measured at a constant temperature, should also be a maximum for the same formulation. This was the trend reported in some of these studies (Morgan *et al.*, 1984; Vallo *et al.*, 1993).

However, for some other epoxy–amine systems, experimental σ_y values exhibited a monotonous decrease from epoxy-rich to amine-rich formulations (Yamini and Young, 1980; Won *et al.*, 1990). This means that the proportionality constant between σ_y and $(T_g - T)$ must vary along the series (β relaxations and physical aging may be invoked as some of the possible factors accounting for the observed trends). For these series, Bowden's theory, with the Burger vector increasing with the amount of hardener, provided a very good fit of experimental data.

e. Conclusions about Structure Effects on Yielding

To summarize, most of the experimental results on the yielding of thermosets may be interpreted by stating that the structure affects σ_y mainly through its influence on T_g. The effects of the chemical structure, secondary relaxations, physical aging, etc., on the proportionality constant are still to be explored.

The macroscopic upper yield stress is lineary related to $T_g - T$ for a given strain rate, and can be adjusted with Argon's, Bowden's, and Kitagawa's models. The influence of strain rate is well represented by the phenomenological Eyring's equation.

Yielding of thermosets is linearly controlled by T_g and the factors that control T_g (Burton and Bertram, 1996) (Fig. 12.9). As demonstrated by Perez and Lefebvre (1995), T_g can include variations of both flexibility and crosslink density, and a generalized diagram can be plotted as $\ln(\sigma_y/G_\infty) = f(T/T_g)$, where G_∞ is the shear modulus at low temperatures (Fig. 12.10).

From a macroscopic point of view, yielding of thermoplastic or thermosetting polymers is similar and depends mainly on $T_g - T$ or T/T_g, as a reduced parameter.

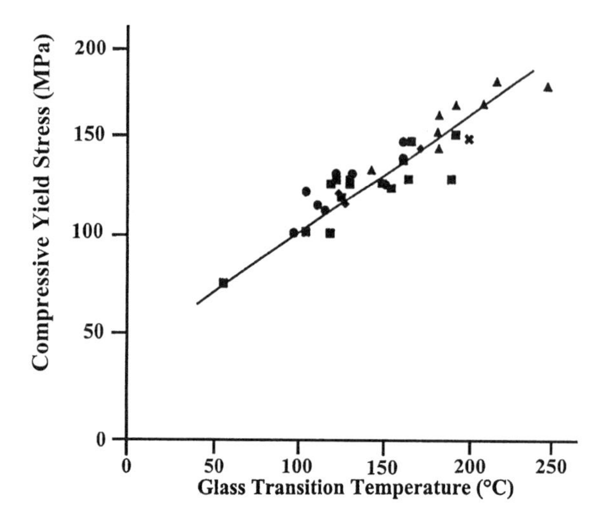

FIGURE 12.9 Compressive yield stress versus T_g for various epoxy resin systems. (Reprinted from Burton and Bertram, 1996 by courtesy of Marcel Dekker Inc.)

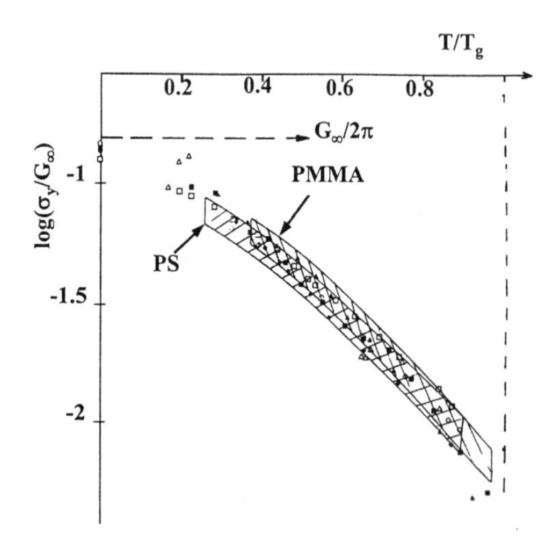

FIGURE 12.10 Evolution of $\ln(\sigma_y/G_\infty)$ versus T/T_g for amorphous thermoplastics (PS, PMMA). The different points correspond to various epoxy networks. (Reprinted from Perez and Lefebvre, 1995 with kind permission of INPL.)

12.4 BULK FRACTURE OF NETWORKS

12.4.1 Static Failure

The fracture energy G_{Ic} of unsaturated polyesters (UP), vinyl esters (VE), and phenolic "resins," is less than 200 J m^{-2} at room temperature. Epoxy networks can exhibit higher values but always lower than those of thermoplastics of similar T_g, as polycarbonate, polyetherimide, or polyphenylene ether.

The fracture energy cannot be related to the failure of chemical bonds which may contribute only with a few J m^{-2}. Furthermore, the possibility of crazing is not allowed in thermosets because fibrils cannot exist due to the high crosslink density. So, in the case of high-T_g crosslinked materials the main source of energy absorption before failure is the yielding of the network. This assumption is obviously valid only above the ductile–brittle transition temperature (Fig. 12.5), where yielding is temperature-dependent.

The basis for the relationship between fracture and yielding is related to crack-tip blunting which causes a decrease of the local stress concentration. So, higher applied loads are required to produce failure. A decrease of σ_y will increase the values of K_{Ic} and G_{Ic}. This is the general trend that is observed; therefore, it is not possible to pretend to increase both the yield stress and the fracture resistance of a neat thermoset. However, this may be accomplished by including a modifier in the formulation (Chapters 8 and 13). The resulting heterogeneous structure promotes new deformation mechanisms that increase the fracture resistance.

Every factor that increases the value of σ_y will produce a decrease in the fracture resistance. Physical aging acts in this direction (Lin *et al.*, 1986; Cook *et al.*, 1999).

The effect of structural parameters on K_{Ic} and G_{Ic} can be summarized as follows: any change in the chemical structure (use of monomers with different molar masses, use of nonstoichiometric formulations, etc.), will produce a variation in T_g; this will directly affect the value of σ_y, which, in turn, will modify the fracture resistance. An increase in T_g will lead to an increase in σ_y, measured at constant temperature, and to a decrease in the fracture resistance. This is why high-T_g epoxy networks exhibit very low values of fracture resistance.

The assessment of relationships between fracture resistance and structural parameters would require analyzing experimental results obtained at constant (T_g – T). Unfortunately, most of the results reported in the literature for materials of different T_g were obtained at room temperature. Conclusions regarding the influence of structural parameters on the fracture resistance mask the direct effect of the network modifications on T_g. Some examples of these studies are discussed in the following sections.

a. Ideal Networks

A linear decrease of K_{Ic} with an increase in crosslink density was reported for model PU based on triisocyanate and diols of various molar masses (Bos and Nusselder, 1994), and for epoxy networks (Lemay *et al.*, 1984). It was suggested that the dilational stress field at the crack tip may induce an increase in free volume and a devitrification of the material. A linear relationship between G_{Ic} and $M_e^{1/2}$ was verified for these systems, although other empiric equations were found in other cases (Urbaczewski-Espuche *et al.*, 1991).

Epoxy networks were synthesized using mixtures of amines of different functionality to vary M_e and the average functionality, $\bar{f}_E$ (Galy *et al.*, 1994; Crawford and Lesser, 1998). In both studies, the increase in crosslink density increased both σ_y and T_g and produced a corresponding decrease in K_{Ic}. A linear relationship between K_{Ic}, measured at room temperature, and crosslink density is shown in Fig. 12.11 (Galy *et al.*, 1994).

To study the influence of chain flexibility, DGEBA, was replaced by DGEBD, a more flexible epoxy monomer, in formulations with the same hardener, 3DCM (Urbaczewski-Espuche *et al.*, 1991). Although the crosslink density increases with the DGEBD fraction (average molar mass of DGEBD = 202 g/mol^{-1}, average molar mass of DGEBA = 345 g/mol^{-1}),

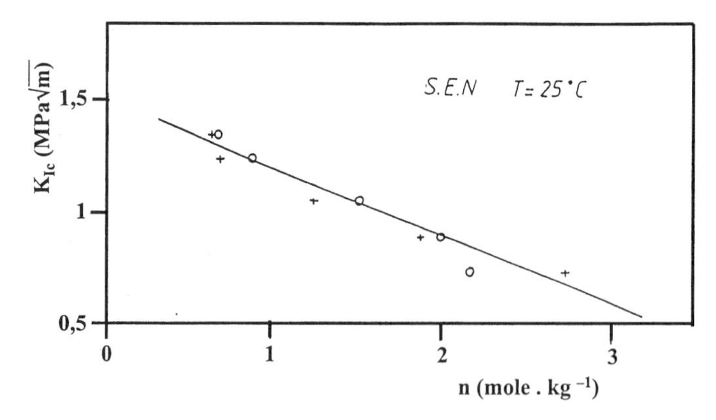

FIGURE 12.11 Evolution o K_{Ic} at 25°C for ideal epoxy networks based on DGEBA./3DCM/MCHA versus crosslink density. (+), theoretical; (o), experimental. (see Table 12.1 for abbreviations). (By permission from Galy *et al.*, 1994, Copyright 2001, ChemTec Publishing.)

the yield stress decreases with the introduction of DGEBD, due to the prevailing effect of chain flexibility over crosslink density (T_g decreases by increasing the DGEBD fraction). A significant increase in G_{Ic}, measured at 25°C, with the DGEBD fraction, was found; it increased from 200 J m^{-2} for DGEBA/3DCM to 700 J m^{-2} for DGEBD/3DCM (Fig. 12.12). The effect of chain flexibility is much greater than the effect of crosslink density.

The effect of chain flexibility is also evidenced by replacing aromatic hardeners by aliphatic ones. This produces an increase in the fracture resistance at room temperature (Kinloch and Young, 1983).

Grillet *et al.* (1991) studied mechanical properties of epoxy networks with various aromatic hardeners. It is possible to compare experimental results obtained for networks exhibiting similar T_g values (this eliminates the influence of the factor $T_g - T$). For instance, epoxy networks based on flexible BAPP (2-2′- bis 4,4′-aminophenoxy phenyl propane) show similar T_g values (~ 170°C) to networks based on 3-3′DDS (diamino diphenyl sulfone). However, fracture energies are nine times larger for the former. These results constitute a clear indication that the network structure does affect the proportionality constant between σ_y and $T_g - T$. Although no general conclusions may be obtained, it may be expected that the constant is affected by crosslink density, average functionality of crosslinks and chain

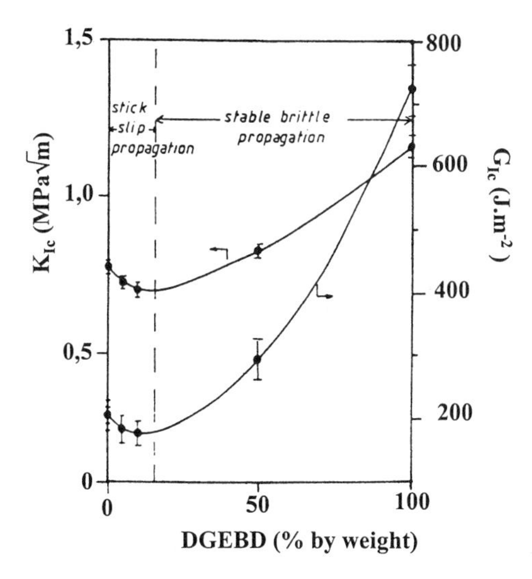

FIGURE 12.12 Critical stress intensity factor (K_{Ic}) and critical energy release rate (G_{Ic}) at 25°C as a function of DGEBD percentage. (Reprinted with permission of SPE from Urbaczewski-Espuche *et al.*, 1991.)

flexibility. But the general trend is that the fracture toughness increases when the yield stress decreases (Fig. 12.13). This is a measure of the ability of plastic deformation of the polymer network.

b. Non-Ideal Networks Prepared from Step Polymerization

In general, an increase in conversion increases T_g and σ_y and decreases K_{Ic} (Amdouni *et al.*, 1990; Jordan *et al.*, 1992). In series prepared with non-stoichiometric compositions, the minimum value of fracture energy is obtained when the stoichiometric ratio of both functionalities is used (Morgan *et al.*, 1984; Won *et al.*, 1990), as expected (maximum values of both T_g and σ_y).

c. Inhomogeneous Networks

Morphological inhomogeneities are present in most polymer networks obtained by chainwise polymerization. In networks produced by polycondensation, such as resols (Chapter 2), reaction species and volatile products phase-separate from the matrix, leading to the formation of voids in the final product. The presence of inhomogeneities leads to stress concentration during mechanical loading, so that yielding and/or fracture can occur at lower loads than with homogeneous materials. Micromechanics is needed to describe the mechanical behavior.

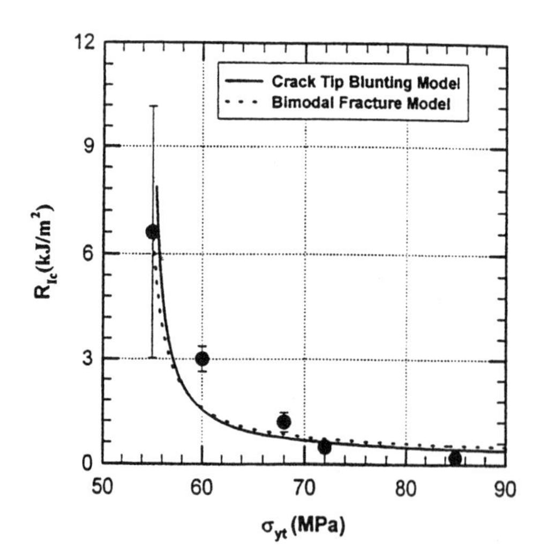

FIGURE 12.13 Fracture energy (G_{Ic} or J_{Ic}) at 25°C versus tensile yield stress for different epoxy networks. (Reprinted with permission of SPE from Crawford and Lesser, 1999.)

For phenolic "resins" of the resols type, numerous voids (a few micrometers in size) are formed during the synthesis. The average diameter and distribution of voids depend strongly on the cure cycle; their presence, together with the high values of crosslink density, explain the low fracture resistance of these networks (Wolfrum *et al.*, 1999).

For acrylate networks, the structure–properties relationships are not yet well established. For systems based on bis-GMA (bis-phenol-A diglycidylmethacrylate)/TEGDMA (Triethylene glycol dimethocrylate), K_{Ic} increased with conversion (Cook, 1991), an effect which is contrary to the one expected. This was explained by the decrease in the fraction of unreacted dimethacrylate. It was also found that by varying the ratio of dimethacrylate/monomethacrylate, K_{Ic} remained constant, $\sim 0.8\,\text{MPa}\,\sqrt{\text{m}}$ (Karmaker *et al.*, 1997). It was suggested that, in these materials, failure can occur in the less crosslinked regions (Simon *et al.*, 1991). Fracture properties of inhomogeneous networks are still an open field.

12.4.2 Influence of Test Variables ($T, \dot{\epsilon}$) on the Fracture of Networks

Viscoelastic models show that an increase in strain rate is equivalent to a decrease in test temperature. This concept may, in principle, be applied to the fracture of thermosets. However, the presence of secondary relaxations (β, γ, etc), can complicate the analysis.

In general, the fracture energy of a network is correlated to its ability for plastic deformation. This means that G_{Ic} decreases with an increase of σ_y. The influence of test variables ($T, \dot{\epsilon}$) on σ_y was analyzed before (Sec. 12.3.4), and can explain the changes observed in G_{Ic} values.

The main concepts are

1. At a constant temperature, the increase of $\dot{\epsilon}$ involves an increase of σ_y and a decrease of K_{Ici} (for initiation). The crack propagation is generally of the stick-slip type (see Fig. 12.4), at low $\dot{\epsilon}$, and becomes brittle-unstable when $\dot{\epsilon}$ increases (Galy *et al.*, 1994). K_{Ica} (for arrest) remains at an almost constant value (Kinloch, 1985).
2. At a constant strain rate, the crack propagation is brittle-stable at low temperatures and becomes of the stick-slip type at high temperatures, with an increase of K_{Ici} (Kinloch, 1985).

The changes in crack propagation types (from stable to stick-slip) are associated with the crack blunting mechanism, which is favored by high temperatures and low strain rates, conditions that decrease σ_y. Nevertheless, these general trends cannot be extended to very high strain rates because a transition from isothermal to adiabatic conditions may occur, due to the heat dissipation and the low thermal conductivity of polymers.

The effects of strain rate and temperature are correlated, and can be modeled (Kinloch and Young, 1983, Kinloch, 1985). For different temperatures and strain rates, G_{Ic} and the time to failure, t_f, were measured. Using the time–temperature superposition principle, shift factors (a_T) applicable to the time to failure t_f, were determine. Shift factors plotted against $(T - T_g)$ are independent of the type of test used (Fig. 12.14). The construction of a typical master curve G_{Ic} versus t_f/a_T is shown in Fig. 12.15 (Hunston *et al.*, 1984). The value of G_{Ic} may be predicted for any strain rate/temperature combination. This model can also be applied to rubber-modified epoxies (See chapter 13).

12.4.3 Impact Strength of Thermosets

Few investigations dealing with impact properties of neat thermosets have been reported because they are too brittle at room temperature and their impact energy is very low (a few $kJ\,m^{-2}$ for unnotched specimens).

The main characteristic of an impact test is the sudden application of a load at a high speed (typically, strain rates are in the range 10^{-1}–$10^{2}\,s^{-1}$.

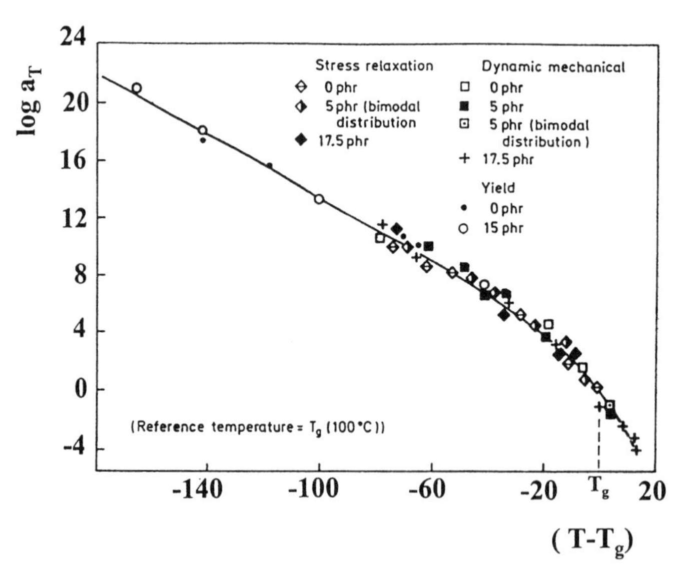

FIGURE 12.14 Values of the shift factor a_T needed for superposing of bulk mechanical data as a function of temperature for neat and rubber-modified epoxy networks. (Hunston *et al.*, 1984. Reprinted with permission from Kluwer Academic/Plenum Press.)

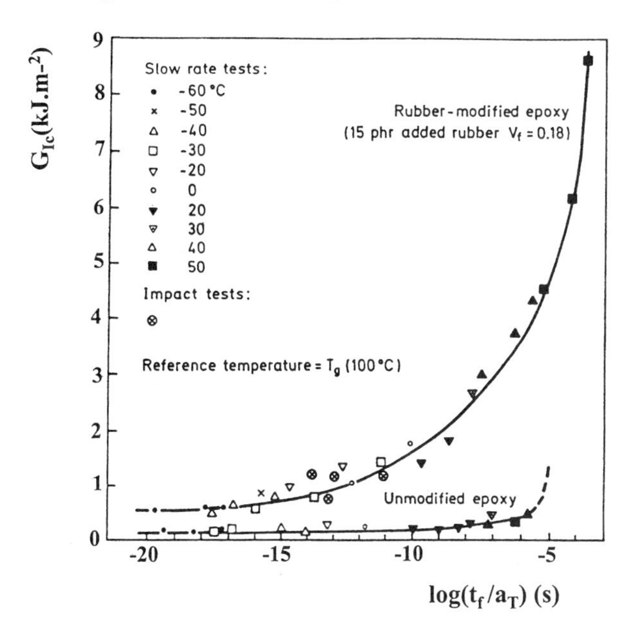

FIGURE 12.15 Fracture energy G_{Ic} versus reduced time of failure (t_f/a_T) for neat and rubber-modified epoxies. (Hunston *et al.*, 1984. Reprinted with permission from Kluwer Academic/Plenum Press.)

a. Tests Description

A pendulum may be used (Charpy, Izod, tensile impact) to determine the work of fracture (Brown, 1999). Instrumented devices provided with piezoelectric transducers are also available; load–time or load–displacement curves can be recorded (Merle *et al.*, 1985), giving as much information as static tests. Servohydraulic or pneumatic setups and falling weight devices are also used. The drop ball test from the US Food and Drug Administration, is especially useful for optical lenses (acrylate networks).

Impact tests may be performed with notched or unnotched specimens, but the results cannot be compared. The presence of a notch induces a great part of the crack initiation energy. The speed of the striker or of the cross-head may be varied from $1\,m\,s^{-1}$ to several $km\,s^{-1}$ for ballistic tests; strain rates may vary in the range of 10^{-1} up to $100\,s^{-1}$. Very often the energy for failure is divided by the effective cross-sectional area or the deformed volume, in order to calculate, respectively, the surface or the volume resilience. Using the compliance formula and simple assumptions, G_{Ic} may be directly calculated (Williams, 1984).

New possibilities appeared with the use of nonbreaking impact tests. The question is how a high-speed striker with a low energy can induce damages and what the residual energy is. These tests are very often performed on composites with thermosetting matrices.

Impact tests can be performed at various temperatures, especially at low temperatures (where there is a combination with the high speed), in order to determine the ductile–brittle transition. This transition is very important for characterizing the polymer behavior, and is determined usually at a constant speed and changing the temperature. Although it is less usual, it is possible to fix the temperature and to vary the speed.

The ductile–brittle transition is clearly related to the yielding behavior of the thermoset in static experiments. (see Fig. 12.5).

At room temperature, well below T_g, a brittle failure is generally observed. The ductile behavior appears when yielding becomes a competitive mechanism of deformation. At high speeds the brittle stress is not too much affected but σ_y depends strongly on $\dot{\epsilon}$ (see Eyring's Eq. (12.9)), producing a shift of the ductile–brittle transition to higher temperatures.

b. Results

As already mentioned, there are very few works dealing with impact properties of thermosets and sometimes they are contradictory, because:

1. Most of the time the network structure is not well defined.
2. The use of conventional tests, without relying on fracture mechanics, leads to a high scattering of experimental results due to the presence of flaws induced by the processing and machining.

An excellent impact resilience at room temperature ($\sim 20\,\mathrm{kJ\,m^{-2}}$), was found for DDM- and BAPP-based epoxy networks. Using the activation energy of the β relaxation, it was shown that, in both cases, networks were impacted above the temperature of the β relaxation at the corresponding strain rate (Grillet *et al.*, 1991).

The impact properties of acrylate networks used in optical lenses have been widely investigated, mainly from an industrial point of view. Due to internal stresses which develop during the reaction, the cure schedule and the heating and cooling rates have to be carefully adjusted. The structural complexity of these networks does not allow us to obtain correlations of the observed impact strength values (Matsuda *et al.*, 1998).

Independently of the strain rate and the test temperature, the toughness of the material depends on its ability to absorb or dissipate energy, and this requires chain mobility. In the glassy state the energy-absorbing mechanism is related to the β relaxation. Therefore, impact resistance increases if the β relaxation is active at the temperature and frequency (some kHz) of impact.

A relationship between the room temperature frequency of the β relaxation, $f_{\beta,RT}$ and toughness was described by Boyer (1976) for thermoplastics, and applied to epoxy networks by Schroeder *et al.* (1987):

$$\log(f_{\beta}, T/f_{\beta,RT}) = (E_{a\beta}/2.3R)(1/T - 1/T_{RT}) \tag{12.11}$$

where $E_{a\beta}$ is the activation energy of the β relaxation. The general trend observed for various epoxy networks is that toughness increases with $f_{\beta,RT}$.

The low values of the impact resistance of thermosets justifies the need to improve them by using modified formulations (Chapters 8 and 13).

12.4.4 Fatigue Failure of Thermosets

a. Introduction

Fatigue or dynamic fatigue that results from cyclic loading can produce damage, with decay of certain properties such as modulus and, eventually, fracture. This phenomenon is very important in the analysis of the long-term behavior of materials, because failure can occur at very low loads, much lower than those required in static tests. Although the fatigue behavior of thermoplastics has been well researched (e.g., see Hertzberg and Manson, 1980), thermosets have only been the subject of a few studies due to their brittleness.

b. Experimental Methods

The main specific characteristic of fatigue experiments is the application of a cyclic stress or strain.

To use the true stress or strain, it is necessary to measure the cross section or the length of the sample continuously and to have a closed-loop feedback on a (hydraulic) setup. Therefore, most of the time, only cyclic loads or displacements are applied onto the sample.

The cyclic load (or strain) is usually a sinusoidal function, but triangle or square functions may be used. It is characterized by its frequency, σ_{max} and σ_{min} (ϵ_{max} and ϵ_{min}, respectively). The fluctuating stress amplitude is $\sigma_a = 1/2(\sigma_{max} - \sigma_{min})$, and σ_m is the average stress.

One of the most important parameters is the value of the stress ratio, $R_{\sigma} = \sigma_{min}/\sigma_{max}$, which can be used to describe all the different tests (Fig. 12.16).

The frequency has to be carefully chosen. It is always attractive to increase the frequency in order to reduce the test time but, especially in the case of polymers, due to the viscoelastic damping and the low thermal

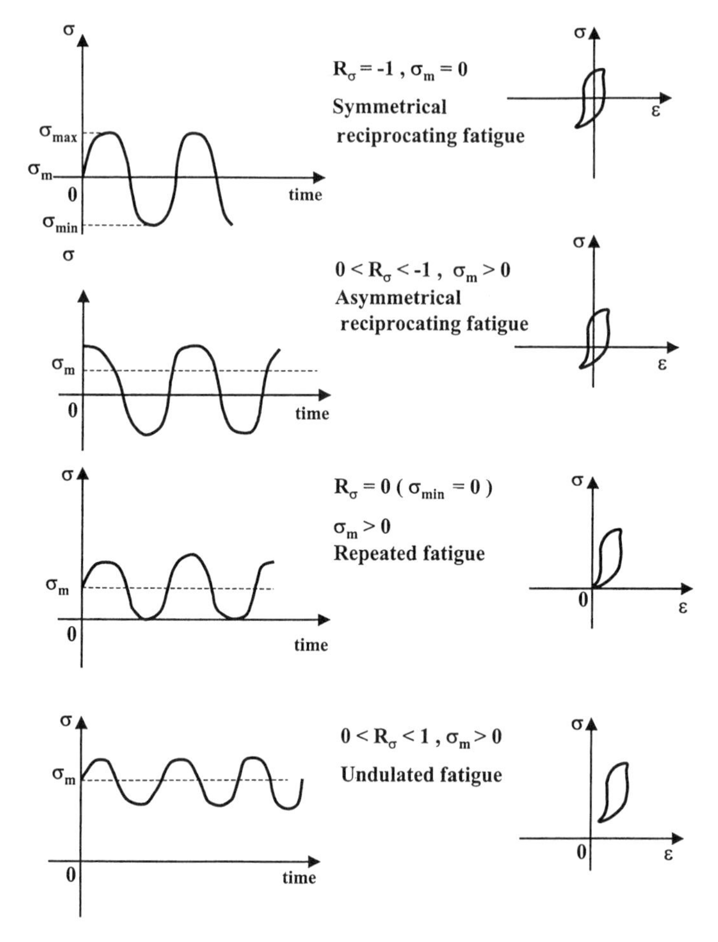

FIGURE 12.16 Evolution of stress versus time for different fatigue tests depending on the stress ratio R_σ and the corresponding stress–strain cycles; σ_m is the average stress.

conductivity, the actual temperature in the sample may be higher than the test temperature. This can induce incorrect lifetime estimations due to crack-tip blunting. The dissipated energy, at an imposed stress, is proportional to the frequency and to the square of the stress amplitude. In certain cases, failure of the sample may be observed in a fatigue test, due to the temperature becoming higher than Tg. In most cases, a frequency of below 5 Hz seems to be a reasonable selection (Brown, 1999).

Rotational machines are used for imposed strains and hydraulic setups are necessary for imposed stresses. Different geometries can be used (tensile, compressive, flexural, etc.). It is better to test samples or parts as close as possible to the actual stress conditions. Using records of real parts or structures, it is possible to perform fatigue experiments with different sequences of σ_a and frequencies.

Two main approaches have been developed in dynamic fatigue.

The Classical Fatigue Curves. Often referred to as S–N curves or Wöhler curves, the amplitude of stress σ_a is plotted versus the number of cycles to failure. A sigmoidal curve, determining a fracture and a nonfracture region, is generated (Fig. 12.17). Sometimes, but not always, a fatigue limit (or endurance limit) is obtained, below which no fracture will occur whatever the number of cycles. This conventional approach gives a total fatigue lifetime without separation between crack initiation and propagation.

The Fracture Mechanics Approach. This was developed to describe the crack growth rate. Paris proposed a law:

$$\frac{da}{dN} = C\Delta K^m \tag{12.12}$$

where a is the crack length and $\Delta K = K_{max} - K_{min}$ is the stress intensity factor range corresponding to σ_{max} and σ_{min}. C and m are constants that depend on the material and test variables (temperature, R_σ, frequency, etc.).

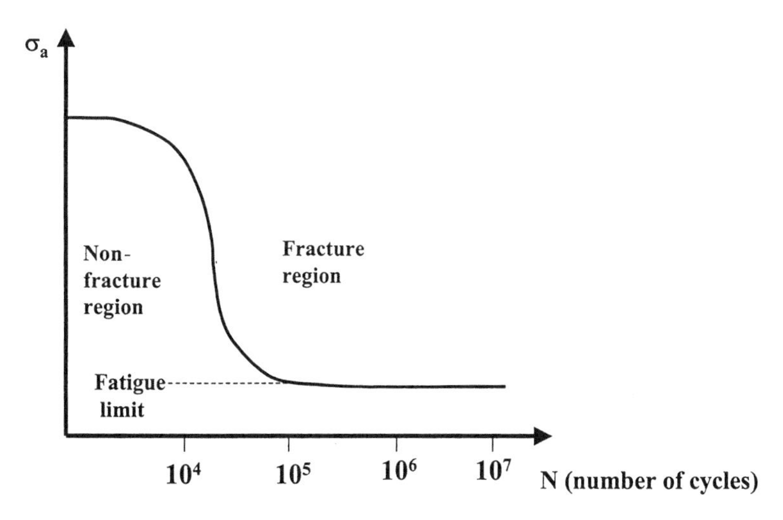

FIGURE 12.17 Typical Wöhler or "S–N" curve for polymers (stress amplitude σ_a versus number of cycles N).

To obtain the fatigue crack propagation rate, da/dN, it is necessary to record the crack length (a) versus time. This can be done by the compliance method (Fischer *et al.*, 1995), using special gauges with many parallel electrical wires which are cut by the propagating crack, or using a traveling microscope or a CCD (charge-coupled device) camera (Rey *et al.*, 1999).

A typical Paris diagram in a log–log scale is shown in Fig. 12.18. The Paris power law is not generally observed over the full ΔK range. For low ΔK values, below a threshold (ΔK_{thres}), the crack cannot propagate. In the subcritical domain (zone I), the Paris law is not observed. If $\Delta K \sim K_{Ic}$ (zone III), the sample will break in a single cycle. Sometimes, the value of ΔK at a given fatigue crack propagation rate ($da/dN = 7.5 \times 10^{-4}$ mm/cycle) is chosen for comparison purposes (Hwang *et al.*, 1989).

c. Structure Effects on Thermosets Fatigue

The Paris power law is generally well verified with neat epoxy thermosets (Fischer *et al.*, 1995; Sautereau *et al.*, 1995; Rey *et al.*, 1999). The exponent varies between 9 and 17. The crack resistance increases continuously when M_e increases; ΔK at $da/dN = 7.5 \times 10^{-4}$ mm/cycle is linearly related to K_{Ic} for numerous networks, as already described for thermoplastics (Hwang *et al.*, 1989; Rey *et al.*, 1999).

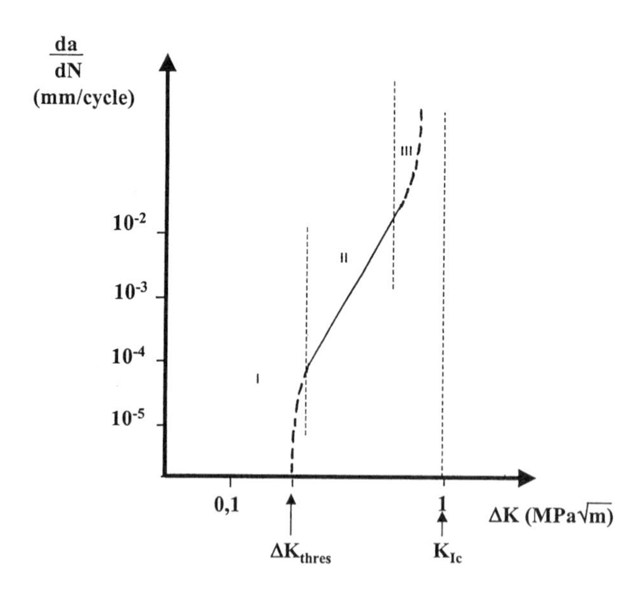

FIGURE 12.18 Typical Paris diagram for thermosets at room temperature.

As K_{Ic} is related to σ_y and T_g, the fatigue crack propagation is also related to the ability of plastic deformation at the crack tip. Fischer *et al.* (1995) proposed a model for fatigue crack propagation where the crack opening displacement is proportional to M_e; then, the logarithm of the exponent m is linearly related to $(\sigma_y/EM_e)^{1/2}$, with a good agreement.

However, in unnotched samples, the fatigue lifetime is governed by crack initiation rather than propagation. S–N curves display an induction time during which no change of elastic properties is observed. It seems that there is no correlation between the induction time and toughness as represented by K_{Ic}. For example, epoxide–amine networks have an endurance limit higher than tough thermoplastics such as polycarbonate or polysulfone (Trotignon *et al.*, 1993). This is probably because in amorphous thermoplastics the endurance limit is closely related to the critical crazing load. Since crazing is inhibited in thermosets, fracture can only result from other mechanisms occurring at a higher stress level.

12.5 CONCLUSIONS

As found for elastic properties (Chapter 10), there are some similarities between the behavior of amorphous linear polymers and thermosets in the domain of large deformations.

The influence of temperature and strain rate can be well represented by Eyring's law; physical aging leads to an increase of the yield stress and a decrease of ductility; the yield stress increases with hydrostatic pressure, and decreases with plasticization effect. Furthermore, it has been demonstrated that σ_y decreases with $T - T_g$ at a constant strain rate. Structure–property relationships display similar trends; e.g., σ_y increases with the chain stiffness through a T_g increase and yielding is favored by the existence of mechanically active relaxations due to local molecular motions (β relaxation).

As for linear polymers, the ultimate properties can be interpreted by a competition between brittle fracture (not very dependent on temperature and $\dot{\epsilon}$, but linked to cohesive energy density) and shear yielding and plastic deformation (see Fig. 12.5).

For thermosets, yielding and fracture are very often measured at room temperature because materials are used at this temperature. To establish structure–properties relationships, these results have to be considered with care and the best way is to measure both σ_y and K_{Ic} at a constant $T_g - T$ in order to take the influence of structure properly into account.

There is a strong competition between thermoplastics and thermosets in the field of engineering composites and it is often claimed that thermoplastics are tougher than thermosets. This assessment cannot be generalized, because many thermosets are tougher than thermoplastics such as polystyrene. If chemical crosslinking plays a role, it must probably reduce the interval between T_B (ductile–brittle transition temperature) and T_g (see Fig. 12.5), but it cannot suppress this region.

The most frequently cited argument favoring the "embrittlement" effect of crosslinking is derived from rubber elasticity theory where the fracture energy is described as proportional to $M_e^{1/2}$, the square root of the chain length between crosslinks. This argument is probably valid in the temperature interval $T_B - T_g$, where plastic deformation can occur (Lemay *et al.*, 1984), but it does not explain why this range of ductility is smaller in thermosets than in linear polymers of similar T_g. The β relaxation probably plays an important role in this domain, but there is not enough experimental data to have a general view of this problem.

The role of morphology in inhomogeneous networks is also not very clear. Is the mechanical behavior under the influence of molecular/macromolecular parameters or is it controlled by stress concentration effects at the boundaries of dispersed domains?

Finally, the most important qualitative difference between amorphous thermoplastics and thermosets is the absence of crazing in the latter. This difference calls for two comments:

1. Crazing inhibition by crosslinking explains the better fatigue resistance of thermosets compared with thermoplastics (this is not valid for semicrystalline thermoplastics such as polypropylene or poly (ether ether ketone).
2. In thermoplastics, structural modifications disfavoring crazing have generally a positive effect on ductility and toughness and these properties are increasing functions of the entanglement density (e.g., physical crosslinking). Why does permanent chemical crosslinking display opposite trends? There is, to our knowledge, no clear answer to this question.

NOTATION

a = crack length, m
a_T = shift factor for time to failure
b = width of sample in three-point bending test, m
C_0 = constant used in the modified von Mises criteria, MPa
C_1 = constant used in the modified von Mises criteria
D = span length of sample in three-point bending test, m
dL = length variation of a sample in a tensile test, m
E = Young's modulus, Pa (GPa)

$E_{a\beta}$ = activation energy of the β relaxation
$f_{\beta,RT}$ = frequency of the β transition at room temperature, Hz
$f_{\beta,T}$ = frequency of the β transition at temperature T, Hz
f_e = amine functionality
$\bar{f}_e$ = average amine functionality
f_c = crosslink functionality
f_g = free volume at T_g
f_M = mechanical free volume
F_{max} = maximum load, N
f_T = thermal free volume
G = shear modulus, Pa (GPa)
G_{Ic} = critical strain energy release rate in mode I, $J\,m^{-2}$
G_∞ = shear modulus at low temperature, GPa
h = thickness of sample in three-point bending test, m
K = stress intensity factor, $MPa\,m^{1/2}$
K_{Ic} = critical stress intensity factor in mode I, $MPa\,m^{1/2}$
K_{Ica} = critical stress intensity factor in mode I for arrest, $MPa\,m^{1/2}$
K_{Ici} = critical stress intensity factor in mode I for initiation, $MPa\,m^{1/2}$
L = length of a sample in a tensile test, m
L_0 = initial length of a sample in a tensile test, m
M_e = molar mass between crosslinks, $kg\,mol^{-1}$
N = number of cycles in fatigue
p = hydrostatic pressure
R = gas constant, $kJ\,kmol^{-1}\,K^{-1}$
R_σ = stress ratio in fatigue
t = time, s
T = temperature, K
T_B = ductile–brittle transition temperature, K
t_f = time to failure, s
T_g = glass transition temperature, K
T_{g0} = glass transition temperature without stress, K
T_{gs} = glass transition temperature under stress, K
V^* = activation volume in Eyring's theory, $m^3\,molecule^{-1}$
x, y, z = main directions

Greek Letters

α_l = expansion coefficient in the liquid state, K^{-1}
α_g = expansion coefficient in the glassy state, K^{-1}
ΔH = (activation energy in Eyring's theory, $kJ\,kmol^{-1}$
ΔK = variation of the stress intensity factor during a fatigue cycle, $MPa\sqrt{m}^{1/2}$
ϵ = strain
$\dot{\epsilon}$ = strain rate, s^{-1}
ϵ_b = strain at break
ϵ_n = nominal strain

ϵ_p = strain at lower yielding point
ϵ_t = true strain
ϵ_y = strain at upper yielding point
μ = coefficient of dependence for yielding stress with hydrostatic pressure
ν = Poisson's coefficient
σ = stress, MPa
$\sigma_1, \sigma_2, \sigma_3$ = principal stresses, MPa
σ_b = stress at break, MPa
σ_n = nominal stress, MPa
σ_p = stress at the lower yield point, MPa
σ_t = true stress, MPa
σ_y = stress at the upper yield point, MPa
σ_{yc} = stress at the yield stress in compression, MPa
σ_{yt} = stress at the yield stress in tension, MPa
τ_{oct} = octahedral shear stress, MPa
τ_y = stress in pure shear, MPa

REFERENCES

Amdouni N, Sautereau H, Gerard JF, Pascault JP, *Polymer*, 31, 1245–1253 (1990).
Argon AS, *Phil Mag*., 28, 839–865 (1973).
Argon AS, Bessonov MI, *Polym. Eng. Sci*., 17, 174–182 (1977).
Bauwens-Crowet C, Bauwens JC, Homes G, *J. Mat. Sci*., 7, 176–183 (1972).
Bos HL, Nusselder JJH, *Polymer*, 35, 13, 2793–2799 (1994).
Bowden PB, Raha S, *Phil. Mag*., 29, 149–166 (1974).
Boyer RF, *Polymer*, 17, 996–1003 (1976).
Brown N, *Mat. Sci. Eng*., 8, 839–850 (1971).
Brown R, *Handbook of Polymer Testing, Physical Methods*, Marcel Dekker, New York, 1999.
Burton LB, Bertram JL. In *Polymer Toughening*, Arends CB (ed.), Marcel Dekker, New York 1996, pp 339–379.
Cook WD, *J. Appl. Polym. Sci*., 42, 1259–1269 (1991).
Cook WD, Mayr AE, Edward GH, *Polymer*, 39, 3725–3753 (1998).
Cook WD, Mehrabi M, Edward GH, *Polymer*, 40, 1209–1218 (1999).
Crawford E, Lesser AJ, *J. Polym. Sci., Part B, Polym. Phys*. 36, 1371–1382 (1998).
Crawford E, Lesser AJ, *Polym. Eng. Sci*., 39(2), 385–392 (1999).
Escaig B, G'Sell C, *Plastic Deformation of Amorphous and Semi-Crystalline Materials*, Les Editions de Physique, Les Ulis (France), 1982.
Fischer M, Martin D, Pasquier M, *Macromol. Symp*., 93, 325–336 (1995).
Galy J, Gérard JF, Sautereau H, Frassine R, Pavan A, *Polym. Net. Blends* 4(3–4), 105–114 (1994).
Girard-Reydet E, Vicard V, Pascault JP, Sautereau H, *J. Appl. Polym. Sci*., 65, 2433–2445 (1997).
Grillet AC, Galy J, Gérard JF, Pascault JP, *Polymer*, 32(10), 1885–1891 (1991).

Hertzberg RW, Manson JA, *Fatigue of Engineering Plastics*, Academic Press, New York, 1980.
Hunston WD, Kinloch AJ, Shaw SJ, Wang SS In *Adhesive Joints*, Mittal KL (ed.), Plenum Press, New York, :1984, pp. 789–807.
Hwang JF, Manson JA, Hertzberg RW, Miller GA, Sperling LH, *Polym. Eng. Sci.*, 29(20), 1477–1487 (1989).
Jordan C, Galy J, Pascault JP, *J. Appl., Polym. Sci.*, 46, 859–871 (1992).
Karmaker AC, Dibenedetto AT, Golberg AJ, *J. Mat. Sci.: Materials in Medecine*, 8, 369–374 (1997).
Kinloch AJ. In *Epoxy Resins and Composites I. Advances in Polymer Science No. 72*, Dusek K (ed.), Springer Verlag, Berlin, 1985, p. 45–68.
Kinloch AJ, Young RJ, *Fracture Behaviour of Polymers*, Applied Science Publishers, London, 1983.
Kitagawa M, *J. Polym. Sci., Polym. Phys.* Ed., 15(9) 1601–1611 (1977).
Lemay JD, Swetlin BJ, Kelley FN, In *Characterization of Highly Crosslinked Polymers*, SS Labana, RA Dickie (eds), ACS Symposium Series No. 243, Washington, 1984, p. 165.
Lin YG, Sautereau H, Pascault JP, *J. Appl. Polym. Sci.*, 32, 4585–4605 (1986).
Liu CH, Nairn JA, *Polym. Eng. Sci.*, 38(1), 186–193 (1998).
McCrum NG, Buckley CP, Bucknall CB, *Principles of Polymer Engineering,* Oxford Science Publishers, New York, 1992.
Mai YM, Powell P, *J. Polym. Sci. , Polym. Phys.*, 29, 785–793 (1991).
Matsuda R, Funae Y, Yoshida M, Yamamoto T, Takaya T, *J. Appl. Polym. Sci.*, 65, 2247–2255; 68, 1227–1235 (1998).
Mayr AE, Cook WD, Edward GH, *Polymer*, 39, 3719–3724 (1998).
Merle G, O YS, Pillot C, Sautereau H, *Polym. Testing*, 5, 37–43 (1985).
Morgan RJ, Mones ET, Steele WJ, *Polymer*, 23, 295–305 (1982).
Morgan RJ, Kong FM, Walkup CM, *Polymer*, 25, 375–386 (1984).
Oyanguren PA, Vallo CI, Frontini PM, Williams RJJ, *Polymer*, 35, 5279–5288 (1994).
Perez J, *Physique et Mécanique des Polymères Amorphes, Techniques et documentations,* Lavoisier, Paris, 1988.
Perez J, Lefebvre JM, In *Introduction à la Mécanique des Polymères*, G'Sell C, Haudin JM (ed.), INPL (France), 1995, pp. 289–318.
Rey L, Poisson N, Maazouz A, Sautereau H, *J. Mat. Sc*i., 34, 1775–1781 (1999).
Robertson RE, *J. Chem. Phys.*, 44, 3950–3956 (1966).
Sautereau H, Maazouz A, Gérard JF, Trotignon JP, *J. Mat. Sci.*, 30, 1715–1718 (1995).
Schroeder JA, Madsen PA, Foister RT, *Polymer*, 28, 929–940 (1987).
Simon G, Allen PEM, Williams DRG, *Polym. Eng. Sci.*, 31(20). 1483–1492 (1991).
Tcharkhtchi A, Faivre S, Roy LE, Trotignon JP, Verdu J, *J. Mat. Sci.*, 31, 2683–2692 (1996).
Tcharkhtchi A, Trotignon JP, Verdu J, *Proceedings of the 18th Discussion Conference 'Mechanical Behaviour of Polymeric Materials*', IUPAC and Czech Acad. Sci., Prague, 20–23 July 1998.

Trotignon JP, Verdu J, Martin Ch, Morel E, *J. Mat. Sci.*, 28, 2207–2213 (1993).
Urbaczewski-Espuche E, Galy J, Gérard JF, Pascault JP, Sautereau H, *Polym. Eng. Sci.*, 31, 1572–1580 (1991).
Vallo CI, Frontini PM, Williams RJJ, *Polymer Gels and Networks*, 1, 257–266 (1993).
Williams JG, *Fracture Mechanics of Polymers*, Ellis Horwood, Chichester, UK, 1984.
Williams JG, Cawood MJ, *Polym. Testing*, 9, 15–26 (1990).
Wolfrum J, Ehrenstein GW, *J. Appl. Polym. Sci.*, 74, 3173–3185 (1999).
Won YG, Galy J, Gérard JF, Pascault JP, Bellenger V, Verdu J, *Polymer*, 31, 1787–1792 (1990).
Wu WL, Bauer BJ, *Macromolecules*, 21, 457–464 (1988).
Yamini S, Young RJ, *J. Mat. Sci.*, 15, 1814–1822 (1980).

13

Yielding and Fracture of Toughened Networks

13.1 INTRODUCTION

In the previous chapter, the relationships between structure and mechanical and fracture properties of neat thermosets were analyzed. It was shown that an increase in thermal resistance (T_g or HDT, heat deflection temperature) and yield stress leads to a decrease in toughness (K_{Ic}), and in impact or fatigue resistance. So the challenge is how to increase toughness without sacrificing thermal and mechanical properties?

The two methods of improving the macroscopic toughness of thermosets are similar to those used for amorphous or semicrystalline thermoplastics: (i) plasticization and (ii) amplification of deformation mechanisms via the generation of a heterogeneous structure.

The plasticizer addition is a relatively simple technique. A miscible low-T_g compound is added to the formulation, so as to produce a decrease in both the glass transition temperature and the yield stress, and a corresponding improvement in the fracture resistance. These drawbacks are very severe for thermosets, and generally this method is not used for toughening purposes.

The most frequently applied methods for improving toughness are the addition of preformed particles or the in-situ formation of dispersed rubbery or thermoplastic particles in the thermoset matrix (Chapter 8).

In Secs. 13.2–13.3 the principles of toughening of thermosets by rubber particles, and the role of morphologies, interfacial adhesion, composition, and structural parameters on the toughening effect are analyzed. Section 13.4 is devoted to the use of initially miscible thermoplastics for toughening purposes. The effect of core-shell rubber particles is discussed in Sec. 13.5 and, in Sec. 13.6, miscellaneous ways of toughening thermosets (liquid crystals, hybrid composites, etc.), are analyzed.

13.2 TOUGHENING OF THERMOSETS

Epoxy networks are the most widely studied materials, due to their well-known chemistry. Consequently, many studies are devoted to epoxy networks as model networks, although the principles and models developed can be applied to other thermosets.

The principles of toughening have been described by Kinloch (1989), Mülhaupt (1990), Huang *et al.* (1993b), and McGarry (1996). The roles of particles during both the initiation and propagation of the crack may be analyzed separately.

13.2.1 Role of the Inclusions in the Initiation Step (Before the Appearance of an Intrinsic Defect or Crack)

a. Modification of the Stress Field

For a single rubber particle in an infinite uniaxial tensile stress field, it was demonstrated that there is a stress concentration effect with a factor around 2, at the particle equator (Fig. 13.1).

This is only valid for particles with a modulus lower than the matrix. On the other hand, in the case of glass beads in polymers the stress concentration occurs at the poles.

If the particle is bonded firmly to the matrix (we will discuss this point later), the initial uniaxial tension stress is changed into a triaxial tension stress field, due to the low rubber incompressibility. The stress field around rubbery particles is not the same as that around a void.

Increasing the concentration of particles (roughly for a volume fraction approaching 10%), the stress concentration effects of neighboring particles can overlap (Fig. 13.2). Therefore, a large volume fraction of the matrix supports an average load higher than the applied load and can yield. This stress concentration effect increases when the volume fraction of dispersed particles increases or the interparticle distance decreases.

As shear yielding is the main deformation mechanism of the network, it is clear that the presence of rubber particles favors the yielding of the

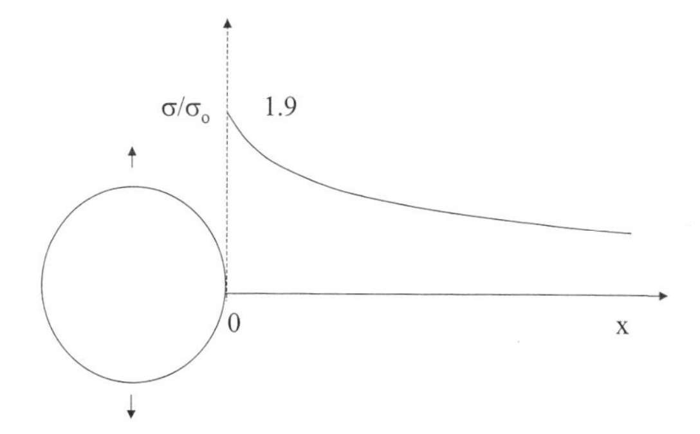

FIGURE 13.1 Stress concentration around a single rubber particle.

matrix. But also, due to incomplete phase separation (Chapter 8), a fraction of rubber remains dissolved in the matrix and contributes to the decrease of T_g and σ_y. Yielding can then occur at lower applied loads.

The introduction of rubber particles increases the fracture energy of the networks at room temperature, but also decreases the temperature of the ductile–brittle transition (Van der Sanden and Meijer, 1993). This ductile–brittle transition is strongly dependent on the nature (and T_g) of the rubber-rich phase and the amount of rubber dissolved in the matrix. The lowest ductile–brittle transition is obtained with butadiene-based copolymers ($T_g \sim -80°C$), compared with butylacrylate copolymers ($T_g \sim -40°C$).

b. Internal Cavitation and Debonding

Due to the difference of expansion coefficients between the particles and the matrix, different kinds of stress fields may be developed (Raghava, 1987). Rubbery and thermoplastic particles are placed in a hydrostatic tension

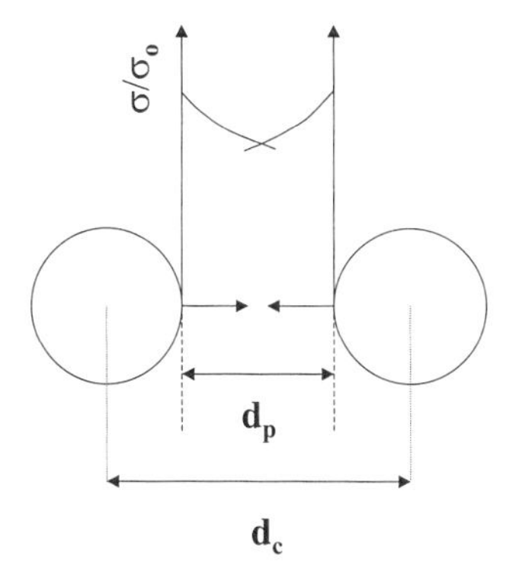

FIGURE 13.2 Stress field overlap between rubber particles.

stress field (8–12 MPa) (Sec. 13.3.2 d), while glass beads are placed in a compressive stress field (15–20 MPa). The ratio of the bulk modulus, $K = E/3(1 - 2\nu)$, over the shear modulus, G, is around 1000 for a rubber and close to 1 for a glassy polymer.

If the adhesion between particles and matrix is good, rubber particles internally cavitate when a load is applied.

If the adhesion is low, debonding at the rubber particle–matrix interface can occur. In both cases voids are formed and this reduces the degree of stress triaxiality in the surrounding matrix and favors the further growth of shear bands.

Internal cavitation was proved by comparison of the initial particle diameter with the diameter measured on a fracture surface (Huang *et al.*, 1993b). An increase of about 20–70% of the initial volume was found, depending on the temperature. This voiding process participates in the energy consumption and is the cause of the stress whitening effect observed on deformed samples.

In the case of thermoplastic particles, because the bulk modulus is equivalent to that of the matrix, no cavitation is observed.

c. Initiation of Matrix Shear Yielding

As discussed in Chapter 12, crazing does not occur in thermosets; therefore, the only possible response of the matrix to a load is to promote localized shear yielding between particles.

A considerable amount of energy is stored in the sample before the appearance of the first crack. In this step, the rubbery particles act – after cavitation or debonding – as triggers for the generation of shear bands in the matrix (Huang and Kinloch, 1992b).

Using finite element stress analysis, Huang *et al.* (1993b) demonstrated that shear bands must appear at 45°, between voids formed in a previous step. As there are many particles, a network of shear bands is generated in the deformed sample (Yee and Pearson, 1986). Their growth generates the appearance of the first crack.

13.2.2 Role of the Particles during Crack Propagation

As a result of the increase in stress and/or strain, shear bands develop in a large fraction of the sample but, at a certain point, a crack appears and starts to propagate. Several mechanisms for energy absorption, associated with the presence of particles, become active during crack propagation.

a. Crack-Bridging Mechanism

The crack-bridging mechanism is illustrated in Fig. 13.3. The particles are stretched between the edges of the propagating crack, increasing the fracture energy. This mechanism needs a good adhesion between matrix and particles. However, because of the very low modulus of rubber, and in spite of the high failure strain, the dissipated energy in such a mechanism is low: ~ 5–10% of the total energy (Kunz-Douglass *et al.*, 1980).

In the case of thermosets toughened with thermoplastics particles (Sec. 13.4), this mechanism may be of a considerable importance because of the intrinsic toughness and/or ductility of these particles.

b. Increase of the Fracture Surface

The presence of particles can modify the fracture surface from a mirror-like surface (for a brittle material), to a rough stress-whitened surface. The roughness can act as a multiplication factor for the absorbed energy.

Sometimes, steps of height (h) are created when the crack jumps over a particle. This leads to the presence of tails issuing from particles, on fracture surfaces. The fracture energy may be expressed by

$$\gamma = \gamma_{\text{matrix}}(1 - \phi_\beta) + \text{k}\frac{\text{h}}{\text{d}_\text{c}} \tag{13.1}$$

where d_c is the interparticle distance (center to center), k is a constant, and ϕ_β is the volume fraction of particles. Furthermore, the particle–matrix decohesion gives an additionnal surface and increases the fracture energy.

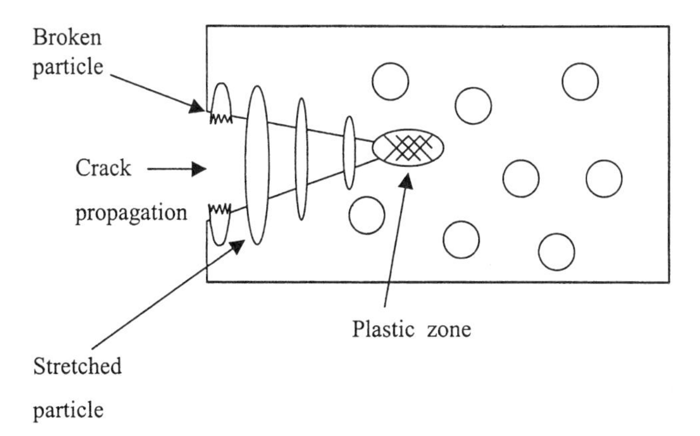

Figure 13.3 Illustration of crack-bridging mechanism.

c. Crack-Front Pinning

This mechanism, very often mentioned in metallic alloys and in filled polymers, can also be considered in the case of low-modulus particles (Fig. 13.4).

The particles, arranged in lines, act as obstacles for the crack front, in the same way as a line of trees constitutes a good protection against the wind. The crack front has to bow locally between particles in order to pass through the line of particles, slowing down the propagation rate.

Lange (1970) gives a quantitative description of the critical energy release rate supplied by this mechanism:

$$G_{Ic} = G_{Ic_{matrix}} + 2\frac{T_L}{d_p} \tag{13.2}$$

where T_L is a constant (called the line tension) and d_p is the interparticle distance (surface to surface). For particles with the same diameter, an increase in their volume fraction leads to a decrease in d_p and an increase in G_{Ic}. On fracture surfaces observed by SEM (scanning electron microscopy), the presence of a crack-pinning mechanism is revealed by features such as river markings.

The crack-pinning mechanism is not very efficient with low-modulus particles such as rubbers. But with stiff thermoplastics (Sec. 13.4), or with high-modulus particles such as inorganic fillers, this mechanism may have an important contribution.

d. Crack Blunting

During crack propagation, macromolecular chains in the vicinity of the crack tip are stretched and broken. The initially sharp crack becomes

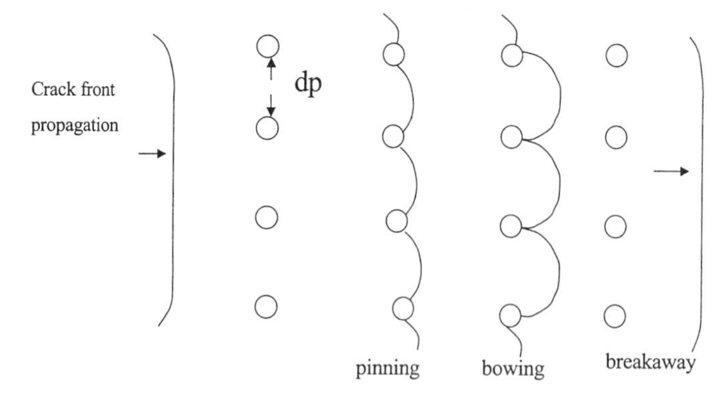

FIGURE 13.4 Scheme of the crack-pinning mechanism: d_p, interparticle distance.

more and more blunted as a result of the formation of a plastic zone and the decohesion of particles. The stress concentration effect at the crack tip becomes lower, and the crack is slowed down and even stopped, for the case of stick-slip propagation (Fig. 12.4).

e. Crack Deflection

During a fracture-mechanical test performed in mode I, the crack propagates in this mode from a macroscopic point of view. But the crack can be deflected locally by the rubbery particles and can also propagate in mode II. As for isotropic materials, G_{IIc} is generally higher than G_{Ic}; an artificial increase of the macroscopic G_{Ic} value will be then evidenced.

f. Conclusion

An improvement in the toughness of thermosets can be favored by rubber or thermoplastic particles, which operate both in crack initiation and propagation mechanisms. The different toughening mechanisms can act simultaneously and can be modeled quantitatively.

13.3 RUBBER TOUGHENING OF THERMOSETS

13.3.1 Fracture Modeling of Rubber-Modified Epoxy Networks

The fracture modeling of rubber-modified thermosets was developed by Huang and Kinloch (1992a), Kinloch and Guild (1996), Huang *et al.* (1993b), and Yee *et al.* (2000).

Huang *et al.* (1993b) proposed a two-dimensional plane strain model, which was successfully used to identify the stress field around the rubbery particles and to simulate the initiation and growth of shear bands between rubbery particles. A model was proposed to quantify the different mechanisms. G_{Ic} of the rubber-modified network was written as

$$G_{Ic} = G_{Icn} + \psi \tag{13.3}$$

where G_{Icn} is the fracture energy for the neat network and ψ is the additional energy dissipated per unit area due to the presence of rubber particles. It is given by

$$\psi = \Delta G_r + \Delta G_s + \Delta G_v \tag{13.4}$$

where ΔG_r is the contribution of particle bridging, ΔG_s is the contribution from plastic shear banding, and ΔG_v is due to plastic-void growth in the matrix. The other possible contributions were not taken into account.

According to Kunz-Douglass *et al.* (1980), ΔG_r may be calculated as

$$\Delta G_r = 4\Gamma_t \phi_\beta$$

ΔG_r is proportional to the tearing energy (Γ_t) multiplied by the rubber volume fraction ϕ_β.

ΔG_s may be expressed as proportional to ϕ_β, σ_{yc} (compressive yield stress), γ_f (fracture strain), the plastic zone size, and the square of the concentration factor, K_m^2. The influence of hydrostatic pressure was taken into account with a modified von Mises criterion (Chapter 12).

ΔG_v is proportional to the void growth and plasticity parameters of the matrix.

The model was successfully applied to rubber-modified epoxy networks, taking into account both the test temperature and the rate effect (Fig. 13.5).

Furthermore, the model makes it possible to separate the contributions of the three toughening mechanisms as a function of temperature (Fig. 13.6). At high temperatures, the crack-bridging mechanism plays a minor role; the void-growth mechanism is very sensitive to temperature and can be completely suppressed at low temperatures. Shear yielding is the main mechanism, except at very high test temperatures where cavitation plays the major role. The contribution of shear yielding depends on the difference between the test temperature and T_g, as discussed in Chapter 12.

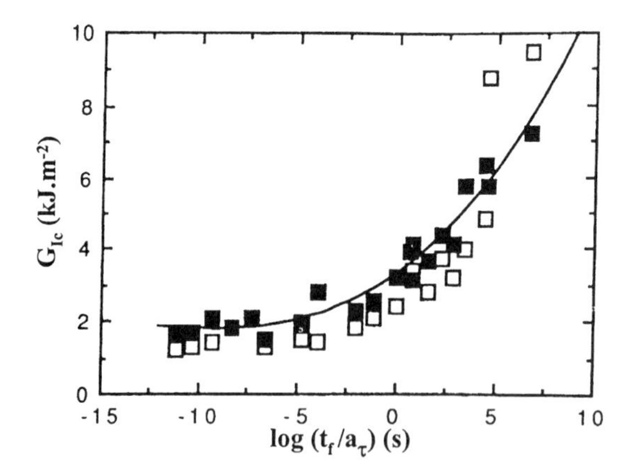

FIGURE 13.5 Comparison between theoretical predictions of Kinloch's model (□) and the experimental results (■) at different test rates and temperatures of a rubber-toughened epoxy. (Huang and Kinloch, 1992a, with kind permission from Kluwer Academic Publisher.)

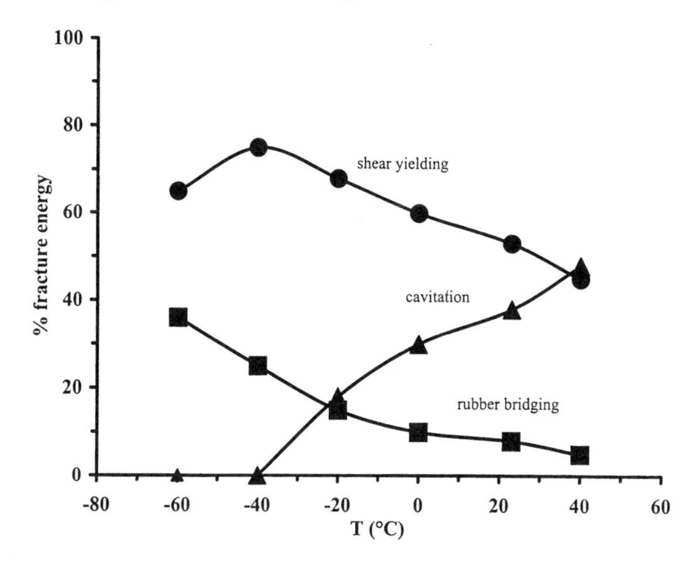

FIGURE 13.6 Relative contributions (%) of the different toughening mechanisms in epoxy networks versus temperature: (■) rubber bridging; (●) shear yielding; and (▲) cavitation. (From the results of Huang *et al.*, 1993b.)

13.3.2 Influence of Network Structure and Morphology on Fracture Properties of Epoxy Networks

As discussed in the previous section, the toughening effect depends both on the matrix, where the shear bands are propagating, and the rubbery phase, which induces cavitation and crack bridging.

In this section, the influence of in-situ formed rubber particles is discussed, while the influence of preformed particles is analyzed in Sec. 13.5.

For epoxy networks modified by liquid reactive rubbers, it is not so easy to discuss these parameters separately, because they are interdependent. For example, an increase in the acrylonitrile content of the carboxy-terminated butadiene acrylonitrile rubber (CTBN) induces a size reduction of the rubbery domains but also a higher miscibility with the epoxy-rich phase, leading to a higher amount remaining dissolved in the matrix at the end of cure (Chapter 8). It is not possible to separate the influence of these two effects on toughness.

a. Influence of the Volume Fraction of Rubber

In a first approximation, G_{Ic} depends linearly on the amount of dispersed phase up to a 20–25% volume fraction (Yee and Pearson, 1986; Verchère *et al.*, 1993; McGarry, 1996).

The use of an in-situ phase-separated rubber produces a decrease in both the Young's modulus and the yield stress (Fig. 13.7). Therefore, high rubber volume fractions cannot be used for structural applications (high stresses, long-time creep, etc.). A stiffness–toughness compromise has to be considered. But, in any case, the initial volume fraction of rubber cannot be higher than ϕ_{crit}, to avoid phase inversion, leading to a rubbery matrix with thermoset inclusions (Chapter 8). Another limiting factor may be the cost of the liquid reactive rubbers.

b. Influence of Particle Size and Particle Size Distribution

It is generally observed that rubber particles are effective for toughening purposes when their sizes are in the 0.1–10 μm diameter range. For a given volume fraction of the rubbery phase, there is a critical particle size below which toughening occurs and above which there is no significant effect. The critical size is related to the interparticle distance, which, in fact, is the main parameter affecting the toughening effect for both thermoplastics (Wu, 1985) and thermosets (Van der Sanden and Meijer, 1993). To keep the

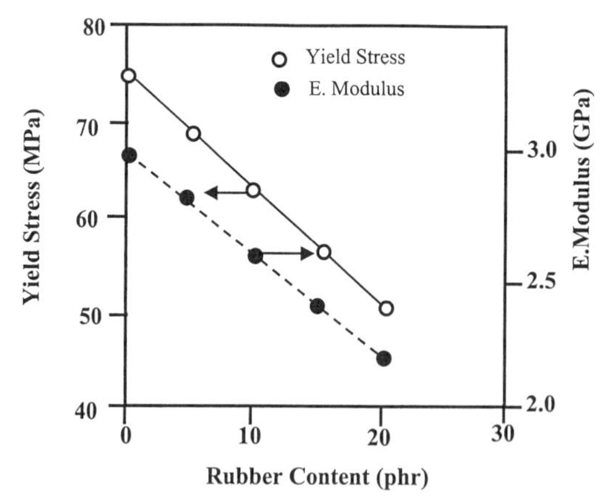

Figure 13.7 Variation of yield stress (σ_y) and Young's modulus (E) for rubber-modified epoxy networks. Rubber = CRBN: carboxy-terminated butadiene acrylonitrile random copolymer. (Reprinted with permission from Pearson, 1993, Copyright 2001. American Chemical Society.)

critical interparticle distance constant, the critical particle size must increase with the volume fraction of dispersed phase.

The effect produced by a rubbery phase is based on the following mechanism:

$$\text{Mechanical loading} \rightarrow \text{cavitation in rubber particles} \rightarrow \text{promotion of shear bands in matrix} \rightarrow \text{toughness improvement}$$

To initiate this mechanism, particles must produce an adequate stress concentration effect: diameters larger than 0.1–0.2 μm are effective for this purposes. Smaller particles cannot store sufficient elastic energy to induce cavitation (Lazzeri and Bucknall, 1993; Dompass and Groeninckx, 1994). The existence of a critical interparticle distance below which a significant toughening effect is observed is based on shear banding being favored by the stress-field overlap between neighboring particles.

The maximum particle size for efficient toughening is in the order of 5–10 μm (Kinloch, 1989; Pearson and Yee, 1991). It has been proved experimentally that larger particles are relatively inefficient (Pearson and Yee, 1991), although they are expected to be active in crack bridging; however, the contribution of this mechanism to rubber toughening is less than 10% of the total fracture energy at room temperature (see Fig. 13.6).

It has been assessed that bimodal particle size distributions consisting of a population of small and large particles may exhibit a better toughening effect than unimodal ones, due to a "synergistic effect" (Chen and Yan, 1992). But this does not seem to be a general trend (Pearson and Yee, 1991; Grillet *et al.*, 1992).

c. Influence of the Matrix T_g

The effect of the matrix T_g on toughness has been analyzed extensively (Levita, 1989; Pearson and Yee, 1989; Van der Sanden and Meijer, 1993). Although the observed effect is usually ascribed to changes in the crosslink density, it is better to regard it as being produced by a change in T_g with respect to the test temperature, T.

Figure 13.8 shows that the toughening produced by a dispersed rubbery phase increases with an increase in the molar mass of the DGEBA (diglycidyl ether of bisphenol A) monomer and a corresponding decrease of the matrix, T_g. As G_{Ic} was measured at room temperature, the main observed effect is the decrease in the shear yield stress of the matrix by lowering its glass transition temperature (a decrease of $T_g - T$ at constant T). This favors shear yielding of the matrix and increases the toughening effect. The same trend is observed for different rubber volume fractions and also under impact conditions (Van der Sanden and Meijer, 1993).

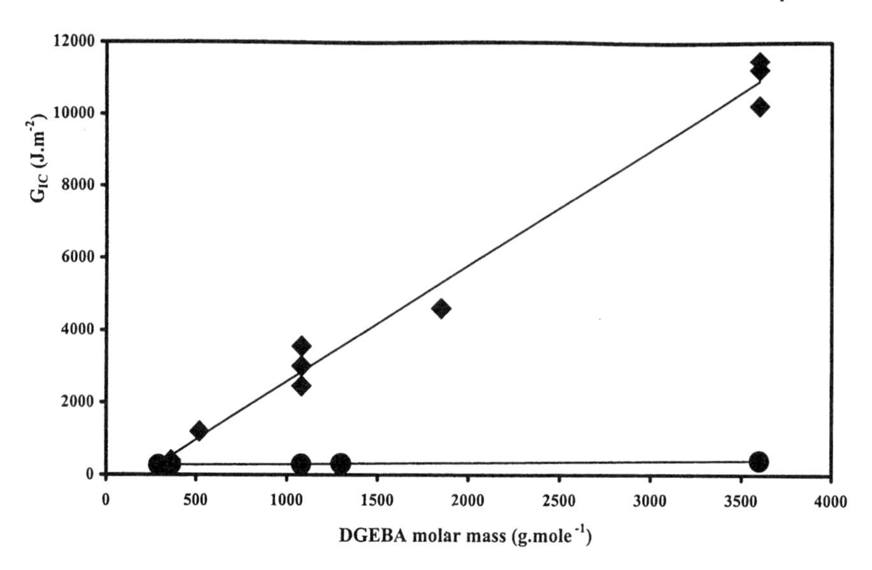

FIGURE 13.8 Fracture energy of epoxy networks cured with DDS (diamino dipheny sulfone) versus the initial DGEBA (diglycidyl ether of bisphenol A): (●) neat systems; (◆) with 10% CTBN (27% AN). Rubber = CTBN: carboxy-terminated butadiene acrylonitrile random copolymer. (Pearson and Yee, 1989 with kind permission from Kluwer Academic Publisher.)

Results presented in Fig. 13.8 could have been interpreted as an effect of crosslink density on toughening. But this is an incorrect concept, because crosslink density can be increased by the use of a low-molar-mass aliphatic diepoxide. This would decrease the matrix T_g and increase its toughenability, in spite of the increase in crosslink density. But also, it may be stated that at the same $T_g - T$, other factors related to the chemical structure, such as sub-T_g relaxations, will play a role on toughening mechanisms.

d. Influence of the Interfacial Adhesion

A threshold of interfacial adhesion between both phases is needed to (a) promote the cavitation mechanism and (b) activate the crack-bridging mechanism. For rubbery particles, the former contributes much more than the latter to the total fracture energy. Adhesion is achieved by the use of functionalized rubbers that become covalently bonded to the matrix. Higher toughness values have been reported by the use of functionalized rubbers (Kinloch, 1989; Huang *et al.*, 1993b). However, these experimental results also reflect the effect of other changes (particle size distribution, amount of dissolved rubber, etc.), produced by introducing functional groups in the rubber (Chapter 8).

A threshold level of interfacial adhesion is also necessary to produce a triaxial tensile state around rubber particles as the result of the cure process. When the two-phase material is cooled from the cure temperature to room temperature, internal stresses around particles are generated due to the difference of thermal expansion coefficients of both phases. If particles cannot debond from the matrix, this stress field magnifies the effect produced upon mechanical loading.

Under these conditions, the hydrostatic pressure, p, around a rubber particle is given by (Raghava, 1987)

$$p = \frac{(\alpha_\alpha - \alpha_\beta)\Delta T}{\frac{1+\nu_\alpha}{2E_\alpha} + \frac{1-2\nu_\beta}{E_\beta}} \tag{13.5}$$

where α is the volumetric thermal expansion coefficient, ν is the Poisson's ratio, E is the Young's modulus, ΔT is the difference between the cure temperature and room temperature, the subscript α represents the matrix, and the subscript β the particle. The thermal expansion coefficient of the rubber is around 600 ppm K^{-1}; that of an epoxy matrix is about 4 times lower. Therefore, rubbery particles are in a triaxial tensile state ($\sim$ 10 MPa), at room temperature.

e. Influence of Rubber Nature: Hyperbranched Polymers (HBP)

Polyester hyperbranched polymers (HBP) with different degrees of branching and functionality can also be used for toughening purposes. They exhibit a compact structure with a high functionality and a low viscosity and can be chemically modified in order to change the initial solubility and the interfacial adhesion. They phase-separate during cure, generating particles in the micrometer range. Particle cavitation is clearly present as a toughening mechanism.

A six-fold increase in G_{Ic} values was obtained for a thermoset based on DGEBF (diglycidyl ether of bisphenol F) containing 5 wt% of an epoxidized hyperbranched polymer (Boogh *et al.*, 1999). At this small concentration, the decrease of the Young's modulus and the glass transition temperature were not so significant. It was suggested that the high toughening capacity of HBP modifiers is induced by the generation of a gradient of properties within the phase-separated particles.

But other studies showed that increasing the amount of HBP led to a decrease in both the Young's modulus and the glass transition temperature;

moreover, a linear polyester of the same chemical structure as the HPB produced a similar toughening effect (Gopala *et al.*, 1999; Wu *et al.*, 1999). Particles with lower sizes than 1 μm gave no increase in toughness, while particles of 2–3 μm gave nearly the same fracture toughness irrespective of molar mass or thermoplastic architecture.

Regarding the use of hyperbranched polymers for toughening purposes, it may be concluded that

1. The hyperbranched architecture does not always afford an advantage in toughness or viscosity compared with low-molar-mass thermoplastic modifiers.
2. The nature of the functional groups of the HBP is a significant factor for the control of viscosity, miscibility with the thermoset precursors, phase separation during cure, and particle–matrix adhesion.

f. Conclusions

Regarding the toughening of epoxy networks, it may be stated that

- Fracture energy is roughly proportional to rubber volume fraction up to phase inversion.
- The optimum particle size lies in the range 0.1–10 μm and depends on the interparticle distance that must be lower than a critical value. The claim about the better effect achieved by the generation of a bimodal distributions of particle sizes has yet to be proved.
- Interfacial adhesion between particles and matrix is necessary and can be achieved using reactive rubbers.
- The toughening effect at room temperature increases with a decrease of the matrix T_g.

13.3.3 Rubber Toughening of Vinyl Ester (VE) and Unsaturated Polyesters (UP)

Different types of modifiers are used in UP formulations to increase toughness and also to decrease shrinkage and improve the surface aspect.

Poly(vinyl acetate) (PVAc) is very often used for "low-profile" applications. At low PVAc contents, the continuous matrix is a polyester network with PVAc inclusions. Increasing the PVAc amount leads first to a bicontinuous structure, and then to a phase-inverted system (Chapter 8). The low-profile action is observed in the concentration range where bicontinuous structures are formed (Pascault and Williams, 2000). However, the fracture energy attains a maximum value for lower PVAc concentrations (Bucknall *et al.*, 1991).

Liquid reactive rubbers were also used for UP and vinyl ester formulations (Suspene *et al.*, 1993; Siebert *et al.*, 1996). Increases in fracture energy and fatigue- crack resistance were reported for some systems, although no significant improvements were observed for some other systems. These different behaviors are probably related to the heterogeneous structure of the matrix (Chapter 7). Toughening mechanisms in three-phase systems are not yet well established.

13.4 THERMOPLASTIC TOUGHENING OF THERMOSETS

13.4.1 Generalities

The major limitation of rubber toughening of thermosets results from the fact that the increase in toughness can be achieved only at the expense of high-temperature performance or of mechanical properties, e.g., a decrease in modulus and yield stress. This can be unacceptable for structural and long-term applications (see Fig. 13.7). A second limitation is the lack of significant success in the toughening of high-T_g networks (see Fig. 13.8).

Following the requests to increase toughness by keeping a high T_g, for several applications (the aerospace industry in particular), high-T_g or semicrystalline thermoplastics (TP) can be used instead of rubbers to modify thermosetting polymers (Hedrick *et al.*, 1985; Pearson, 1993; Hodgkin *et al.*, 1998; Pascault and Williams, 2000).

The thermoplastic-rich phase may be separated in the course of polymerization (Sec. 13.4.2) or can be incorporated as a dispersed powder in the initial formulation (Sec. 13.4.3). A strong drawback of the in situ-phase separation for processing purposes is the high viscosity of the initial solution which results from the much higher average molar mass of the TP compared with the liquid rubbers. Also, for the same reason, the critical concentration ϕ_{crit} has a smaller value (phase inversion is observed at smaller concentrations of modifier).

Different TPs have been used to modify thermosets, such as poly(ether sulfone) (PES), polysulfone (PSF), poly(ether ketone) (PEK), polyether imide (PEI), poly(phenylene oxide) (PPO), linear polyimides, polyhydantoin, etc. (Stenzenberger *et al.*, 1988; Pascal *et al.*, 1990, 1995; Pascault and Williams, 2000).

The major trend observed is a modest increase in K_{Ic} or G_{Ic} by the introduction of initially miscible thermoplastics. This improvement is obtained without any loss in stiffness and thermal properties. Some very high improvements in K_{Ic}, claimed by some authors, are due to phase inversion, leading to a thermoplastic matrix with thermoset particles.

Specific quantitative models have been developed for the TP toughening of thermosets (Yee *et al.*, 2000).

13.4.2 Influence of Structural Parameters on the Toughening of Thermosets by Initially Miscible TPs

a. Volume Fraction

As in the case of rubber particles it was demonstrated that the fracture energy is roughly proportional to the volume fraction of TP rich phase, both in the case of epoxy/PEI networks (Bucknall and Gilbert, 1989) and bismaleimide networks toughened with various TPs (Stenzenberger *et al.*, 1988). This improvement was evidenced in the range of volume fractions below ϕ_{crit}.

b. Morphology of the Dispersed Phase

In the range where bicontinuous morphologies are obtained, K_{Ic} may attain a maximum local value (but it then increases again for phase-inverted structures). The presence of this maximum seems to be related to the formation of an incipient phase-inverted structure that exhibits a lack of adhesion between both phases. In these cases, bicontinuous or double-phase morphologies lead to a better fracture resistance (Girard-Reydet *et al.*, 1997).

The effect of TP particle size is controversial. While crack pinning is favored by small particles, large particles can promote particle bridging.

c. Nature and Molar Mass of the TP

The chemical nature of the TP affects its miscibility with the thermoset precursors and, consequently, the phase-separation process. On occasions, it can also modify the cure kinetics.

A change in the molar mass of the thermoplastic will affect the location of the critical concentration in the phase diagram (ϕ_{crit} will shift to lower values as the average molar mass of the TP increases, Chapter 8). At a constant TP volume fraction, an increase in its molar mass may bring the material to the phase-inverted region, an effect that will produce a significant increase in the fracture resistance. Trends reported in the literature regarding the effect of the molar mass of the TP on fracture resistance (Hedrick *et al.*, 1985) should be re-analyzed using the corresponding phase diagrams.

Apart from the modification of the phase diagram, an increase in the average molar mass of the TP will increase the viscosity of the initial solution (an undesired effect), but will increase the energy dissipated in the crack-bridging mechanism (a desired effect).

d. Interfacial Adhesion and Interface Modifications

From a mechanical point of view, a critical level of adhesion (chemical or physical) is required between TP and thermoset TS phases to insure stress and strain transfers and to favor ductile stretching of TP particles in the crack-bridging mechanism (Pearson, 1993; Hodgkin *et al.*, 1998).

For some thermoplastics such as PEI, a toughening effect is observed even in the absence of reactive end groups (no covalent bonds are formed with the matrix). Possibly, physical interactions are strong enough to produce an adequate adhesion level for toughening purposes.

Another approach for improving adhesion is to introduce a small amount of a third component that may simultaneously be dissolved in the TP and form chemical bonds with the thermoset. One example is the addition of CTBN in a PSF–DGEBA–DDS system, where CTBN is preferentially dissolved in the PSF phase but can react with the epoxy groups, promoting adhesion between both phases (Pascault and Williams, 2000).

The use of block copolymers, with each one of the blocks exhibiting miscibility with each one of the phases increases the adhesion level but also promotes dramatic changes in the final morphologies (Girard-Reydet *et al.*, 1999).

When TP-modified thermosets are used as matrix for composites, the resulting morphologies can be strongly affected by the presence of glass or carbon fibers (Varley and Hodgkin, 1997).

13.4.3 Toughening by TP Powders

The use of preformed nonmiscible TP powders enables the TP load to be increased in the final material while keeping a continuous thermoset matrix. The final morphology is more easily controlled than in the in-situ phase-separation process.

High-T_g thermosets (epoxies, cyanate esters, etc.), may be toughened by adding nonmiscible high-T_g amorphous or semicrystalline TP powders, such as PA6 or 12 (Zeng *et al.*, 1993; Cardwell and Yee, 1994; Girodet *et al.*, 1996), polyimides (Pascal *et al.*, 1994), or PBT Poly (butyleneterephtalate) (Kim and Roberston, 1992). To avoid (partial) miscibility, the cure has to be performed below the melting temperature of the semicrystalline modifier. The dissolution of the TP leads to an improved toughness but a lower T_g (Lennon *et al.*, 2000b).

The TP particles can be ground from bulk samples to a size in the range of 10–30 μm. TP particles based on PA6, PA12, or PA6–12 and with a very narrow size distribution can be synthesized directly by anionic polymerization of the corresponding lactam (Chapter 8).

The adhesion between phases and the dispersion state seem to play a very important role. To improve wetting behavior, dispersion, and adhesion by grafting reactions, a plasma treatment ($O_2/N_2/NH_3$) of the TP powders can be performed in a fluidized bed (Lennon *et al.*, 2000a).

TP powders can be also be used for mode II-interlaminar fracture improvement of laminates (Zeng *et al.*, 1993). TP particles spread between layers largely improve the plastic deformation. G_{IIc} varies linearly with the TP volume fraction (Zeng *et al.*, 1993).

13.4.4 Conclusions

Regarding the toughening of thermosets by the addition of a TP as a modifier, it may be stated:

1. TPs may provide some toughening for high-T_g networks without loss in thermomechanical properties; the effect is modest below phase inversion and may increase significantly when bicontinuous morphologies are formed.
2. Crack bridging by TP particles seems to be the main energy-dissipating mechanism.
3. For the in-situ phase-separation process, an increase in the TP molar mass modifies the phase diagram (it lowers ϕ_{crit}), increases the viscosity – which is a drawback for processing and impregnation of composites – but increases the fracture energy associated with the crack-bridging mechanism.
4. The introduction of preformed TP particles enables us to increase the volume fraction of dispersed TP particles with a corresponding toughness increase.

13.5 TOUGHENING BY CORE-SHELL RUBBER (CSR) PARTICLES

Core-shell rubber (CSR) particles are prepared by emulsion polymerization, and typically exhibit two or more alternating rubbery and glassy spherical layers (Lovell 1996; Chapter 8). These core-shell particles are widely used in thermoplastics, especially in acrylic materials (Lovell, 1996), and have also been used to modify thermosets, such as epoxies, cyanates, vinyl ester resins, etc. (Bécu *et al.*, 1995).

13.5.1 Design of the Core-Shell Rubber Particles

The emulsion free-radical polymerization carried out in different steps ensures a precise control of the particle size and particle size distribution. The particle diameter can be adjusted between 100 nm and 1000 nm, with a low polydispersity (generally less than 1.1) (Chapter 8). Rubber particles with sizes lower than 100 nm are ineffective for toughening purposes (Sec. 13.3.2b).

The usual requirement is a rubbery core and an outer shell of a glassy polymer. The rubbery core is generally based on polybutadiene or butylacrylate or copolymers, the choice affecting strongly the ductile–brittle transition temperature.

The crosslinking of the rubber is important to maintain the structural integrity during the mixing and curing processes. In order to favor internal cavitation of the rubbery core and a good stress transfer, grafting of the inner shell with the rubbery core is necessary.

The outer glassy shell is designed to prevent coalescence of rubbery particles during synthesis and drying, and to insure a good interface with the matrix. The shell is usually based on styrene/acrylonitrile or styrene/methyl methacrylate copolymers. Depending on the acrylonitrile or methyl methacrylate content, the compatibility between the shell and the matrix can be varied over a wide range.

The glassy shell does not need to be crosslinked, since it is grafted to the crosslinked rubbery core. Sometimes the shell can be crosslinked and some reactive functions (such as glycidyl methyl methacrylate) may be grafted onto the surface at the end of the polymerization step.

The main problem arising with core-shell toughening is obtaining a good state of dispersion. This can be favored by chemical grafting on the surface, or the use of emulsifiers or solvents. A good dispersion increases the particle surface in contact with the matrix, producing a considerable increase in viscosity (Maazouz *et al.*, 1994). This may be a drawback for the use of core-shell particles. Twin-screw extruders or ultra high-speed stirrers were used with success to produce a good dispersion of the CSR particles. Moreover, the use of very small particles and the possibility of matching their refractive index offers the possibility of having transparent, toughened materials, which is not generally possible with the use of rubbers or TP particles.

13.5.2 Specific Toughening Mechanisms with CSR Particles

Sue (1992, 1996a,b) studied the fracture behavior of high-performance epoxies and thermosets toughened with acrylic core-shell particles. Using the technique of four-point bending test with two notches, together with microscopy observations, he found that a specific mechanism called croiding (contraction of crazing and voiding), took place. This is a craze-like damage, which is also denoted as dilatation bands. Although this mechanism was described for systems modified with CSR particles, there is no reason why it should not be present in systems modified with rubber particles.

The sequence of events arising during croid formation is summarized in Fig. 13.9. At the crack tip, due to a high triaxial stress field, internal

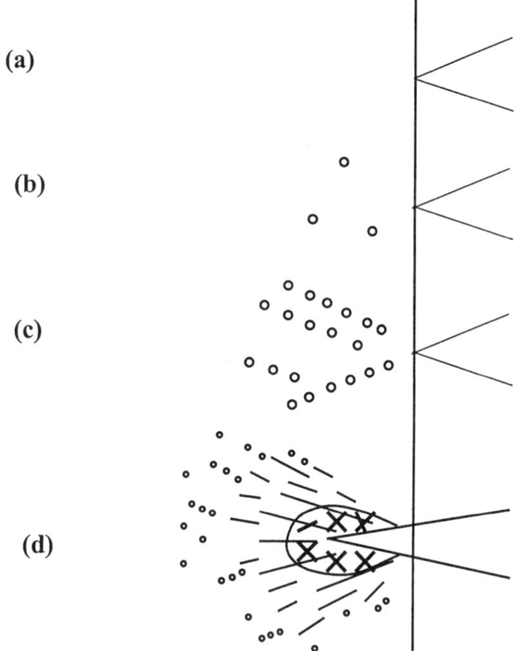

FIGURE 13.9 Sequence of events in a croid formation. (a) Initial state at the crack tip. (b) Cavitation of the rubber particles due to loading head of the crack tip. (c) Cavitation of rubber particles near the already cavitated particles due to stress–concentration effect. The croid is forming. (d) Croids are propagating ahead of the crack and inside the craze-like damaged zone; many shear bands develop between cavitated rubber particles. (Sue, 1992 with kind permission from Kluwer Academic Publisher.)

cavitation of the rubbery cores takes place. This leads to the cavitation of neighboring CSR particles and formation of lines and arrays, with a characteristic shape in transmission electron micrographs (TEM) (Sue, 1992). These arrays are roughly perpendicular to the normal stress direction, like the crazes in thermoplastics.

As in rubber toughening, the cavitated particles induce shear yielding of the matrix, producing a plastic zone smaller than the cavitated zone. These mechanisms induce a significant toughening effect. Should matrix yielding precede the cavitation of CSR particles, the croiding mechanism would be suppressed.

13.5.3 Influence of CSR Structural Parameters

As a result of their highly versatile chemical structures and morphologies, thermosetting polymers toughened with CSR particles allow us to obtain a very large improvement in fracture resistance. The addition of 24% of a commercial CSR to a brittle epoxy matrix ($G_{Ic} \sim 200\,J\,m^{-2}$) enables us to increase G_{Ic} to $1130\,J\,m^{-2}$ (Bécu *et al.*, 1997); but the use of especially designed CSR particles has led to a G_{Ic} of $1400\,J\,m^{-2}$ (Bécu-Longuet *et al.*, 1999). The toughening of epoxy networks seems to be at least as good as that observed using reactive liquid rubbers, compared at the same volume fractions, but without any effect on the matrix glass transition temperature (Qian *et al.*, 1995).

Core-shell particles have been also introduced in other thermosets with significant success. In vinyl esters (VE), the fracture toughness was increased from $0.9\,MPa\sqrt{m}$ for the pure VE up to 1.5 or $2.5\,MPa\sqrt{m}$ using less than 10% CSR particles. The best toughening effect was obtained using CSR particles with a polybutadiene core (Roberts *et al.*, 1998); not only static toughness but also impact strength and fatigue resistance are improved by using CSR particles (Bécu *et al.*, 1997).

The main conclusions about the influence of CSR structural parameters are

1. A convenient particle size is required for cavitation (Dompass and Groeninckx, 1994). In an epoxy matrix cured with dicyandiamine, the maximum toughness was obtained using CSR particles with a diameter of 450 nm, which corresponds to an optimum interparticle distance of 500 nm (Bécu-Longuet *et al.*, 1999).
2. The effect of covalently bonding the shell surface to the matrix was studied (Sue *et al.*, 1996a). It is possible to introduce glycidyl methacrylate (GMA) in the last step of the CSR synthesis. GMA

contains epoxy functions, which can react with an epoxy matrix and so improve the interfacial adhesion. The GMA introduction can also favor the dispersion of CSR particles in the reactive mixture. However, the toughening effect was shown to be practically independent of the GMA content (up to 30%). This implies that the chemical bonding between the shell and the epoxy matrix is not a critical parameter for toughening purposes. Physical interactions via interdiffusion or partial solubility seem to be sufficient to reach an adequate adhesion strength, higher than the cavitational strength of the CSR particles.

3. The influence of the shell thickness has also been reported (Sue *et al.*, 1996a). By varying the shell thickness from 15% to 35% of the total thickness, the dispersion of CSR particles changed from a random to a well-dispersed but locally clustered distribution. The latter produced a higher toughening effect, probably due to the promotion of a crack-deflection mechanism, in addition to the other toughening mechanisms.

13.5.4 Influence of Matrix T_g on Toughness and Toughening Mechanisms

For a series of epoxy phenolic novolac prepolymers cured with bisphenol A, imidazole, or DDS and toughened with 10 per CSR particles, the crosslink density was varied over a wide range of M_e from 200 g mol^{-1} up to 2000 g mol^{-1} (Lu, 1995). In this series an increase in M_e is related to a decrease in

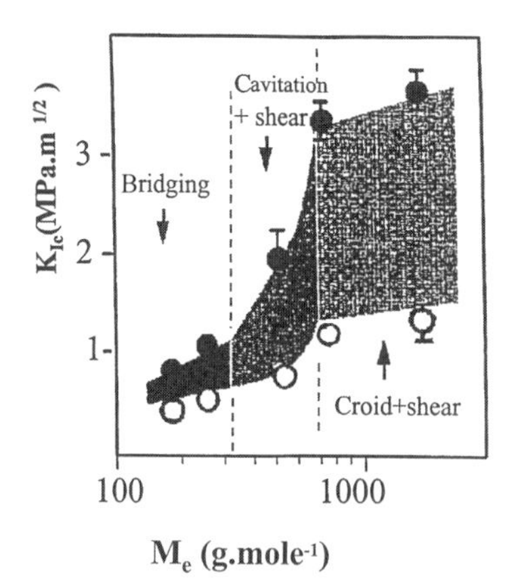

FIGURE 13.10 Toughness (K_{Ic}) and toughening mechanisms of epoxy networks modified with 10 phr CSR versus molar mass between crosslinks, M_e: (○) neat systems; (●) with 10 phr CSR. (Adapted from Lu, 1995.)

T_g and a corresponding decrease in the yield stress of the matrix. As shown in Fig. 13.10, K_{Ic} increased from about $0.5\,MPa\sqrt{m}$ up to $3\,MPa\sqrt{m}$, with an increase in M_e. This toughness enhancement was connected to changes in the toughening mechanisms from bridging for low M_e values, to cavitation plus shear yielding at intermediate M_e values, to croiding plus shear for yielding for high M_e values.

13.5.5 Toughening with Preformed Rubber Particles

It is not possible to produce a dispersion of rubber particles in the thermoset precursors due to their agglomeration. It is possible, however, to synthesize a stable emulsion or suspension of rubber particles in one of the monomers. These particles, stabilized by copolymers and surfactants, may be considered as a limiting case of CSR particles when the shell thickness tends to zero. The use of dispersed acrylic rubbers (Sue *et al.*, 1996a and Ashida *et al.*, 1999) and poly(dimethyl-siloxane) (PDMS) emulsions (Rey *et al.*, 1999), have been reported.

With 10 wt% of dispersed acrylic rubbers (average diameter $0.4\,\mu m$), the fracture energy of an epoxy network was largely improved. G_{Ic} increased from $180\,J\,m^{-2}$ for the neat matrix to $420\,J\,m^{-2}$ for the modified formulation. The addition of 10 wt% CTBN led to a value of $240\,J\,m^{-2}$ but, with the addition of 10 wt% CSR particles, increased the value to $490\,J\,m^{-2}$.

Functionalized PDMS particles were dispersed in a DGEBA prepolymer, and a curing agent (dicyandiamide) was added (Rey *et al.*, 1999). The particle size (average diameter $1\,\mu m$) remained stable for different concentrations and dispersion techniques (twin-screw extruder or high-speed stirrer). K_{Ic} showed a maximum value for 8 wt% PDMS ($1.36\,MPa\sqrt{m}$ compared with 0.93 for the pure matrix and 1.23 for a 15 wt% PDMS). An improvement was also observed in fatigue crack propagation (Fig. 13.11). The Paris law was obeyed and a clear correlation was found between K_{Ic} values and the fatigue resistance of notched samples.

The use of an emulsion of PDMS particles is attractive because its low viscosity makes it suitable for preimpregnation of fibers to produce composite materials. A possible drawback is the emulsion stability which has to be controlled in order to avoid particle coalescence.

13.5.6 Conclusions

A high degree of toughening may be attained by the use of CSR particles as modifiers of thermosetting polymers. This is because several adjustable parameters – the chemical structure and size of the core; the number, chemical structure, and thickness of shells; and the possibility of crosslinking

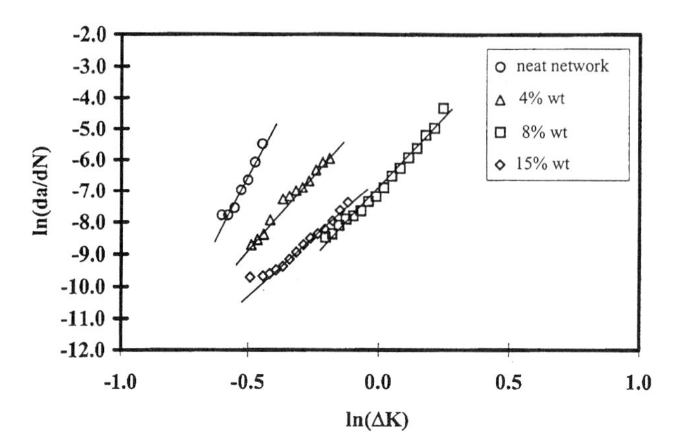

FIGURE 13.11 Paris diagrams for fatigue crack propagation of poly(dimethylsiloxane)-modified epoxy networks. (Rey *et al.*, 1999 with kind permission from Kluwer Academic Publisher.)

and grafting core and shells – can be designed during their synthesis to obtain a significant increase in fracture energy.

Twin-screw extruders or high-speed stirrers may be used to obtain good dispersions in the thermoset precursors. The main drawback is the possibility of producing agglomerates of particles during storage or processing.

13.6 MISCELLANEOUS TOUGHENING AGENTS

13.6.1 Use of Liquid-Crystalline Polymers (LCP)

Liquid-crystalline polymers (LCP) can be used in thermoset systems as initially miscible modifiers that phase-separate during cure.

Carfagna *et al.* (1992) used a block copolyester (LCP) blended with a thermoplastic amorphous polyacrylate. It was extruded and spun under various conditions to obtain filaments, which were then dissolved in the epoxy prepolymer to produce a homogeneous phase. A curing agent (a diamine) was added and the LCP phase-separated during cure to form isotropically dispersed microfibers, with a very high aspect ratio (50). The addition of only 2 wt% of LCP increased the fracture toughness by 20% (fibers acts as crack stoppers), without sacrificing T_g or modulus.

Functionalized LCP may be used as comonomers in thermoset formulations. Upon cure, liquid-crystalline thermosets (LCT) are obtained, with a morphology characterized by the presence of liquid-crystalline dispersed domains.

Examples of these formulations are systems based on a difunctional LC epoxy monomer (diglycidyl ether of 4-4′-dihydroxy-α-methylstilbene), cured with methylene dianiline (Ortiz *et al.*, 1997). The generation of liquid-crystalline microdomains (smectic or nematic) in the final material required their phase-separation before polymerization or at low conversions. This could be controlled through the initial cure temperature. Values of G_{Ic}, ($kJ\,m^{-2}$) were 0.68 (isotropic), 0.75 (nematic), and 1.62 (smectic). The large improvement produced by the smectic microdomains was attributed to an extensive plastic deformation.

Sue *et al.* (1997) reported results for the same LC epoxy monomer cured with various hardeners. K_{Ic} values could be increased up to $1.89\,MPa\,m^{1/2}$. Observations of fracture surfaces indicated that crack bridging, crack branching, and crack deflection were the main toughening mechanisms.

13.6.2 Rubber plus Thermoplastic – Toughened Networks

The use of liquid reactive rubbers to toughen thermosets leads to a decrease of both T – g and the Young's modulus (see Fig. 13.7). Thermoplastics can be added to these formulations to improve thermal and mechanical properties.

Romano *et al.* (1994) introduced both CTBN and a linear phenoxy thermoplastic in an epoxy formulation. The resulting morphologies were strongly dependent on heating rates. They included subdomains and interconnected zones. A combination of 30 phr phenoxy and 15 phr CTBN, associated with a slow cure schedule, greatly improved the peel shear strength.

Woo *et al.* (1994) studied a DGEBA/DDS system with both polysulfone and CTBN. The thermoplastic/rubber-modified epoxy showed a complex phase-in-phase morphology, with a continuous epoxy phase surrounding a discrete thermoplastic/epoxy phase domain. These discrete domains exhibited a phase-inverted morphology, consisting of a continuous thermoplastic and dispersed epoxy particles. The reactive rubber seemed to enhance the interfacial adhesive bonding between the thermoplastic and thermosetting domains. With 5 phr CTBN in addition to 20 phr polysulfone, G_{Ic} of the ternary system showed a 300% improvement ($700\,J\,m^{-2}$ compared with $230\,J\,m^{-2}$ for the neat matrix).

These studies were based on an empirical search of compositions and cure schedules directed to applications. Phase-separation processes in ternary reactive systems, including the possibility of controlling the generated morphologies and the relationships with mechanical properties, is still an open field.

13.6.3 Rubber plus Fillers – Toughened Networks (Hybrid Composites)

To balance some of the drawbacks produced by the rubber toughening of thermosets, inorganic fillers that increase modulus and yield stress can be added to generate hybrid composites. Inorganic fillers such as glass beads, alumina, or silica – with high values of modulus and strength – are frequently included in thermoset formulations.

A combination of alumina and core-shell particles led to an increase in G_{Ic} of about 16 times over the value of the neat epoxy, without any decrease in T_g (Geisler and Kelley, 1994).

Others studies dealt with liquid reactive rubbers in combination with solid glass beads (Maazouz *et al.*, 1993; Azimi *et al.*, 1995) or with hollow glass spheres (Azimi *et al.*, 1996). A synergistic toughening effect in both static fracture energy and fatigue crack propagation was evidenced. The effect was higher than the predictions arising from the rule of mixtures for each one of the modifiers. This improvement was explained by the greater localized plastic deformation ahead of the crack front and by the efficiency of the crack-pinning mechanism due to the presence of inorganic fillers. It was proposed that the overlap of the stress field due to the dispersion of glass beads promotes the cavitation of the rubbery particles in the vicinity and then favors the shear yielding mechanism (Pearson and Yee, 1991). When the glass particles were silane-treated to increase the interfacial adhesion, the fracture energy was increased even further.

A crosslinked rubber may be synthesized at the surface of the glass beads to produce a core-shell structure (glassy core and rubbery shell). Thermosets modified with these particles showed a strong toughening effect for an optimum thickness of the rubbery shell (Amdouni *et al.*, 1992).

Different ways of toughening thermosets may be employed; such as the use of rubber block copolymers (Mülhaupt and Buchholz, 1996) or introduction of microgels (crosslinked microparticles) (Funke *et al.*, 1998).

It is possible to design tailored thermosets for specific applications, by varying the amount and the nature of the inorganic fillers.

13.7 CONCLUSIONS

Thermosets can be toughened by different procedures. The aim is to increase toughness by keeping the good intrinsic properties of the matrix, such as modulus, T_g, and solvent resistance. The use of plasticizers is thus not recommended, and the main concept for toughening purposes is the introduction of a second phase (rubbery or/and thermoplastic) capable of generating energy-dissipation mechanisms that increase fracture resistance. Apart from the different mechanisms associated with the presence of a dispersed phase, particles should promote shear yielding of the matrix, which is the main deformation mechanism of highly crosslinked networks.

The use of initially miscible liquid reactive rubbers is very efficient for the toughening of bulk materials and adhesives. Deformation mechanisms involve internal cavitation of particles and a subsequent shear yielding of the matrix. The fracture energy increases linearly with the rubber volume fraction. However, the amount of rubber that may be dispersed in the thermoset without producing phase inversion is limited by the critical volume fraction. For particles smaller than 100–200 nm, no toughening effect is evidenced. Rubber-toughened thermosets exhibit a good impact strength and a low ductile–brittle transition temperature. The main drawbacks of these formulations are the difficulties in toughening high-T_g matrices, the decrease in thermal properties due to partial rubber solubility, the decrease in modulus and yield stress, and the sensitivity to oxidation (thermal or photochemical) in the case of butadiene-based rubbers.

To overcome some of these disadvantages, initially miscible thermoplastics can be used to toughen moderately high-T_g networks while maintaining good mechanical and thermal properties. Crack bridging, which is favored by the intrinsic toughness of the thermoplastics, is a significant toughening mechanism. But there remains the problem of controling morphologies arising from the in-situ phase-separation process.

The introduction of preformed particles constitutes another important possibility for toughening thermosets. TP powders or core-shell rubber particles may be used. Morphologies can be designed and controlled, and thermal properties kept invariable. Significant increases in the fracture energy may be attained. Cavitation of the rubber cores associated with croiding remains the triggering mechanism for shear yielding. The main drawback to using preformed particles is the high viscosity of the initial dispersions and the possibility of producing agglomeration of particles during processing or long-term storage.

The combination of both soft and rigid particles permits the design of tailored materials with specific properties; but, in any case, the formulations must be suitable for conventional processing.

NOTATION

d_c = center to center interparticle distance, m
d_p = surface to surface interparticle distance, m

E = Young's modulus, Pa (GPa)
G = shear modulus, Pa (GPa)
G_{Ic} = critical strain energy release rate in mode I, $J\,m^{-2}$
G_{Icn} = critical strain energy release rate in mode 1 for the neat network, $J\,m^{-2}$
G_{IIc} = critical strain energy release rate in mode II, $J\,m^{-2}$
h = height of a step between crack fronts, m
K = bulk modulus, Pa (GPa)
K_{Ic} = critical stress intensity factor in mode I, $MPa\,m^{1/2}$
K_m = stress concentration factor
M_e = average molar mass between crosslinks, $kg\,mol^{-1}$
M_n = number-average molar mass, $kg\,mol^{-1}$
p = hydrostatic pressure, Pa
T_g = glass transition temperature, K

Greek Letters

α = expansion coefficient, K^{-1}
γ = fracture energy of a composite material, $J\,m^{-2}$
γ_{matrix} = fracture energy of the matrix, $J\,m^{-2}$
γ_f = fracture strain
Γ_t = tearing energy, $J\,m^{-2}$
ΔG_r = energy contribution of particle bridging, $J\,m^{-2}$
ΔG_s = energy contribution of plastic shear banding, $J\,m^{-2}$
ΔG_v = energy contribution of voiding, $J\,m^{-2}$
ΔT = temperature variation, K
ν = Poisson's coefficient
σ = stress, Pa
σ_{yc} = compressive yield stress, MPa
ϕ_{crit} = critical volume fraction for phase inversion
ϕ_β = volume fraction of the β phase
ψ = additional energy dissipated in rubber-modified networks, $J\,m^{-2}$

Subscripts

α, β = relative to α or β phase

REFERENCES

Amdouni N, Sautereau H, Gérard JF, *J. Appl. Polym. Sci.*, 45, 1799–1810 (1992); 46, 1723–1735 (1992).
Ashida T, Katoh A, Handa K, Ochi M, *J. Appl. Polym. Sci.*, 74, 2955–2962 (1999).
Azimi HR, Pearson RA, Hertzberg RW, *J. Appl. Polym. Sci.*, 58, 449–463 (1995).
Azimi HR, Pearson RA, Hertzberg RW, *Polym. Eng. Sc.*, 36(18), 2352–2365 (1996).
Bécu L, Sautereau H, Maazouz A, Gérard JF, Pabon M, Pichot C, *Polym. Adv. Tech.*, 6, 316–325 (1995).

Bécu L, Maazouz A, Sautereau H, Gerard JF, *J. Appl. Polym. Sci.*, 65, 2419–2431 (1997).
Bécu-Longuet L, Bonnet A, Pichot C, Sautereau H, Maazouz A, *J. Appl. Polym. Sci.*, 72, 849–858 (1999).
Boogh L, Petterson B, Manson JAE, *Polymer*, 40, 2249–2261 (1999).
Bucknall CB, Gilbert AH, *Polymer*, 30, 213–221 (1989).
Bucknall CB, Partridge IK, Philips MJ, *Polymer*, 32(5,) 786–790 (1991).
Cardwell BJ, Yee AF, *Polymer Mat. Sci. Eng.*, 70, 254–255 (1994).
Carfagna C, Nicolais L, Amendola E, Carfagna C Jr, Filippov AG, *J. Appl. Polym. Sci.*, 44, 1465–1471 (1992).
Chen TK, Jan YH, *J. Mat. Sci.*, 27, 111–121 (1992).
Dompass D, Groeninckx G, *Polymer*, 35, 4743–4765 (1994).
Funke W, Okay O, Joos-Müller B, *Adv. Polym. Sci.*, 136, 139–234 (1998).
Geisler B, Kelley FN, *J. Appl. Polym. Sci.*, 54, 177–189 (1994).
Girard-Reydet E, Vicard V, Pascault JP, Sautereau H, *J. Appl. Polym. Sci.*, 65, 2433–2445 (1997).
Girard-Reydet E, Sautereau H, Pascault JP, *Polymer*, 40, 1677–1687 (1999).
Girodet C, Espuche E, Sautereau H, Chabert B, Ganga R, Valot E, *J. Mat. Sci.*, 31, 2997–3002 (1996).
Gopala A, Wu H, Xu J, Heiden PA, *J. Appl. Polym. Sci.*, 71, 1809–1817 (1999).
Grillet AC, Galy J, Pascault JP, *Polymer,* 33(1), 34–43 (1992).
Hedrick JL, Yilgör I, Wilkes GL, McGrath JE, *Polym. Bull.*, 13, 201–208 (1985).
Hodgkin JH, Simon GP, Varley RJ, *Polym. Adv. Tech.*, 9, 3–10 (1998).
Huang Y, Kinloch AJ, *J. Mat. Sci.*, 27, 2753–2762; 2763–2769 (1992a).
Huang Y, Kinloch AJ, *Polymer*, 33(6), 1330–1332 (1992b).
Huang Y, Kinloch AJ, Bertsch RJ, Siebert AR, In *Toughened Plastics I – Science and Engineering*, Riew CK, Kinloch AJ (ed.), *Adv. Chem. Ser.* No. 233, ACS, Washington, DC, 1993a, pp. 189–210.
Huang Y, Hunston DL, Kinloch AJ, Riew CK, In *Toughened Plastics I – Science and Engineering*, Riew CK, Kinloch AJ (ed.), *Adv. Chem. Ser.* No. 233, ACS, Washington, DC, 1993b, pp. 1–35.
Kim J, Roberston RE, *J. Mat. Sci.*, 27, 3000–3009 (1992).
Kinloch A,. In *Rubber-Toughened Plastics*, Riew CK (ed.), *Adv. Chem. Ser.*, No. 222, ACS, Washington, DC, 1989, pp 67–91.
Kinloch AJ, Guild FJ, In *toughened Plastics II – Novel Approaches in Science and Engineering*, Riew CK, Kinloch AJ (ed.), *Adv. Chem. Ser.*, No. 252, ACS, Washington, DC, 1996, pp. 1–32.
Kunz-Dunglass S, Beaumont PWR, Ashby MF, *J. Mat. Sci.*, 15, 1109–1123 (1980).
Lange FF, *Phil. Mag.* 22, 983–989 (1970).
Lazzeri A, Bucknall CB, *J. Mat. Sci.*, 28, 6799–6808 (1993).
Lennon P, Espuche E, Sage D, Gauthier H, Sautereau H, Valot E, *J. Mat. Sci.*, 35, 49–55 (2000a).
Lennon P, Espuche E, Sautereau H, Valot E, *J. Appl. Polym. Sci.*, 77, 857–865 (2000b).

Levita G. In *Rubber-Toughened Plastics*, Riew CK (ed.), *Adv. Chem. Ser*, No. 222, ACS, Washington, DC, 1989, pp. 94–118.

Lovell PA, *Trip*, 4(8), 264–272 (1996).

Lu AF, PhD thesis No. 1391, EPFL, Lausanne, Switzerland, 1995.

Maazouz A, Sautereau H, Gérard JF, *J. Appl. Polym. Sci.*, 50, 615–626 (1993).

Maazouz A, Sautereau H, Gérard JF, *Polym. Bull.*, 33, 67–74 (1994).

McGarry FJ, In *Polymer Toughening*, Arends CB (ed.), Marcel Dekker, New York, 1996, pp. 175–188.

Mülhaupt R, *Chimia*, 44(3), 43–52 (1990).

Mülhaupt R, Buchholz U, In *Toughened Plastics II – Novel approaches in Science and Engineering*, Riew CK, Kinloch AJ (ed.), *Adv. Chem. Ser.*, No. 252, Washington, DC, 1996, pp. 75–94.

Ortiz C, Kim R, Rodeghiero E, Kramer EJ, Ober CK, *Proceedings: Deformation Yield and Fracture of Polymers*, Cambridge, 1997, pp.125–128.

Pascal T, Mercier R, Sillion B, *Polymer*, 31, 78–83 (1990).

Pascal T, Bonneau JL, Biolley N, Mercier R, Sillion B, *Polym. Adv. Tech.*, 6, 219–229 (1994).

Pascault JP, Williams RJJ, In *Polymer Blends: Volume I: Formulation and characterization of Thermoset-Thermoplastic Blends*, Paul DR, Bucknall CB (eds), John Wiley & Sons, New York, 379–415 (2000).

Pearson RA, In *Toughened Plastics I: Science and Engineering*, Riew CK, Kinloch AJ, (eds), *Adv. Chem. Ser.*, No. 233, ACS, Washington, DC, 1993, pp. 405–425.

Pearson RA, Yee AF, *J. Mat. Sci.*, 24, 2571–2580 (1989).

Pearson RA, Yee AF, *J. Mat. Sci.*, 26, 3828–3844 (1991);

Qian JY, Pearson RA, Dimonie VL, El-Aasser MS, *J. Appl. Polym. Sci.*, 58, 439–448 (1995).

Raghava RS, *J. Polym. Sci.*, Part B, 25, 1017–1031 (1987).

Rey L, Poisson N, Maazouz A, Sautereau H, *J. Mat. Sci.*, 34, 1775–1781 (1999).

Roberts K, Simon G, Cook W, Burchill P, *Proceedings of the IUPAC World Polymer Congress*, Gold Coast, 1998, p. 487.

Romano AM, Garbassi F, Braglia R, *J. Appl. Polym. Sci.*, 52, 1775–1783 (1994).

Siebert AR, Guiley CD, Kinloch AJ, Fernando M, Heihnsbrock EPL, In *Toughened Plastics II – Novel approaches in Science and Engineering*, Riew CK, Kinloch AJ (ed.), *Adv. Chem. Ser.*, No. 252, ACS, Washington, DC, 1996, pp. 151–160.

Stenzenberger HS, Römer W, Herzog M, König P, *Sci. Adv. Mat. Proc. Eng. Series*, 33, 1546 (1988).

Sue HJ, *J. Mat. Sci.*, 27, 3098–3107 (1992).

Sue HJ, Garcia-Meitin EI, Pickelmann DM. In *Polymer Toughening* Arends CB (ed.), Marcel Dekker, New York, 1996a, pp. 131–174.

Sue HJ, Yang PC, Puckett PM, Bertram JL, Garcia-Meitin EI, In *Toughened Plastics II – Novel Approaches in Science and Engineering*, Riew, CK, Kinloch AJ (ed), *Adv. Chem. Ser.* No. 252, ACS, Washington, DC, 1996b, pp. 161–175.

Sue HJ, Earls JD, Hefner RE Jr, *Proceedings: Deformation Yield and Fracture of polymers*, Cambridge, 1997, p.129–132.

Suspene L, Yang YS, Pascault JP, In *Toughened Plastics I: Science and Engineering*, Riew CK, Kinloch AJ (ed.), *Adv. Chem. Ser.*, No. 233, ACS, Washington, DC, 1993, pp. 163–188.

Van der Sanden MCM, Meijer HEH, *Polymer*, 34(24), 5063–5072 (1993).

Varley RJ, Hodgkin JH, *Polymer*, 38, 1005–1016 (1997).

Verchère D, Sautereau H, Pascault JP, Moschiar SM, Riccardi CC, Williams RJJ, In *Toughened Plastics I – Sciencea nd Engineering*, Riew CK, Kinloch AJ (ed.), *Adv. Chem. Ser.* No. 233, ACS, Washington, DC, 1993, pp. 335–363.

Woo EM, Bravenec LD, Seferis JC, *Polym. Eng. Sci.*, 34(22), 1664–1673 (1994).

Wu S, *Polymer*, 26, 1855–1863 (1985).

Wu H, Xu J, Liu Y, Heiden PA, *J. Appl. Polym Sci* 72, 151–163 (1999).

Yee AF, Pearson RA, *J. Mat. Sci.*, 21, 2462–2474 (1986).

Yee AF, Du J, Thouless MD, In *Polymer Blends*, Vol. 2, Paul DR, Bucknall CB (eds), John Wiley & Sons, New York, 2000, pp. 225–267.

Zeng S, Hoisington M, Seferis JC, *Polymer Composites*, 14(6), 458–466 (1993).

14

Durability

14.1 INTRODUCTION

Aging can be defined as a slow and irreversible (in use conditions) variation of a material's structure, morphology and/or composition, leading to a deleterious change of use properties. The cause of this change can be the proper material's instability in use conditions or its interaction with the environment (oxidation, hydrolysis, photochemical, radiochemical, or biochemical reactions, etc.).

Aging becomes a difficult problem to study in practice, because it proceeds too slow in use conditions (typical lifetime of years). It is then necessary to make accelerated aging tests to build kinetic models that describe the time changes of the material's behaviour, and to use these models to predict the durability from a conventional lifetime criterion. Indeed, the pertinence of the choice of accelerated aging conditions, the mathematical form of kinetic model, and lifetime criterion has to be proved. Empirical models are highly questionable in this domain because they have to be used in extrapolations for which they are not appropriate.

In the ideal case, an aging study would involve the following steps:

(a) Identification of aging mechanisms through physical and analytical observations.

(b) Kinetic modeling based on the previously established mechanistic scheme (diffusion law, chemical kinetics, etc.).
(c) Prediction of use properties from the structural state using polymer physics.

Among polymers, thermosets are especially difficult to study for many reasons: structural complexity, making difficult the chemical analysis, lack of rigorous tools to investigate the macromolecular structure; lack of physical theories to interpret the change of properties (e.g., embrittlement) against structural changes.

These difficulties are again increased, sometimes considerably, by the fact that thermosets are generally used in composites or as adhesives, e.g., in applications where aging can result also from a change of the interfacial properties and in which certain key properties, e.g., the rubbery modulus, are practically inaccessible.

Despite these difficulties, the practical importance of durability in composite or adhesive applications, has given rise to a vast amount of literature in this field during the past decades. Most of the studies deal with two main aging cases:

1. "humid aging" in liquid media (boats, pipes, tanks, etc.) or in wet atmospheres (aerospace structural parts, (e.g., helicopter blades)
2. "thermal aging" at high temperatures in processing as well as in use conditions (engine parts, electrical insulations, etc.).

Both domains constitute the main sections of this chapter.

Photochemical or radiochemical aging are not examined here for many reasons:

1. The literature on thermosets in these domains is very scarce. Most authors use theoretical concepts and experimental methods elaborated for linear polymers, and extensive reviews are available (Ranby and Rabek, 1975).
2. For photochemical aging, it is well known that photooxidation affects only a thin superficial layer directly exposed to solar radiation – a few dozens of micrometers in the case of epoxies (Bellenger and Verdu, 1983). Thus the aging mode cannot control the material's lifetime in most cases (composites, adhesives), except for applications such as, for example, varnishes of automotive bodies (Bauer et al., 1992).
3. Radiochemical aging has very specific applications. The uses of thermosets in nuclear engineering have been growing. Most of the data on radiochemical aging of thermosets, available at the beginning of the 1990s, has been reviewed (Wilski, 1991).

14.2 AGING RESULTING FROM WATER ABSORPTION

14.2.1 Reversible and Irreversible; Physical and Chemical Aging

A simple and very useful method of studying humid aging consists of exposing an initially dry sample in a medium at constant temperature and relative humidity (or water activity in immersion) and recording weight changes. Various types of gravimetric curves can be obtained; the most frequent ones are shown in Fig. 14.1.

In case (a), an equilibrium is reached. It can be considered here that there are only reversible physical interactions between the polymer and water. Drying leads to a curve that is practically a mirror image of the absorption curve. The behavior of the material can be characterized by two quantities: the equilibrium water concentration W_∞, which characterizes the polymer affinity for water (hydrophilicity), and the duration t_D of the transient, which is sharply linked to the sample thickness L and to a parameter characteristic of the rate of transport of water molecules in the polymer – the diffusion coefficient D.

In cases (b) and (b′) there is no equilibrium; the mass increases continuously or decreases after a maximum, which indicates the existence of an irreversible process – chemical, hydrolysis, or physical, damage; microcavitation increases the capacity of the material for water sorption. The experimental curves having the shape of curves (b) or (b′), indicate that the irreversible processes induce significant mass changes in the timescale of diffusion. When the irreversible processes are significantly slower than diffusion, the behavior shown by curves (c) or (d) is observed. The sorption equilibrium is reached at $t \sim t_D$, and a plateau can be observed in the curve

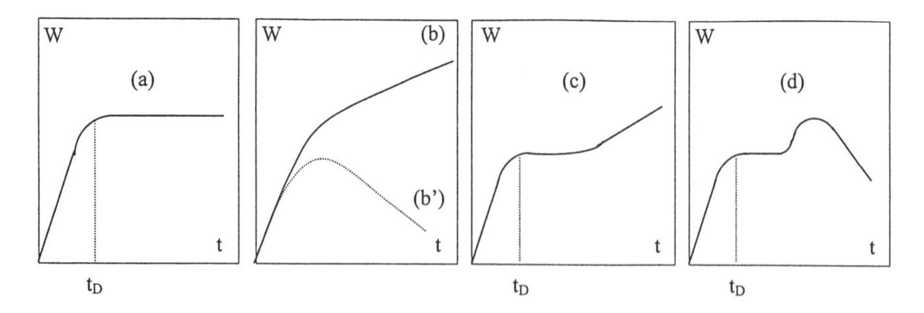

FIGURE 14.1 Shape of the kinetic curves of mass change in the most frequent cases of humid aging.

before irreversible changes become significant.

Since both reversible and irreversible processes are influenced in distinct ways by temperature and water activity, the first step of a humid aging study consists of searching for the conditions (T, RH, sample thickness) in which both phenomena can be clearly decoupled, as in Figs 14.1c and d. The interpretation of experimental results and the modeling of the kinetics of property changes would be difficult or even impossible if physical characteristics such as W_∞ and t_D (or better D) were not known.

To distinguish between physical and chemical aging, one needs analytical data on structural changes (chain scission by hydrolysis, decrease of crosslink density, evolution of small molecules resulting from degradation). Visual and microscopic observations enable us to detect damage. In the most complicated case, chemical degradation and mechanical damage are sharply coupled (osmotic cracking). The following section is devoted to physical, reversible, water–polymer interactions and water diffusion.

14.2.2 Hydrophilicity

Polymer hydrophilicity can be judged in the cases (a), (c), or (d) of Fig. 14 .1. It is defined as the affinity of a polymer for water, which can be quantified by the equilibrium mass gain, W_∞, determined in standard conditions, e.g., in a saturated atmosphere from a sorption experiment. W_∞ depends on the vapor pressure or activity of water and on the temperature. It varies, typically from 0 to 10% in most networks.

a. Influence of Vapor Pressure or Activity of Water

W_∞ is usually obtained by weighing:

$$W_\infty = \frac{\text{Asymptotic mass} - \text{initial mass}}{\text{Initial mass}} \tag{14.1}$$

The water equilibrium mass fraction is given by

$$m_\infty = \frac{W_\infty}{1 + W_\infty} \tag{14.2}$$

For rough estimations, one can use $m_\infty \approx W_\infty$.

The water equilibrium concentration is expressed by

$$C_\infty = \frac{\rho_w}{0.018}\left(\frac{W_\infty}{1 + W_\infty}\right) \qquad (\text{mol m}^{-3} \text{ if } \rho_w \text{ is expressed in kg m}^{-3}) \tag{14.3}$$

where ρ_w is the density in the wet state; C_∞ is linked to the water vapor pressure, p, by a more or less complex law depending on the sorption mechanism. For a low-to-moderate hydrophilic behavior, it may be assumed that Henry's law is valid, at least in a first approximation:

$$C_\infty = S \times p \tag{14.4}$$

where S is the solubility coefficient.

Thus, in the domain of validity of Henry's law, the equilibrium concentration is proportional to the relative hygrometry (at a given temperature).

Immersion in pure water must lead to the same result as in a saturated atmosphere. If water contains solutes, its vapor pressure decreases. The corresponding equilibrium concentration, linked to the water activity, is proportional to the water vapor pressure. In other words, the water equilibrium concentration is a decreasing function of the solute concentration: salt water is less active than pure water.

Most of these aspects of water-sorption equilibrium correspond to the equality of chemical potentials of water in the medium and in the polymer. The consequences of this principle are illustrated by the experiment of Fig. 14.2, where an interface is created between water and a nonmiscible liquid (oil, hydrocarbon, etc.), and a polymer sample is immersed into the organic liquid. It can be observed that, despite the hydrophobic character of the surrounding medium, the sample reaches the same level of water saturation as in direct water immersion or in a saturated atmosphere. What controls the water concentration in the polymer is the ratio C/C_s of water concentrations in the organic phase, where C_s is the equilibrium concentration, which can be very low but not zero. In other words, hydrophobic surface treatments can delay the time to reach sorption equilibrium but they cannot avoid the water absorption by the substrate.

Let us remind ourselves that the saturated vapor pressure, p_s, increases with temperature. In a first approximation it can be written:

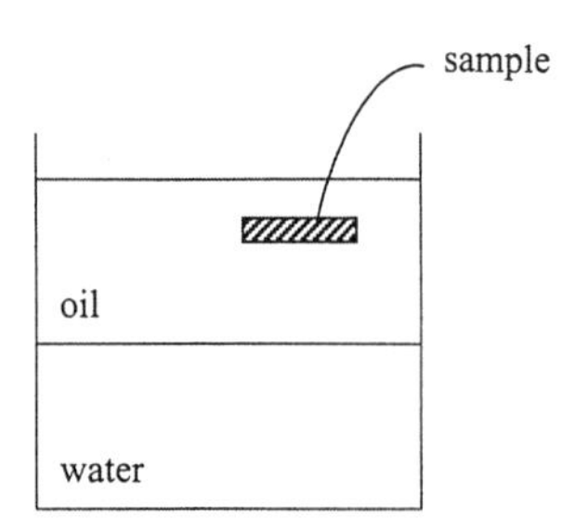

FIGURE 14.2 Water equilibrium in a nonaqueous medium immiscible with water.

$$p = p_0 \exp - \frac{H_w}{RT} \tag{14.5}$$

where $\ln p_0 = 25.33$ (p_0 in p_a) and $H_w = 42.8\,kJ\,mol^{-1}$.

On the other hand, S obeys an Arrhenius law:

$$S = S_\sigma \exp - \frac{H_s}{RT} \tag{14.6}$$

where H_s is the heat of dissolution. In the case of water, H_s is negative (the dissolution process is exothermic), and ranges generally from $-25\,kJ\,mol^{-1}$ (polymers of low polarity) to $-50\,kJ\,mol^{-1}$ (highly polar polymers). Then,

$$C_\infty = Sp = S_0 p_0 \exp - \frac{(H_s + H_w)}{RT} \tag{14.7}$$

For highly polar polymers, $H_s < -H_w$, and the equilibrium concentration is a decreasing function of temperature. This is often found in the most hydrophilic networks, based, for example, on the aromatic amine – aliphatic diepoxide of diglycidyl ether of butane diol (DGEBD) type (Tcharkhtchi *et al.*, 2000), or on particular polyimides (Hilaire and Verdu, 1993).

For many usual, moderately polar networks, such as epoxides of the diglycidyl ether of bisphenol A (DGEBA) diamine type, or vinyl esters, $H_s \sim -H_w$, so that the equilibrium concentration appears almost temperature-independent. For most of the less polar networks such as polyesters or anhydride-cured epoxies, C_∞ (or W_∞) increases slightly with temperature: $\Delta W_\infty/\Delta T \approx 0.01–0.02\%\ K^{-1}$ between 20 and 50°C.

Except, eventually, for networks of very low polarity, $-H_s$ is close to H_w, so that the above equation can be well approximated by

$$\frac{1}{C_\infty}\frac{\Delta C_\infty}{\Delta T} \approx -\frac{H_s + H_w}{RT^2} \tag{14.8}$$

This opens the way for a rapid determination of H_s from experimental values of C_∞ at two different temperatures.

b. Influence Of Structure

On the basis of observations made on limited structural series, certain authors (e.g., Adamson, 1980) suggested that water absorption would occur by occupancy of the available "free volume" by water molecules. Despite its seductive intuitive character, this theory fails to explain why free-volume rich substances such as silicone rubbers, crosslinked polyethylene, or simply liquid aliphatic hydrocarbons are hydrophobic. Furthermore, experimentally determined apparent heat of dissolution values (H_s) and plasticization effects generally agree well with theoretical predictions, so that there is no reason to assume that water is not homogeneously dissolved into the polymer matrix, except, in the case of a macroscopic porosity.

From a global analysis of W_∞ values (Table 14.1), it is clear that hydrophilicity is an increasing function of the polarity of the groups contained in the polymer and their concentration:

$-S-$; $-CF_2-$; $Si(CH3)_2$; $-CH_2-$; $-CH(CH_3)-$; $-C(CH_3)_2-$ are almost hydrophobic groups

$-SO_2-$; $C(=O)$; $-O-$; $C(=O)-O$; $>N-$ are moderately hydrophilic groups

$(CO)_2N-$; $-COOH$; $-C(=O)-NH-$; $-OH$ are highly hydrophilic groups

Since the unrelaxed bulk modulus, K_u e.g., determined by ultrasonic propagation velocity measurements, is a good measure of the cohesive energy density, CED ($K_u \sim 11$ CED; Chapter 10), and CED gives a good indication of the overall material's polarity, one can expect a correlation between K_u and W_∞. This is shown in Fig. 14.3 for the amine–epoxy and styrene–vinyl ester networks. The following relationship is found:

$$W_\infty = 1.63(K_u - 4.5) \quad (W_\infty \text{ in } \%, K_u \text{ in GPa}) \tag{14.9}$$

The intercept (4.5 GPa) corresponds essentially to the nonhydrophilic dispersive component of cohesive energy, which does not vary very much from one polymer to another.

Predictions of W_∞ could be performed using global (Hildebrand) or partial (Hansen) solubility parameters. but these are very difficult (and perhaps impossible) to determine accurately from solvent–sorption experiments, so that this way is not realistic. The best experimental approach is, in our opinion, using the ultrasonic modulus.

Empirical calculations of W_∞ from the CRU structure can give relatively good results provided that sufficiently large structural units are considered to take into account eventual intramolecular interactions (Bellenger *et al.*, 1988). However, their practical interest for predicting the behavior of new systems is relatively limited: the molar contribution of the most hydrophilic groups, e.g., hydroxyl groups, is not an integer, which means that

TABLE 14.1 Tentative classification of network hydrophilicities

Family	Examples	W_∞ (293 K)
Very low hydrophilicity	Polydimethylsiloxane Polyethylene Polystyrene co divinyl benzene	Typically < 0.5%
Low hydrophilicity	Many styrene crosslinked UP Some styrene crosslinked vinylesters Some anhydride crosslinked epoxies	0.5–1.5% Increase with ester concentration
Moderate hydrophilicity	Some vinyl esters Amine crosslinked epoxies of relatively low crosslink density Some polyimides	1.5–3% Increases with alcohol or amide concentration
High hydrophilicity	Amine crosslinked epoxies of high crosslink density (TGAP, TGMDA) Many polyimides	> 3%

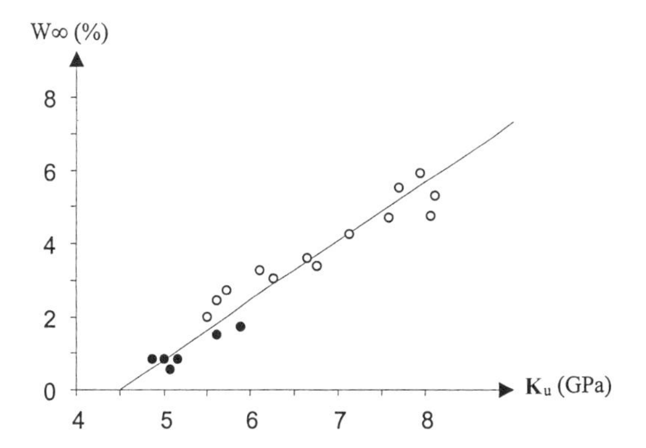

FIGURE 14.3 Correlation K–W_∞ between the bulk modulus and the equilibrium water concentration. o Amine crosslinked epoxies. • Vinylesters and polyesters.

certain groups are able to fixate a water molecule and others are not. A theory for predicting the fraction of "active" groups is necessary.

Recent studies showed that, in a given structural series in which the main variable is the OH concentration, the molar contribution of a particular OH group is apparently an increasing function of the OH concentration. A possible explanation of this result is that in the polymer–water complex, water is generally doubly or triply bonded. Thus, a hydrophilic site would be composed of two, eventually three, neighboring polar groups. The average distance between polar groups is generally too high to allow such a concerted process. But the spatial distribution of these groups is more or less heterogeneous, so that there is a more or less important fraction of these groups sufficiently close to form an hydrophilic site. Thus it is possible to explain why OH groups have a weak molar contribution to water absorption in polyesters ($[OH] \leq 0.5\,\mathrm{mol\,kg^{-1}}$) (Bellenger *et al.*, 1990), a medium molar contribution in amine-crosslinked epoxies ($4 < [OH] < 8\,\mathrm{mol\,kg^{-1}}$) (Bellenger *et al.*, 1988); and a high molar contribution in linear polymers such as those soluble in water – poly(vinyl alcohol), poly(acrylic acid), etc. – in which $[OH] > 10\,\mathrm{mol\,kg^{-1}}$. It explains also why, in epoxide–amine structural series differing by the amine/epoxide molar ratio, W_∞ increases pseudo-parabolically with the OH concentration (Tcharkhtchi *et al.*, 2000)

14.2.3 Diffusion

Solvent transport in organic polymer matrices is usually depicted as a two-step mechanism. The first step is the dissolution of the solvent in the superficial polymer layer. This process, which can be considered almost instantaneous in the case of water, creates a concentration gradient. The second step is the diffusion of the solvent in the direction of the concentration gradient. This process may be described by a differential mass balance (often called Fick's second law), which, in the unidimensional case, may be written as

$$\frac{\partial C}{\partial t} = D\frac{\partial^2 C}{\partial x^2} \tag{14.10}$$

where D is the diffusion coefficient and x the coordinate along the sample's thickness (L).

The resolution of this differential equation gives

$$\frac{C}{C_\infty} = 1 - \frac{8}{\Pi^2}\sum_{n=0}^{\infty}\frac{1}{(2n+1)^2}\exp\left(-\frac{\Pi^2(2n+1)^2}{L^2}Dt\right) \tag{14.11}$$

At short times, typically when $C \leq 0.5\,C_\infty$, this function can be well approximated by

$$\frac{C}{C_\infty} = \frac{4}{\sqrt{\Pi}}\left(\frac{Dt}{L^2}\right)^{1/2} \tag{14.12}$$

It is thus usual to plot C against $\sqrt{t}$. The linearity of the curve in its initial part can be considered as a validity criterion for the Fick's law. The slope of the linear part lets us determine the coefficient of diffusion:

$$D = \frac{\Pi}{16}L^2\frac{\Delta(C/C_\infty)}{\Delta(\sqrt{t})} \tag{14.13}$$

A characteristic time of diffusion, t_D, defined as the duration of the transient, can be arbitrarily taken at the intersection between the tangent at the origin and the asymptote (Fig. 14.4). This leads to

$$t_D = \frac{\Pi}{16}\frac{L^2}{D} \approx \frac{L^2}{5D} \tag{14.14}$$

Typical values of diffusion coefficients are 10^{-12}–$10^{-13}\,m^2\,s^{-1}$ at 20–50°C. The diffusion time t_D is about 1 day to 1 week for samples of 1 mm thickness, and 1 year for samples of 1 cm thickness.

Here, the best way to accelerate aging is to decrease the sample thickness (when possible). Except in very highly hydrophilic materials, D is independent of the relative hygrometry or water activity, but is an increasing function of temperature:

$$D = D_0 \exp -\frac{H_D}{RT} \tag{14.15}$$

where H_D is often of the same order of magnitude but opposite sign to H_s; $H_D \sim 20$–$70\,kJ\,mol^{-1}$

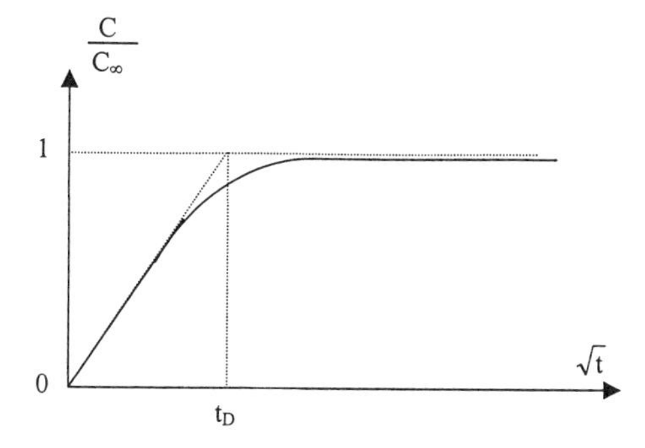

FIGURE 14.4 Shape of a Fickian diffusion curve. Definition of the characteristic time of diffusion.

The structure–diffusivity relationships are not clearly established. A rough correlation between the diffusivity and the reciprocal of hydrophilicity is usually found. For example, for networks absorbing less than 1.5% water, such as unsaturated polyesters and anhydride-cured epoxies, the coefficient of diffusion is often higher than $5 \times 10^{-13}\,m^2\,s^{-1}$ at 20–50°C. In contrast, for epoxies cured by amines absorbing more than 2% water, the coefficient of diffusion is of the order of $10^{-13}\,m^2\,s^{-1}$ at 20–50°C, and it tends to decrease with W_∞ in most of the structural series.

Diffusion kinetics were studied on both sides of T_g for epoxy samples based on an aliphatic diepoxide (DGEBD) cured by an aromatic diamine (diethylenetoluenediamine: ETHA). Surprisingly, there is no clear discontinuity of D and S around T_g (Tcharkhtchi *et al.*, 2000). All these observations suggest that, in these systems, the diffusion rate of water is controlled by the strength of the polymer–water hydrogen bond, as in the following scheme:

$$W + P_1 \rightarrow [W - P_1]$$

$$[W - P_1] \rightarrow W + P_1 \qquad (I)$$

$$W \rightarrow \text{ jump to } P_2 \qquad (II)$$

$$W + P_2 \rightarrow [W - P_2] \qquad \text{etc}\ldots$$

where W is the water molecule, P_1 and P_2 are neighboring polymer hydrophilic sites, and [W–P] is the polymer–water complex. The whole kinetics would be thus governed by the dissociation of the water–polymer complex (I) rather than by the rate of activated jumps from one site to another (II).

14.2.4 Physical Consequences of Water Absorption

a. Plasticization

Using the classical hypotheses of the free volume theory, the glass transition temperature for a polymer (p) and solvent (s) solution, with a volume fraction of solvent, v, is given by (Kelley and Bueche, 1960):

$$T_g = \frac{(1 - v)_p T_{gp} + v\alpha_s T_{gs}}{(1 - v)\alpha_p + v\alpha_s} \qquad (14.16)$$

where T_{gp} and T_{gs} are the glass transition temperatures of the dry polymer and the solvent ($T_{gs} \sim 100$–150 K for water, depending on authors), and α_p and α_s are the respective coefficients of free volume expansion ($\alpha = \alpha_l - \alpha_g$).

This relationship can be simplified assuming the validity of the Simha–Boyer rule, $\alpha T_g =$ constant, which leads to

$$\frac{1}{T_g} = \frac{1}{T_{gp}} + AV \text{ where } A = \frac{1}{T_{gs}} - \frac{1}{T_{gp}} \tag{14.17}$$

A numerical application of this relationship is given in Table 14.2. These results are in good agreement with experimental data. They show that the plasticizing effect increases:

- With the equilibrium water content. It is almost insignificant in low-polarity networks such as unsaturated polyesters or anhydride-cured epoxies.
- With the initial glass transition temperature T_{gp} of the polymer in the dry state. It is limited in low-T_g epoxies and vinyl esters, but important in high-T_g amine-cured epoxies.

b. Swelling

The swelling mechanism is well established in rubbers (Flory, 1943). The expansion force generated by the solvent penetration is equilibrated by the entropic force linked to chain stretching. Little is known, in contrast, for the case of glassy polymers where plasticization effects are not sufficient to induce a devitrification. Swelling can be defined by

$$\frac{V}{V_0} = 1 + \psi \tag{14.18}$$

where V_0 and V are the sample volumes in the dry and wet (equilibrium) states, respectively; ψ is generally lower than the value predicted from the hypothesis of additivity of polymer and water volumes. This means that water penetration in the polymer is accompanied by some contraction.

To our knowledge there is no theory that predicts efficiently the swelling ratio from W_∞ and the network characteristics.

Table 14.2 Calculated plasticization effect in saturation conditions for some industrial networks

Network	φ	T_{gp}(K)	A (K^{-1})	T_g(K)	ΔT_g(K)
UP–styrene	0.010	375	5.7×10^{-3}	367	8
Flexibilized epoxy–anhydride	0.008	340	5.4×10^{-3}	335	5
Vinyl ester	0.015	395	5.8×10^{-3}	382	13
Aliphatic epoxy–amine	0.040	270	4.6×10^{-3}	257	13
Aromatic epoxy-amine[a]	0.025	460	6.2×10^{-3}	429	31
Aromatic epoxy-amine[b]	0.060	520	6.4×10^{-3}	433	87

[a]DGEBA–DDM network.
[b]TGMDA–DDS network.

c. Dielectric Properties

Water is highly polar. Its penetration in a polymer induces an increase of both components ϵ' and ϵ'' of the complex dielectric constant (Chapter 6), an increase of the conductivity and a decrease of the dielectric strength. Microdielectric sensors can be used to monitor water diffusion in thick samples, especially in composites (Kranbuehl, 2000).

14.2.5 Mechanical Consequences of Water Absorption

a. Plasticization Effects

In the domain where the material is ductile or semiductile, the yield stress $\sigma_y$is linked to the glass transition temperature T_g by

$$\sigma_y = C_k(T_g - T)$$

where C_k is a constant (Kambour, 1984; Chapter 12). Plasticization, which decreases T_g, leads to a decrease of σ_y. This effect can be important when a material that is not very far from its glass transition point is submitted to loads. In this case, water absorption can induce yielding. The data of Table 14.2 show that σ_y can decrease by more than 30%. A significant increase of creep compliance can be expected in such a case.

A softening can be observed at temperatures ranging typically from T_g (dry) to (T_g (wet) – 50 K). Far from T_g there is no significant influence of water (at moderate concentrations) on stiffness.

b. Swelling Effects

Differential swelling induces a strain ϵ

$$\frac{\Delta V}{V} = 3\epsilon$$

Thus, concentration gradients during the transient of diffusion (before equilibrium is reached) induce stresses. These stresses can be predicted using the following approach:

(a) Establish the diffusion law. From this law, calculate the water concentration C at any time and any point of the sample.
(b) From equilibrium experiments, determine the relationship between the concentration C and the degree of swelling $\Delta V/V$.
(c) From the data obtained in (a) and (b), calculate the volume change at any point.
(d) From the profile of volume change, calculate the strain profile.

(e) From the behavior law $\sigma = f(\epsilon, t, C)$, calculate the stress. Indeed, the effect of sorbed water on the behavior law has to be established under equilibrium conditions.

The results in terms of stress distribution along the thickness are shown in Fig. 14.5.

In extreme conditions, e.g., essentially at temperatures typically higher than 60°C, where diffusion is fast and generates strong concentration gradients and where the yield stress is sufficiently low, water absorption can induce damage in medium to high hydrophilic networks.

14.3 AGING DUE TO HYDROLYSIS

14.3.1 The "Ideal" Case of Hydrolysis

Hydrolytic processes can be analyzed by comparison with an ideal case, which can be defined by the following hypotheses:

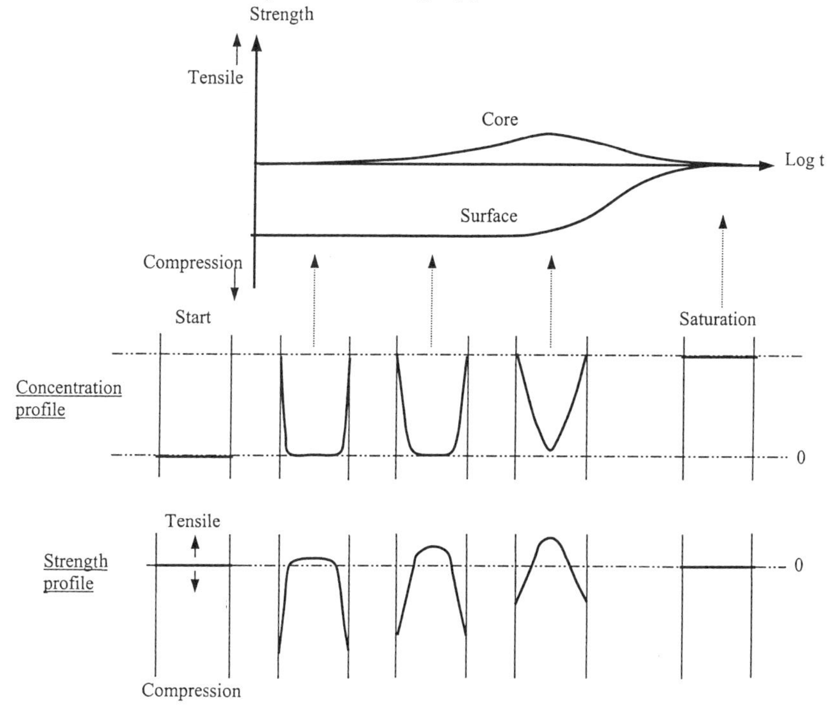

FIGURE 14.5 Stress distribution at various stages of water absorption and desorption.

(a) Hydrolysis is homogeneous at all dimension scales. It is not diffusion-controlled (no degradation gradient along the thickness).

(b) Hydrolysis ($E + W \rightarrow$ chain scission) obeys a second-order kinetic law:

$$\frac{dn}{dt} = k[E]C \tag{14.19}$$

where n is the number of hydrolysis events (moles) per unit volume, E is the concentration of reactive groups, C is the water concentration, and k is the rate constant that depends only on temperature. C is an increasing function of the hygrometric ratio in a first approximation:

$$C \sim C_0\left[\frac{RH}{100}\right]$$

For a reversible process ($E + W \leftrightarrow A + B$, where A and B are reaction products), the kinetic law may be written as

$$\frac{dn}{dt} = k[E]C - k'[A][|B] \tag{14.20}$$

An equilibrium would be then observed for

$$\frac{[A_\infty][B_\infty]}{[E_\infty][W_\infty]} = \frac{k}{k'} \tag{14.21}$$

(c) Hydrophilicity changes due to hydrolysis are negligible in the conversion range of interest. This means that $C = \text{constant}$, so that a pseudo-first-order rate constant, $K = kC$, may be defined.

(d) A pertinent end-life criterion may be defined, corresponding to a particular conversion (e.g., to a given structural state independent of exposure conditions): $n = n_f$. Then,

$$\frac{dn}{dt} = K[E] = K\big([E]_0 - n\big), \quad \text{which leads to} \quad \ln\frac{[E]_0 - n}{[E]_0} = Kt \tag{14.22}$$

The lifetime may be thus calculated as

$$t_f = \frac{1}{K}\ln\frac{[E]_0 - n_f}{[E]_0} = \frac{1}{kC}\ln\frac{[E]_0 - n_f}{[E]_0} \tag{14.23}$$

$$t_f = \left[\frac{1}{k_0 \exp -\frac{H_H}{RT}}\right]\left[\frac{1}{C_0\left(\frac{RH}{100}\right)}\right]\left[\ln\frac{[E]_0 - n_f}{[E]_0}\right] \tag{14.24}$$

The lifetime t_f may be thus calculated as the product of three terms, respectively dependent on T, RH (or water activity), and polymer structure.

Generally, embrittlement occurs at relatively low conversions ($n_f \ll [E]_0$), so that

$$n \approx [E]_0 Kt \quad \text{and} \quad t_f \approx \frac{1}{kC}\frac{n_f}{[E]_0} \tag{14.25}$$

Hydrolysis must appear in principle as a pseudo-zero-order process, except if the end-life conversion n_f/E_0 exceeds largely 10%, which seems unlikely. In linear polymers, the number of chain scissions n can be determined from molar mass measurements. In thermosets, it is considerably more difficult to determine. Some possible ways are discussed in the following section.

14.3.2 Determination of The Conversion Degree in Thermosets

Infrared or NMR Determination of [E]. These methods have been used for instance in unsaturated polyesters or in vinyl esters (Ganem *et al.*, 1994); however, they are not very sensitive at low conversions.

Rubber Elasticity. This method is based on the hypothesis that Flory's relationship can be applied to thermosets, which is only a rough approximation (Chapter 10):

$$G = RT\nu_e \tag{14.26}$$

where G is the shear modulus at temperature T (above T_g), and ν_e is the concentration of elastically active network chains. For a network initially free of dangling chains, each chain scission leads to the disappearance of one elastically active chain, for networks in which the crosslink functionality is $\varphi \geq 4$, and three elastically active chains for networks in which the crosslink functionality is $\varphi = 3$ (Fig. 14.6). Then,

$$\nu_e = \nu_{e0} - n \quad \text{for } \varphi \geq 4 \tag{14.27}$$

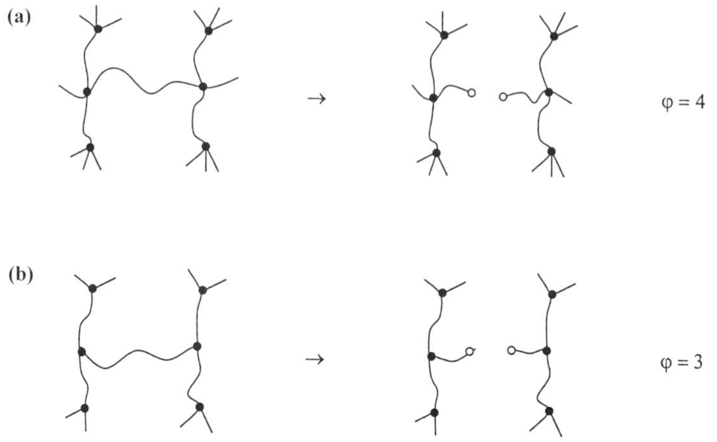

FIGURE 14.6 Destruction of elastically active network chains resulting from a chain scission in the case of tetrafunctional (a) and trifunctional nodes (b).

and

$$\nu_e = \nu_{e0} - 3n \quad \text{for } \varphi = 3 \tag{14.28}$$

This gives

$$\Delta\nu_e = \frac{1}{RT}(G_0 - G) \quad \text{for } \varphi \geq 4 \tag{14.29a}$$

and

$$\Delta\nu_e = \frac{1}{3RT}(G_0 - G) \quad \text{for } \varphi = 3 \tag{14.29b}$$

In the case of unsaturated polyesters, nondegraded samples made from a prepolymer of molar mass M and a styrene mass fraction s have a chain-ends concentration $b = [2(1 - s)/M]\rho$, where ρ is the density. If ν_e is the actual concentration of elastically active network chains, an ideal network would be obtained by "welding" each chain end to another one, leading to

$$\nu_{e0} = \nu_e + \frac{b}{2} \tag{14.30}$$

A particular polyester network may be considered as a model of degraded "ideal" polyester based on a prepolymer having an infinite molar mass with a number of chain scissions equal to $\Delta\nu_e = b/2$. Thus, polyester samples differing by the initial prepolymer molar mass can be used to calibrate rubber-elasticity measurements.

This approach can be illustrated by unsaturated polyesters based on an almost equimolar combination of maleate and phthalate of propylene glycol, crosslinked by styrene (45 wt%) (Mortaigne *et al.*, 1992). Six samples differing by the prepolymer molar mass were analyzed. The chain-ends concentration, b, was determined by volumetric analysis of alcohols and acids in the initial reactive mixture. Then, the system was cured, elastic measurements were made in the rubbery state at $T_g + 30^\circ C$, and the shear modulus G' was plotted against chain-ends concentration (Fig. 14.7). The following relationship was obtained:

$$G' = 14 - 5 \times 10^{+3} b \tag{14.31}$$

where G' is in MPa and b is in $\text{mol}\,g^{-1}$. Equtation (14.31) applied to degradation (each chain scission creates 2 chain ends) leads to

$$\Delta \nu_e = \frac{1}{2}\left(\frac{\Delta G'}{510^{+3}}\right) = 10^{-4} \Delta G' \tag{14.32}$$

where $\Delta G'$ is the decrease of rubbery modulus (in MPa) for a number of chain scissions $\Delta \nu_e$ ($\text{mol}\,g^{-1}$).

14.3.3 Structure–Reactivity Relationships

The simplified expression for lifetime is written as

$$t_f = \frac{1}{kC}\frac{n_t}{[E]_0} \tag{14.33}$$

The lifetime is a decreasing function of hydrophilicity (C), which is generally directly linked to the concentration of reactive groups E_0.

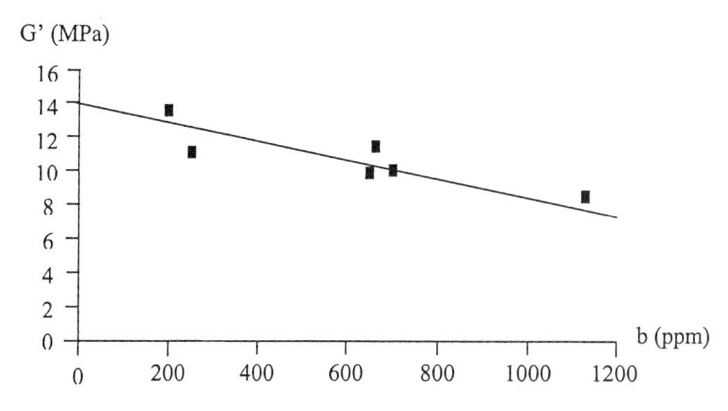

FIGURE 14.7 Rubbery modulus against dangling chains concentration in styrene crosslinked unsaturated polyester. (After Mortaigne *et al.*, 1992.)

For example, unsaturated polyesters based on neopentyl glycol are more stable than those based on propylene glycol. In series differing only by the diol nature, a good correlation between the hydrophilicity and the hydrolytic stability is generally observed, as illustrated by the example of Fig. 14.8. The pseudo-first-order rate constant of hydrolysis ($K = k[E]_0$), was plotted against equilibrium water concentration for two linear polyester families: the first one based on isophthalic acid (IPA) and the second one based on maleic anhydride (MAA). For a given family, the hydrophilicity depends essentially on the diol structure, the order being always: neopentyl glycol (NPG) < propylene glycol (PG) < diethylene glycol (DEG).

The intrinsic reactivity of the ester, which essentially influences k, plays an important role. The following order of increasing stability has been established: reacted fumarate < orthophthalate < isophthalate ≪ methacrylate. It can be seen in Fig. 14.8 that maleates are about 10 times more reactive than isophthalates (Belan 1997).

Practitioners know well that isophthalates are more stable than orthophthalates, and that vinyl esters (methacrylates) are at least one order of magnitude more stable than unsaturated polyesters (UP). The weak point of UP is the fumarate unit, which is, however, necessary for the crosslinking .

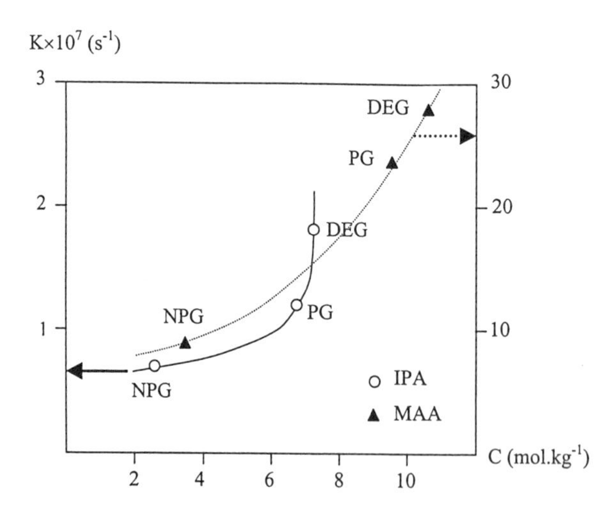

FIGURE 14.8 Pseudo-first-order hydrolysis rate constant at 70°C against equilibrium water concentration for isophthalate (○) and maleate (▲) polyesters. The abbreviations are given in the text.

An estimation of ester concentrations and rate constants for some polyesters and vinyl esters crosslinked by styrene is given in Table 14.3. To compare with linear polymers: $r(100^\circ C) \sim 8 \times 10^{-5}\,mol\,kg^{-1}\,day^{-1}$, so that $K \sim 2 \times 10^{-5}\,day^{-1}$ for bisphenol A polycarbonate ($[E]_0 \sim 4\,mol\,kg^{-1}$), and $r(100^\circ C) \sim 8.45 \times 10^{-3}\,mol\,kg^{-1}\,day^{-1}$, so that $K \sim 1.6 \times 10^{-3}\,day^{-1}$ for poly(ethylene terephthalate) ($[E]_0 \sim 5.20\,mol\,kg^{-1}$).

It appears that the ester reactivity towards hydrolysis is governed by the chemical environment of the ester group rather than by the macromolecular architecture (there is no systematic difference between linear polymers and networks).

In polyesters, the existence of an osmotic cracking process (see below), gives importance to other structural factors, especially the concentration of dangling chain ends.

14.3.4 Lifetime Criterion

For linear polymers, the following lifetime criterion can be proposed:

$$n_c = \frac{1}{M_{crit}} - \frac{1}{M_{n0}} \tag{14.34}$$

where M_{crit} is the critical molar mass (entanglement threshold) and M_{n0} is the initial molar mass. The toughness drops suddenly two or three orders of magnitude at M_{crit}, which thus corresponds to a ductile–brittle transition. In thermosets, the problem is considerably more complicated because we lack information on the molecular mechanisms of fracture. Furthermore, most

TABLE 14.3 Characteristics of the hydrolysis kinetics at 100°C, 100% HR for styrene crosslinked unsaturated polyester and vinyl esters. (After Ganem *et al.*, 1994.

Material	E_0 (mol kg^{-1})	K (day^{-1}) × 1000
UP[a]	6.35	35.0
VE[b]	0.3–3.0	0.36–1.4[c]

[a]Unsaturated polyester: maleate/phthalate (1/1) of propylene glyconol and neopenty glycol crosslinked by 38%, by weight, of styrene.
[b]Vinyl esters of various types based on DGEBA dimethacrylates crosslinked by 23–50%, by weight, of styrene.
[c]K is clearly correlated to E_0 ($K \sim (0.6 \pm 0.15)\,E_0$), because hydrophilicity increases with E_0 (1 hydroxyl per methacrylate).

thermosets are initially brittle. In this case, linear elastic fracture mechanics can be applied. The ultimate stress, σ_f, may be related to the modulus E by

$$\sigma_f = \gamma E \tag{14.35}$$

where γ is a parameter that essentially depends on the defects geometry: typically, $0.03 \geq \gamma \geq 0.02$ in unaged polymers.

Aging does not modify the value of E significantly (except in the close vicinity of T_g), so that aging effects on fracture properties are relatively low in the brittle regime, except if defects are created. This is the case in polyesters (osmotic cracks), where durability is controlled by this process rather than by chain scission or any other structural change at the molecular or macromolecular scale.

Thus, there are two ways to select an end-life criterium for thermosets:

- arbitrary choice in the case where there is no cracking
- cracks appearance in the other cases.

It remains then to establish the relationship between the hydrolysis conversion and the crack initiation, which is not obvious.

14.3.5 Some "Nonideal" Cases of Hydrolytic Aging

a. Diffusion-Controlled Hydrolysis

The kinetic equations analyzed in the previous sections are valid for a spatially homogeneous hydrolysis. However, it is possible that above a certain boundary in the reactivity (K) and sample thickness (L) map, all the water molecules penetrating in the sample are consumed in superficial layers, so that aging becomes heterogeneous (Fig. 14.9).

A simplified theory of these processes has been established (Audouin *et al.*, 1994). It leads to a simple scaling law for the prediction of the thickness of the degraded layer (TDL):

$$\mathrm{TDL} \approx \sqrt{\frac{D}{K'}} \tag{14.36}$$

where K' is the first-order constant relative to water, for which the consumption rate is given by

$$\frac{dC}{dt} \approx -k[E]_0 C \quad \text{and} \quad K' = k[E]_0 \tag{14.37}$$

There is a limiting thickness, $L_c \sim 2\mathrm{TDL}$, such that for $L < L_c$, aging is homogeneous, whereas for $L > L_c$, it is heterogeneous (the degradation rate is lower in the core than in the superficial layers). From

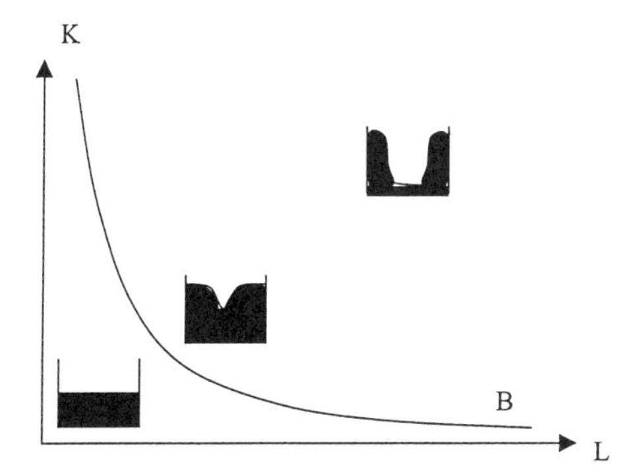

FIGURE 14.9 Non diffusion controlled and diffusion controlled domains in a reactivity–thickness (K–L) graph. The thickness profiles of degradation at various points on both sides and at various distances of the boundary B are shown.

$$D = D_0 \exp - \frac{H_D}{Rt} \quad \text{and} \quad K' = K_0' \exp - \frac{H_R}{T} \tag{14.38}$$

it results that

$$TDL \approx (TDL)_0 \exp - \frac{H_L}{RT} \tag{14.39}$$

$$H_L = \frac{1}{2}(H_D - H_R) \tag{14.40}$$

H_D is of the order of 20–70 kJ mol^{-1}; H_R is generally higher (50–100 kJ mol^{-1}), so that H_L is generally negative. This means that the thickness of the degraded layer is generally a decreasing function of temperature, as is experimentally observed in many cases. A practical consequence of this result is that care must be taken in the choice of the combination of sample thickness and exposure conditions, for accelerated aging.

b. Nonzero-Order Processes

Sometimes, the experimental $n = f(t)$ curve is not linear in its initial part: for example, in the case of polycarbonate, where the rate of chain scission decreases rapidly in the early period of exposure and tends towards a constant value, r_∞, so that: $n = n_l + r_\infty t$. In such cases, there is no other way to explain this behavior than to assume that the material contains very unstable groups (in a concentration close to $n = n_l$) that degrade rapidly. When they are consumed, the system adopts a "normal" pseudo-zero-order behavior.

c. Random and Nonrandom Hydrolysis – Mass Changes

Gravimetric studies of hydrolysis are especially interesting because they combine simplicity and richness of information provided that a kinetic model is available to interpret experimental curves. This model can be built from the following approach: let us consider a network containing initially $[E]_0$ reactive linkages, b_o dangling chains (each one containing initially β reactive groups sufficiently near to the chain end to give a small, extractable molecule by hydrolysis), and γ_0 free extractable molecules (monomers), such as unreacted glycols, etc. The rate of scission of elastically active chains (or dangling chains far from their extremity) is

$$\frac{dn}{dt} = K\big([E]_0 - \beta b\big) \tag{14.41}$$

where b is the actual concentration of dangling chains. These processes lead to two new chain ends, so that

$$\frac{db}{dt} = 2K\big([E]_0 - \beta b\big) = 2K\beta\left(\frac{[E]_0}{\beta} - b\right) \tag{14.42}$$

$$\ln\frac{[E]_0/\beta - b_0}{[E]_0/\beta - b} = 2K\beta t \quad \text{and} \quad b = \frac{[E]_0}{\beta} - \left(\frac{[E]_0}{\beta} - b_0\right)\exp -2K\beta t \tag{14.43}$$

The corresponding mass gain m^+ due to water incorporation into the network is $18 \times 10^{-3}\ \text{kg mol}^{-1}$ of hydrolysis event:

$$\begin{aligned}\frac{dm^+}{dt} &= 18 \times 10^{-3}\frac{dn}{dt} = 18 \times 10^{-3}\big([E]_0 - \beta b\big)\\ &= 18 \times 10^{-3} K\big([E]_0 - \beta b_0\big)\exp -2K\beta t\end{aligned} \tag{14.44}$$

The rate of chain scission leading to extractable molecules is

$$\frac{dn'}{dt} = K'\beta b = K'\big[[E]_0 - \big([E]_0 - \beta b_0\big)\exp 2K\beta t\big] \tag{14.45}$$

The corresponding mass loss, m^- is thus given by

$$\frac{dm^-}{dt} = -(M_{nb} - 18)10^{-3}\frac{dn'}{dt} \tag{14.46}$$

where M_{nb} is the number-average molar mass of extractable molecules in g mol^{-1}. The whole mass change is calculated from

$$\frac{dm}{dt} = \frac{dm^+}{dt} + \frac{dm^-}{dt} = 18 \times 10^{-3} K\big([E]_0 - \beta b_0\big) \exp -2K\beta t - K'$$
$$(M_{nb} - 18)10^{-3}[E]_0 + K'(M_{nb} - 18)10^{-3}\big([E]_0 - \beta b_0\big) \exp -2K\beta t \tag{14.47}$$

In the case of a true random process, all the groups are equireactive, so that $K = K'$, and

$$10^3 \frac{dm}{dt} = -K(M_{nb} - 18)[E]_0 \left[1 - \left(\frac{M_{nb}}{(M_{nb} - 18)} \frac{([E]_0 - \beta b_0)}{[E]_0}\right) \exp -2K\beta t\right] \tag{14.48}$$

The curve $m = f(t)$ exhibits a maximum for

$$t_{max} = \frac{1}{2K\beta} \ln\left[\frac{M_{nb}}{(M_{nb} - 18)} \frac{([E]_0 - \beta b_c)}{[E]_0}\right] \tag{14.49}$$

since $M_{nb} \gg 18\,g\,mol^{-1}$ and $[E]_0 \gg \beta b_0$, the following approximation may be used:

$$t_{max} \approx \frac{1}{2K\beta}\left(\frac{18}{M_{nb}} - \frac{\beta b_0}{[E]_0}\right) \tag{14.50}$$

Integration of dm/dt gives

$$10^3 m = m_0 + m_1 + m_2 \tag{14.51}$$

where m_0 is the integration constant corresponding to the initially extractable products (γ_0)

$$m_1 = -\frac{1}{2\beta} M_{nb}\big([E]_0 - \beta b_0\big) \exp -2K\beta t \tag{14.52}$$

is the term corresponding to a transient and depending on the concentration of dangling chains.

$$m_2 = -K(M_{nb} - 18)[E]_0 t \tag{14.53}$$

is the term corresponding to a steady state, which corresponds to hydrolysis events in elastically active chains.

The mass loss curve is expected to have a linear asymptote of slope

$$dm_2/dt = -K(M_{nb} - 18)[E]_0 \tag{14.54}$$

$$\frac{dm_2}{dt} t_{max} = \frac{1}{2\beta}(M_{nb} - 18)\left(\frac{18[E]_0}{M_{nb}} - \beta b_0\right) \tag{14.55}$$

This equation allows us to check the validity of the approach if M_{nb} is known (analytical study of the extractable compounds).

There are at least two cases where experimental results are not consistent with the above approach. In both cases, hydrolysis remains homogeneous but the weight-loss rate is higher than predicted. This means that the hydrolysis rate is higher on chain ends than on internal network segments: $K' > K$. Many possible causes can be proposed for such a behavior:

- Terminal groups are effectively more reactive than internal ones, e.g., the case of a "depolymerization." The reason is generally that the end group (e.g., an acid) influences the reactivity of the next reactive group through inductive or mesomeric effects or by some catalytic effect involving a cyclic intermediate ("backbiting").
- Terminal groups are not more reactive than internal groups but reverse reaction is highly favored, owing to the low mobility of the network chains. The following scheme can be written:

$\sim$ A-B $\sim + H_2O \rightarrow \sim$ [A-OH HB] $\sim_{cage}$ (k_H) primary hydrolysis

$\sim$ [A-OH HB] $\sim_{cage} \rightarrow \sim$ [A-B] $\sim + H_2O$ (k_r) cage "recombination"

$\sim$ [A-OH HB] $\sim_{cage} \rightarrow \sim$ AOH + HB $\sim$ (k_D) diffusion out of the cage

The yield of the hydrolysis reaction depends on the competition between recombination and diffusion. The rate of diffusion (k_D) depends sharply on the molar mass of the fragments, since the diffusivity of a given molecular species is a decreasing exponential function of its molar volume (Van Krevelen, 1990). Thus one can imagine a process in which all the groups A–B are equireactive (k_H constant), but in which the yield of small molecules (high k_D) is higher than predicted from the hypothesis of a random process.

The problem of osmotic cracking is well known for polyester boat and swimming pool users. After a more or less long time of exposure in water, blisters and cracks appear at the surface. Their formation can be detected sometimes gravimetrically (Fig. 14.10). It can be shown that the induction time of osmotic cracking, t_b, is (Gautier *et al.*, 1999):

- a decreasing function of temperature
- a decreasing function of the concentration of catalytic residues and monomers (γ_0)
- a decreasing function of the concentration of reactive groups ($[E]_0$)
- a decreasing function of the concentration of dangling chains (b_0)
- a decreasing function of the polymer reactivity towards hydrolysis (K).

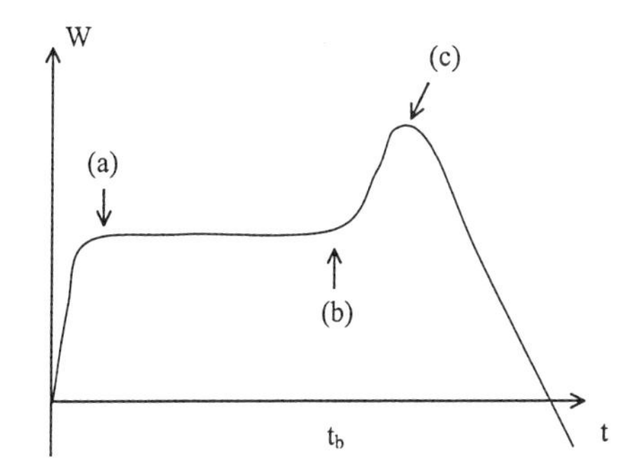

FIGURE 14.10 Shape of a mass variation–time curve in the case of osmotic cracking: (a) end of the diffusion transient, (b) onset of cracking, and (c) percolation threshold of the crack network.

Osmotic cracking is influenced by the same factors as weight loss. The following mechanistic scheme can be proposed:

(a) Organic molecules initially present or generated by hydrolysis accumulate in the matrix, their concentration increases continuously because their loss rate by diffusion is very low compared with their formation rate.
(b) When the concentration of these organic molecules reaches its equilibrium value, the system becomes oversaturated. Then, the excess of organic molecules demixes and forms a separated phase comprising microdomains dispersed into the matrix.
(c) These microdomains are very hydrophilic, owing to the chemical nature of the small molecules (glycols, diacids). They absorb water. Then, the polymer layer between a given microdomain and the water bath behaves as a semipermeable membrane since it is considerably more permeable to water than to organic molecules (these molecules can be considered trapped in the microdomains).
(d) Osmosis takes place and leads to an osmotic pressure build-up into the microdomains.
(e) When the osmotic pressure becomes higher than a certain threshold value linked to the polymer fracture toughness, the microdomain becomes a propagating crack.

14.4 THERMAL AGING

14.4.1 Introduction, Definitions

When exposed to high temperatures, organic polymers undergo irreversible structural changes linked to their proper instability or to their interaction with atmospheric oxygen. These structural changes, especially at the macromolecular scale (crosslinking, chain scissions), lead to changes of use properties.

Let us consider, for a given application, a critical property P (initial value P_0) and a critical value of this property P_F (end-life criterion). At a given time t, the property value is P_t and the corresponding aging index ξ may be defined by

$$\xi = \frac{P_t - P_F}{P_0 - P_F} \tag{14.56}$$

Then, when ξ changes its sign, failure becomes probable and the material can no longer be used with a sufficient degree of security.

From experimental results and chemical kinetics considerations, the function $P_0 = f(t)$ may be established, and the lifetime, t_F, determined by

$$t_F = f^{-1}(P_F) \tag{14.57}$$

where f^{-1} is the reciprocal function of f.

Let us now consider a set of isothermal exposure experiments at various temperatures $T_1, T_2, T_3, \ldots, T_n$, giving the corresponding lifetime values $t_{F1}, t_{F2}, t_{F3}, \ldots, t_{Fn}$. In a (t_F, T) plot, the corresponding points form a curve that is the thermal stability ceiling (TSC) of the polymer (Fig. 14.11).

From a time–temperature equivalence principle (see below), any material history may be represented by an isothermal equivalent, corresponding to a point U (t_u, T_u) in the (t, T) graph. If the point U is below the TSC curve, the material will not undergo failure in the particular conditions. In contrast, if the point U is above the TSC curve, the material will undergo failure because its index ξ will change sign.

In Fig. 14.12 we show two TSC curves: one for a typical commodity polymer (PVC, PP, PMMA, etc. and one for a thermostable polymer (aromatic polyimide). Users are essentially interested in the maximum temperature at which the material works for a long period. This temperature (for a duration of about 10 years) will be 50–100°C for the commodity polymer and 200–250°C for the thermostable polymer. Processing practitioners are rather interested in the maximum temperature at which the material can be processed for a duration of, typically, a few minutes. This temperature is of the order of 180–250°C for commodity polymers and 350–450°C for ther-

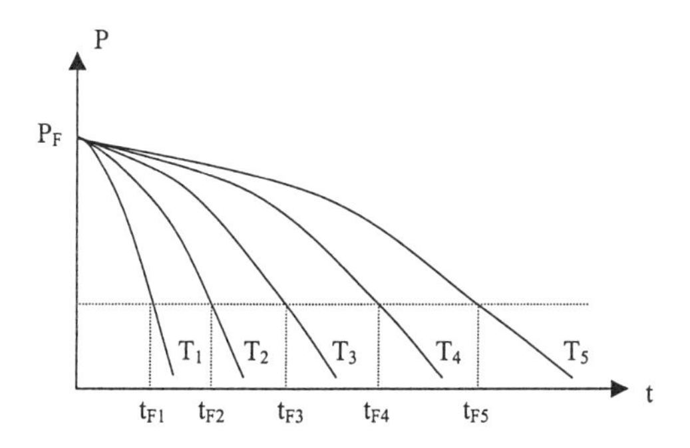

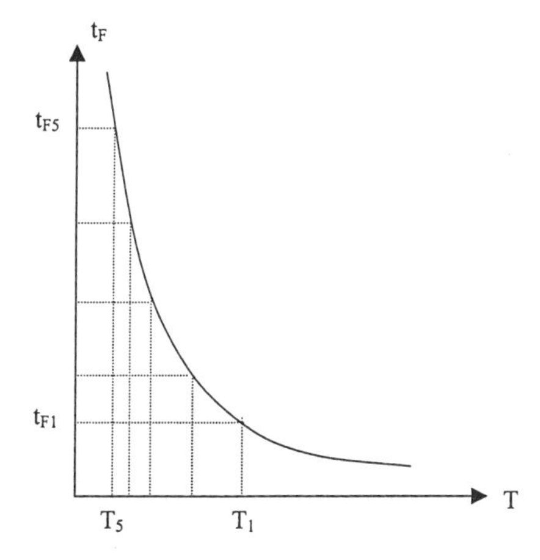

Figure 14.11 Principle of determination of the thermal stability ceiling (TSC) of a polymer (see text).

mostable polymers. All the commercial polymers, including thermosets, range between these boundaries.

Some very important features of thermal aging cannot be easily understood without a kinetic reasoning. In the following section, a simple kinetic model is postulated, which, in spite of its oversimplification, allows us to understand most of the general properties of thermal aging. The structure–thermal stability relationships are briefly examined and then some practical aspects are presented.

14.4.2 A Basic Kinetic Model to Begin the Study of Polymer Aging

The proposed model was formerly used to describe oxidation kinetics (Bolland, 1946), but can be extended to a general thermal degradation

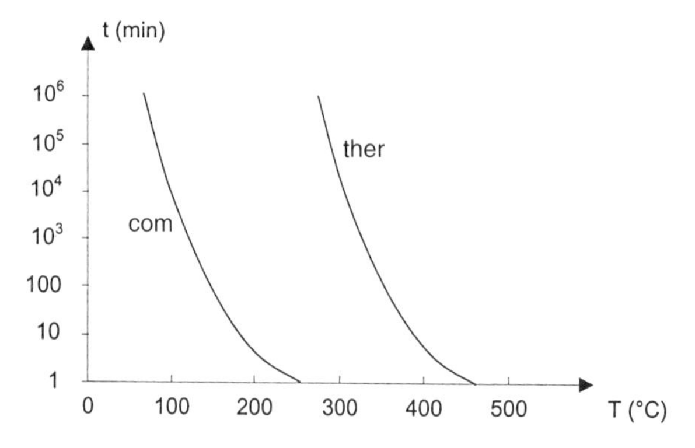

FIGURE 14.12 Shape of thermal stability ceiling curves for a commodity polymer of the PVC–PP type (com) and a thermostable polymer of the aromatic polyimide type (ther).

kinetics, including polymer thermolysis. The model starts from the hypothesis that all the main reactions are radical chain processes initiated by the polymer thermolysis (reaction 14.I below) or the hydroperoxide decomposition (reaction 14.I′).

Hydroperoxides that are initially present can result from the radicals formed in thermolysis or in side reactions produced during synthesis or processing. Since the chain oxidation generates hydroperoxides (reaction 14.III), it is expected to be autoaccelerated and needs very low initial POOH concentrations to occur (Audouin *et al.*, 1995):

14.I $\quad PH\ (\text{polymer}) \rightarrow 2P^0 + \alpha V \qquad (k_1)$

14.I′ $\quad \lambda POOH \rightarrow 2P^0 + \alpha' V' (k_1'),$

with $\lambda = 1$ (unimolecular) or 2 (bimolecular)

14.II $\quad P^\bullet + O_2 \rightarrow PO_2^\bullet \qquad (k_2)$

14.III $\quad PO_2^\circ + PH \rightarrow PO_2H + P^\circ \qquad (k_3)$

14.IV $\quad P^\circ + P \rightarrow \text{inact prod} \qquad (k_4)$

14.V $\quad P^\circ + PO_2^\circ \rightarrow \text{inact prod} \qquad (k_5)$

14.VI $\quad PO_2^\circ + PO_2^\circ \rightarrow \text{inact prod} + O_2 \qquad (k_6)$

V and V′ are volatile molecules that can be formed also in termination steps, but it is kinetically equivalent (except in very special cases) to consider that they are formed only in initiation. P° radicals can abstract hydrogens:

14.VII $\quad P^\circ + PH \rightarrow PH + P^\circ$

But this transfer reaction does not influence the whole kinetics. It can simply explain the relatively high mobility of the P° radical and then the high value of the termination rate constant, k_4. P° radicals can also undergo rearrangements – e.g., β scissions:

$$\sim\!\overset{H}{\underset{R}{C}}\!-CH_2-\underset{R}{CH}-CH_2\!\sim \;\longrightarrow\; \sim\!\overset{\circ}{\underset{R}{C}}\!-CH_2-\underset{R}{CH}-CH_2\!\sim \;\longrightarrow\; \sim\!\underset{R}{C}\!=\!CH_2 + \overset{\circ}{\underset{R}{C}}H-CH_2\!\sim$$

(P°) (P'°)

It is considered in a first approach that both P° and P'° radicals are equally reactive in terminations or oxidations, so that this reaction – as hydrogen abstraction – does not modify the whole kinetics. The oxygen addition to radicals is very fast, $k_2 \sim 10^8\text{–}10^9\,\mathrm{l\,mol^{-1}\,s^{-1}}$, so that there is practically no competitive process when O_2 is in sufficient concentration to scavenge all the P° radicals.

An Arrhenius plot of the rate constants is shown in Fig. 14.13. The reactions 14.II, 14.IV, 14.V, and 14.VI have a low activation energy. Their rate constants are in order: $k_2 > k_4 > k_5 \gg k_6$. The activation energies are in the reverse order. The hydrogen abstraction has an activation energy depending on the C–H dissociation energy but is typically of the order of $50\,\mathrm{kJ\,mol^{-1}}$. The hydroperoxide decomposition (k_1') has an activation energy of the order of $100\,\mathrm{kJ\,mol^{-1}}$. The polymer thermolysis (k_1) has an activation energy of the order of $200\text{–}400\,\mathrm{kJ\,mol^{-1}}$.

From Fig. 14.13, POOH decomposition, e.g., oxidation, will predominate at low temperatures, whereas polymer thermolysis will predominate at high temperatures. This means that there will be enormous differences of

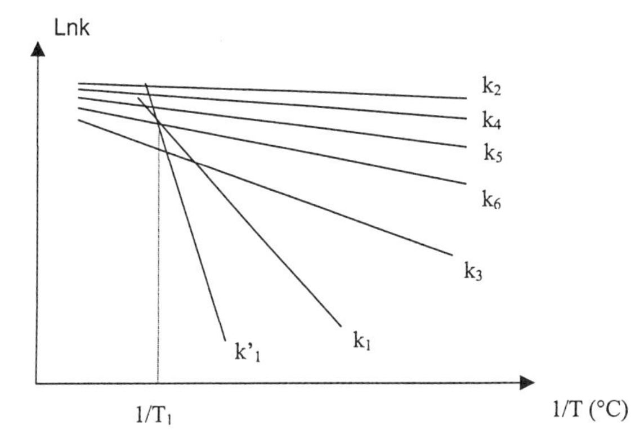

FIGURE 14.13 Shape of the Arrhenius plot of rate constants.

lifetimes between O_2 and neutral atmosphere exposures at 150°C, and little or no differences at 500°C.

At low temperatures in air, typically below 250°C, the polymer thermolysis (reaction 14.I) of thermostable polymers can be neglected. We are essentially in the presence of an oxidation chain process initiated by hydroperoxide decomposition. The main characteristics of this process are the following (Audouin *et al.*, 1995):

- Relatively low activation energy (typically 100–150 kJ mol^{-1}).
- Existence of a steady-state regime in which the global oxidation rate does not depend on the initiation rate constant.
- Nonmonotonic mass variation (Fig. 14–14a), well predicted by the kinetic model (Rychly et al., 1997).
- Influence of oxygen concentration C, which in a first approach can be approximated by a hyperbolic function,

$$r_{Oa} = \frac{dC}{dt} = \frac{aC}{1 + bC} \tag{14.58}$$

where a and b are constants that depend on the elementary rate constants, which can be determined from experiments made at various oxygen pressures; C is linked to the O_2 partial pressure p by Henry's law, $C = Sp$, where S is in the order of 10^{-3} mol l^{-1} Pa^{-1} (Van Krevelen, 1990). This means that at low oxygen concentrations ($C \ll b^{-1}$), oxidation is a pseudo-first-order reaction, $r \sim aC$, whereas at high oxygen concentrations ($C \gg b^{-1}$), oxidation is a pseudo-zero-order reaction, $r \sim a/b$ independent of C (Fig. 14.14b).

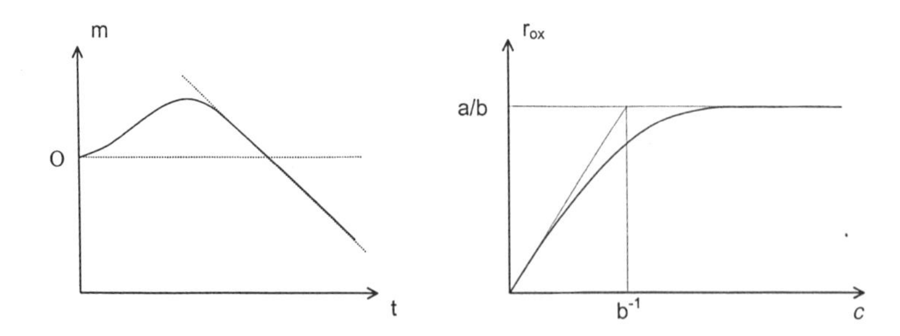

Figure 14.14 Characteristics of the oxidation kinetics at low temperature. (a) Shape of mass variation curves; (b) oxidation rate versus O_2 concentration.

a. Diffusion-Controlled Oxidation (Audouin *et al.*, 1994)

Let us consider a layer of thickness dx, located at a distance x of the sample surface. The local rate of variation of the O_2 concentration can be written as

$$r(C) = r\ (\text{diffusion}) - r\ (\text{consumption}) \tag{14.59}$$

The differential equation for unidirectional diffusion (diffusion by sample edges neglected) may be written as

$$\frac{\partial C}{\partial t} = D\frac{\partial^2 C}{\partial x^2} - \frac{aC}{1 + bC} \tag{14.60}$$

Solving this equation in a steady state, $\partial C/\partial t = 0$, leads to the thickness concentration profile, $C = f(x)$, and the local oxidation rate

$$r(x) = aC(x)/(1 + bC(x))$$

A local conversion may be calculated as

$$Q(x) = \int_0^t r(x)dt \tag{14.61}$$

The predicted $Q = f(x)$ can be compared with experimental data on oxidation thickness profiles established by spectroscopic imaging (FTIR, Raman, NMR, ESR, TOF SIMS), microhardness profiling, measurements made in microtome sections, or simply visual observations on microphotographs of polished samples (in general, oxidation leads to yellowing).

Simple calculations show that, as in the case of hydrolysis (Sec. 14.3.5), the thickness of the oxidized layer (TOL) can be approximated by

$$\text{TOL} \sim \sqrt{\frac{D}{k_{0x}}} \tag{14.62}$$

where k_{0x} is the pseudo-first-order constant of oxidation ($k_{0x} = r_{0x}/C$). In most cases, TOL ranges between 0.01 and 1 mm.

As

$$D = D_0 \exp -H_D/RT \text{ and } k_{0x} = k_{0x} \exp -\frac{H_{0x}}{RT}$$

then

$$\text{TOL} = (\text{TOL})_0 \exp -\frac{H_L}{RT} \tag{14.63}$$

$$H_L = \frac{1}{2}(H_D - H_{0x}) \tag{14.64}$$

Since, generally, $H_{0x} > H_D$, H_L is negative, which means that TOL is a decreasing function of temperature, as systematically observed (see, e.g., Lehuy *et al.*, 1991, for anhydride-cured epoxies).

Accelerated aging by a simple temperature variation not only increases the reaction rate but also modifies the spatial distribution of degradation events. Caution must be taken, therefore, in the interpretation of experimental results or in the comparison of data from various sources.

The thermal stability ceiling will show various slope changes when increasing temperature, corresponding to transitions between the following regimes:

1. The nondiffusion-controlled oxidation regime (homogeneous oxidation along the thickness). This regime is only observable on thin samples, typically ≤ 1 mm.
2. The diffusion-controlled, hydroperoxide-initiated, oxidation regime (oxidation is restricted to a superficial layer). The shape of the distribution of oxidation products depends on the O_2 solubility in the polymer (Fig. 14.15).
3. The diffusion-controlled, initiated by polymer thermolysis (see below), oxidation regime.

A complete investigation leading to a nonempirical lifetime prediction model would thus involve the following steps:

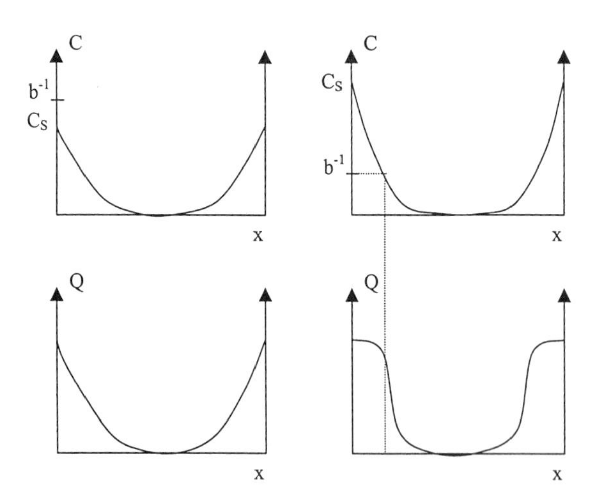

Figure 14.15 Shape of the thickness profiles of O_2 concentration (top) and degradation products concentration (bottom) in the case where the equilibrium O_2 concentration C_s is lower than b^{-1} (left) or higher than b^{-1} (right).

1. O_2 permeation measurements to determine the O_2 solubility, S, and diffusivity, D, in the polymer matrix, in the temperature range of interest.
2. Oxidation experiments on thin films (typically less than 200 μm), at various O_2 pressures, to determine the coefficients a and b (in the nondiffusion-controlled regime). Note that

$$\frac{1}{r} = \frac{1}{aSp} + \frac{b}{a} \tag{14.65}$$

   Thus, plots of the reciprocal oxidation rate against reciprocal O_2 pressure must be straight lines.
3. Thermal degradation experiments in neutral atmosphere to determine if polymer thermolysis has (or has not) to be taken into account in the model.
4. Kinetic modelling: calculation of the thickness distribution of oxidation products.
5. Experimental checking by analyzing thermally degraded samples.

Examples of this approach have been given recently for polybismaleimides (Colin *et al.*, 2000).

Remark: In composites, there are some complications especially due to

- the anisotropic character of diffusion
- the eventual contribution of the interphase to oxygen transport and degradation.

b. Anaerobic Thermal Degradation

A relatively simple model could be derived from the above standard scheme, suppressing O_2:

14.I	$PH \rightarrow 2P^\circ + \alpha V$	(k_i)
14.VIII	$P^\circ \rightarrow P^\circ + \beta V$	(k_p)
14.IV	$P^\circ + P^\circ \rightarrow$ inact prod	(k_t)

Volatiles can result from the initiation event or from a propagation one, e.g., zip depolymerization or simply radical rearrangement. The following equations result:

$$[P^\circ] = [P^\circ]_\infty \frac{1 - \exp -Kt}{1 + \exp -Kt} \quad \text{where} \quad [P^\circ]_\infty = \sqrt{\frac{k_i[PH]}{k_t}} \text{ and}$$

$$K = 4\sqrt{k_i[PH]k_t} \tag{14.66}$$

$$\frac{d[PH]}{dt} = -k_i[PH] \rightarrow [PH] = [PH]_0 \exp -k_i t \tag{14.67}$$

and

$$\frac{dV}{dt} = \alpha k_i[PH] + \beta k_p[P^\circ] = \alpha k_i[PH]_0 \exp -k_i t + \beta k_p[P^\circ]_\infty \frac{1 - \exp -Kt}{1 + \exp -Kt} \tag{14.68}$$

The rate of mass change is given by

$$\frac{dm}{dt} = \frac{M_v}{p} \frac{dV}{dt} \tag{14.69}$$

where M_V is the average molar mass of volatile molecules and ρ is the polymer density.

By comparing the rate constants for the evolution of volatile species and of substrate destruction, one can assess the contribution of initiation and propagation to the mass loss.

Chain scission (embrittlement) would result essentially from reaction 14.VIII, if radicals P° undergo β scission, whereas crosslinking (T_g increase) would result essentially from radical coupling (reaction 14.IV).

Boundary conditions can take into account the existence of a nonreactive fraction (char residue).

In more complicated cases, the model must take into account the existence of many distinct steps. This is not difficult if the steps are well separated as, for example, in epoxies (Fig. 14.16), but becomes complicated when the steps overlap.

14.4.3 Structure–Stability Relationships

Before any examination of structure–property relationships in the field of thermal aging, it is essential to understand that there is no one universal stability criterion, valid for all polymers under all circumstances. This is illustrated by the following examples.

"Intrinsic" thermal stability. This could be judged, for example on the basis of determinations of the rate of substrate consumption d[PH]/dt in a neutral atmosphere. According to this criterion, polyethylene, which begins to decompose at about 450°C, would appear more stable than

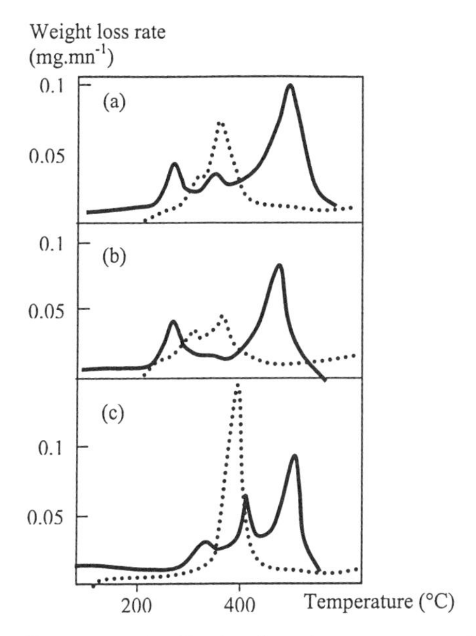

FIGURE 14.16 Derivative thermogram of some aromatic epoxies: (a) DGEBA–DDE (b) DGEBA–DDM, and (c) DGEBA–DDS networks under oxidative (—) and nonoxidative (···) conditions, Reprinted from Bellenger *et al.* 1984, copyright 2001 with permission from Elsevier Science.

most of the common thermostable polymers, which is somewhat paradoxical!

Weight Loss Criteria. Let us consider three possible thermogravimetric behaviors, as plotted in Fig. 14.17. It is clear that polymer (I) reacts rapidly but without weight loss; it behaves as an oxygen absorber. From a weight loss criterion, it would be judged more stable than (II) and (III), which is not necessarily true. As regards polymers (II) and (III), the hierarchy of the respective stabilities depends on the selected criterion.

Mechanical Criteria. There is a big difference in the behavior of initially ductile and initially brittle materials. Ductility is sharply linked to the macromolecular scale structure, whereas in brittle materials (or more generally in the brittle regime for any polymeric material), the properties (including ultimate ones) depend essentially on the molecular scale structure and on the size of eventual defects. This difference can be easily illustrated in the case of an amorphous linear polymer, but the reasoning would be the same for a thermoset.

Let us consider a homopolymer of initial molar mass M_{n0} (molar mass of the monomer unit m_0) and a critical molar mass (entanglement threshold

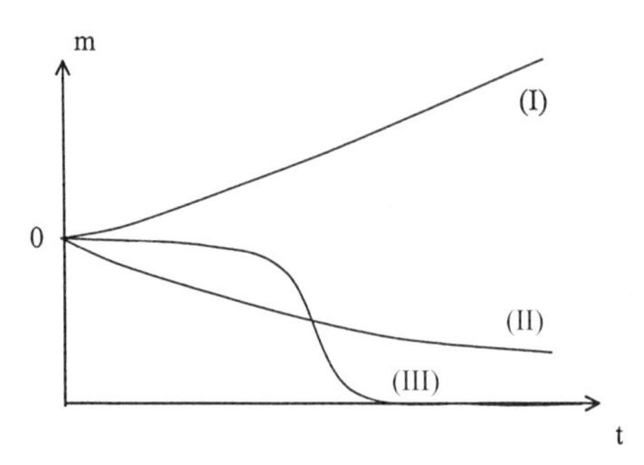

FIGURE 14.17 Different thermogravimetric behaviors.

at which embrittlement occurs), M_{crit}. The number of chain scissions to reach embrittlement is thus

$$n_{crit} = \frac{1}{M_{crit}} - \frac{1}{M_{n0}} \tag{14.70}$$

so that $n_{crit} < 1/M_{crit}$. The initial number of monomer units per mass, n_0, is given by

$$n_0 = \frac{1}{m_0} \tag{14.71}$$

The conversion, x_{crit} (proportion of broken monomer units), is calculated as

$$x_{crit} = \frac{n_{crit}}{n_0} < \frac{m_0}{M_{crit}} \tag{14.72}$$

m_0 is usually of the order of $100\,g\,mol^{-1}$ and M_{crit} of the order of $10^4\,g\,mol^{-1}$, so that x_{crit} is usually of the order of 1% or lower. Such a conversion can be considered negligible from the point of view of the molecular scale structure, whereas it is catastrophic for properties that depend on the macromolecular scale structure, especially the plastic deformation responsible for ductility. This explains why:

1. In an initially ductile polymer, failure properties (ultimate elongation, fracture toughness, impact resistance) decrease rapidly during a chain-scission aging process, whereas elastic and yield properties are practically unaffected at the embrittlement point.
2. For a given rate of chain scission, the failure properties of an initially brittle polymer decrease at a considerably slower rate than for an initially ductile polymer.

3. For an initially ductile polymer, the rate of decay of failure properties decreases abruptly after the embrittlement point.

These differences are illustrated by Fig. 14.18.

It is clear that if a relative end-life criterion (e.g., $p_f = P_0/2$) is chosen, the ductile polymer will systematically appear less stable than the brittle polymer. However, from the practical point of view, this distinction has no sense.

Influence of Sample Thickness and O_2 *Diffusion.* As shown above, the overall conversion of thermal degradation can depend of the sample thickness in the diffusion-controlled regime. Thus, stability comparisons are only valid for samples of comparable thickness. Let us now compare two polymers of glass transition temperatures T_{g1} and T_{g2}, oxidized at a temperature T, such that $T_{g1} < T < T_{g2}$. Even if the intrinsic oxidation rates are equal, polymer 1 will appear more unstable than polymer 2 because oxygen diffusion is faster above than below T_g. The thickness of the oxidized layer will be higher for polymer 1 than for polymer 2.

From the above examples it is apparent that a pertinent classification of thermal stabilities is not obvious.; however, some clear structure–stability relationships can be derived from experimental results. The most important characteristic to be found is the mechanism by which thermal aging proceeds: a chain process or a step process. The most unstable polymers are those that undergo degradation chain processes, for example:

- oxidation of hydrocarbon polymers
- depolymerizations (polyoxymethylene, polymers with tetrasubstituted carbons such as poly(methyl methacrylate))

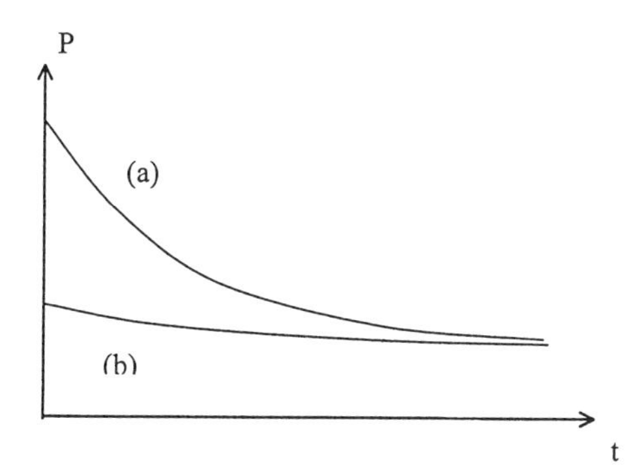

FIGURE 14.18 Shape of the time variation of a fracture property of an initially ductile (a) and initially brittle (b) polymer.

- zip side-chain reactions as RH eliminations from $-(CH_2-CHR)_n-$ (poly(vinyl chloride), poly(vinyl alcohol), poly(vinyl acetate)).

These chain reactions are generally terminated by a bimolecular combination of two propagating radicals, so that their whole rate depends on two quantities: the initiation rate (initiation being generally the most expensive step in terms of energy) and a combination of propagation, k_p, and termination, k_t, rate constants, generally of the form $k_p/\sqrt{k_t}$, that depend sharply on the type of radicals and molecular mobility. For example, in the case of oxidation, k_p decreases in the series:

ter CH > sec CH_2 > prim-CH_3

And k_t varies in the opposite sense:

primary radicals $-CH_2-O_2^\circ$ are about 1000 times more efficient in terminations than tertiary $>C-OO^\circ$ radicals.

Thus, one expects the following order of stability:

$-C(CH_3)_2 > -CH_2- > -CH(CH_3)-$

Double bonds or electronegative atoms in the α position have a destabilizing effect on the propagation rate:

$$-\overset{|}{\underset{|}{C}}-CH_2-\overset{|}{\underset{|}{C}}- < -CH_2-CH=CH-;\ -O-CH_2- \text{ or } >N-CH_2-$$

Aromatic groups are extremely stable and react only at high temperatures where they can undergo H abstraction, OH° and PO° additions, etc.

The initiation efficiency and the thermal stability in step-by-step reactions are essentially linked to the dissociation energy, E_d, of the weakest bond. In the frame of oxidation processes, it is clear that the weakest bond is the O–O bond of peroxides (140 kJ mol^{-1}, against typically 350 kJ mol^{-1} for a –C–C– aliphatic bond, and $\sim$ 380 kJ mol^{-1} for a C–H bond in an aliphatic methylene). This explains the quasi-exclusive role played by peroxides in the initiation of chain oxidation at low temperatures.

In chain processes (except perhaps oxidation initiated by unimolecular hydroperoxide decomposition, Audouin *et al.*, 1995), irregular structures containing weak bonds (as the classical example of hydroxyl chain ends in polyoxymethylene) can play a very important role, especially in the case of a long kinetic chain length, because their decomposition largely controls the whole degradation kinetics. In step-by-step processes, on the contrary, one expects that the contribution of a given "weak point" to the whole degradation is in the order of its molar fraction. The whole behavior of the polymer does not depend of the nature and presence of irregular structures. There are two important consequences of this situation:

1. There is an enormous difference between a nonstabilized and a stabilized commodity polymer (for instance polyoxymethylene) undergoing a chain-degradation process. Stabilization can occur by various routes, e.g., suppression of weak points; interruption of the propagation by a competitive process; increase of the termination rate using a radical scavenger such as carbon black, etc.

   In contrast, there are no big differences (although sometimes they can be significant) between two polymers with distinct irregular structures undergoing a step-by-step degradation process. These polymers are difficult or impossible to stabilize and the effect of an eventual stabilization is not so marked as in the case of chain degradation.
2. The knowledge of weak points, which can need important analytical studies, is useful for chain degradation but useless for step degradation.

Specific Case of Thermosets

Crosslinking has no specific direct effect on thermal degradation: crosslinks can be either weak points (e.g., tertiary carbons in polyester or anhydride-cured epoxies) or thermostable structural units (e.g., trisubstituted aromatic rings in phenolics, certain epoxies, or certain thermostable polymers). Indirect effects can be observed essentially above T_g: crosslinking reduces free volume and thus decreases O_2 diffusivity. It also prevents melting, which can be favorable in burning contexts.

The most general and specific aspect of thermal aging of polymer networks is the existence of a "postcuring" effect, often predominant in the early time of exposure (Fig. 14.19). The curves exhibit a maximum (for the ultimate stress of unsaturated polyesters, P can increase by more

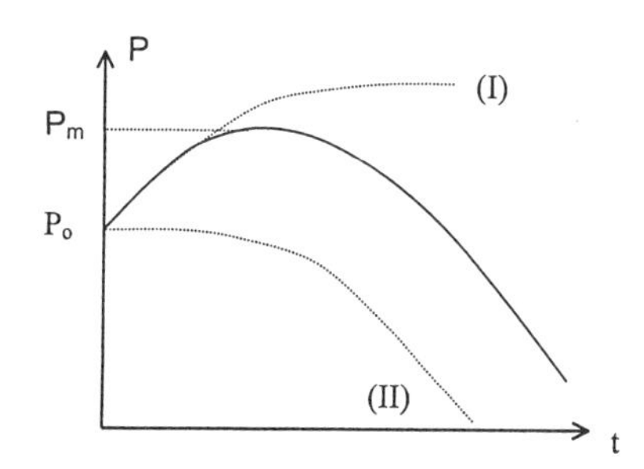

FIGURE 14.19 Shape of the time variation of a thermomechanical property, e.g., the ultimate stress or T_g for a thermoset.

than 30%) and then a continuous decrease. They result from the combination of two processes: postcure (I) and thermal degradation (II); the former is generally faster than the latter one. Indeed postcure effects depend on cure conditions but in practical cases, they are rarely reduced to zero. It is important to note that thermal aging cannot be considered as a sequence: first postcure and then thermal aging both begin at the same time, but postcure predominates in the early period of exposure.

Unsaturated polyesters can undergo degradation by oxidation of their aliphatic segments, decarboxylation of esters and partial depolymerization of polystyrene chains.

Anhydride-cured epoxies can be also degraded by oxidation (leading to very easily hydrolysable anhydride groups):

$$\mathrm{CH{-}O{-}\underset{\underset{O}{\|}}{C}{-}E} \sim\sim + O_2 \; - \; \mathrm{\underset{\underset{O}{\|}}{C}{-}O{-}\underset{\underset{O}{\|}}{C}{-}E} \sim\sim$$

Cycloaliphatic hardeners containing double bonds can participate in oxidative crosslinking processes responsible for embrittlement.

$$\text{(cyclohexenyl)} + PO_2 \longrightarrow \text{(cyclohexyl)}{-}O{-}O{-}P \longrightarrow \text{(cyclohexyl)}\begin{matrix} O{-}O{-}P \\ O{-}O^{\circ} \end{matrix} \text{ ETC...}$$

Epoxies are sensitive to oxidation of their aliphatic segments:

The reactivity of the α aminomethylene increases with the electron density on nitrogen, i.e., with the amine reactivity in the crosslinking reaction.

Amine-crosslinked epoxies also undergo dehydration at temperatures higher than 200–250°C:

$$\sim CH_2{-}\underset{\underset{OH}{|}}{CH}{-}CH_2{-}N\langle \; \longrightarrow \; \sim CH_2{-}CH{=}CH{-}N\langle \; + H_2O$$

Chain scission can also occur:

$$-CH_2{-}\underset{\underset{OH}{|}}{CH}{-}CH_2{-}N\langle \; \longrightarrow \; -CH_2{-}C\begin{matrix}{/\!/}O \\ \diagdown H\end{matrix} + CH_3{-}N\langle$$

The isopropanol segment $-CH_2-CH(OH)-CH_2-$ can always be considered as the weakest point of amine-crosslinked epoxies, both in oxidation and in thermal degradation.

Phenolic networks are interesting because they can be considered as a polymeric antioxidant:

OH $\xrightarrow{PO_2^{\circ}}$ O° + PO_2H $\longrightarrow$ O $\longrightarrow$ O_2 etc...

They exhibit a relatively high resistance to thermal aging but a tendency to discolor owing to the formation of highly conjugated chromophores. In high-temperature degradation they give often high char contents.

Thermostable networks offer a very large diversity of thermal aging behaviors. Their design obeys the following rules:

- avoid aliphatic structures, that are very sensitive to oxidation and are not thermostable
- or use aliphatic groups only in ladder structures in which a chain scission does not suppress an elastically active network segment (e.g., polyimides of the PMR 15 type).

14.4.4 Practical Aspects of Thermal Aging: Lifetime Prediction

a. Test Conditions

1. Sample Thickness. It is convenient to study at least two sample thicknesses: one ($\sim 100\,\mu m$) to determine the kinetic parameters of the non-diffusion-controlled kinetics, the other one ($\geq 1\,mm$) to determine the distribution of oxidation products along the thickness.

2. Isothermal versus Programmed Temperature Experiments. Isothermal experiments are preferable to nonisothermal experiments, owing to the simplicity of kinetic modeling. Let us remind ourselves, however, that the time to reach thermal equilibrium (from ambient temperature) is of the order of $tau_T \approx L^2/\alpha_T$, where L is the sample thickness and α_T is the thermal diffusivity (generally of the order of $10^{-7}\,m^2\,s^{-1}$). At high temperatures, thermal degradation can be so fast that it becomes non-negligible during the thermal transient. In this case, programmed temperature experiments are preferable.

3. Characterization of the Conversion of Degradation Reactions. There are many ways to characterize thermal aging:

- Measurement of the residual mechanical properties, which avoids amking hypotheses on the relationships between the structural state and the mechanical behavior and gives the best lifetime value but is not necessarily easy to use as a kinetic function.
- Gravimetry, using initially dry samples to avoid interferences between water desorption and degradation. Gravimetric data can be useful if they are treated through a kinetic model as previously described. End-life gravimetric criteria are generally highly arbitrary.
- Gas evolution, which can be used globally in the same way as gravimetry, with the advantage of considerably higher sensitivity. However caution must be taken with very high sensitivities because they can lead to the detection of nonrepresentative phenomena. Analysis of evolved gases by GC–MS can, indeed, give interesting information on degradation mechanisms but in complex structure materials such as commercial thermosets, spectral changes are not easy to interpret.
- Thermal analysis (DSC, DMTA, DETA, etc.) gives valuable information on eventual changes of glass transition temperature, which can be interpreted in terms of crosslinking/chain-scission processes.

b. Problems of Lifetime Prediction

Any problem of lifetime prediction in the field of polymer aging is highly complex owing to the high number of structural and environmental variables and the complexity of their interrelationships. Its complete, nonempirical resolution would need a big volume of experiments and years of investigation. In such conditions, a given aging study must be a compromise between economical and technological constraints. In the most favorable cases, the degradation mechanisms will be identified and the approaches presented above may be used to predict lifetimes. In less favorable cases, more empirical approaches may be used, the most popular one being the use of the Arrhenius relationship. Examples of its application are common in the literature. In such complex processes, the Arrhenius equation actually works as an empirical model as illustrated by the following examples.

Sequential Processes. In many cases (anhydride cured-epoxies, amine-cured epoxies, etc.), thermal degradation curves, for instance gravimetric ones, exhibit two distinct stages (Lehuy *et al.*, 1991). The first one corresponds to a relatively fast, pseudo-first-order process, whereas the second one arises from a relatively slow, pseudo-zero-order phenomenon (Fig. 14.20). The curve can be fitted with the following equation:

$$m = m_1\ (1 - \exp -K_1 t) + K_2 t$$

In general, K_1 and K_2 will have different activation energies, E_1 and E_2, so that the whole process will not follow the Arrhenius equation. If an arbitrary end-life criteria m_f is selected, the lifetime t_f will exhibit an apparent activation energy that depends on m_f, according to Fig. 14.21.

Chain Oxidation in the Case of O_2 *Excess (Unimolecular Initiation).* The oxidation rate varies with time, according to a relationship of the type (Audouin *et al.*, 1995):

$$r = \frac{k_p^2 (PH)^2}{k_t} \left(1 - \exp - \frac{k_i t}{2}\right)^2 \qquad (14.73)$$

The rate r does not obey the Arrhenius law. Its apparent activation energy is E_i – activation energy of k_i, the initiation rate constant – at the beginning of exposure, and then it decreases progressively and tends toward $E_\infty = 2E_p - E_t$ (the activation energy of k_p^2/k_t), in steady state.

Chain Oxidation in Diffusion-controlled Conditions. As shown:

$$r = \frac{aC}{1 + bC}$$

where a and b are functions of elementary rate constants. Since both a and b obey the Arrhenius law, r will not, except far from the transition between both extreme regimes $C \gg b^{-1}$ or $C \ll b^{-1}$).

Every time that the whole degradation rate is an algebraic sum of terms that follow the Arrhenius equation or a function of such a sum, it

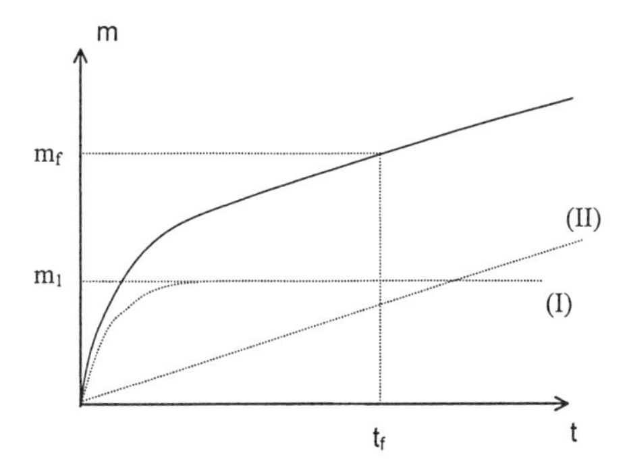

FIGURE 14.20 Schematization of a sequential process involving a pseudo first order (I) and a pseudo zero order (II) component.

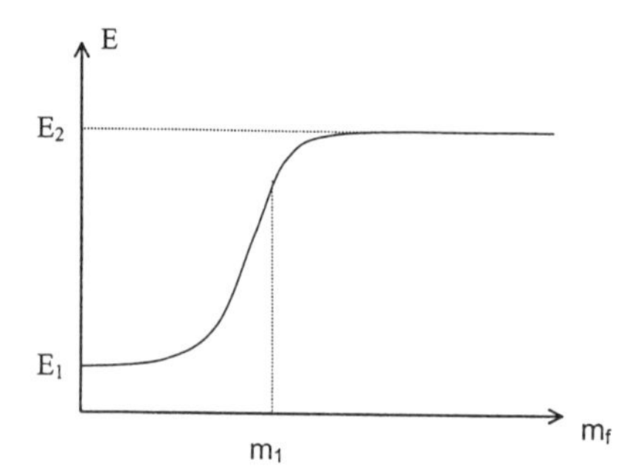

FIGURE 14.21 Schematization of the dependence of apparent saturation energy with the endlife criterium for a sequential process in which $E_1 < E_2$

will not obey this equation. Considering that lifetime prediction results from an extrapolation on a logarithmic time scale (errors on kinetic parameters are largely amplified by the extrapolation), the above examples emphasize the need for a true kinetic analysis of the aging process and of relatively long accelerated aging tests, in order to reduce the extrapolation risks.

In any case, the region of accelerated aging must be limited by physical transitions (melting, glass transition), since the kinetic parameters undergo practically unpredictable discontinuous variations at these transitions, as illustrated by the example of oxidation (amide growth in an amide-cured epoxy) (Fig. 14.22).

Another nontrivial problem is the choice of a polymer property and the end-life criteria. Indeed, the best solution consists of selecting a property whose relationships with structure have been previously established. This allows us to move nonempirically from structure (predicted by chemical

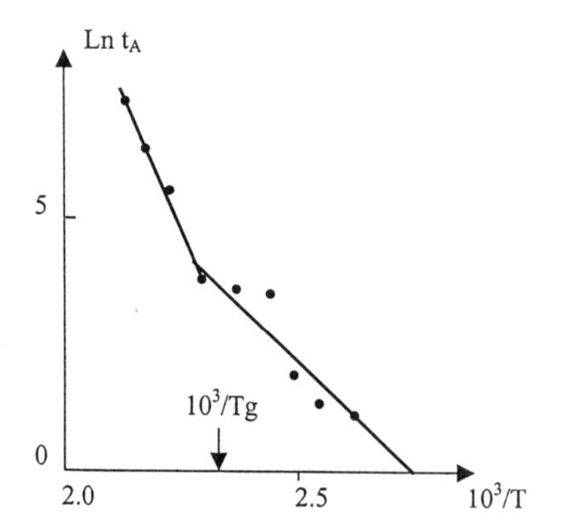

FIGURE 14.22 Arrhenius plot for amide growth in the oxidation of amine crosslinked epoxy films.

kinetics), to property and vice versa.

However, this problem is especially difficult in the case of thermosets, for the most important practical case of embrittlement. As a matter of fact, there is (to our knowledge), no widely accepted theory capable of predicting the fracture properties as a function of the number of chain scissions or crosslinks created by aging. This research domain is wide open and the problem must be, for the moment, treated empirically.

14.5 CONCLUSIONS

Three very important aging modes of thermosets have been examined:

1. Physical aging resulting from water absorption is probably the most widely studied process. The structure–property relationships in this field are relatively well understood, except for diffusivity, which seems to depend sharply on the strength of polymer–water interactions. The resulting plasticization is also relatively well understood, but the research field on swelling in the glassy state still remains open.
2. Chemical aging resulting from water absorption (i.e., hydrolysis) has not been as widely studied as physical aging. It is relatively well understood at the molecular scale (chemical mechanisms). But macromolecular (kinetics of decrease of the elastically active chain concentration) and mechanical aspects (effect of chain scissions on mechanical properties) are far from being elucidated.
3. Thermal aging has also been widely studied for its technological interest, but most of the studies are empirical or semiempirical (using, e.g., the Arrhenius law and an arbitrary end-life criteria, from mass–loss kinetic curves). Powerful conceptual tools (radical chain kinetics, coupled diffusion–reaction schemes, etc.) are available to avoid or, at least, to reduce significantly the empiric approaches.

Finally, there are no systematic differences in the aging mechanisms of linear and network polymers. What is different is essentially (i) the degree of difficulty of analytical studies, especially determinations of the crosslink density and (ii) the lack of theories (in thermosets) linking the network structure to properties, especially ultimate mechanical properties. There is a need for basic research in this area to reach a full understanding of aging effects on use (especially fracture) properties.

NOTATION

b = concentration of dangling chains, $mol\,kg^{-1}$
$C(C_\infty)$ = concentration (equilibrium concentration), $mol\,m^{-3}$
C_k = Kambour 's constant, $Pa\,K^{-1}$
D = diffusion coefficient, $m^2\,s^{-1}$
$[E]$ = substrate concentration (hydrolysis), $mol\,m^{-3}$
E = Young's modulus, Pa
G = shear modulus, Pa
H = activation energy, $J\,mol^{-1}$
k = elementary rate constant, $l\,mol^{-1}\,s^{-1}$
K = first-order rate constant, s^{-1}
K_u = ultrasonic bulk modulus, Pa
L = sample thickness, m
m = mass, kg
m_0 = molar mass of the monomer unit, $kg\,mol^{-1}$
M = molar mass, $kg\,mol^{-1}$
M_{crit} = critical molar mass (embrittlement), $kg\,mol^{-1}$
n = number of chain scissions per mass unit, $mol\,kg^{-1}$
p = H_2O or O_2 pressure, Pa
P = property
Q = conversion of the oxidation process
r = reaction rate, $mol\,m^{-3}\,s^{-1}$
R = gas constant, $J\,mol^{-1}\,K^{-1}$
RH = relative humidity, %
s = styrene mass fraction
S = solubility coefficient, $mol\,m^{-3}\,Pa^{-1}$
t, t_D, t_F = time; time to reach equilibrium; lifetime, s
T = temperature, K
T_g = glass transition temperature, K
TDL = thickness of the degraded layer (hydrolysis), m
TOL = thickness of the oxidized layer, m
TSC = thermal stability ceiling
v = volume fraction
V = volume, m^3
W = water mass uptake

Greek Letters

α = coefficient of free volume expansion, K^{-1}
β = number of hydrolyzable bonds in a dangling chain
ϵ = strain
ρ = density, $kg\,m^{-3}$
$\sigma, \sigma_y \sigma_f$ = stress; yield stress; ultimate stress in brittle regime, Pa
φ = crosslink functionality
ψ = volume gain due to swelling

REFERENCES

Adamson MJ, *J. Mater. Sc.*, 15, 1736–1745 (1980).

Audouin L, Langlois V, Verdu J, De Bruijn, JLM, *J. Mater. Sci.*, 29, 569–583 (1994).

Audouin L, Gueguen V, Tcharkhtchi A, Verdu J, *J. Polym. Sci. Polym. Chem. Ed.*, 33, 921–927 (1995).

Bauer DR, Mielewski DF, Gerlock JL, *Polym. Degrad. Stab.*, 38, 57–68 (1992).

Belan F, Bellenger V, Mortaigne B, Verdu J, *Polym. Degrad. Stab.*, 56, 301–310 (1997).

Bellenger V, Verdu J, *J. Appl. Polym. Sci.*, 28, 2599-2609 and 2677–2688 (1983).

Bellenger V, Fontaine E, Fleishman A, Saporito J and Verdu J., *Polym. Deyrad. Stab.* 9, 195–2908 (1984).

Bellenger V, Verdu J, Morel E, *J. Mater. Sci.*, 24, 63–68 (1988).

Bellenger V, Mortaigne B, Verdu J, *J. Appl. Polym. Sci.*, 41, 1225–1233 (1990).

Bolland J.L.B, and Gee G. *Trans Farad. Soc.* 42, 236–252 (1946)

Colin X, Marais C and Verdue J. To appear in *J. Appl. Polym. Sci.* 2001 (2000).

Flory PJ, Rehner J, *J. Chem. Phys.*, 11, 521–526 (1943).

Ganem M, Mortaigne B, Bellenger V, Verdu *J, Polym. Netw. Blends,* 4, 115–122 (1994).

Gautier L, Mortaigne B, Bellenger V, Verdu J, *Polymer*, 41, 2481–2490 (1999).

Hilaire B, Verdu J, *Water Sorption Characteristics of Some Polynadimides*. Conf. presentation at STEPI 3, Montpellier, France, June 1993.

Kambour RP, *Polym. Comm.*, 24, 292–295 (1983).

Kelley FN and Bueche F, *J. Appl. Polym. Sci.*, 1, 549–556 (1961).

Kranbuehl D, In *Recent Developments in Durability. Analysis of Composite Systems*, Cardon A, Fukida H, Reifsnider K, Verchery G (eds), Balkema (pub). Rotterdam, 2000, pp. 413–420.

Lehuy HM, Bellenger V, Paris M, Verdu J, *Polym. Degrad. Stab.*, 35, 77–86 (1991); 37, 171–179 (1991).

Mortaigne B, Bellenger V, Verdu J, *Polym. Netw. Blends*, 2, 187–195 (1992).

Ranby B, Rabek J, *Photodegradation, Photooxidation and Photostabilization of Polymers*, Wiley, New York, 1975.

Rychly J, Matisova-Rychla L, Csomorova K, Achimsky L, Audouin L, Tcharkhtchi A, Verdu J, *Polym. Degrad. Stab.*, 58, 269–274 (1997).

Tcharkhtchi A, Bronnec S, Verdu J, *Polymer*, 41, 5777–5785 (2000).

Van Krevelen DW, *Properties of Polymers*, 3rd edn, Elsevier, Amsterdam, 1990.

Wilski H, In *Radiation Effects in Polymers*, Clough RL, Skalaby SW (eds), *ACS Symp. Ser.*, 475–500 (1991).

Conversion Factors

Frequently found non-SI unit	SI unit	Conversion
atmosphere, atm	$N \cdot m^{-2}$ or Pascal, Pa	1 at = 98.07, 10^3 $N \cdot m^{-2}$ bar 1 atm = 0.9807 bar
calorie, cal	Joule, J	1 cal = 4.184 J
centipoise, cP	Pascal second, $Pa \cdot s$	1 cP = 1.01, 10^{-3} $Pa \cdot s$
dyne	Newton, N	1 dyn = 1, 10^{-6} N
erg	Joule, J	1 erg = 10^{-7} J
foot, ft	meter, m	1 ft = 0.3048, 10^{-1} m
foot pound-force, $lb \cdot ft$	Joule, J	1 $lb \cdot ft$ = 1.356 J
inch, in.	meter, m	1 in = 0.2540, 10^{-2} m
kilopound, kp	Newton, N	1 kp = 9.807 N
kilogram-force per centimeter squared	$N \cdot m^{-2}$ or Pascal, Pa	1 kgf cm^{-2} = 9.807, 104 $N \cdot m^{-2}$
centimeter, cm	—	
mil (= 0.001 in.)	micrometer, μm	1 mil = 25.4 μm, 2.54×10^{-5} m
pound per square inch, PSI	$N \cdot m^{-2}$ or Pascal, Pa	1 PSI = 6.897, 10^3 $N \cdot m^{-2}$

Index